T0295494

CLIMATE CRISIS, ENERGY VIOLENCE

CLIMATE CRISIS, ENERGY VIOLENCE

MARY FINLEY-BROOK

STEPHEN METTS

ELSEVIER

ACADEMIC PRESS
An imprint of Elsevier

Academic Press is an imprint of Elsevier
125 London Wall, London EC2Y 5AS, United Kingdom
525 B Street, Suite 1650, San Diego, CA 92101, United States
50 Hampshire Street, 5th Floor, Cambridge, MA 02139, United States

Notices
Knowledge and best practice in this field are constantly changing. As new research and experience broaden our understanding, changes in research methods, professional practices, or medical treatment may become necessary.

Practitioners and researchers must always rely on their own experience and knowledge in evaluating and using any information, methods, compounds, or experiments described herein. In using such information or methods they should be mindful of their own safety and the safety of others, including parties for whom they have a professional responsibility.

To the fullest extent of the law, neither the Publisher nor the authors, contributors, or editors, assume any liability for any injury and/or damage to persons or property as a matter of products liability, negligence or otherwise, or from any use or operation of any methods, products, instructions, or ideas contained in the material herein.

ISBN: 978-0-12-819501-7

For information on all Academic Press publications visit our website at
https://www.elsevier.com/books-and-journals

Publisher: Mica Haley
Acquisitions Editor: Kathryn Eryilmaz
Editorial Project Manager: Naomi Robertson
Production Project Manager: Omer Mukthar
Cover Designers: Jess Irish, Matthew Limbert

Typeset by TNQ Technologies

Working together
to grow libraries in
developing countries

www.elsevier.com • www.bookaid.org

Dedication

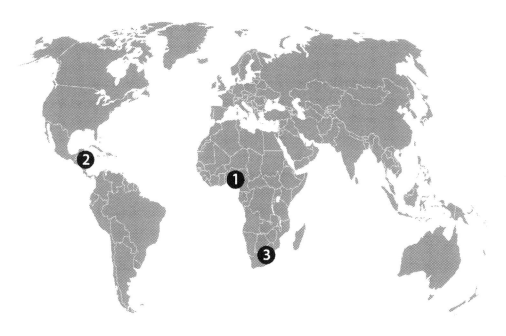

Defenders & Locations:	1 - Ken Saro-Wiwa (Nigeria)
	2 - Berta Caceres (Honduras)
	3 - Fikile Ntshangase (South Africa)
Themes:	Frontline defender; Climate warrior
Subtheme:	Collective action

They tried to bury her, but didn't know she was a seed.[1]

---

[1] Morrill, C. (2021, June 26). As key suspect tried, encampment demands justice for Berta Caceres. Common Dreams.

Honoring frontline defenders

Across the globe, exposure to danger is not equally shared. Regions wracked by political violence often double as hotspots of energy violence.[2] Amidst turmoil, frontline defenders stand up to protect their communities and lands from egregious harms. Systemically intimidated and criminalized, their very words and actions have too many times proven fatal. The energy corporations responsible for these deaths use carefully crafted social responsibility campaigns to clock their culpability, enabling and intensifying patterns of energy violence.[3] Three exemplary frontline defenders, each whose very life was violently taken from them, represent the urgent call to break free of violent energy systems and build a future imbued with energy justice.

Kenule Beeson "Ken" Saro-Wiwa (1941—95)

In the 1989 short story *Africa Kills Her Sun*, Kenule Beeson Saro-Wiwa foreshadowed his execution in 1990. "Ken" Saro-Wiwa, a Nigerian writer and environmental activist from the Ogoni ethnic group in the Niger Delta, helped start an organization called Movement for the Survival of the Ogoni People (MOSOP)[4] to fight for Ogoni rights and end oil colonialism.[5] In a gruesome military tribunal in 1995, Saro-Wiwa and eight other activists called the Ogoni Nine were found guilty of fake charges and executed. Saro-Wiwa's last words were "...the struggle continues." Posthumously, Saro-Wiwa won the Right Livelihood Award and Goldman Environmental Prize for his leadership.

[2] See Global Witness (2012—21) reports in Recommended Readings at the end of this Dedication.
[3] Roth, S. (2020, November 23). The fossil fuel industry wants you to believe it is good for people of color. *LA Times*.
[4] Movement for the Survival of the Ogoni People. (n.d.). Ogoni news and resources.
[5] Movement for the Survival of Ogoni People. (1991, August 26). Ogoni Bill of Rights Presented to the Government and People of Nigeria.

Ken Saro-Wiwa. *(Image credit: Goldman Environmental Prize.)*

The murder of the Ogoni Nine brought awareness of energy violence to an international stage. Saro-Wiwa's family continued pressure for reform.[6] Decades of legal cases seeking amends for oil damages and physical violence in Ogoniland helped seed current corporate social responsibility trends, yet accountability and transparency remain inadequate.

Fikile Ntshangase (1957—2020)

South African activist Fikile Ntshangase fought for territorial rights against the company, Tendele Coal.[7] Ntshangase refused to relocate from her ancestral lands and for this was murdered in her home after receiving death threats. Ntshangase vocally opposed coal expansion as an officer with the Mfolozi Community Environmental Justice Organization (MCEJO). Men shot the elder six times in front of her grandson and his friends.

---

[6] EarthRights International. (n.d.). Wiwa v. Royal Dutch Shell: Getting away with murder: Shell's complicity with crimes against humanity in Nigeria.

[7] Greenfield, P. (2020, October 23). South African environmental activist shot dead in her home. *The Guardian.*

Fikile Ntshangase. *(Image credit: Zelda Ann Hintsa.)*

South Africa's coal regions operate as sacrifice zones, where ethnic minorities and the poor suffer the most. If South Africans speak out like Fikile Ntshangase did, who will protect them?

Berta Caceres (1973—2016)

Berta Caceres, a Honduran Indigenous Lenca activist, was murdered in her home at the hands of security personnel with direct ties to the Honduran military and the hydroelectric company behind the Agua Zarca Dam. While fossil fuels are the focus of this book, renewable energy projects are not exempt from violence. Power inequalities across all energy types foment violent environments.

Wake up, humanity! There is no more time.[8] *¡Viva Berta!*

Berta Caceres. *(Image credit: Goldman Environmental Prize.)*

---

[8] Berta Caceres quoted in Castellanos, A. & Pine, A. (2020, July 29). Berta Caceres in her own words. Toward freedom.

Berta's legacy includes cofounding the Council of Honduran Popular and Indigenous Organizations (COPINH), an antiracist, antipatriarchal 'grassroots' network (i.e., of ordinary people, generally volunteers) who work with Native communities, and especially women and girls, to defend their land, subsistence, and human rights.[9] Berta's daughter continues to advance these objectives.

Try this

Research, discussion, and act: Use Global Witness Reports (2012—20) to discuss the impact and challenges of frontline defenders.[10] What can solidarity networks do to communicate their messages and reverse violence?

Defense of life

There has been an alarming upward trend of assassinations of environmental activists.[11] Whether called land defenders, community leaders, or environmentalists, these change-makers are part of local struggles to protect a common good. A **collective**—cooperative initiative—in the instances above emerges from the grassroots, meaning a social movement from humble origins. Placemaking envisions the world *as it can be*. This process puts in motion geographic imaginaries that remake our relationships to one another and Earth.

Modern societies have not protected the 'next seven generations' as Indigenous societies have consistently advised.[12] Now with no time to spare, frontline defenders beckon us forward, offering three guiding principles: (1) build from the base, (2) harness collective power, and (3) create a more just world.

Climate warriors

An exemplary youth leader who found her voice expressing outrage at the treatment of her people is artist Selina Leem,[13] a Marshallese climate activist.[14] At age 14, Leem was

---

[9] Council of Popular and Indigenous Organizations of Honduras. (2018, March 22). 25 years of struggle and Revolution.

[10] Global Witness. (n.d.). Environmental defenders annual report archive.

[11] Global Witness. (2020, September 13). Last line of defense.

[12] Borrows, J. (2008). *Seven generations, seven teachings: Ending the Indian Act*. National Centre for First Nations Governance.

[13] Staff Writer. (2022, May 16). Sea change: Music helps students grasp climate risk facing island nations. UR Now.

[14] Dreher, A. (2021, March 17). How the legacy of nuclear testing in the Marshall Islands still affects Spokane's Marshallese community. The spokesman-review.

the youngest delegate at the United Nations Framework Convention on Climate Change (UNFCCC) Conference of Parties (COP).[15] Leem gave an emotional message in 2015 with thanks for the UN commitment to 1.5° in international negotiations.[16] Just 5 years later, in a TED talk, Leem speaks of how she emerged as a **climate warrior**:[17] a person committed to combating climate change.

Youth mobilization is expanding all over the world. Bold youth like Ugandan environmental activist Vanessa Nakate highlight global inequity:

> *All we really want is a livable and healthy planet… Is that too much to ask? Not to destroy our only home and have a small group of people benefit from our pain and suffering.*[18]

An invitation

A preponderance of scientific evidence points to a deepening climate crisis, yet hope is not lost: today we are collectively equipped to limit and remediate energy's past and ongoing ecological and social harms. Worldmaking—the uplifting work of collaborative problem solvers—is apparent throughout this book. Each of its chapters amplifies activists and organizations, encouraging readers to learn more through illustrated case studies as well as additional source materials.

As much as this book documents global patterns of energy violence, it is also an invitation to take part in "**the Great Turning**"—a broad and deep transformation from exploitative and self-destructive business as usual to the emergence of mutualistic, life-sustaining systems that are forming and being recovered to heal vast harm.[19] It is an appeal from frontline defenders and their communities to build collective power within everyday struggles for energy justice.[20]

All our global challenges require collaboration, none more so than the transformation to just energy systems. *What part do you best play?* Dr. Anaya Elizabeth Johnson recommends asking oneself: (1) What am I good at? (2) What is the work that needs doing? and (3) What brings me joy?[21] The three intersecting answers to these fundamental questions lead to an ideal starting point for you to fulfill your unique calling.

[15] Leem, S. N. (2021, October). TED talk: Climate change isn't a distant threat, it is our reality.

[16] Leem, S. (2015, December 14). Marshall Islands 18-year-old thanks UN for climate pact. Climate Change News.

[17] Leem, S. N. (2021). TED talk: Climate change isn't a distant threat, it is our reality.

[18] Nakate, V. (2021). *A bigger picture: My fight to bring a new African voice to the climate crisis*. HarperCollins.

[19] Kelly, S., & Macy, J. (2021). The great turning: Reconnecting through collapse. In J. Bendell, & R. Read (Eds.), *Deep adaptation: Navigating the realities of climate chaos* (pp. 197–210).

[20] Sze, J. (2020). Environmental justice in a moment of danger. University of California Press.

[21] Ibid.

Vocabulary

1. climate warrior
2. collective
3. "the Great Turning"

Recommendations

Books

Lakhani, N. (2020). *Who killed Berta Caceres?: Dams, death squads, and an indigenous defender's battle for the planet.* Verso.

Menton, M., & Le Billon, P. (Eds.). (2021). *Environmental defenders: Deadly struggles for life and territory.* Earthscan.

Nakate, V. (2021). *A bigger picture: My fight to bring a new African voice to the climate crisis.* HarperCollins.

Okome, O. (Ed.). (2000). *Before I am hanged: Ken Saro-Wiwa, literature, politics, and dissent.* Africa World Press.

Reports

Global Witness. (2023). *Decade of defiance.*
Global Witness. (2022). *Standing firm.*
Global Witness. (2021). *Last line of defense.*
Global Witness. (2020). *Defending tomorrow.*
Global Witness. (2019). *Enemies of the state?.*
Global Witness. (2018). *At what cost?.*
Global Witness. (2017). *Defenders of the earth.*
Global Witness. (2016). *On dangerous ground.*
Global Witness. (2016). *How many more?.*
Global Witness. (2014). *2002−2013 deadly environment.*
Global Witness. (2013). *A hidden crisis.*

Contents

Introduction: Manufacturing ignorance

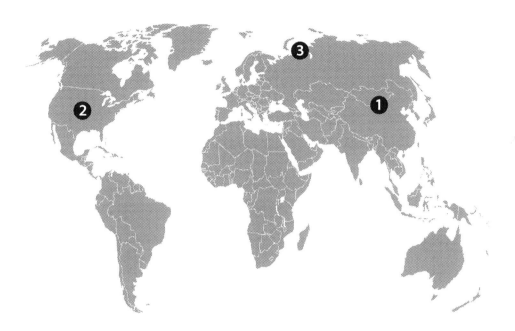

Locations: 1 - Inner Mongolia - China
            2 - United States
            3 - Yamal Peninsula - Russia

Themes: Energy transformation; Power; Violence

Subthemes: Deception playbook; Fast violence; Slow violence

Why study energy violence?

Energy, the capacity to do work, takes a number of different forms (i.e., heat energy, kinetic energy, chemical energy, electric energy). Although energy has been used throughout human history, it did not become a major topic of international public discourse until the early 1970s. It remains a slippery, misunderstood topic. This book primarily addresses **fossil fuels**, a group of energy sources (i.e., coal, oil, gas) formed when plants and organisms were subject to intense heat and pressure over millions of years. Since fossil fuels are composed of organic hydrocarbons located in finite basins, they are subject to depletion. **Renewable energy** is collected from resources naturally replenished on a human timescale, including sunlight, wind, rain, tides, waves, and geothermal heat. These abundant, live generating resources can form the basis of an **energy commons**: sustainable, equitable and distributed energy with associated technologies apart from fossil fuels which operate narrowly as market commodities to be traded, bought and sold.

We highlight a structural and pragmatic distinction between **energy transition** (moving from fossil fuels to renewables) and comprehensive **energy transformation** featuring social, ecological, and economic change beyond fuel switching alone. Since energy is so pervasive as a social and material building block, its transformation opens new opportunites. Absent transformation, **energy colonialism** extends the past's status quo, a spatially uneven process with high costs (e.g., pollution, resettlement) in areas of production while exporting benefits (e.g., electricity, profit).[1] Broadly speaking, **colonialism** is a harmful market-based institution reliant on exploitative and cruel extraction to benefit another area. A first step of colonialism is often control over land to exploit both resources and labor.[2]

Today's energy systems are largely controlled by the private sector marked by outsized profits that remain unquestioned even as many struggle to pay ever-increasing utility bills. Fossil fuel interests tend to reinforce and take advantage of structural inequality while promoting the sector as racially sensitive and inclusive. This subterfuge is one of many forms of energy violence. **Violence** involves the "intentional use of physical force or power, threatened or actual, against oneself, another person, or against a group or community, which either results in or has a high likelihood of resulting in injury, death, psychological harm, maldevelopment, or deprivation."[3]

The fossil fuel industry dominates the "playing field," creating disparity amongst private and public interests. The concept of **energy violence** draws attention to repression and disinformation throughout the sector as well as preventable fatalities and systematic practices in specific projects and companies. Decision-makers can knowingly drive irreversible harm that diminishes the quality of life of groups with less power.[4]

Slow violence emerges from structural inequalities ultimately resulting in damage to oneself, another person, or against a group or community. Slow "'everyday" violence (i.e., pollution, poverty, racism, exploitation, etc.) is by nature an ongoing, lengthy

---

[1] Finley-Brook, M., & Thomas, C. (2013). Renewable Energy and Human Rights Violations: Illustrative Cases from Indigenous Territories in Panama, in *The New Geographies of Energy: Assessment and Analysis of Critical Landscapes,* Zimmerer, K. Ed. New York: Routledge. pp. 162–171.

[2] Táíwò, O. O. (2022). *Reconsidering Reparations.* Oxford University Press.

[3] World Health Organization. (2014, January 9). Global status report on violence prevention 2014.

[4] Britton-Purdy, J. (2016, March 21). The violent remaking of Appalachia. *The Atlantic.*

process. The related concept of "slow death" exposes the severity of potential conse-quences from delayed suffering.[5] Many energy types also contribute to patterns of punc-tuated **fast violence**, such as acutely harmful events involving fires, explosions, blow outs, or collapse. There generally tends to be more notice of fast violence that is dramatic and visible, often with an obvious and direct material cost. Fast violence captures media attention, though sometimes only momentarily. Yet slow violence that pervades throughout poverty or toxic exposure is deadly to greater numbers.

Violence is ordinary and extraordinary.

It is everywhere and geographically uneven.

Through its outsized power, the energy sector holds sway over both social and political norms. **Power** is generally the ability to do or act, or the capability of doing or accomplish-ing something. Having "power over" others includes energy companies taking advantage of near-monopoly control, and political campaign contributions designed to influence pol-iticians. When power is not shared, evidence often exists of both fast and slow violence to vulnerable groups who have been kept from decision-making processes. In its darkest form, **necropolitics**, or death politics, involves the power to decide who lives and dies.[6] Across the petrochemical industries, toxic exposures to pollution are highly uneven, marked by corridors and clusters of slow and fast violence.[7] **Exposure** involves the amount and frequency with which a chemical substance comes into contact with a person.

In our modern era, violence toward environmentalists is not uncommon.[8] Lucrative en-ergy projects in repressive countries rely on private security, police, military, and paramilitary operations to violate human rights as a means of control.[9] The term climate colonialism highlights connections between historical and contemporary oppression further com-pounded by patterns of climate disruption, whether in terms of where harm is most concen-trated (i.e., Most Affected Peoples and Areas [MAPA]; the global South) or in terms of who suffers (minorities, women, children, elderly, disabled, etc.).[10] **Short-termism** is driven by a narrow quantitative focus and investment protocol to boost immediate profit regardless of its costs to long-term security. Over decades, energy companies have embraced mechanization

---

[5] Nixon, R. (2011). *Slow Violence and the Environmentalism of the Poor*. Harvard University Press; Berlant, L. (2007). Slow death (sovereignty, obesity, lateral agency). *Critical inquiry*, 33(4), 754–780.

[6] Mbembe, A. (2019). *Necropolitics*. Duke University Press.

[7] Davies, T. (2018). Toxic space and time: Slow violence, necropolitics, and petrochemical pollution. *Annals of the American Association of Geographers*, 108(6), 1537–1553.

[8] Menton, M., & Le Billon, P. (Eds.). (2021). *Environmental defenders: Deadly struggles for life and territory*. Routledge.

[9] Finley-Brook, M. (2019). Extreme Energy Injustice and the Expansion of Capital in *Organized Violence and the Expansion of Capital*. Paley, D. & Granovsky-Larsen, S. Eds. University of Regina Press. pp. 23–47.

[10] Sultana, F. (2022). The unbearable heaviness of climate coloniality. *Political Geography*, 99, 102638.

in order to downsize employment costs. In the context, pay-for-performance incentives permeate industry workplaces regardless of social and ecological damages.

While both nations and individuals may try to calculate an escape from intensifying global ecological crises, no one is immune, and many face the worst impacts. Climate violence is accentuated by persistent power inequities between the global South and North. Frontline populations experience heightened psychological and physical harm. **Frontline** refers to proximity and exposure to harm, often considered a "last line of defense", even as the most extreme exposures occur here as a result of regulatory failure.[11] Injustice of the climate crisis is rooted in the fact that those who suffer the greatest losses and damage have not generated the emissions nor benefited from their release. To underscore the point, the 54 countries of the African continent constitute 15% of the world's total population but have produced less than 4% of its global greenhouse emissions in 2021. These very same countries are now consistently experiencing debilitating losses from extreme weather.[12]

Decolonizing energy

Inherent to all, clean air should be an obvious universal human right.[13] Though less essential, access to electricity is also a basic right. Households lacking electrification significantly struggle with living conditions, particularly compared to those with access.[14] **Energy poverty** occurs when people can't afford utilities or where there is no access to the electrical grid. Reliance on dirty kerosene, diesel, or biomass with high particulate emissions is more common in low-wealth areas. In such households, women and girls can suffer from higher respiratory disease due to poor indoor air quality in kitchens where unfair gender distribution of domestic chores can be found. **Energy burden**, or the portion of money or time paid for energy, means that poor households pay a greater portion of their total income per watt compared to wealthy households.[15] With high energy burden, resources are siphoned from low-wealth communities for inflated utility rates that advantage corporate executives, shareholders, and complicit politicians.

In a global context, **climate change** refers to long-term shifts in temperatures and weather patterns. Whiles some variance is natural, since the 1800s human activities have been the main driver of climate change. This is primarily due to the burning of fossil fuels (i.e., coal, oil, and gas), which produces **greenhouse gases** (GHGs) (i.e., gases that trap heat in the atmosphere: carbon dioxide, methane, nitrous oxide, and fluorinated

---

[11] Global Witness. (2020, September 13). Last line of defense.

[12] Lakhani, N. (2022, November 8). Climate crisis will have huge impact on Africa's economies, study says. *The Guardian*.

[13] Mbembe, A., & Shread, C. (2021). The universal right to breathe. *Critical Inquiry*, 47(S2), S58–S62.

[14] Hughes, M. (2018, December 10). Why access to energy should be a basic human right. *Forbes*.

[15] National Association for the Advancement of Colored People. (2017, March). Lights out in the cold.

gases).[16] Recent reports of the Intergovernmental Panel on Climate Change (IPCC) all demonstrate without exception the need to end fossil fuels immediately.[17] Despite the panel's urgent declarations, new investment and activity in fossil fuel extraction continues apace; indeed, it is intensifying across particular fuels, notably oil and gas.

Politicians have defended the interests of fossil fuel executives at the expense of the workers who have powered the energy industry. As operations have become increasingly mechanized thereby displacing workers, the increased use of fossil fuels through mechanization has only bolstered the **GHG intensity** of the energy sector wherein more carbon dioxide equivalent (CO_2e) emissions to produce 1 kilowatt of energy per hour occur, all while lowering job opportunities.

Expensive fossil fuel infrastructure represents sunk costs, deterring investment in renewables. The world needs low-carbon energy from solar, wind, wave, geothermal, and other renewable sources. Ugandan climate activist Vanessa Nakate, founder of the Rise Up Movement, succinctly states,

…we need to immediately stop digging and burning fossil fuels…stop funding fossil fuel infrastructure now. Anything less is inexcusable.[18]

Today's infrastructure expansion, like climate governance, operates from the same colonial footing that has led to climate crisis itself. "Just" transition (Conclusion) to sustainable and equitable sources of energy requires alternatives that are structurally different from business-as-usual.[19] Researchers performing simulations of future hurricanes have found it easier to restore power in more decentralized infrastructure that is strategically dispersed than in large, centralized hubs.[20]

Coupled with fossil energy intensification, decades of political inaction reinforce old and new harms, disregarding fixed temporal, spatial and conceptual boundaries.[21]

[16] United Nations. (n.d.) What is climate change?
[17] Intergovernmental Panel on Climate Change (IPCC) (2021). Sixth Assessment Report.
[18] Nakate, V., Kaim,V., & Gibson, E. (2021). Walking the Talk. *Finance & Development.*
[19] Overland, I. (2019). The geopolitics of renewable energy: Debunking four emerging myths. *Energy Research & Social Science, 49*, 36–40.
[20] Budryk, Z. (2023, January 23). Federal study calls for rooftop solar panels to meet Puerto Rican renewable energy goals. *The Hill.*
[21] Morin, E., & Kern, A. B. (1999). *Homeland Earth: A manifesto for the new millennium.* New York: Hampton Press; Morello-Frosch, R., Pastor, M., Sadd, J., 2001. Environmental justice and southern California's "riskscape" the distribution of air toxics exposures and health risks among diverse communities. Urban Aff. Rev. 36, 551–578; Cain, L. R., & Hendryx, M. (2010). Learning Outcomes among Students in Relation to West Virginia Coal Mining: an Environmental Riskscape Approach. *Environmental Justice, 3*(2), 71–77.

Political negligence has allowed an altered Earth marked by crises with mounting economic costs. Steeped in politics, mitigation delay has serious societal repercussions as the toll of extreme weather events, crop failures and resource conflict mount. State and global economies both face accelerating disruptions beyond just "normal" economic benchmarks like inflation and recession.

Climate crisis

In our current geologic Holocene epoch, a more precise yet unofficial designation **Anthropocene** is used to denote human activity inexorably linked to measurable global environmental change, including climate disruption.[22] In this period, particular locations experience a form of **double exposure**—the double impact of climate change and poverty—or even triple exposures, given racism, sexism, and other forms of persistent structural violence. Fossil fuel economies themselves, as well as their operations, are uneven—socially, economically and spatially—resulting in climate harms that exacerbate existing vulnerabilities and exposures.

World leaders have demonstrated ineptness and ineffectuality in the face of existential crisis, ignoring **climate science** (climatology), the scientific study of Earth's climate, typically defined as weather conditions averaged over a period of at least 30 years. The failing of politicians is marked by willful ignorance as well as disregard for the suffering and death of others—a pathological necropolitics buoyed by denialism and disinformation.

> **A cruel irony of climate disruption**
>
> **is those who suffer the greatest**
>
> **usually contribute the least emissions.**

Ecological debt is an accumulated obligation for restitution after exploitation of resources and degradation of the environment occurs, most pronounced in poor countries. Decades ago affected countries began to ask for reparations; the topic of reparative justice remains a sticking point in climate negotiations, particularly in contentious deliberations of **loss and damage,** which is defined by a "polluters pay" principle. Loss refers to that which is gone forever and cannot be brought back, such as people who die or species that go extinct. Damages refer to financial and societal costs and harms which may be recouped, rebuilt, or replaced, at least partially, depending on reslience and investment.

A spatial approach to energy is useful to determine responsibilities for loss and damage. Time too is a prominent factor of ecological debt accumulation across generations. Young people today face loss and damage from emissions produced both now and by past generations. However, timely justice and mitigation is continually withheld, in part due to disparity of resource

---

[22] Lewis, S. L., & Maslin, M. A. (2015). Defining the anthropocene. *Nature, 519*(7542), 171—180.

flow based on power inequities. Broadly, **flow** is movement during a time period, like an electrical circuit. The material flows and discursive influences of the energy sector cross borders and spread around the globe. **Connectivity** connotes interaction: it occurs through geographical features linked to each other functionally, spatially, or logically. When analyzing energy flows, connectivity shows how features such as hubs or nodes (i.e., a drill pad, power plant, transmission tower, or end user) are linked to one another. While local energy systems traditionally created infrastructure paths designed to obtain the shortest route in terms of Euclidean (straight line) distance from one site to another, our globalized world relies upon supply circuits that are complex and frequently non-linear. Today's fossil energy sector has amassed power in and through its materiality, as well political influence, to such a degree that its created juggernaut threatens the very interconnectivity of our social and ecological systems. The supply chains on which we now rely face increasing disruptions only exacerbated by climate change.

Methane madness

Methane as a GHG pollutant deserves particular attention because of its high potency as a destructive force to the atmosphere over the next decades. In this book, methane's particular temporal and spatial impacts are termed **methane rift**, as measurement, regulation, emissions, and energy violence commingle in uniquely destructive ways. Derived from the Marxian concept of metabolic rift—generalized ecological crisis at the hands of capitalism—methane rift shares characteristics of carbon rift where emissions drive climate warming, creating an existential rupture between humans and the biosphere.[23] GHGs are generally invisible to the naked eye, and methane (CH_4) is no exception. **Leakage**, a spatial process associated with flow beyond boundaries, means pollutants like methane escape from drill pads and pipeline infrastructure. These "fugitive" emissions are built into the system with costs paid by ratepayers. There is little incentive for companies to change, since methane emissions are inadequately regulated and harms have been externalized from the consumer price of fossil fuels.

Methane is poorly measured and intentionally downplayed throughout the energy sector, particularly in fossil gas,[24] although also in hydropower[25] and coal.[26] **Fossil gas**

---

[23] Clark, B., & York, R. (2005). Carbon metabolism: Global capitalism, climate change, and the biospheric rift. *Theory and Society, 34,* 391–428.

[24] Alvarez, R. A., Pacala, S. W., Winebrake, J. J., Chameides, W. L., & Hamburg, S. P. (2012). Greater focus needed on methane leakage from natural gas infrastructure. *Proceedings of the National Academy of Sciences, 109*(17), 6435–6440; Howarth, R. W. (2020). Methane emissions from fossil fuels: exploring recent changes in greenhouse-gas reporting requirements for the State of New York. *Journal of Integrative Environmental Sciences,* 1–13.

[25] Kemenes, A., Forsberg, B. R., & Melack, J. M. (2007). Methane release below a tropical hydroelectric dam. *Geophysical Research Letters, 34*(12).

[26] Warmuzinski, K. (2008). Harnessing methane emissions from coal mining. *Process Safety and Environmental Protection, 86*(5), 315–320.

is generally referred to as "natural" gas.[27] Industry uses this name to suggest that gas is clean. Selling gas as "natural" ignores **hydraulic fracturing** (fracking), when unconventional fossil gas is produced from shale and other types of sedimentary rock formations by forcing water, chemicals, and sand down a well under high pressure (Chapter 8). Fracking increases gas flow and raises yield and profit in the short term yet it uses "forever chemicals" and leads to the production of technologically enhanced naturally occurring radioactive materials (TENORMs) with serious long-term damage to local communities. A comprehensive cradle-to-grave strategy (i.e., life cycle analysis) with data from independent experts exposes gas as a problematic fossil fuel.[28] Fossil gas produces hazardous pollutants with health risks including asthma, heart attacks and strokes, preterm delivery, low birth weight, cancer, and early death.[29] Exposure to methane itself has been linked to increases in cardiovascular disease.[30] Fracking processes produce a variety of hazardous substances including heavy metals, radioactive materials, and volatile organic compounds (VOCs). Air pollutants emitted by gas extraction sites and gas-powered electricity generation plants include hydrogen sulfide (H_2S), nitrogen oxides (NO_x), ozone (O_3), particulate matter (PM), and sulfur dioxide (SO_2).[31] Fossil gas is a major contributor to anthropogenic climate change. Methane leaks cause heating on a relatively brief timescale.[32]

Coal, oil, and gas production are major global sources of methane, much of it wasted. Fugitive methane mitigation is considered a "low hanging fruit" of climate action. Reducing methane pollution is achievable because valuable gases escape, meaning

[27] Clairemont, N. (2022, June 2). Word of the Week: 'Gas'. *The Washington Examiner.*

[28] Colborn, T., Kwiatkowski, C., Schultz, K., & Bachran, M. (2011). Natural gas operations from a public health perspective. *Human and ecological risk assessment: An International Journal, 17*(5), 1039–1056; McKenzie, L. M., Witter, R. Z., Newman, L. S., & Adgate, J. L. (2012). Human health risk assessment of air emissions from development of unconventional natural gas resources. *Science of the Total Environment, 424,* 79–87; Tollefson, J. (2013). Methane leaks erode green credentials of natural gas. *Nature, 493*(7430), 12; Greiner, P. T., York, R., & McGee, J. A. (2018). Snakes in The Greenhouse: Does increased natural gas use reduce carbon dioxide emissions from coal consumption? *Energy Research & Social Science, 38,* 53–57.

[29] Partin, M. R. (2020). The Health, Safety, Climate and Economic Risks of Fossil Gas Extraction and Use. Sierra Club/MN 350.

[30] Mendoza-Cano, O., Trujillo, X., Huerta, M. et al. (2023). Assessing the relationship between energy-related methane emissions and the burden of cardiovascular diseases: a cross-sectional study of 73 countries. *Sci Rep* 13, 13515. https://doi.org/10.1038/s41598-023-40444-7

[31] Nordgaard, C. L., Jaeger, J. M., Goldman, J., Shonkoff, S. B., & Michanowicz, D. R. (2022). Hazardous air pollutants in transmission pipeline natural gas: An analytic assessment. *Environmental Research Letters.*

[32] Brandt, A. R., Heath, G. A., & Cooley, D. (2016). Methane leaks from natural gas systems follow extreme distributions. *Environmental Science & Technology, 50*(22), 12512–12520.

controls can pay for themselves by capturing resources.[33] Super emitters with massive releases are responsible for a significant portion of methane, so first targeting negligent operations has an impact.[34] A small number of firms—the most intensive producers of GHGs—are most blameworthy for climate change,[35] while the industry itself clusters at the top of the most destructive global industries.[36] Historically, the sector has been very proficient at avoiding responsibility, but recent independent monitoring innovations show great promise, and are already providing increased transparency.[37]

This book highlights regulatory gaps driving ecological crises; for example, regulations designed to first measure then curb methane flares and leaks remain largely weak and voluntary. We have passed ecological thresholds faster than predicted due to reckless fossil fuel extraction. We approach several **tipping points** with alarming speed—benchmarks by which changing climate could push parts of the Earth's system into abrupt or irreversible change. After reaching a tipping point (Fig. 1), additional inputs surpass limits and will push the biosphere into a new state.[38] Tipping points portend a critically challenging future.[39]

Climate change is not a linear process. Tipping points can further propel us beyond global averages 4−5°C higher than pre-industrial temperatures.[40] A core theme in the climate crisis is the factor of time. GHG emissions released in the past remain in the atmosphere contributing to climate change. We have waited past time when small or easy GHG reductions could achieve necessary emission reductions. Any luxury of taking incremental steps is now past due after decades of insufficient action. The window to maintain a habitable Earth is rapidly closing. Experiencing climate disruption, we must simultaneously gather extra resources and energy to act against it.

[33] UN Economic Commission for Europe. (2017). UNECE gas experts discuss methane emissions.

[34] Grant, D., Jorgenson, A., & Longhofer, W. (2020). *Super Polluters: Tackling the World's Largest Sites of Climate-Disrupting Emissions.* Columbia University Press.

[35] South, N. (2015). Anticipating the Anthropocene and greening criminology. *Criminology & Criminal Justice, 15*(3), 270−276.

[36] Erickson, P., van Asselt, H., Koplow, D., Lazarus, M., Newell, P., Oreskes, N., & Supran, G. (2020). Why fossil fuel producer subsidies matter. *Nature, 578*(7793), E1−E4.

[37] Erland, B. M., Thorpe, A. K., & Gamon, J. A. (2022). Recent Advances Toward Transparent Methane Emissions Monitoring: A Review. *Environmental Science & Technology, 56*(23), 16567−16581.

[38] McSweeney, R. (2020, February 10). Explainer: Nine 'tipping points' that could be triggered by climate change. *Carbon Brief.*

[39] Armstrong McKay, D. I., Staal, A., Abrams, J. F., Winkelmann, R., Sakschewski, B., Loriani, S., .. & Lenton, T. M. (2022). Exceeding 1.5 C global warming could trigger multiple climate tipping points. *Science, 377*(6611), eabn7950.

[40] Steffen, W., Rockström, J., Richardson, K., Lenton, T.M., Folke, C., Liverman, D., Summerhayes, C. P., Barnosky, A.D, Cornell, S.E., Crucifix, M., Donges, J.F., Fetzer, I., Lade, S.J., Scheffer, M., Winkelmann, R., and Schellnhuber, H.J. (2018) Trajectories of the Earth System in the Anthropocene. *Proceedings of the National Academy of Sciences (USA)*, DOI: 10.1073/pnas.1810141115.

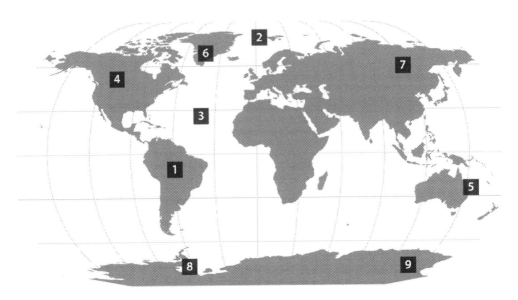

Figure 1 Examples of global tipping points. *(Adapted from: CodeOne, DWikiMan via Wikimedia Commons.)*

A warming threshold of 1.5° or less is now understood as a relatively safe target to avoid tipping points.[41] Yet we face moving beyond the critical 1.5° threshold by as early as 2027;[42] in fact, each of the 12 months of 2023 consecutively breached the threshold.[43] While the specific impacts of further warming past 1.5° remain unclear,[44] if heating continues on a trajectory to "Hothouse Earth" (Fig. 2) life as we know it now would be severely disrupted, facing potential extinction.

A tipping points framework has also been used in social analysis. For example, with the Dakota Access Pipeline (DAPL) resistance (Chapter 7), concerned Native populations and allies were able to draw enough attention to spur a global movement and disrupt construction. In 2014, Indigenous youth sought to grab attention and provoke a rippling response

---

[41] Livingston, J. E., & Rummukainen, M. (2020). Taking science by surprise: The knowledge politics of the IPCC Special Report on 1.5°. *Environmental Science & Policy, 112*, 10–16.

[42] World Meteorological Organization (2023, May 17). Global temperatures set to reach new records.

[43] *First year-long breach of 1.5 degrees Celsius could be more eduring without accelerated action by world leaders.* Union of Concerned Scientists. (n.d.). https://www.ucsusa.org/about/news/first-year-long-breach-15-degrees-celsius-could-be-more-enduring-without-accelerated

[44] Hood, M. (2023, May 31). 1.5C of warming is too hot for a just world: study. Phys.org.

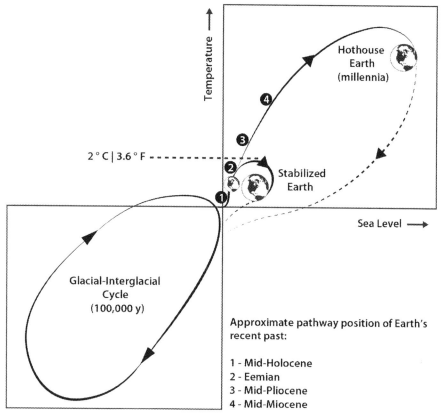

Figure 2 Hothouse Earth. *(Adapted from: Trajectories of the Earth System in the Anthropocene, W. Steffen, et al., 2018.)*

bringing waves of support to the Standing Rock movement to stop the DAPL.[45] Since leaders have failed to act prudently, we are now in a race with a social tipping point—as has been identified in the civil rights movement—when enough people stand up and transformative change ensues. Groups like Just Stop Oil in the UK and *Ultima Generazione* (Last Generation) in Italy engage in disruptive acts to bring attention to climate breakdown. **Civil disobedience** is the active, professed refusal to obey certain laws, demands, orders, or commands. Nonviolent civil disobedience has historically been a way to achieve mass social change when all other avenues have failed (e.g., women's right to vote in the international Suffragette Movement; the US civil rights movement to end segregation).[46]

History has shown upsurges in mobilization are coupled with transformation. As individuals and collectives demand change, leveraging their agency to shift the attention of

[45] Táíwò, O. O. (2022). *Reconsidering Reparations*. Oxford University Press.
[46] Malm, A. (2021). *How to blow up a pipeline*. Verso Books.

political representatives, collective "people power" can alter the balance towards the reversal of GHG emissions.[47]

How do we remove roadblocks to climate action?

Stark cognitive dissonance persists today between our knowledge about climate change and action to limit its disruption after decades of false equivalence—acting as if belief in climate change was a "perspective" rather than fact. Media outlets granted equal air time to scientists and deniers, in a show of "both-sidism." Ads and paid infomercials from industry promote fossil fuel disinformation and **climate denialism**—the dismissal or unwarranted doubt that contradicts the scientific consensus on climate change, including the extent to which humans have caused it.[48]

Manufacturing ignorance

The production of ignorance, the study of which is known as agnotology, involves deliberate manufacturing and dissemination of misinformation.[49] Misinformation is widespread in the energy sector. For example, a common denial tactic includes spreading the pervasive myth that environmentalism equates to job loss and economic downturns. To the contrary, independent analysis shows renewable alternatives are excellent for employment generation while building forward and backward linkages to multiply jobs in local economies.

We live in a time of both local and a global information deficit. Post-truth normalcy encourages **environmental illiteracy**, or the lack of ability to make informed decisions pertaining to issues including pollution and natural resource use. Instead of relying on science, leaders co-opted by industry create ineffective targets (i.e., carbon-zero by 2050) and false solutions (i.e., carbon capture and storage).

Epistemic justice requires accurate access to information and fair production of knowledge. Instead we live in a time of epistemological privilege where having money allows for access to produce knowledge, creating unfair advantages in the production and dispersal of truth. The way forward becomes opaque following multi-million-dollar campaigns aimed to obfuscate. This exacerbates a collective public **nonresponse**, referring to the absence of a verbal or written reply, however subtle, to climate change. Populations cognizant of the science of climate change cannot come to terms with rapid shifts in their daily lives. There is both (1) a preference to not see and name it and (2) "not

---

[47] Otto, I. M., Donges, J. F., Cremades, R., Bhowmik, A., Hewitt, R. J., Lucht, W., .. & Schellnhuber, H. J. (2020). Social tipping dynamics for stabilizing Earth's climate by 2050. *Proceedings of the National Academy of Sciences*, 117(5), 2354–2365.

[48] Carroll, S. B. (2020, November 8). The Denialist Playbook. *Scientific American*.

[49] Proctor, R.N., & Schiebinger, L. Eds. (2008). *Agnotology: The Making and Unmaking of Ignorance*. Stanford University Press.

knowing *how* to know" such that people use "cultural narratives to deflect disturbing information and normalize a particular version of reality in which 'everything is fine.'"

Informed nonresponse to climate chaos
is reproduced in our everyday lives.

Informed nonresponse, or collective lies (i.e., shared mistruths or partial realities), are used when it is difficult to face the truth. This is what occurred after the "Great Bleaching" of the Great Barrier Reef in Australia in 2016 as even the national tourism industry refused to acknowledge or criticize coal extraction or the ties between coral bleaching and climate change. Meanwhile, tourists coming to the Great Barrier Reef post pictures of beautiful fish in remnants of dead reefs in a type of collective amnesia without knowing how much richer and full of life the world once was.[50] A **shifting base syndrome** can occur in the absence of any point of comparison, allowing an assumption of normality.

Climate disruption is a **hyperobject**: something that is distributed across space and time to such an extent that our minds struggle to comprehend it. Because of this, this book employs helpful heuristics such as space, scale, and power to communicate patterns and connections between locations. Heuristic examples belonging to power structures often feature **censorship**, the restriction of collections, displays, dissemination, and exchange of information, opinions, ideas, and imaginative expression. Laws silencing science have become common[51] while weather channel coverage of disasters generally ignores or minimizes discussion of climate change.[52] Fossil fuel companies have aggressively channeled their power into marketing and lobbying their own version of the **deception playbook** (Fig. 3) developed first by tobacco and fast food industries.

Common industry tactics include deploying "spin masters" from public relations departments, who pretend to be honest brokers sharing growth forecasts or seemingly sound environmental strategies. Google and other internet search engines commonly return manufactured market greenwash.[53] Unable to distinguish sound research results from purchased ads appearing as top-of-page results, public interest is undermined.[54] In similar fashion, the industry front group Energy In Depth couples its prolific ad spaces with an extensive pro-fossil fuel blog disguised as a journalism outlet.[55]

[50] Ritter, D. (2018). *The Coal Truth*. UWA Publishing.
[51] Sabin Center for Climate Climate Law. (n.d.) Silencing Science Tracker.
[52] Hewett, F. (2021, July 2). Media Coverage Of Climate Change Is Improving. But That Alone Won't Stamp Out Disinformation. *WBUR*.
[53] Asher-Schapiro, A. (2022, November 2). Gaming Google: Oil firms use search ads to greenwash, study says. *Reuters*.
[54] McIntrye, N. (2022, January 5). Fossil fuel firms among biggest spenders on Google ads that look like search results. *The Guardian*.
[55] Kaufman, A. C. (2022, March 5). A natural gas giant is waging a sneaky war on a minor Colorado climate policy. *Huffington Post*.

Tactic	Description
anti-regulation	• limit climate action to private sector constructs, like carbon credits • focus on individual responsibility • allege that environmental laws and civil rights codes usurp individual rights and personal freedoms
polarization	• slander or disparage scientists and advocates • argue consensus is false (i.e., hire a contrarian scientist, use minor examples or side issues that seem to not support the consensus) • employ either-or binaries (i.e., jobs vs owls); divide-and-conquer
disinformation	• reify false solutions that maintain fossil energy • promulgates doubt about transformative change • vilify critics (i.e., alarmist, unrealistic, exclusive) • promote fear about economic downturn; assert economic reliance on status quo without admitting inequality • downplay true costs, such as public health expenses • defame alternative paths • impede evidence (i.e., nondisclosure agreements)
distraction	• put social advocates or environmentalists on the defensive with frivolous investigations or lawsuits • create elaborate philanthropic campaigns • design media campaigns about offsets • greenwash (i.e., exaggerate commitments)

Figure 3 The deception playbook

Climate anxiety

A 2021 global study found that four in 10 people ages 16−25 feared bringing a child into the world due to the climate crisis.[56] Climate anxiety is adding to an already existing mental health crisis.[57] One form of energy violence is the psychological trauma caused by climate change, made worse due to feeling helpless and betrayed by global leaders.[58]

---

[56] Harvey, F. (2021, September 14). Four in 10 young people fear having children due to the climate crisis. *The Guardian.*

[57] Sullivan, T. (2022, April 4). 'Climate anxiety' is on the rise: Here's how leaders can address the mental health crisis. *Health Evolution.*

[58] Hickman, C., Marks, E., Pihkala, P., Clayton, S., Lewandowski, E. R., Mayall, E. E., .. & van Susteren, L. (2021). Young people's voices on climate anxiety, government betrayal and moral injury: A global phenomenon. *Government Betrayal and Moral Injury: A Global Phenomenon.*

Children particularly are at risk of unfair physical harm: pollutants have some of the most severe and long-term consequences on developing organs in young bodies. Comprehensive prevention of pollution is of upmost urgency as harms to fetuses, babies, and youth are often irreversible.[59] While minors should have legal protections for their well-being as trusts of the state, this responsibility has been abnegated, leading to lawsuits from nonprofit groups like Our Children's Trust (OCT) seeking redress of withheld protections.[60]

Supporting youth contributes to collective resilience.[61] With fresh perspectives and original viewpoints, youth leadership has been instrumental, whether as ecologists, activists, or inventors. The climate movement is a space for young leaders, especially those from MAPA whose families face a polycrisis of climate breakdown woven with other challenges, to be heard and to organize equitable solutions.[62] As it can be hard to discern exactly how and where to act with the greatest urgency, one objective of this book is to help readers dissect the energy sector across their very own geographies and from their own informed perspectives to best engage climate challenges both local and global. A "toolbox" approach provides action-oriented methods while engaging in critical reflection. A series of exercise prompts within each chapter are followed by an extended practice guide (Praxis) aimed to integrate the book's theoretical framework with as series of concrete learning approaches.

The climate is changing. Why aren't we?

Assumptions we've taken for granted, experiences we've acquired, skills we've learned may no longer prove adequate to a rapidly changed world. Energy disinformation campaigns will continue to undermine the public's best interests, even faced with existential threats. In this situation, **discontinuity**—a break from the past—should be not only be expected, but prepared for by accelerating and honing our educational foundation in energy transformation.

Social paralysis benefits those who are most responsible for bringing us to this precipice— the energy industry—as climate action threatens their profit margin. Two strategies are at play across climate obstruction: (1) delay or weaken global targets so there is less regulatory pressure; (2) embrace voluntary markets to undercut and distract from transformative change.

We must unflinchingly face the direness of the situation
without collapsing into cynicism and paralysis.

---

[59] Steingraber, S. (2010). *Living downstream: An ecologist's personal investigation of cancer and the environment.* Da Capo Press.

[60] Our Children's Trust. (n.d.). Youth v Gov.

[61] Rousell, D., & Cutter-Mackenzie-Knowles, A. (2020). A systematic review of climate change education: Giving children and young people a 'voice' and a 'hand' in redressing climate change. *Children's Geographies, 18*(2), 191−208.

[62] Staff. (2021, November 15). In the fight against climate change, young voices speak out. *New York Times.*

Without clear and equal access to information, not all communities know how or why to fight back against fossil industries that deliver yet more toxic burdens and climate disruption.[63] Such is the story of an Argentine shantytown, known as "Flammable," that hosts Shell Industries and a series of associated chemical plants. People living in neighborhoods with refineries often feel confused about their exposures, the pollution's spatial distribution, and its concrete impacts.[64] A **nonview**, or a form of blindness to the situation, can develop regarding the sources and effects of toxicity. Ignorance transforms into self-doubt regarding the extent of contamination and creates divisions and stigmas regarding who is "really contaminated." Residents are constantly waiting for further testing that will "truly" demonstrate the effects of pollution. They are also waiting for an "always imminent" state relocation plan and a settlement with a company that will "allow us to move out." While waiting, already vulnerable residents submit to a further damaging reality—a slow violence unidentified by its lack of information.

Soft denialism involves believing in climate change but continuing to act as if it were not real. Contradictions like this lead to the emergence of psychological defenses, such as disavowal (Conclusion): a subtle but powerful form of soft denialism whereby impacts aren't fully recognized due to subconscious psychological defenses. Disavowal can be described as keeping only one eye open to climate breakdown. Self-censorship occurs frequently when knowledgeable people choose to not talk about climate change in the presence of deniers.

Try this

Reflection and discussion: in your perspective, what is the most common form of environmental illiteracy? What underlying biases are linked to this misunderstanding?

Hyper greenwash

The term greenwashing was coined in the 1980s to refer to dishonest portrayals of positive environmental acts with inflated claims.[65] For decades many companies have been pronouncing their operations as "green" without actually making products in a sustainable way. **Greenwashing** is generally understood as intentional environmental deception to embellish environmental positives or obscure negatives. Signature acts include using fluffy

---

[63] Auyero, J., & Swistun, D. A. (2009). *Flammable: Environmental suffering in an Argentine shantytown.* Oxford University Press.

[64] Auyero, J., & Swistun, D. (2007). Confused because exposed: Towards an ethnography of environmental suffering. *Ethnography, 8*(2), 123–144.

[65] Watson, B. (2016, August 20). The troubling evolution of corporate greenwashing. *The Guardian.*

language and suggestive pictures (Fig. 4). The chapters referred to in the figure feature deeper explorations of each greenwashing example.[66–70]

We have included detailed discussion of US greenwashing as the US is a trend setter in coal oil and gas violence as well as industry disinformation. Nonetheless, Saudi Aramco greenwashing of oil[71] and Russian Gazprom depictions of gas[72] engage in comparable forms of ecological deception. Around the world, energy firms have superior access to media and communication resources, which they use to obfuscate with strategies of deception, including sophisticated greenwashing, which serves to distract from criticisms of social and ecological failings. Previously, greenwashing was considered relatively benign—a normal skill set for public relations teams serving large energy and utility companies.[73] Today, greenwashing opearates as repressive control over knowledge at the very moment science clearly necessitates collective climate action.

As environmental mistruths pushed by companies are exposed, and more people push for more effective regulations, companies have responded with yet more complex deception. In 2021, the National Association for the Advancement of Colored People (NAACP) identified 10 forms of fossil fuel deception (Fig. 5)[74].

Try this

Research and analysis: *Look at examples of energy sector advertisements or media campaigns. Do they demonstrate fossil fueled foolery discussed in* Fig. 5?

---

[66] Levantesi, S. (2022, June 6). Climate deniers and the language of climate obstruction. Desmog.

[67] Price of Oil. (2022, August 11). Madness is the Method: How Cheniere is Greenwashing its LNG With New Cargo Emissions Tags.

[68] Saudi Green Initiative. (n.d.). Championing climate action at home and abroad.

[69] Carpenter, S. (2020, August 4). After Abandoned 'Beyond Petroleum' Re-brand, BP's New Renewables Push Has Teeth. Forbes.

[70] Crunden, E. A., & Wittenberg, A. (2021, December 10). Toxicologist who belittled PFAS risks resigns from EPA role. E&E News.

[71] Aramco. (n.d.). Circular Carbon Economy.

[72] Gazprom. (2021). We are the future: Gazprom Groups Sustainability Report 2021.

[73] Futerra Sustainability (2008, July 8). The ten signs of greenwash.

[74] National Association for the Advancement of Colored People. (2021). Fossil Fueled Foolery 2.0.

Deception	Descriptions	Examples
green by association	baseless virtue signaling	photo advertisement of a gas guzzling vehicle in pristine nature
lack of definition	incomplete; vagueness	'cleaner' burning, 'lower' emissions fuels, 'low' carbon future[66]
'smoke and mirrors'	misleading; exaggerating	justification of coal-to-gas transition based on carbon dioxide while ignoring other GHG emissions, such as methane
bait and switch	distracts by drawing attention to the good while ignoring or hiding the bad	offsets, monetary donations for conservation groups or charities without correcting behavior or reducing harm (i.e., Mountain Valley Pipeline **Chapter 9**)
not cradle-to-grave	lacking Life Cycle Assessment (LCA)	coal ash as 'beneficial' reuse (**Chapter 4**)
hidden tradeoffs	impact paradox - negatives occur along with improvements	2022 US Inflation Reduction Act, which tied support for renewables to a decade of expansion of fossil fuels (**Conclusion**)
rally behind a low standard	race to the bottom; cheap, not holistic	LNG cargo emissions tags[67]
rally behind a slightly improved standard	making huge claims while only being incrementally better	'Certified Sustainable Gas' (**Chapter 8**); best in class
lesser of two evils	justifying a bad choice by comparing with a worse option	'clean' coal vs regular coal (**Chapter 4**)
worshiping false labels	taking complex concepts and oversimplifying them while elevating to a fetish	net-zero (**Chapter 5**),[68] carbon positive, carbon neutral, zero emission
reluctant enthusiast	will support renewable energy when profitable, but keep all energy sources' on the table	name change of British Petroleum to Beyond Petroleum, with reversal[69]
minimizing harm	using the minimal standards that exist in law as proxies	for-hire expert witnesses, consultants that sully the reputation of regulatory agencies or undercut their authority[70]
green product from a dirty company	a majority of operations pollute heavily, but one or more cleaner alternative is marketed	AES Corp's solar farm in Guayama, Puerto Rico (**Conclusion**)
smokescreen	irrelevant or misleading claims	Frackademia (**Chapter 8**): industry financed research
inflation of scores on scorecards	overly positive grading schematics for social or ecological scorecards	when ESG numerical scores are used to suggest sustainability, while significant problems remain

Figure 4 Forms of green deception.

1	Invest in efforts that undermine democracy
2	Finance political campaigns & pressure politicians
3	Fund scientists and scientific research institutions to publish biased research
4	Say government regulations hurt the economy and low-income communities
5	Deny or understate the harms polluting facilities cause to people and the environment
6	Deflect responsibility – shift blame to communities they pollute
7	Co-opt community leaders and organizations and misrepresent the interests and opinions of communities
8	Exaggerate the level of job creation and downplay the lack of quality and safety in jobs
9	Praise false solutions while claiming that real solutions are impractical, impossible, or harmful for BIPOC (Black, Indigenous, People of Color) and poor communities
10	"Embrace" renewables to control the New Energy Economy[74]

Figure 5 Fossil fueled foolery.

Extreme energy hazardscapes

Global energy **infrastructure** is comprised of both public and private distribution and transmission networks for energy and electricity. Infrastructure is a material display of economic and political power[75]—what infrastructure is built where exhibits a visible expression of surrounding social norms.[76] Physical structures in the energy sector, such as power plants and pipelines, are representations of modern culture.[77] The prefix infra- (below, beneath, or within) suggests a hidden yet essential characteristic.[78] After construction, infrastructure tends to become unseen, "out of sight, out of mind," unless an emergency, like a fire or explosion, occurs.

[75] Pasternak, S., Cowen, D., Clifford, R., Joseph, T., Scott, D. N., Spice, A., & Stark, H. K. (2023). Infrastructure, jurisdiction, extractivism: keywords for decolonizing geographies. *Political Geography*, *101*, 102763.

[76] Bridge, G., Özkaynak, B., & Turhan, E. (2018). Energy infrastructure and the fate of the nation: Introduction to Special Issue. *Energy Research & Social Science*, *41*, 1—11.

[77] Blok, A., Nakazora, M., & Winthereik, B.R. (2016). Infrastructuring Environments. *Science as Culture*. 25: 1—22; Beuret, N. (2017). Counting carbon: Calculative activism and slippery infrastructure. *Antipode*, *49*(5), 1164—1185.

[78] Carse, A. (2012). Nature as infrastructure: Making and managing the Panama Canal watershed, *Social Studies of Science*. 42: 539—563.

Analysis of "siting" (Chapter 9), meaning to fix or build something in a particular location, is deficient unless attention is given to the place selected for installing new infrastructure, including its cultural uses and ethnic histories. A **place** is a location with emotional affect and personal meaning. Places are unique, humanized landscapes that are often important to social identity. Energy companies don't tend to recognize the importance of places. Their exclusionary goal is to move resources from one hub to another as efficiently and profitably as possible.

Companies invested in competitive fossil fuel markets seek to conquer frontiers with the placement of infrastructure on mountains, frozen tundra, platforms, and in coastal areas with recurring patterns of life-threatening storms. **Extreme energy**, or high-intensity, ecologically destructive extractive energy sector operations, is rarely defined, although it's risk-laden and growing in prevalence. One of the few decisive sources, specifically addressing extreme oil, describes it as hydrocarbons that should have remained in the ground but were driven into the world economy by capitalist pressures to extract.[79] Fig. 6 highlights several common patterns found in extreme energy projects, while recognizing not all elements exist in every project.[80]

Examples of types of energy that are frequently "extreme" include tar sands oil extraction, deep water drilling, coal seam gasification, and fracking used to extract oil and gas. However, any type of energy can be extreme. This includes even renewable energies like

Risk Patterns	Description
intensive technologies	• require vast natural, financial, and technical resources to extract and process energy (minimal net gain)
ecologically destructive	• harm at small and large scales (i.e., proximate and distant locations)
informal or illegal environments	• propensity for leveraging politics to advance inequality and injustice
spatial disjuncture	• extreme locations (i.e., deep underground, in the Arctic); • interruptions of social spaces (i.e., polluting industry proximate to residential areas)
inappropriate siting	• placement in high-risk locations • disregard for social, cultural or ecological value

Figure 6 What is extreme energy?

[79] Carroll, W. K. (Ed.). (2021). *Regime of obstruction: How corporate power blocks energy democracy*. AU Press.
[80] Finley-Brook, M. (2019). Extreme Energy Injustice and the Expansion of Capital in *Organized Violence and the Expansion of Capital*. Paley, D., & Granovsky-Larsen, S. Eds. Regina, SK: University of Regina Press. pp. 23–47.

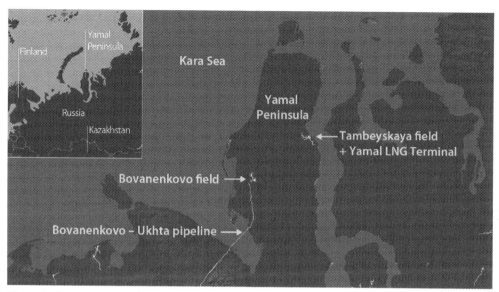

Figure 7 Extreme energy footprint, Yamal Peninsula, Russia. *(Sources: NASA Black Marble Nighttime Lights (NTL), 2016. Pipeline features extracted via OpenStreetMap (Overpass Turbo API).)*

solar, as seen with the ultra-mega parks in India[81] or fast solar in China.[82] Various components of extreme energy may be found at the same locations, creating hotspots with higher cumulative or comprehensive risk, as is the case with the liquified natural gas (LNG) facility sited at Russia's Yamal Peninsula. Extreme energy crystallizes in the Yamal LNG Terminal Project (Fig. 7 above).

Yamal roughly translates as "End of the Land" in the language of the Indigenous Nenets, and this Arctic site is risky for a massive LNG facility.[83] A common signpost of extreme energy is the desire to push physical and technical limits. Yamal LNG requires complex mitigation strategies such as thermal stabilization units (Fig. 8) to protect from permafrost.[84]

Approximately 80% of "Russian" fossil gas is located in the Yamalo-Nenets Autonomous Okrug, an autonomous territory. Russian sponsorship of Yamal LNG on Indigenous land is an act of internal colonialism. State allocation of land for oil and gas

---

[81] Chari, M. (2020, September 21). How solar farms fuel land conflicts. *Mint*.

[82] Staff. (2021, December 21). Human cost of China's green energy rush ahead of Beijing winter Olympics. NDTV.

[83] Cherepovitsyn, A., & Evseeva, O. (2021). Parameters of Sustainable Development: Case of Arctic Liquefied Natural Gas Projects. *Resources*, 10(1), 1.

[84] Forbes, B. C., Stammler, F., Kumpula, T., Meschtyb, N., Pajunen, A., & Kaarlejärvi, E. (2009). High resilience in the Yamal-Nenets social—ecological system, west Siberian Arctic, Russia. *Proceedings of the National Academy of Sciences, 106*(52), 22041—22048.

Figure 8 Stabilization units required for extreme temperature modulation. *(Image credit: Gazprom.)*

development disrespects territorial rights of the Nenets, creating displacement and a shortage of pasture space for reindeer herding activities.[85] When Yamal LNG was constructed, it was Russia's largest energy project. A railroad was built to connect to the remote area with state subsidies for a seaport and airport. Environmental ramifications abound from steel, concrete, and iron being placed across the taiga.[86] Craters erupt as high methane gas pushes through melting ground, making construction even more dangerous as terrain shifts can cause infrastructure ruptures with explosions, leaks, or spills.[87] At the time of writing, Russia is expanding additional LNG sites made possible by the melting of the Arctic (Chapter 10).

As seen from space, many extreme energy sites appear as malignant, expanding cancers—these sacrificed zones exhibit severely diminished present and future ecological services (Fig. 9). For example, after mountaintop removal for surface coal mining, rivers and valleys across millions of acres are marred indefinitely. Water may become toxic for decades on end in energy wastelands. The concept of **hazardscape** draws attention to risk associated with hazards and harms in energy sector landscapes now proliferating, significant contributors to the Anthropocene. Hazardscape builds from the core concept of

[85] Gorbuntsova, T., Dobson, S., & Palmer, N. (2019). Diverse geographies of power and spatial production: Tourism industry development in the Yamal Peninsula, Northern Siberia. *Annals of Tourism Research, 76*, 67–79.

[86] Chuvilin, E., Stanilovskaya, J., Titovsky, A., Sinitsky, A., Sokolova, N., Bukhanov, B., .. & Badetz, C. (2020). A Gas-Emission Crater in the Erkuta River Valley, Yamal Peninsula: Characteristics and Potential Formation Model. *Geosciences, 10*(5), 170.

[87] Evseeva, O. O., & Cherepovitsyn, A. E. (2019). An approach to assessment of sustainability of the large-scale Russian liquefied natural gas project. *Topical Issues of Rational Use of Natural Resources, 2*: 608.

Unconventional (Fracking) oil & gas well pads - Permian Basin, Texas, USA

Haerwusu Surface Coal Mine, Jungar Banner, Inner Monogolia, China

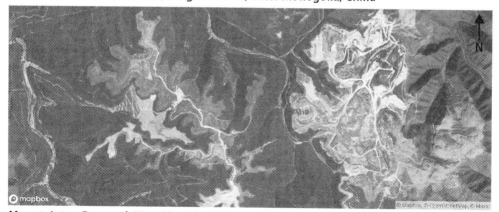

Mountaintop Removal, West Virginia, USA

Figure 9 Landscapes of extreme energy. *(Map Image credit: © Mapbox, © OpenStreetMap, © Maxar.)*

landscape, which at its base is an ideological way of "seeing": a viewshed that expresses style, significance, and ideology.[88] Landscapes materialize power.[89]

Hazardscapes depict the unprecedented disruptions of the Anthropocene.

Hazardscapes serve to challenge understandings of safety.[90] For example, the Motiva Petroleum Refinery hazardscape, located in Port Arthur, Texas, at the terminus (end point) of the Keystone XL Pipeline, illustrates how companies normalize toxic infrastructure within existing violent energy landscapes. Energy hazardscapes frequently exhibit mission creep, referring to gradual expansion beyond an original scope over time, spawned by initial success.

Situated at the southern Louisiana and Texas state border along Sabine Lake, the Motiva Petroleum Refinery in Port Arthur shares a shoreline position with three expansive LNG terminals, resulting in a region-wide hazardscape. While touted as a "green" option when compared to oil via industry greenwashing, LNG brings its own harms and dangers (Chapter 10). Taken together, these facilities compound the long history of localized pollution, increasing both the number and types of exposure for residents in Port Arthur, many of whom live remarkably close to the refinery. For years, children in Port Arthur have developed rare cancers and leukemias. High mortality rates have been documented since the 1980s,[91] as seen in other refinery locations.[92] As early as the 1970s, studies revealed high lead exposures in this area.[93]

In 1902, the Texas Company, later known as Texaco, constructed the initial refinery. In 1989, Saudi Refining secured 50% ownership. In 1998, a joint venture under the name Motiva Enterprises conjoined Texaco/Saudi Refining with Shell Oil. After expansion in 2007, Motiva Port Arthur became the largest US refinery. The large plant lies proximate to residential areas (Fig. 10); notably, only 20% of Port Arthur residents are

[88] Mitchell, D. (2005). Landscape. In Sibley, D., Atkinson, D., Jackson, P., & Washbourne, N., Eds. *Cultural Geography: A Critical Dictionary of Key Concepts*. London: I.B. Taurus, p, 50.

[89] Cosgrove, D. (2008). Geography is Everywhere: Culture and Symbolism in Human Landscapes. In Oakes, T.S., & Price, P. L., Eds. The Cultural Geography Reader. New York: Routledge, 176—185. Palgrave, London.

[90] Berg, J., & Shearing, C. (2018). Governing-through-harm and public goods policing. *The Annals of the American Academy of Political and Social Science*, 679(1), 72—85.

[91] Thomas, T. L., Waxweiler, R. J., Moure-Eraso, R., Itaya, S., & Fraumeni, J. J. (1982). Mortality patterns among workers in three Texas oil refineries. *Journal of Occupational Medicine*, 24(2), 135—141; Wen, C. P., Tsai, S. P., McClellan, W. A., & Gibson, R. L. (1983). Long-term mortality study of oil refinery workers I. Mortality of hourly and salaried workers. *American Journal of Epidemiology*, 118(4), 526—542.

[92] Schnatter, A. R., Wojcik, N. C., & Jorgensen, G. (2019). Mortality Update of a Cohort of Canadian Petroleum Workers. *Journal of Occupational and Environmental Medicine*, 61(3), 225.

[93] Eads, E. A., & Lambdin, C. E. (1973). A survey of trace metals in human hair. *Environmental Research*, 6(3), 247—252.

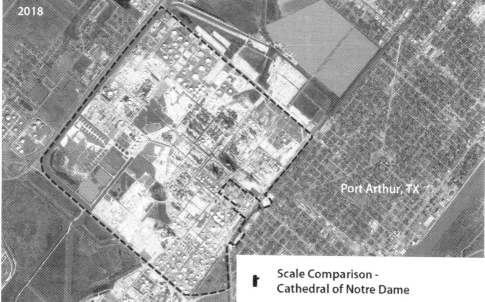

Figure 10 Motiva refinery hazardscape. *(Sources: Google Earth, © Maxar Technologies, Texas General Land Office, NASA. Motiva Boundary extracted via OpenStreetMap (Overpass Turbo API).)*

white. The area is also vulnerable to flooding and extreme weather events, including hurricanes. These events not only cause weather-related damages, but increase exposure to pollutants produced by the refinery when these events impact the refinery's facilities and waste containment areas. In communities of color like Port Arthur, there is usually less assistance in rebuilding following these extreme events.[94]

The Gulf region, where Port Arthur is located, hosts some of the largest petrochemical facilities (Chapter 5) in the US, contributing to a "cancer belt," or a strip of toxic hazardscapes extending into Texas from Louisiana's notorious "cancer alley"—an area increasingly dubbed "death alley."[95]

The US cancer belt is just one of many global toxic hotspots where environmental racism (Chapter 7) causes disproportionate harm for people of color as a specific form of environmental injustice. While pressure for social responsibility and ecological sustainability mounts in prosperous locations, low-wealth areas and communities of color continue to be unfairly burdened with dangerous pollutants coupled with environmental risk. For example, repeating and intensifying Gulf Coast hurricanes spread toxic petrochemicals. Disasters happen too frequently for communities to fully recover between events, as was the case for Lousiana residents unable to repair homes between serial hurricanes in 2020 and 2021 (i.e., Laura, Delta, Ida) compounding their exposure. Such tragedies bleed together across space and time.

Mapping power

As fossil energy has concentrated power over time, its relationships across politics and utilities, down individual neighborhoods, are marked by extreme inequities. This creates a literal regime of obstruction as types and modalities of corporate power are stratified yet highly orchestrated as shown in Fig. 11. Each of these modalities can further be classified according the various levels of transparency noted in Fig. 12.

A particularly effective technique known as a **power map** (Praxis 1) can be developed in order to effect change in power relationships. This visual tool can identify the best targets to promote social change, making it a useful analytical method to show influence. This participatory approach helps groups of impacted populations determine how to leverage change, even when faced with seemingly overwhelming odds. "Following the money" (Chapter 5) is another critically informative technique used to uncover relationships between money and politics that exist throughout the energy sector.

---

[94] Bullard, R. D., & Wright, B. (2012). *The wrong complexion for protection: How the government response to disaster endangers African American communities.* NYU Press.

[95] Coalition Against Death Alley. (2019).

Power Type	Definition	Energy Company Example
operational	decision-making	chain of command
strategic	control	board of directors; majority shareholders
allocative	availability and conditions of capital firms depend on	financial institutions (i.e., banks, insurers, asset managers, hedge funds, etc.)
instrumental	influence agendas	lobbying, campaign finance
structural	maintain the agenda, make the rules	threaten capital withdrawal or energy insecurity/blackout if regulation or decarbonization threatens profits
discursive	shaping norms, values and beliefs	corporate social responsibility (CSR), Environment, Social, Governance (ESG), corporate citizenship

Figure 11 Modalities of corporate power.

Modalities of power	Brief description	Examples
visible	observable decision making	visible and definable aspects of political power: the formal rules, structures, authorities, institutions and procedures of decision-making as well as the informal everyday politics of resistance
hidden	defining the agenda	control over who gets a seat at the decision-making table, what is on the agenda, and who has access to information (i.e., data in/equity)
invisible	shaping meaning and what is acceptable	ability to shape beliefs, sense of self, acceptance of the status quo; socialization perpetuates exclusion and inequality by defining what is normal or acceptable

Figure 12 Forms of power.[96]

Worldmaking

To move past fossil energy towards systemic transformation entails envisioning, referred to as **worldmaking**—to imagine and build places around us in ways the world can be, beyond how the world has been. Worldmaking is particularly important for young people to become involved in the creation of a planet that is different from the one they are inheriting.[97] The Earth is in flux and places are constantly becoming. We are being challenged to imagine and transform what exists, to make it better in spite of the challenges.[98]

---

[96] Gaventa, J. (2006). Finding the spaces for change: a power analysis. *IDS Bulletin*, 37(6), 23–33.

[97] Sze, J. (2020). *Environmental Justice in a Moment of Danger*. University of California Press.

[98] Pellow, D.N. (2018). *What is Critical Environmental Justice?* Wiley.

Indigenous writer Elizabeth Archuleta suggests we embrace an ethos of responsibility: our calling (moral duty) to address oppression individually and collectively (Chapter 7).[99]

Those with access to resources, power, and information can promote more ethical economies, without reproducing history's oppressions. **Capitalism** simply defined is an economic system in which private individuals or businesses own trade and industry as a means to generate profit. In a more equitable world, with better balance of power and fairer access to resources, capitalism might not be as harmful. The classism, racism, sexism, and increasingly fascism that thrive under current practices are part of a global system that has carried us to a precarious edge where choices are stark between systemic transformation or ecological overshoot.

To date, so-called "green" capitalist interventions remain shallow through subtle or incremental shifts, all the while maintaining, even increasing, exploitation and extraction. Slightly **"greener" capitalism** is not sustainable, ecologically or socially sound, and thus is not deserving of the label "green." Similarly, "sustainable development" was recognized as an outright oxymoron more than a decade ago.[100] The current pace of growth is not feasible indefinitely whereby markets undermine sustainability and perpetrate ecocide.

Systems seeking perpetual growth are inextricably linked to first extraction zones, followed by disposal zones for dumping waste and pollution.[101] Frontline communities continue to pay for a system with their very lives. As harms from profit-driven growth are evidenced in both data and lived experience, capitalism is increasingly open to criticism. Standards of living and life expectancy in capitalist countries are falling, not rising. The same can be said for most socialist countries. **Socialism** is a political and economic theory of social organization which advocates for the means of production, distribution, and exchange to be owned and regulated by the people. Governments have allowed and facilitated environmental overshoot whether communist, socialist, or capitalist.

A variety of societal and economic worldviews generate alternative frameworks. Two ideologies directly tied to ecological repair are ecosocialism and ecofeminism (Fig. 13), both of which have experienced a surge in popularity over the past decade. A critique of unbridled capitalism has emerged in the global South in post-extractivism, which complements a burgeoning degrowth movement with origins in Europe.

---

[99] Archuleta, E. (2006). "I Give You Back": Indigenous Women Writing to Survive. *Studies in American Indian Literature, 18*(4), 88–114.

[100] Redclift, M. (2005). Sustainable development (1987–2005): an oxymoron comes of age. *Sustainable development, 13*(4), 212–227.

[101] Manbiot, G. (2019, April 25). Dare to declare capitalism dead - before it takes us all down with it. *The Guardian.*

	'Greener' Capitalism	Ecosocialism	Ecofeminism	Post-extraction degrowth
popularity	hegemonic	alternative	alternative	alternative
perceived value of nature	-nature is cheap, exploitable or exchangeable -privatization and commodification are desirable	-nature is interconnected with humans -red-green alliances build sustainable non-exploitative exchanges	-humans should not oppress nature -gender, racial, and other forms of injustice must stop	-nature is finite
perceived value of society	-encourages individualized social hierarchy of power -views people with low resources as exploitable or exchangeable	-encourages collective power -attention to improving lives of workers and poor people	-inequality is ethically and functionally incorrect -sexism augments ecological harm	-prioritizes social well-being over corporate profits, over-production and excess consumption

Figure 13 Worldviews.

The **degrowth** process is designed to intentionally reduce inequality and improve well-being, a replacement of hypercompetitive capitalism leading to ecological overshoot.[102] Given global inequality, it would be insensitive to universally expect degrowth;[103] a more nuanced reading could be agnostic to growth as a goal (i.e., agrowth). Instead of fetishizing "stuff," degrowth focuses on relationships, giving attention to differences in time, space, and geography. **Post-extractivism** emerged in Latin America as an anti-imperial and anti-colonial agenda for more equitable social relations and a healthier environment.[104] These movements toward **ecosocialism** recognize intersections between social and ecological well-being, advocating for the subordination of both state and market to society (i.e. the "people").[105] Ecosocialism has roots in many traditions. In political spheres, it is poised as a critique of capitalism as harmful to society and nature. Ecosocialism in the 21st century includes Ecuador's constitutional reforms for *buen vivir* (roughly translated as "living good") and rights of nature (RoN) (Chapter 6). Grassroots

---

[102] Hickel, J. (2021). What does degrowth mean? A few points of clarification. *Globalizations, 18*(7), 1105—1111.

[103] Löwy, M. (2018). Why ecosocialism: for a red-green future. *Great Transition Initiative, 1*, 1—13.

[104] Acosta, A. (2017). Post-extractivism: from discourse to practice—reflections for action. In *Alternative pathways to sustainable development: Lessons from Latin America* (pp. 77—101). Brill Nijhoff.

[105] Löwy, M. (2018, December 19). Why Ecosocialism? A discussion of the case for a red-green future. *Climate and Capitalism*.

ecosocialism exists in *Via Campesina*, a local-to-global network of subsistence farmers advancing food sovereignty and agrarian justice.[106] These examples make up part of global efforts to block corporations from "toxic trespass," and remind us that there are viable alternatives based in life-affirmative practices. Likewise, **ecofeminism** is concerned with the complex interrelationship between environmental degradation and various forms of oppression, such as gender inequality and human domination of nature. New waves of ecofeminists are decidedly younger and more international—like Ugandan youth leader Vanessa Nakate fighting for women's education as part of climate justice.

Everyone deals with crises differently based on their history, resources, and mental state. Amongst its various ramifications, the climate crisis challenges us to reflect on how we can do better individually and as members of both our local and global communities. Disasters are not new. While we search for newly effective means to communicate the scope and severity of the climate challenge without increasing trauma, we can also incorporate important lessons for facing uncertainty from past traditions.[107]

This book's objectives

The infrastructures we live among are rooted in the past, influential in the present, and determine collective futures. This book aims to demonstrate "textbook" cases of harm in the energy sector using an empirical, science-based approach highlighting (1) spatial and geographic evidence and (2) knowledge from frontline communities. A core objective is to provide a vocabulary and framework for analyzing energy violence as a basis for participatory engagement and active learning. Chapter 1 focuses on research methods—systematic, verifiable, and repeatable practices to document evidence and produce knowledge. Each subsequent chapter addresses fossil fuel (either coal, oil, or gas) and is delivered within a unifying structure (Fig. 14).

Creating energy democracy in a shared energy commons is not a spectator sport. Chapters provide applied and critical activities (i.e., power mapping) to build analytical and practical skills to engage in energy research and climate action. While each chapter first addresses energy violence through evidence, each part also identifies groups and individuals involved in transformative change. Active learning and ongoing engagement are encouraged in a **Praxis** guide following the book's conclusion. "Try this" prompts throughout chapters challenge readers to compare and contrast locations in the book with places they experience firsthand. A **living laboratory** refers to a physical or virtual space in which to test and

---

[106] Brownhill, L., & Turner, T. E. (2019). Ecofeminism at the Heart of Ecosocialism. *Capitalism Nature Socialism, 30*(1), 1–10.

[107] Ghosh, A. (2018). *The great derangement: Climate change and the unthinkable.* Penguin.

Sections	Chapter Subheadings
1	Energy Violence
2	Spatial Distribution
3	Temporal Analysis
4	Illustrative Cases
5	Recommended Resources

Figure 14 Structure of Chapters 2—10.

Number	Objective
1	Dissect industry disinformation and agendas behind it;
2	Provide educational tools for deep, critical engagement (i.e., decolonization, anti-racism, anti-sexism);
3	Encourage active participation in energy democratization;
4	Engage in economic, environmental and climate justice as intellectual and social frames for informed, transformative climate action

Figure 15 Pedagogical objectives.

solve challenges faced by society at large.[108] Both authors are active in public scholarship at scales from local to global. Sections within Chapters 1, 3, and 7—9 in particular, draw from participatory action research. Fig. 15 shows an overview of operational place-based approaches for energy research and systemic transformation.

Violence and misuse of power are present in the energy sector around the world. Readers are encouraged to apply both concepts and approaches from the book to their own relationships with energy structures locally and globally. The book's illustrative examples cover the globe across 27 countries, designed to encourage inquiry and engagement rather than perform the impossible task of exemplifying all global locations. Many of the featured cases are unfolding in real time. Energy transformations surround us.

Summary

We are fortunate if we learn about energy violence from a book rather than living on the frontlines of an extreme energy hazardscape. Coming to terms with climate change isn't easy, particularly for youth who have done little to cause the damage but must live with

---

[108] Hossain, M., Leminen, S., & Westerlund, M. (2019). A systematic review of living lab literature. *Journal of Cleaner Production*, *213*, 976—988.

expensive and deadly consequences. Reducing climate disruption while contributing to equity and justice is relevant and urgent. Teaching about energy violence demonstrates important patterns, relationships, and illustrated cases to encourage additional inquiry and action to challenge corporate power and greenwashing, reduce climate disruption, and engage in new worldviews and worldmaking.

Vocabulary

1. Anthropocene
2. capitalism
3. censorship
4. civil disobedience
5. climate change
6. climate denialism
7. climate science
8. connectivity
9. colonialism
10. deception playbook
11. degrowth
12. discontinuity
13. double exposure
14. ecofeminism
15. ecosocialism
16. energy
17. energy burden
18. energy colonialism
19. energy commons
20. energy poverty
21. energy transformation
22. energy transition
23. energy violence
24. environmental illiteracy
25. exposure
26. extreme energy
27. fast violence
28. flow
29. fossil fuels
30. fossil gas
31. frontline
32. GHG intensity
33. greener" capitalism
34. greenhouse gases (GHGs)

35. greenwashing
36. hazardscape
37. hydraulic fracturing
38. hyperobject
39. infrastructure
40. landscape
41. leakage
42. living laboratory
43. loss and damage
44. methane
45. methane rift
46. necropolitics
47. non-response
48. non-view
49. place
50. post-extractivism
51. power
52. power map
53. recognition
54. renewable energy
55. shifting base syndrome
56. short-termism
57. slow violence
58. soft denialism
59. tipping point
60. violence
61. worldmaking

Recommended

Barca, S. (2020). *Forces of reproduction: Notes for a counter-hegemonic anthropocene.* Cambridge University Press.

Bullard, R. D., & Wright, B. (2012). *The wrong complexion for protection: How the government response to disaster endangers African American communities.* NYU Press.

Jacobson, M. Z. (2023). *No miracles needed: how today's technology can save our climate and clean our air.* Cambridge University Press.

Klein, N. (2007). *The shock doctrine: The rise of disaster capitalism.* Macmillan.

Lockwood, J. A. (2017). *Behind the carbon curtain: The energy industry, political censorship and free speech.* University of New Mexico Press.

Oreskes, N., & Conway, E. M. (2011). *Merchants of doubt: How a handful of scientists obscured the truth on issues from tobacco smoke to global warming.* Bloomsbury Publishing.

Proctor, R. N., & Schiebinger, L. (Eds.). (2008). *Agnotology: The making and unmaking of ignorance.* Stanford: Stanford University Press.

Schmelzer, M., Vetter, A., & Vansintjan, A. (2022). *The future is degrowth: A guide to a world beyond capitalism.* Verso.

Wenzel, J., & Yaeger, P. (2017). *Fueling culture: 101 words for energy and environment.* Fordham Univ Press.

CHAPTER 1

Research methods

Contents

Climate Crisis, Energy Violence
ISBN 978-0-12-819501-7,
https://doi.org/10.1016/B978-0-12-819501-7.00012-6

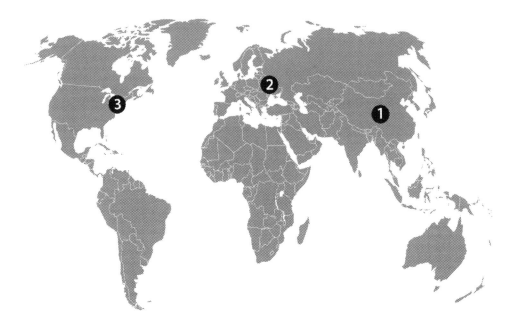

Locations: 1 - China
             2 - Ukraine
             3 - United States

Themes: Spatial Analysis, Energy Violence and Deathprint

Subthemes: Proximity, Exclusion, Connectivity and Segmentation

Energy is a fundamental medium and metric.

Space and power

Space is a social product. It is not simply "there," a neutral container waiting to be filled, but is instead a dynamic, human-constructed means of control, and hence of domination and power.[1] Supply disruption and social tensions from the COVID-19 pandemic reinforce how communities and countries are interconnected and interdependent.

---

[1] Lefebvre, H. (1991). *The production of space*. Verso.

Space is multidimensional, non-homogenous, relative and often discontinuous.[2]

Critical analysis of power (Introduction) and space helps inform **praxis**, the process by which a theory, lesson, or skill is enacted, embodied, or realized. Praxis involves engaging, applying, exercising, realizing, or practicing ideas. Similarly, "critical making" highlights the productive and transformative qualities of critical engagement as an invitation to embrace **justice** ("just" behavior or treatment; getting what one deserves) as an important unifying objective. To create a more equitable and livable future means understanding and influencing power. Critical academic fields such as political ecology[3] often employ a **power lens**,[4] meaning to use the construct of power to view or comprehend social and ecological patterns. Power has been insufficiently addressed in much of the prior research in the energy sector.[5] Most power analysis has privileged the most obvious form of power: direct, visible power like coercion, also commonly identified as "power over." To have **power over** someone means having the ability to make them do something they would not do otherwise. Focusing solely on abuse of power, however, may overlook building power at the base and creating spaces for participation. Relationships involving **power with** (as opposed to power over) allow participants to pursue common goals, such as social equity and a livable future.

While power over space can be temporary or relatively fixed, the political spaces of the energy sector extend beyond formal state relations. Formal spaces include those mandated by government institutions as recognized under legal code. Formal spaces represent a small portion of decision-making in the energy sector. Political spaces can also be informal or fluid (not fixed in place) and include political discourses and community practices as well as courtrooms, legislative chambers, or boardrooms. Fig. 1.1 breaks down the organizational construct of space, and the examples suggest expansive horizontal, vertical, and digital coverage.

People experience space differently based on nationality, gender, race, religion, income, and/or other characteristics. Native populations often experience current places

[2] Harvey, D. (2009). *Social justice and the city*. University of Georgia Press.

[3] Sultana, F. (2021). Political ecology 1: From margins to center. *Progress in Human Geography, 45*(1), 156−165.

[4] Wood, B., Baker, P., & Sacks, G. (2022). Conceptualising the commercial determinants of health using a power lens: a review and synthesis of existing frameworks. *International Journal of Health Policy and Management, 11*(8), 1251−1261.

[5] Fuchs, D., Di Giulio, A., Glaab, K., Lorek, S., Maniates, M., Princen, T., & Røpke, I. (2016). Power: the missing element in sustainable consumption and absolute reductions research and action. *Journal of Cleaner Production, 132*, 298−307.

Types of Space	Description	Examples
absolute	objective	geographical coordinates (latitude and longitude)
relative	subjective	homeland; place
metaphorical	virtual or imagined	cyberspace; dystopia

Figure 1.1 Categories of space.

as dystopia.[6] A child born during a war or famine may never experience their family's homeland. Experiencing an absolute space with clear borders (i.e., an uncontested country) can be distinct for a citizen of France versus one from Afghanistan or Ukraine. Location matters. Quantitative **spatial analysis** highlights patterns of relationality within and between locations and allows for solving complex location-oriented problems (i.e., the power of where). **Geographic Information Science** (GIS) excels in objective, quantifiable spatial data and analysis in both vector and raster models. These two models both represent real-world phenomena; they differ, however, in their fundamental composition. In the case of the vector model, points, lines, and polygons are used to demarcate objects within coordinate space. National boundaries, discrete coordinate point locations, and linear features such as pipelines are typically represented by the vector model. The raster model utilizes a grid of pixels—rectangular cells—to best represent surfaces such as elevation, terrain, and land cover; captures the Earth's surface through **remote sensing**; and tracks concentrations of materials or substances, such as spills or toxic plumes.

Energy is spatial

Spatial distribution involves relationships between areas (i.e., linkage, corridor, or proximity between zones, clusters, or fragments). Analysis of spatial patterns like this can help document environmental and social implications but the data selected must fit the context of the place.[7] There are limits to what can be answered depending on the quality and precision of spatial data. Even with adequate data, excessive fixation with quantifiable absolute space can contribute to bureaucratic and regulatory violence. Spatial and social marginalization (i.e., environmental racism, voter disenfranchisement, gerrymandering) forge negative quality-of-life outcomes.

[6] Whyte, K., Caldwell, C., & Schaefer, M. (2018). Indigenous lessons about sustainability are not just for "all humanity." In *Sustainability: Approaches to environmental justice and social power*, 149–179.
[7] Monmonier, M. (2018). *How to lie with maps*. University of Chicago Press.

Spatial **proximity** (i.e., nearness), such as residential location at the fenceline of a polluting facility, will increase the likelihood of a person developing respiratory disease if that infrastructure releases harmful air emissions.[8] Yet space is seldom a fixed "container" or constant attribute. Never has that been clearer than with the 1986 Chernobyl nuclear disaster in Ukraine. Though the country has resurfaced as a hot geopolitical conflict zone nearly 4 decades later tied to fossil gas (Chapter 10), Ukraine remains a complex space with its nuclear meltdown and radioactive drift. The following discussion of the radioactivity levels from the Chernobyl disaster demonstrates that, while it is common in many situations that a proximate resident will have higher risk from contamination,[9] exposures do not always follow fixed, concentric patterns.

There are two ways of determining exposure—uniform and irregular. Uniform involves simplified proximity indicators that assume exposure increases as the distance decreases; often, concentric rings of various radii (e.g., 1, 5, 10, 30 km) are drawn around the source to identify exposure groups. Fig. 1.2 shows Chernobyl's exclusion zones: the areas evacuated and where certain activities continue to be prohibited following the nuclear meltdown.

As seen in Fig. 1.2, the present-day exclusion zone is a highly irregular shape, unlike the uniform radius of the original 30 km exclusion zone. Moreover, wind, water flow, terrain, and landcover all influence the movement of toxic elements and radioactive levels.[10] Whether air pollution arises from stationary power plants or is dispersed like fugitive dust, spatial relationships like proximity and flow help define risk. Fig. 1.3 depicts a singular power plant relative to other social and ecological resources, since infrastructure does not exist in isolation. Cumulative emissions found in a hotspot with high ecological burden due to various pollution sources are known as a **cluster** (i.e., a group of similar things positioned or occurring closely together).

Energy infrastructure often causes **fragmentation**, which is a process of a space or area being broken into small or separate parts, such as when contiguous habitat gets divided into isolated patches. Forest fragmentation leads to biodiversity loss and increases

[8] Maantay, J., Chakraborty, J., & Brender, J. (2010). Proximity to environmental hazards: Environmental justice and adverse health outcomes. In *Strengthening environmental justice research and decision making: A symposium on the science of disproportionate environmental health impacts* (pp. 17−19). Environmental Protection Agency (EPA).

[9] Huang, Y. L., & Batterman, S. (2000). Residence location as a measure of environmental exposure: A review of air pollution epidemiology studies. *Journal of Exposure Science & Environmental Epidemiology, 10*(1), 66−85.

[10] Mappes, T., Boratyński, Z., Kivisaari, K., Lavrinienko, A., Milinevsky, G., Mousseau, T. A., ... & Watts, P. C. (2019). Ecological mechanisms can modify radiation effects in a key forest mammal of Chernobyl. *Ecosphere, 10*(4), e02667.

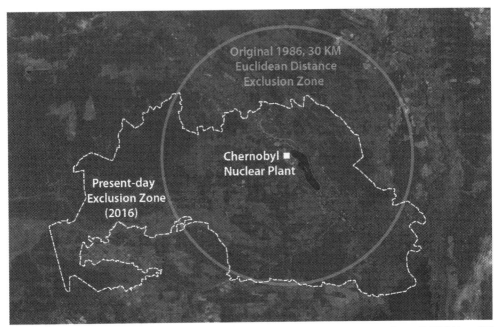

Figure 1.2 Chernobyl exclusion zone. *(Sources: Esri, Maxar, Earthstar Geographics, CNES/Airbus DS, USDA FSA, USGS, Aerogrid, IGN, IGP and the GIS User Community. Exclusion Zone extracted via Open-StreetMap (Overpass Turbo API).)*

invasive plants and pathogens, among other consequences. Similarly, social fragmentation has repercussions such as the disintegration, collapse, or breakdown of norms of behavior or relationships. Community conflict over fossil fuel infrastructure can last for years.

Try this

Research and discuss: *How does energy influence your life? Identify and describe both the benefits and impacts from access to electricity, its economic advantages as well as pollution exposures (i.e., asthma) and climate change.*

Spatializing infrastructure

Energy can be hard to grasp at a global scale, making it helpful to begin locally in an area you know well. Readers need to have the tools to delineate the patterns of energy violence in order to engage in energy democracy. Material or physical connectivity exists in associated spaces of an energy **vector**: a structure or system transferring energy across

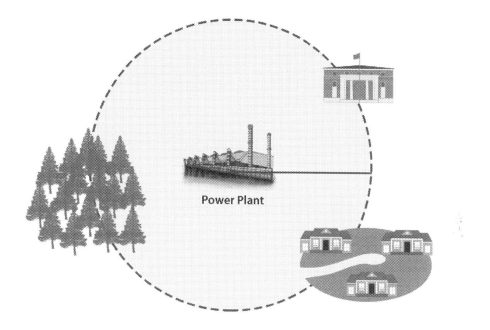

Energy Proximity

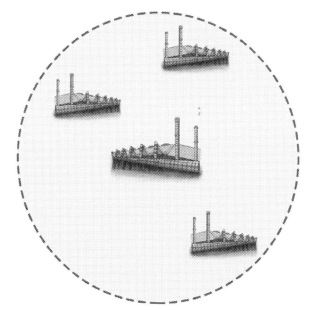

Clusters & Hotspots

Figure 1.3 Energy is spatial.

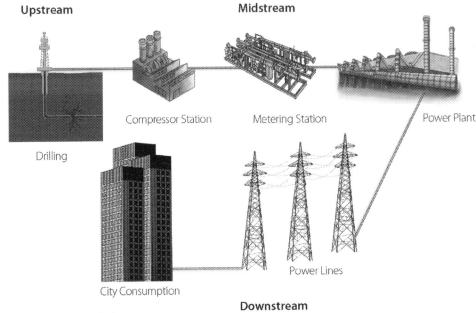

Flow & Connectivity

Figure 1.4 Energy vector.

space and time.[11] Fig. 1.4 above depicts these basic constructs as the core building blocks of energy's material form.

Energy vectors connect a series of spaces. The oil and gas industry's energy vector is usually divided into three major components: upstream, midstream, and downstream. **Upstream** exploration and production activities involve searching for, extracting, and producing crude oil or gas, such as drilling operations to bring raw resources to the surface. The **midstream** sector involves transmission networks, whether by pipeline, rail, barge, tanker, or truck. Midstream aggregation facilities may include basic processing or purification before merging materials for transport. The **downstream** sector includes refining as well as selling, distributing products, and consumption. Downstream facilities include petrochemical plants, oil refineries, gas distribution companies, and retail outlets (i.e., gas stations).

In Fig. 1.5, each element or facility undergoes its own independent bureaucratic review cycle, thus fragmenting emissions into discrete segments rather than considering the total consequences of their cumulative impacts.

---

[11] Krajačić, G., Martins, R., Busuttil, A., Duić, N., & da Graça Carvalho, M. (2008). Hydrogen as an energy vector in the islands' energy supply. *International Journal of Hydrogen Energy, 33*(4), 1091–1103.

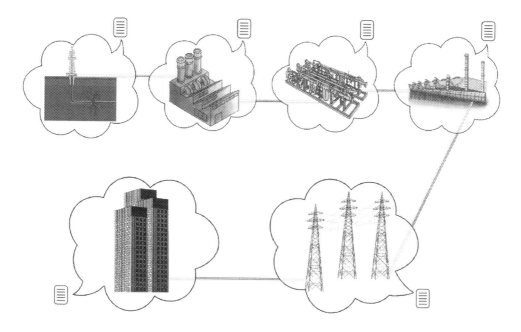

Segmentation & Fragmentation

Figure 1.5 Segmentation and fragmentation.

The US Federal Energy Regulatory Commission (FERC) states segmentation is impermissible—an applicant is prohibited from breaking one large project into components where each piece may be considered to have few impacts to avoid responsibility for broader consequences. Yet energy projects are commonly broken down. This **segmentation** is regulatory manipulation (see Fig. 1.5). Segmentation creates a low regulatory floor for large infrastructure spread across multiple sites. Each smaller piece receives approval by suggesting it can have "insignificant" impacts and damages can be easily offset, even when the larger system has very significant cumulative implications.

Try this

Research and analyze*: how does an energy vector supply your region? Analyze an energy site, hub, or corridor using Google Earth, OpenStreetmap, or Sentinel Hub Playground.*

Time matters

The factor of time is also central to understanding energy.[12] **Temporal analysis** examines change over time, covering either short or long temporal spans. For example, changes in weather conditions are usually studied over a period of at least 30 years even though climate scientists have hundreds of years of data.[13] In briefer time scales, time influences the power of impacted communities. Public comment in air and water permitting is reviewed in a truncated period (i.e., 30 days) before moving to the next phase. Afterward, prior decisions are fixed or extremely hard to reverse, even when new information emerges about risk. Processes of both fast and slow violence involve elements of time.[14] Slow regulatory violence in permitting can be accelerated strategically. Timeframes can be accelerated to take power away. For example, fast track, expedited reviews mean participatory planning is limited or absent.

Earlier placements of infrastructure typically do not consider the full extent of potential hazards that can accumulate over time. Subsequent placements show how installments cluster in space across time by building on the time-bound, limited site decision-making of earlier placements while likely intensifying risk with each new addition. For example, while the 9–11 bombings of the Pentagon and Twin Towers were both calamitous and symbolic, a strike on the Indian Point Energy Center (Fig. 1.6), located approximately 35 miles northward of New York City, could have caused an exponentially higher death count. Compounding risk, a proximate explosion at Algonquin gas pipeline could cause a similar, large-scale disaster.

The Indian Point nuclear facility, currently undergoing long-term decommissioning (2021), is located just hundreds of feet from three large-diameter midstream gas pipelines that have operated successively over decades, all with potential blast radii impact in close proximity to the nuclear site. A **blast radius**, or damage radius, is the distance from an explosive source in the event of an accident. For pipelines, it is estimated with a mathematical formula involving gas pressure and pipe diameter; yet precise distance is known only after the accident, since various factors can influence exact impact intensity and range.

In the case of Indian Point, a blast event was deliberated extensively by multiple agencies, yet experts concluded that "modern construction" coupled with heightened "high consequence area" (HCA) inspections were adequate to mitigate risk.[15] This

[12] Pasqualetti, M. J. (2013). The geography of energy and the wealth of the world. In *The new geographies of energy: Assessment and analysis of critical landscapes* (p. 270).

[13] Marcott, S. A., Shakun, J. D., Clark, P. U., & Mix, A. C. (2013). A reconstruction of regional and global temperature for the past 11,300 years. *Science, 339*(6124), 1198–1201.

[14] Sandlos, J., & Keeling, A. (2016). Toxic legacies, slow violence, and environmental injustice at Giant Mine, Northwest Territories. *Northern Review,* (42), 7–21.

[15] Staff. (2020, April 27). NRC: No danger to Indian Point from natural gas pipeline. *Nuclear Newswire.*

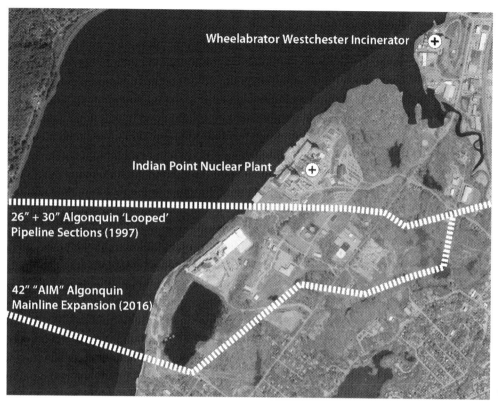

Figure 1.6 Indian Point's compounded risks. *(Sources: Esri, Maxar, Earthstar Geographics, CNES/Airbus DS, USDA FSA, USGS, Aerogrid, IGN, IGP and the GIS User Community.)*

exhibits regulatory violence, a concept discussed in more detail below and in Chapter 3, because assessors rely on formulas and industry assumptions to assert there is "low enough" risk rather than seeking to prevent harm. Regulatory floors become standards to avoid culpability in lawsuits after carefully selected experts postulate risk as low. Yet, infrastructure projects are often riddled with unknown variables, poor historical records, cost-cutting, and simple human error. Furthermore, siting or placing two or more things together, such as with the **co-location** of various industrial sites in a single area like at Indian Point, can multiply risks.

At the core, time is essential to understanding energy violence and the climate crisis: yesterday's GHG emissions have decades-long repercussions. Of particular importance, methane emissions have 86 times the potency of carbon dioxide averaged over the first 20 years, and 35 times potency across 100 years.[16] GHGs produced or eliminated in the

---

[16] Intergovernmental Panel on Climate Change. (2013). *Climate change 2013: The physical science basis.*

next few years will make a critical difference in what climate scientists now deem "uncharted territory" where "time is up".[17] The most heavily polluting parts of fossil fuel infrastructure (refineries, compressor stations, liquefaction trains, cracker plants) are not only flagrant emitters forcing global GHG, but often sited in degraded areas, compounding risk and harm. Placing new energy infrastructure (especially high-pressure, flammable components) can amplify risk with a domino effect, such as a series of blasts or fires across infrastructure. **Synergistic effects**—when the sum of the whole is greater than each of the parts—mean that different toxic pollutants or multiple exposures compound, harming local communities while often raising global GHG emissions. Similarly, climate change itself is a **threat multiplier**: it extends and exacerbates risks and harms, including increased flooding that can spread contamination. Extreme weather contributes to fast energy violence.

Failing infrastructure becomes stressed by climate change. For instance, dams that breach during hurricanes and extreme flooding events exhibit histories of structural instability that then become tragedy, such as the Edenville Dam collapse in 2020.[18] A decade prior, FERC issued violation notices for imperative redesign and repair measures. In 2018, FERC terminated the operating license due to inability to pass Probable Maximum Flood (PMF) tests.[19] Following dam collapse, toxic exposure downstream was as much of an emergency as the collapse when Dow Chemical treatment ponds overflowed (Fig. 1.7).

Due to unequal spatial distribution of facilities and inadequate responses where vulnerable people reside, poor neighborhoods often suffer most from hazards, such as toxic floods.[20] In another example in Houston, Texas, following Hurricane Harvey, the storm spread contamination from energy facilities into vulnerable neighborhoods and local water bodies.[21]

As a threat multiplier, climate disruption exacerbates other drivers of insecurity making precarious situations worse. Aggregate climate impacts are staggering: there were more than 7000 extreme weather events since 2000, a major increase over the previous

[17] Ripple, W., Wolf, C., Gregg, J. W., Rockstrom, J., Newsome, T., Law, B., Marques, L., Lenton, T., Xu, C., Huq, S., Simons, L., & King, Sir D. A. (2023). The 2023 state of the climate report: Entering uncharted territory. *BioScience*, biad080.

[18] DiChristopher, T., & Rooney, K. (2018, September 21). Duke Energy says dam breached at North Carolina plant and coal ash may be flowing into Cape Fear river. *CNBC*.

[19] Tabuchi, H. (2020, May 5). Dam failure threatens a Dow chemical complex and superfund cleanup. *New York Times*.

[20] Kiaghadi, A., & Rifai, H. S. (2019). Physical, chemical, and microbial quality of floodwaters in Houston following Hurricane Harvey. *Environmental Science & Technology, 53*(9), 4832–4840.

[21] Stone, K. W., Casillas, G. A., Karaye, I., Camargo, K., McDonald, T. J., & Horney, J. A. (2019). Using spatial analysis to examine potential sources of polycyclic aromatic hydrocarbons in an environmental justice community after Hurricane Harvey. *Environmental Justice, 12*(4).

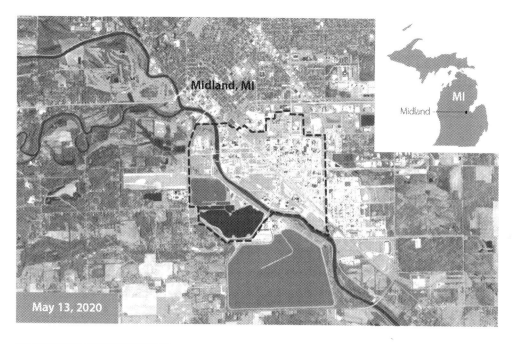

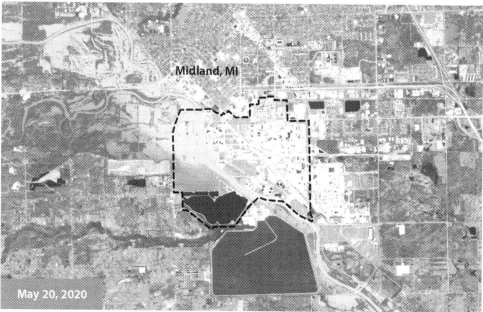

– – – **Dow Chemical Plant (Approximate)**

Figure 1.7 Flood event at Dow Chemical superfund site. *(Sources: PlanetScope © Planet Labs, Inc., Down Chemical Plant Boundary extracted via OpenStreetMap (Overpass Turbo API).)*

2 decades. From 2000 to 2019, there were 7348 major disasters recorded, claiming 1.23 million lives, affecting 4.2 billion people (more than half the world's population) and causing approximately $2.97 trillion in economic losses.[22] Not doing everything possible to mitigate GHGs is climate violence, particularly since those who suffer the most are already vulnerable.

Try this

Research, analyze, and share: *Climate change is often said to be a threat magnifier or multiplier: explore the ways this happens in the energy sector. Share specific examples.*

Environmental injustice

Environmental injustice is a display of power over space. People of color frequently experience greater harm. For example, African Americans are disproportionately vulnerable in natural or human-induced disasters, such as public health emergencies, toxic contamination, industrial accidents, and bioterrorism threats.[23] Significant burden falls on those living at or near pollution sources and sinks. **Sources** are areas of contamination origin, while **sinks** are zones that hold or store contamination. The distribution of both sources and sinks—and whether they are growing or shrinking in particular geographical and social contexts—is most often unfair: time and again the pattern is a disproportionate burden to those without power, and inequitable benefit to those with privilege.

Hotspots and blindspots

A toxic **hotspot**, or a cluster of spatial phenomena like pollution (Chapters 6 and 9), show how impacted communities can experience cumulative and synergistic effects from multiple polluting industries over decades or even centuries. **Hotspot analysis** uses spatial analysis and mapping techniques to highlight risks from proximity to contamination clusters.[24] While pollution from all sources adds up, research suggests that a small number of the worst offenders make up an unfair share of pollution. In addition to attending to these big emitters, we can make significant gains by drawing into focus

---

[22] Beitsch, R. (2020, October 12). Climate change a factor in most of the 7000 natural disasters over the last 20 years: UN report. *The Hill*.

[23] Bullard, R. D., & Wright, B. (2012). *The wrong complexion for protection: How the government response to disaster endangers African American communities*. NYU Press.

[24] Morath, S. J., Hamilton, S., & Thompson, A. (2021). Plastic pollution litigation. *Natural Resources & Environment*; Djenohan, Z. (2020). Making way for unjust enrichment in environmental justice litigation. *Loyola Law Review*, 67, 223.

our **blindspots**—areas of significant environmental damage that have been overlooked by the public and policymakers. Some examples are spin-offs of fossil gas, such as vinyl, plastic, liquefied natural gas (LNG), and blue hydrogen (Chapter 10).

Formosa Plastic in St. James Parish, Louisiana, is a well-known example of environmental injustice[25] and represents both a hotspot due to clustering and a blindspot due to the production of chemicals for use in plastic and vinyl. Branded "The Sunshine Project," the complex of 14 facilities with 10 plants produces chemical byproducts such as ethylene glycol, polyethylene, and polypropylene. This toxic situation came to international attention because of the work of local organizations Rise St. James, the Bucket Brigade, and others. Organizations brought plastic pollution that local inhabitants live with to a meeting with lobbyists—and were accused of terrorizing and faced a 15-year prison sentence.[26] This is just one example of how anti-terrorism laws are employed to dissuade environmental activism. Other examples of criminalization of dissent are discussed in Chapter 7.

Companies like Formosa work with little oversight and relative immunity in most countries in spite of the toxicity of their manufacturing sites. Formosa's record internationally is alarming. A famous event occurred in Vietnam in 2016 when Formosa Ha Tinh Steel, built by the Taiwanese corporation Formosa Plastics, discharged industrial waste illegally into the ocean,[27] killing thousands of fish.[28] The disaster also cost human lives: divers, fishers, and seafood-consumers got sick and some died.[29]

Products that companies like Formosa produce are widely utilized and highly popular. A life cycle analysis (LCA) - which factors all inputs and outputs, from a product's materials extraction to final disposal - finds high pollution and GHG emissions making vinyl unsuitable as a construction product in spite of low cost and convenience. The need to move "Beyond Plastic" is known worldwide.[30] Plastic garbage covers vast areas, with concern growing about the huge amount of microplastics found in croplands,

[25] Blanks, J., Abuabara, A., Roberts, A., & Semien, J. (2021). Preservation at the intersections: Patterns of disproportionate multihazard risk and vulnerability in Louisiana's Historic African American Cemeteries. *Environmental Justice, 14*(1), 1—13.

[26] *Democracy Now.* (2020, June 29). Louisiana activist's face 15 years for "terrorizing" oil lobbyist with a box of plastic pollution.

[27] Fan, M. F., Chiu, C. M., & Mabon, L. (2022). Environmental justice and the politics of pollution: The case of the Formosa Ha Tinh Steel pollution incident in Vietnam. *Environment and Planning E: Nature and Space, 5*(1), 189—206.

[28] Fan, M. F., Chiu, C. M., & Mabon, L. (2022). Environmental justice and the politics of pollution: The case of the Formosa Ha Tinh Steel pollution incident in Vietnam. *Environment and Planning E: Nature and Space, 5*(1), 189—206.

[29] Mollman, S. (2016, June 30). A Taiwanese steel plant caused Vietnam's mass fish deaths, the government says. *Quartz.*

[30] Beyond Plastics. (2021). *The new coal: Plastics and climate change.*

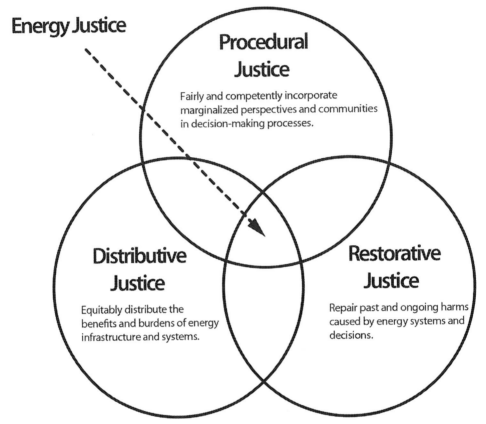

Figure 1.8 Energy justice. *(Adapted from: The Emerging Potential of Microgrids in the Transition to 100% Renewable Energy Systems, R. Wallsgrove, et al., 2021.)*

oceans, and homes. Occupational hazards for workers with common fossil fuel-based products we regularly use, like polyester, are discussed in Chapter 8. Laborers in manufacturing sites experience cancers and respiratory ailments among other fatal or debilitating conditions.

Environmental justice means that no group of people should have to bear a disproportionate share of negative consequences resulting from policies, decisions, and actions; and benefits should be shared and create restorative and reparative actions. It necessitates fair treatment, meaning processes are non-discriminatory and equitable for all people regardless of race, color, national origin, or income. It also requires meaningful involvement, which is the guarantee that impacted and vulnerable residents have a realistic opportunity to participate in the full cycle of the decision-making process of environmental regulations and policies. The three spheres of environmental justice—distributive, procedural, and restorative justice—also align in energy justice (Fig. 1.8 above).[31]

Environmental justice challenges power structures and criticizes mal-distribution of resources and risks.[32] Restorative justice requires recognition of difference and reparation for historical and present-day structures of violence. Energy justice is not a process that can be done piecemeal—even breaking it down into three components suggests that respect for difference and recognition of marginalized groups only pertains to one area, which is what commonly happens when states fail to incorporate diversity, equity, inclusion, and justice (DEIJ) into all components. The 4Ds of **energy justice**—decarbonization, democratization, decolonization, decentralization—form an interwoven structure since each process can't be done alone.[33]

To achieve energy justice requires understanding of **intersectionality**, a concept brought first to the fore by African American feminist scholar Kimberle Crenshaw,[34] which highlights intensification of harm with two or more forms of discrimination, whether based on class, race, nationality, religion, age, sexual orientation, illness, or other status. Context like this is essential for action-oriented praxis to re-frame policies for more equitable solutions and requires paying careful attention to power dynamics to see clearly who is excluded or left out.[35] A related international concept is that of double exposure, where there is magnified vulnerability of those who experience both climate change and poverty. People who live close to stationary sources of pollution, particularly low-income households who cannot afford to relocate, are frontline communities (Fig. 1.9). This concept connotes spatial injustice in social relations of domination and oppression.

Critical geographies of energy organize and connect social constructs, like race, with ecological components across location and time. Polluting facilities are frequently situated in low-income areas, particularly communities of color, as is the case with the

[31] Finley—Brook, M., & Holloman, E. L. (2016). Empowering energy justice. *International Journal of Environmental Research and Public Health, 13*(9), 926.

[32] Dehm, J. (2022). Environmental justice challenges to international economic ordering. *American Journal of International Law, 116*, 101—106.

[33] de Onis, C. M. (2021). *Energy Islands: Metaphors of power, extractivim and justice in Puerto Rico*. University of California Press.

[34] Crenshew, K. (1989). Demarginalizing the intersection of race and sex: A Black feminist critique of antidiscrimination doctrine, feminist theory and antiracist politics. *University of Chicago Legal Forum, 140*, 139—67; Adewunmi, B. (2014). Kimberlé Crenshaw on intersectionality: "I wanted to come up with an everyday metaphor that anyone could use." *New Statesman, 2*

[35] Sultana, F. (2021). Climate change, COVID-19, and the co-production of injustices: A feminist reading of overlapping crises. *Social & Cultural Geography, 22*(4), 447—460.

Figure 1.9 Frontline community opposition, Cancer Alley, 2018. *(Image credit: Julie Dermansky.)*

community proximate to Entergy Gas Plant in New Orleans shown in Fig. 1.9 above.[36] Racism has often meant dark-skinned populations were disproportionately impacted, as if they had "the wrong complexion for protection."[37]

The consolidation of press under a few major corporations[38] directly relates to the insufficient coverage and censorship of environmental injustices and causes.[39] Some scholars downplay political censorship or the buyout of academic research, either because they benefit or because it has become normalized to such an extent that they are complacent.[40] Others may falsely reassure themselves that while those less fortunate may suffer

[36] Taylor, D. (2014). *Toxic communities: Environmental racism, industrial pollution, and residential mobility.* NYU Press.
[37] Bullard, R. D., & Wright, B. (2012). *The wrong complexion for protection: How the government response to disaster endangers African American communities.* NYU Press.
[38] Chamberlain, D. (2018). Why local democracy needs a healthy local media. *The Political Quarterly,* 89(4), 719–721.
[39] Park, D. J. (2021). *Media reform and the climate emergency: Rethinking communication in the struggle for a sustainable future.* University of Michigan Press.
[40] Lockwood, J. A. (2017). *Behind the carbon curtain: The energy industry, political censorship and free speech.* University of New Mexico Press.

from climate disruption, they and their loved ones will miraculously escape harm. This deceptive ideation is a psychological defense to make one feel better—yet it is a false exceptionalism. Unsubstantiated optimism will erode quickly as supply shortages intensify and more people find themselves lacking medicines and food items that they previously took for granted.

Methods in this book help readers see dangerous blindspots connected to fossil fuels, including those in our own thought patterns. It also brings to the fore the many vibrant grassroots responses all over the world, recognizing that people are organizing and mobilizing for transformative change, even in spite of mainstream media coverage that is censored and limited.[41] Independent press, nonprofit organizations, and social media make these efforts known even without coverage in corporate-owned outlets. Every reader of this book uses fossil fuels in some capacity—most of us all day, every day—which means we share responsibility for action.

When marginalized impacted communities fight expansion of fossil fuels, they are often told a permit or expansion is a "done deal" and not to waste their efforts. In reality, mobilized communities have been able to stop large harmful projects. For example, in St. James Parish, Louisiana, a local community fought to stop a giant new petrochemical plant and won. Located along the 80-mile stretch of the Mississippi River known as Cancer Alley, residential communities—mostly Black—are interspersed among factories, pipelines, and tank farms. Against the odds, the grassroots organization Rise St. James stopped Formosa's proposed plant. They demonstrated unfair harm, erasure of African American history, and cumulative toxic burden. These are common geographic expressions of **racial capitalism**, defined as the process of deriving social and economic value from the racial identity of another person. This concept asserts that racialized exploitation and capital accumulation are mutually reinforcing.

The example of St. James, like other communities in Cancer Alley, shows how advantages and disadvantages extend and consolidate through time. Impacted communities experience clusters of burden, yet governmental codes have largely avoided addressing cumulative impacts.[42] **Cumulative impacts** from multiple sources magnify over time and space, and documenting this synergy creates a more realistic depiction of **body burden**: the pollution in people and individually the amount or concentration of chemicals that can be detected in the human body signifying toxic exposure. Health risks may be synergistic, producing combined effects greater than the sum of the individual or separate exposures. Cumulative impacts also correspond with wealth, investment, and social impacts.[43] Privilege often brings with it additional advantages when doors open and

[41] Burch, E. (2021). A sea change for climate refugees in the South Pacific: How social media—not journalism—tells their real story. *Environmental Communication, 15*(2), 250–263.

[42] Tollefson, J. (2022, June 2). How science could aid the US quest for environmental justice. *Nature*.

[43] Táíwò, O. O. (2022). *Reconsidering reparations*. Oxford University Press.

savings accrue. Disadvantages may also augment such that lost education or poor health lead to greater hardship. These unequal patterns can repeat across generations.

Unfortunately, many environmental scientists have not prioritized intersectional thinking, nor have they considered unequal power dynamics of institutions. This leads to biased and incomplete delivery of policy recommendations and unmet environmental justice mandates. Environmental practitioners, like engineers and consultants, frequently lack understanding of requirements for meaningful participation and informed consent. In practice, technical roles often reinforce maldistribution. Captured politicians reinforce industry paradigms generated from public relations offices. Experts often defend objective methods and theories riddled with bias. Technocrats lose focus and miss how segmented assessments (per chemical or per facility rather than all parts), siloed ways of thinking in state permits (i.e., air vs. water vs. soil) and permitting agencies advance ecological harm and social injustice.

Climate change presents itself differently around the globe and population vulnerabilities range from moderate to very high. The urgency may feel different and the resources to get out of harm's way are vastly disparate.[44] Whatever the reason for **displacement**— the forced loss of land and resources—it is difficult to remedy and seldom justified. The majority of people pushed from their homes by climate change remain within the borders of their country. However, climate change and disaster displacement include cross-border movement in times of political and economic tension.[45] Pacific islands like Tuvalu have been forced to initiate foreign colonies of "expats."[46]

Ecological debt is an accumulated obligation for restitution from wealthier countries after exploitation of resources and degradation of the environment in poor countries. The climate crisis will create two classes: those who can flee and those who cannot.[47] Eighty percent of disabled people live in low- and middle-income countries, many of which are vulnerable to climate change.[48] When disasters occur, disabled people experience greater risks before displacement, translating into additional challenges during the displacement process. Early warning systems are often inaccessible. Disabled people

---

[44] Pearson, T. W. (2017). *When the hills are gone: Frac sand mining and the struggle for community*. Minnesota Press; Pearson, T. W. (2016). Frac sand mining and the disruption of place, landscape, and community in Wisconsin. *Human Organization, 75*(1), 47–58.

[45] United Nations High Commissioner for Refugees. (n.d.). *Climate change and disaster displacement*.

[46] Farbotko, C., & Lazrus, H. (2012). The first climate refugees? Contesting global narratives of climate change in Tuvalu. *Global Environmental Change, 22*(2), 382–390.

[47] Gleik, A. (2021, July 7). The climate crisis will create two classes: Those who can flee, and those who cannot. *The Guardian*.

[48] United Nations High Commissioner for Refugees. (2021). *Disability, displacement and climate change*.

may also face heightened protection risks including discrimination, exploitation, and violence in disaster response contexts.[49]

Try this

Research and discuss: What frontline communities live near you? How do they experience structural violence, including ecological or generational debt?

Decision-making power

Today's **energy regimes** (carbon economy + top-down state) were built from historical political economies.[50] A regime is a system or planned way of doing things, especially one imposed from above, in some cases authoritarian. The centralized nature of the energy sector makes state institutions appear as a "regime." Processes of **productive exclusion** mean that the penetration of capital from outside an area or sector reduces local or domestic access to resources,[51] a pattern that thrives under capitalism. This exclusion can occur from simple operations to more complex supply circuits. The energy sector is historically and presently laden with examples of productive exclusion[52]—the playing field for the penetration of capital is grossly uneven, dependent on the financial power of particular investors.

Capitalism (Introduction) contributes to **uneven development**, a systemic process by which the power relations are translated into spatial forms, some areas prospering while others stagnate or decline due to inequity in access to power and resources across space. Today's centralized economies replete with state-run energy operations continue to rely on hierarchies of power and prioritized access to critical resources. Socialism advocates for the means of production, distribution, and exchange to be owned and regulated by the people (Introduction), which is rare in current state-run energy operations. For example, state-run firms controlling the energy sector of China and South Africa lack transparency and exacerbate social inequalities.

Political decision-making spaces are too often hierarchical and unequal rather than collaborative and transformative. Inclusive spaces are more accessible to those with limited mobility, health, and childcare constraints. Fig. 1.10 discusses formal political

[49] Fjord, L. (2007). Disasters, race, and disability: [Un]seen through the political lens on Katrina. *Journal of Race & Policy, 3*(1): 46–65.
[50] Seow, V. (2022). *Carbon technocracy: Energy regimes in modern East Asia.* University of Chicago Press
[51] McKay, B., & Colque, G. (2016). Bolivia's soy complex: The development of 'productive exclusion'. *The Journal of Peasant Studies, 43*(2), 583–610.
[52] Bridge, G., Barr, S., Bouzarovski, S., Bradshaw, M., Brown, E., Bulkeley, H., & Walker, G. (2018). *Energy and society: A critical perspective.* Routledge.

Type of space	Description	Examples
closed	decisions by a set of actors behind closed doors	-executive order -court ruling
invited	transient; must be held open	-written comment periods -permit hearings with public comment
claimed or created	mobilization to change the status quo	-blockadia (i.e., tree sits,[53] climate camps[54]) -community-managed electrical grid[55]

Figure 1.10 Power spaces.

spaces governed under legal code and informal political spaces, like blockades and collaboratives[53,54,55].

Claimed spaces can be emancipatory. Yet with poor distribution of costs and benefits, energy and utility project sites often become **contested spaces**, meaning locations or places struggle over power to shape realities and perceptions. Community solar grids, such as in Puerto Rico (US) (Chapter 9), form autonomous local power networks to take control of their own energy.

Production of space can manifest and reproduce injustice … or justice.

As effective climate action has stalled for decades, more people are protesting even as states act to dissuade climate activism by increasing financial punishments and jail time for violations (Chapter 7). Protest activities include blockades (Chapter 7) like the Yellow Finch tree sit in Virginia (US) where a tree platform and surrounding camp formed a liminal space as a 2 year, self-governing site of resistance to block construction of the interstate Mountain Valley Pipeline.

To rewire the violent energy sector, a growing number of "regular folks" are turning to nonviolent civil disobedience.[56] **Direct action** is the use of strikes, demonstrations, blockages, or other public forms of protest to achieve demands, such as stopping destructive industries.[57] Most of these social movements have been non-violent and grassroots.[58]

[53] Ludwig, M. (2021, April 12). Appalachian pipeline blockade ends with arrests after 932 days. *Truthout*.
[54] Klimacamp. Bei Wein. (n.d.). *What is climate camp?*
[55] La Rosa, M. (2019, September 19). Step by powerful step, citizens lead Puerto Rico into its solar future. *NACLA*.
[56] Sovacool, B. K., & Dunlap, A. (2022). Anarchy, war, or revolt? Radical perspectives for climate protection, insurgency and civil disobedience in a low-carbon era. *Energy Research & Social Science, 86*, 102,416.
[57] Táíwò, O. O. (2022). *Reconsidering reparations*. Oxford University Press.
[58] Malm, A. (2021). *How to blow up a pipeline*. Verso.

Regulatory 'scale'	Sites of struggle, spaces of participation
local/regional	body home property parcel neighborhood turf city/town subnational region (i.e., province, district, county)
national	country ethnic territory or homeland
international/global	supranational region bilateral/multilateral agency quasi-governmental institution[63]

Figure 1.11 Overlapping sites, spaces, and scales of participation.

While protest builds, companies lobby for bills for criminalization of protests,[59] whether in rural[60] or urban areas.[61] Risks from protest are not equal: people of color are at increased risk for criminalization, surveillance, and police violence.[62]

In complex energy projects or climate policy, it can be challenging to identify power structures. One heuristic device to help delineate decision-making roles is **scale** (i.e., socially defined governance or institutional levels). While state scales and formal government processes are often the focus of energy research, the sector relies on capillary power at many levels. Fig. 1.11 above details how different scales and spaces overlap.[63]

Privatization of energy has been a trend for decades. The high costs and poor service experienced in most areas are a result of regulatory violence in our electrical systems. **Investor-owned utilities** (IOUs)—private enterprises acting as public utilities—are widespread and consolidating, which is concerning given how this foments vulnerability

---

[59] Alvarez, C. H., Theis, N. G., & Shtob, D. A. (2021). Military as an institution and militarization as a process: theorizing the US military and environmental justice. *Environmental Justice, 14*(6), 426–434; Rasch, E. D. (2017). Citizens, criminalization and violence in natural resource conflicts in Latin America. *European Review of Latin American and Caribbean Studies/Revista Europea de Estudios Latinoamericanos y del Caribe,* (103), 131–142.

[60] Graddy-Lovelace, G. (2021). Leveraging law and life: Criminalization of Agrarian movements and the Escazú agreement. In *Our extractive age* (pp. 94–113). Routledge.

[61] Dillon, L., & Sze, J. (2016). Police power and particulate matters: Environmental justice and the spatialities of in/securities in US cities. *English Language Notes, 54*(2), 13–23.

[62] Cordon, G. (2022, June 17). Government accused of making 'hostile environment' for peaceful protest. *Independent.*

[63] Quasi refers to the fact that agencies like the UN are a hybrid with formal state roles and non-governmental components.

to climate change,[64] terrorism, ransom, or sabotage. For example, across the US Pacific Coast there have been climate-related blackouts during prolonged forest fires.[65] Blackouts were announced by utility companies as "public safety power shut off events." This wording sought to make it seem like companies had acted in public interest when in fact the situation emerged from a combination of neglect of local customer needs and of global planetary boundaries. IOUs leave **captive ratepayers** no choice in utility providers because of what is essentially a monopoly status (i.e., monopsony— only one supplier). Ratepayers often feel stuck with IOUs focused on return on investments and shareholder dividends.[66] Viewing energy as a commons, public power and electric cooperatives are alternatives to IOUs (Conclusion). Distributed renewable energy and storage systems have more equitable foundations and infrastructures.[67]

The strategic technique known as **counter-mapping** challenges state and corporate territorial assumptions with localized, community-based data. A local community in Virginia (Chapter 9) utilized counter-mapping techniques to dismantle their misrepresentation by state and corporate actors, overthrowing a critical state permitting process. **Community-based participatory action research** (CBPAR) involves directly impacted communities collaborating with scientists and practitioners who center local knowledge and concerns in their investigative design and implementation.

Representation—who is speaking or acting on behalf of a group, territory, or institution—is part of anti-colonial struggle.[68] Counter-mapping and counter-narratives are essential to shifting persistent tropes. In practice, counter-mapping is a fight over how space is represented. Pulling out layers of participation and building spatial consciousness allows people to understand how, in many instances, a problematic degree of control has been handed to a small number of powerful companies that often have too close relationships with a government and/or individual politicians. To understand and attend to the role of non-state actors, it is helpful to introduce the notion of **energy governance**. This notion is not to be confused with the narrower purview of government, which attends primarily to the formal structures and actions of state officials. Energy governance involves decision-making around energy by formal governments as well as non-state actors—these include the private sector and civil society, meaning a broad range of not-for-profit organizations working in the areas of education, religion, social justice, and more.

[64] Baker, S. (2021). *Revolutionary power. An activist's guide to the energy transition.* Island Press.

[65] *Oil and Gas Journal.* (2020, June 27). Shell restarts damaged Bintulu GTL plant.

[66] Táíwò, O. O. (2022, January 24). Toward an Energy Democracy. *Intelligencer.*

[67] McNamara, W., Passell, H., Montes, M., Jeffers, R., & Gyuk, I. (2022). Seeking energy equity through energy storage. *The Electricity Journal, 35*(1), 107,063.

[68] Bosworth, K., & Chua, C. (2021). The Countersovereignty of Critical Infrastructure Security: Settler-State Anxiety versus the Pipeline Blockade. *Antipode.*

Try this

Research, analyze and counter-map: *Examine justice in your energy grid. Analyze time and space to delineate how processes unfolded and how costs and benefits are distributed. Develop counter-maps to show alternative claims and possibilities.*

Deathprint

Slow violence seldom makes headlines as it can be harder to document than fast violence.[69] Air pollution deaths are only recently being tied to nearby harmful facilities. An energy **deathprint**, defined as the mortality rate per kilowatt hour (kWh) of an energy source (Fig. 1.12), shows deathprint comparisons across major fuel types.

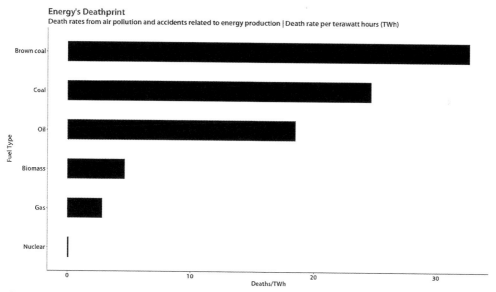

Figure 1.12 Basic comparison of energy-related deaths. *(Source: Markandya & Wilkinson (2007); Sovacool et al. (2016); UNSCEAR (2008; & 2018), accessed via Our World in Data.)*

---

[69] Ahmann, C. (2018). "It's exhausting to create an event out of nothing": Slow violence and the manipulation of time. *Cultural Anthropology, 33*(1), 142–171.

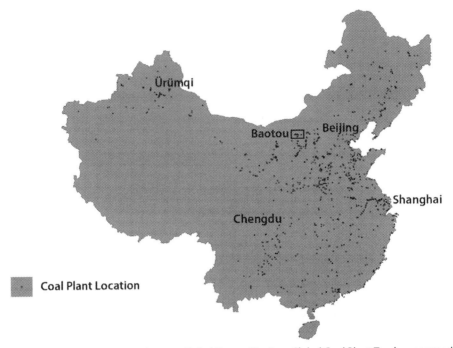

Ürümqi

Baotou Beijing

Shanghai

Chengdu

Coal Plant Location

Figure 1.13 Coal plants in China. *(Source: Global Energy Monitor, Global Coal Plant Tracker, accessed 11/8/2020.)*

This deathprint does not effectively tally deaths across the full life cycle of a particular energy's infrastructure. It does, however, suggest that coal is by far the deadliest fuel.[70] Yet location matters: coal in the US is less deadly than the global average, while coal in China has the highest deathprint. China is the location where half of the world's coal is used (Fig. 1.13 above),[71] often for the manufacturing of products then shipped around the globe.

Emission statistics from fossil fuels are often aggregated, masking local disparities and uneven regional responses. Further, proximity alone is an insufficient risk indicator as power plants vary in both the type of fuels combusted as well as the types of pollution controls used.[72] Mismanagement of toxic residues contributes to a public health emergency

[70] Conca, J. (2016, September 30). The 'deathprint' of energy grapples with the power of regulation. *Forbes.*

[71] Wang, J., Wang, R., Zhu, Y., & Li, J. (2018). Life cycle assessment and environmental cost accounting of coal-fired power generation in China. *Energy Policy, 115,* 374–384.

[72] Dai, W., Dong, J., Yan, W., & Xu, J. (2017). Study on each phase characteristics of the whole coal life cycle and their ecological risk assessment—A case of coal in China. *Environmental Science and Pollution Research, 24*(2), 1296–1305.

in hotspots.[73] A future deathprint can and should additionally quantify fatal impacts attributed directly and indirectly to climate disruption for those now living as well as those yet born.[74]

Sacrifice zones

Inequity in **spatial planning**—designed arrangement of living, working, and environmental conditions at a wide range of spaces and scales—creates material expressions of power. Baotou, the largest city in China's Inner Mongolia, demonstrates spatial unevenness: a multitude of coal plants feed the city's intensive industrial complex in an otherwise remote region.

It's impossible to tell where Baogang refineries end and the city begins.

Baotou, described as the "worst place on earth,"[75] situated northwest of Beijing (Fig.1.13), gained this title largely because of its deposits of rare earth minerals. Massive pipes erupt from the ground and there is a constant stream of diesel-belching coal trucks connecting to the city's coal plants (Fig. 1.14).

A myriad of electronic devices with components made in Baotou span the globe. Because of our collective technology lust, Baotou has become a **sacrifice zone**: a geographic area permanently impaired by environmental damage or economic disinvestment. Sacrifice zones are most commonly found in low-wealth and minority communities.

Try this

Discuss*: Air pollution in urban hubs of China receive attention more often than rural clusters like Baotou's Baogang refineries complex. Is it "good" spatial planning to concentrate pollution in remote places? Is this fair?*

The costs of air pollution are growing. Research suggests one in five deaths globally are linked to fossil fuel combustion. Existing legal standards don't fully protect health,[76]

---

[73] Kravchenko, J., & Lyerly, H. K. (2018). The impact of coal-powered electrical plants and coal ash impoundments on the health of residential communities. *North Carolina Medical Journal, 79*(5), 289–300.

[74] Pearce, J. M., & Parncutt, R. (2023). Quantifying global greenhouse gas emissions in human deaths to guide energy policy. *Energies, 16*(16), 6074.

[75] Maughan, T. (2015, April 2). The dystopian lake filled by the world's tech lust. *BBC*.

[76] Environmental Protection Agency (EPA). (2020). *Integrated Science Assessment (ISA) for ozone and related photochemical oxidants*; Vohra, K., Vodonos, A., Schwartz, J., Marais, E. A., Sulprizio, M. P., & Mickley, L. J. (2021). Global mortality from outdoor fine particle pollution generated by fossil fuel combustion: Results from GEOS-Chem. *Environmental Research, 195*, 110754.

Figure 1.14 Coal plants cluster, city of Baotou, China. *(Sources: Esri, Maxar, Earthstar Geographics, CNES/Airbus DS, USDA FSA, USGS, Aerogrid, IGN, IGP and the GIS User Community. Global Energy Monitor, Global Coal Plant Tracker, accessed 11/8/2020.)*

particularly for the vulnerable.[77] Power plants can create cross-border damages.[78] Europe's "Toxic 30" (Fig. 1.15) shows deaths extend beyond the country where coal facilities are based.

While research and media identifying coal's preventable fatalities are common,[79] experts also draw attention to the fact that fossil gas is deadly as well.[80] Even cooking indoors with gas stoves has been shown to contribute to disease.[81] Disinformation

---

[77] Milman, O. (2021, July 29). Three Americans create enough carbon emissions to kill one person, study finds. *The Guardian*.

[78] HEAL, WWF European Policy Office, Climate Action Network Europe & Sandbag. (2016). *Europe's dark cloud: How coal burning countries are making their neighbors sick*.

[79] Yun, X., Meng, W., Xu, H., Zhang, W., Yu, X., Shen, H., ... & Tao, S. (2021). Coal is dirty, but where it is burned especially matters. *Environmental Science & Technology*.

[80] Conca, J. (2018, January 25). Natural gas and the new deathprint for energy. *Forbes*.

[81] Michanowicz, D. R., Dayalu, A., Nordgaard, C. L., Buonocore, J. J., Fairchild, M. W., Ackley, R., ... & Spengler, J. D. (2022). Home is where the pipeline ends: Characterization of volatile organic compounds present in natural gas at the point of the residential end user. *Environmental Science & Technology*; Gorski, I., & Schwartz, B. S. (2019). Environmental health concerns from unconventional natural gas development. In *Oxford research encyclopedia of global public health*.

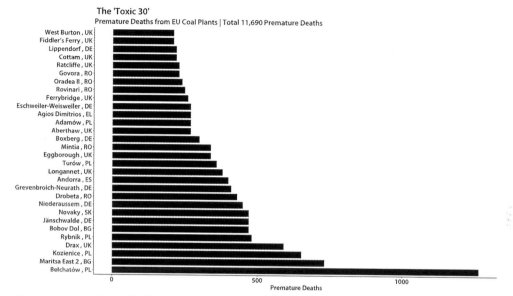

The 'Toxic 30'
Premature Deaths from EU Coal Plants | Total 11,690 Premature Deaths

Figure 1.15 The Toxic 30. *(Source: Europe's Dark Cloud Report, WWF European Policy Office, Sandbag, CAN Europe and HEAL, Brussels, Belgium, 2016.)*

(Introduction) is so systematic and widespread in the fossil fuels industry that it has become self-perpetuating. For example, poorly informed energy pundits focus on carbon dioxide emissions but not on methane, a potent greenhouse gas also released from fossil fuel operations, whether they be coal, oil, or gas. If we do not identify or correctly measure a problem—like methane leaks from fossil gas contributing to rapid climate change—the problem can easily be ignored.

Evidence alone is no match for the power of vested interests.

With strong **vested interests**—personal or especially financial stakes in a situation or undertaking—officials at top levels of government prioritize and exempt energy companies in order to increase their profits while suppressing critical opposition. Energy consumers are often manipulated with fear of scarcity, and threat of blackouts designed to keep the public constantly paying to build new fossil fuel infrastructure to the benefit of energy developers and investors.

Embodied energy violence

The concept of a deathprint, while instructive, has significant informational gaps, as fatality alone is not inclusive of many significant harms. For example, harm also includes non-lethal but life-altering consequences, including slander, intimidation, and anticipatory stress.[82] There can be labor abuse in any manufacturing sector; even the solar industry has its own toxicity in terms of silicosis from sand mining and from contaminating materials in necessary wires and batteries. As with many green options, it is tempting to assume that fixes that address one area are the "solution"; yet, it is necessary to understand the complete picture as there are often tradeoffs between solutions and harms. Having accurate data matters. As a core example, coal, oil, and gas ignore or underestimate methane emissions and thus have miscalculated climate damage that might have tipped the scale toward renewables decades sooner. Heating induced by fossil gas methane emissions causes permafrost regions to thaw, in turn releasing the methane that has been left frozen, encapsultated within natural features. This is an example of "warming feeding warming." Positive or **amplifying feedback loops** appear when the product of a reaction leads to an increase in that reaction—these feedbacks can accelerate rapidly until they "run away."[83]

Energy violence creates particularly harmful consequences for those experiencing multiple forms of harm, as damages can compound or multiply. In Nigeria (Chapter 6), oil's harm is nothing short of a catastrophe after decades of spills and flares. The word "violence" fits with energy systems because processes continue past the point of discovery. The definition of violence from the World Health Organization (WHO) is the intentional use of physical force or power, threatened or actual, against oneself, another person, or against a group or community, that either results in or has a high likelihood of resulting in injury, death, psychological harm, maldevelopment, or deprivation.[84] Violence can be subdivided differently to highlight its various aspects like structural, legal, or regulatory violence (Fig. 1.16). The chapters referred to in Fig. 1.16 provide additional information regarding violence.[85]

The physical body is central to our understanding of violence.[86] Bodies commit harmful acts and violence is enacted on bodies. Nonetheless, body—violence linkages are not simple, in part due to subconscious biases linked to **embodiment**: awareness and visibility of corporeal expression, including how people engage the world through their physical bodies and how they are perceived based on characteristics such as race,

[82] Finley—Brook, M., Williams, T. L., Caron-Sheppard, J. A., & Jaromin, M. K. (2018). Critical energy justice in US natural gas infrastructuring. *Energy Research & Social Science, 41*, 176—190.

[83] Macy, J., & Johnstone, C. (2022). *Active hope: How to face the mess we're in with unexpected resilience and creative power.* New World Library.

[84] Rutherford, A., Zwi, A. B., Grove, N. J., & Butchart, A. (2007). Violence: A glossary. *Journal of Epidemiology & Community Health, 61*(8), 676—680.

[85] Bell, S. E., Fitzgerald, J., & York, R. (2019). Protecting the power to pollute: Identity co-optation, gender, and the public relations strategies of fossil fuel industries in the United States. *Environmental Sociology, 5*(3), 323—338.

[86] Tyner, J. (2012). *Space, place, and violence: Violence and the embodied geographies of race, sex and gender.* Routledge.

Type of Violence	Definition	Examples
physical	bodily harm (**Chapter 2**)	-murder and assault
psychological	anguish or threat (**Chapter 9**)	-trauma, intimidation or stress
structural	inequality without accountability (**Chapter 2**)	-energy colonialism -racism, sexism, classism, etc. -rural-urban disparity
legal	manipulation of legislative weaknesses or loopholes (**Chapter 6**)	-protracted lawfare -counter-suits to deter justice -bankruptcy to avoid cleanup
regulatory	sabotage of policies to limit or impede effective legislation (**Chapter 3**)	-regulatory rollbacks -grandfathering -loopholes -unresolved occupational hazards
bureaucratic	ineffective state actions that misdiagnose, misdirect and distract (**Chapter 3**)	-technical criteria poorly aligned with risk -toxic hotspots
lateral	when those who feel or are powerless, direct their dissatisfaction toward another, themselves, or those less powerful (**Chapter 10**)	-divide and conquer -cooptation[85]
territorial	dispossession of marginalized groups or households (**Chapter 10**)	-loss of property rights -land maldistribution
gender	harm directed at an individual based on biological sex or gender identity (**Chapter 7**)	-sexual assault and rape -disappearances

Figure 1.16 Intersecting forms of energy violence.

gender, size, and age.[87] Energy embodiment becomes a literal process with microplastics, hydrocarbons, and fossil fuel pollutants entering our bodies daily.

**Inability to breathe is a metaphor and
a material reality of state violence and ongoing racism.**

---

[87] Hall, J. M. (2012). Revalorized black embodiment: Dancing with Fanon. *Journal of Black Studies, 43*(3), 274–288.

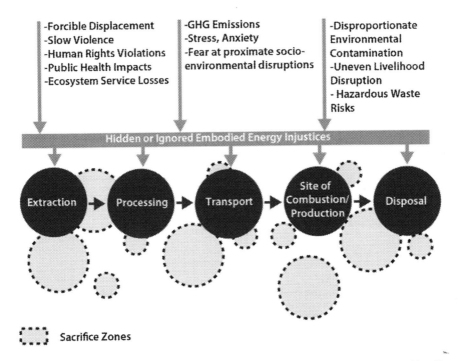

- Forcible Displacement
- Slow Violence
- Human Rights Violations
- Public Health Impacts
- Ecosystem Service Losses

- GHG Emissions
- Stress, Anxiety
- Fear at proximate socio-environmental disruptions

- Disproportionate Environmental Contamination
- Uneven Livelihood Disruption
- Hazardous Waste Risks

Hidden or Ignored Embodied Energy Injustices

Extraction → Processing → Transport → Site of Combustion/Production → Disposal

Sacrifice Zones

Figure 1.17 Embodied energy life cycle. *(Adapted from: Embodied energy injustices: Unveiling and politicizing the transboundary harms of fossil fuel extractivism and fossil fuel supply chains, N. Healy, et al., 2019.)*

Examining embodiment in energy requires considering the intersections between race and gender.[88] The concept of **embodied carbon** (which includes embodied methane) symbolically represents carbon dioxide (CO_2) and other GHG emissions associated with materials and construction processes throughout the supply chain. Like "blood diamonds," violent energy carries within it a plethora of unethical ramifications from its use, as shown in LCA (also known as **cradle-to-grave**) covering all inputs and outputs from the start to end of life (Fig. 1.17 above).

Traditionally overlooked, there is growing evidence of psychological harm from stress and conflict during energy development.[89] Structural violence is a backdrop to other forms of harm. Whether emerging from slow violence or fast violence, or some

---

[88] Healy, N., Stephens, J. C., & Malin, S. A. (2019). Embodied energy injustices: Unveiling and politicizing the transboundary harms of fossil fuel extractivism and fossil fuel supply chains. *Energy Research & Social Science, 48,* 219–234.

[89] Pearson, T. W. (2017). *When the hills are gone: Frac sand mining and the struggle for community.* Minnesota Press.

combination of both, crises create opportunities for change, while showing the need to move away from the status quo.

Try this

Reflect and analyze: *Is deathprint a useful calculation? What are its strengths? How could it be improved? Does the concept of energy violence help do enough to expose other concerns? Why or why not?*

Energy imaginaries

Fossil fuel companies lobby for subsidies and against climate action to ensure record profits.[90] Renewable, alternative energy solutions can only truly thrive when governments stop providing regulatory advantages and subsidies to coal, oil, and gas, and their byproducts. We require new modes of engagement to face unprecedented changes. We are all connected, with the climate crisis being an obvious example.

Extreme heat frays everything.[91]

We must shift power to lead upward. In reality, energy is our commons.[92] As we draw on our narrative imagination we extend social visions and thoughts about what is possible to meet our challenges. **Embodied research** unites body and mind in investigative practices in physicality and praxis. Community-based science or **street science** is data and evidence gathered by regular people without professional training. CBPAR is increasingly common in environmental and climate research as a growing number of scientists realize the need to work with directly impacted communities.

Due to its complex broad-reaching implications, energy sector analysis requires **mixed methods research** involving two or more research methodologies. Narrow analysis in one discipline, such as economics, chemistry, or physics, can impede cross-disciplinary understanding necessary for energy transformation in the Anthropocene. Science and **quantitative research methods** are required, but there are also benefits from **qualitative research methods** involving the collection and analysis of non-numerical data to understand concepts, opinions, or experiences—this should especially involve direct communication with frontline populations. Helpful primary sources include interviews, focus groups, and participant-observation.

This book is for problem-solvers.

---

[90] Bailout Watch. (2021, August 20). *6 bailed out polluters already spent $15M this year lobbying for subsidies and against climate action.*
[91] Wallace–Wells, D. (2020). *The uninhabitable earth: Life after warming.* Tim Duggan Books.
[92] Dawson, A. (2022). *People's Power: Reclaiming the Energy Commons.* OR Books.

Our shared need for community-controlled off-grid and island-grid solutions to increase the resiliency and economic power of rural areas is well demonstrated. Yet one-size-fits-all solutions will never address the challenges we face. While community-based solutions are positive on many fronts, we can't afford to fall into the "local trap"[93], the risk of assuming the local scale is always preferential to larger, often global scales. Local energy provides important benefits but is not a panacea, as an arrangement in which resources or decisions are controlled locally can also be ecologically unsustainable or socially unjust. The outcomes of a given scalar arrangement are dependent on the political agendas of those empowered by the arrangement. For example, mass mobilization of organized labor in large-scale utilities can lead to sustainability in cities and for workers in very different ways than distributed community-run solar power.[94] Energy **prosumers** (consumers involved in production) can imagine new forms of energy ownership and control based on the needs and sources of each location and governance body. Many options exist to improve social and economic benefits from energy production/distribution and reduce harm.

Geographical imaginaries are not merely representations of space; instead, they represent a place for ways that produce knowledge and construct meaning (i.e., placemaking, worldmaking). Our understanding of climate (i.e., atmospheric knowledge) and our present, collective inability to effectively respond to the looming crisis speaks to the longevity of colonial spaces and imaginaries. We need to form new spatial imaginaries and social practices for equitable **placemaking (and worldmaking)**— collaborative acts to achieve the potential of a place, not only as it has been but also *as it can be*. It is necessary to reinvigorate the collective narrative imagination, seeking to avoid societal chaos through degrowth's intentional, equitable reduction framework.[95] In a future of energy transition and climate mitigation, reparations for past harm (Conclusion) will play a critical role in establishing equity and justice. Reparations need to be built into energy's international finance to address specific histories of forced labor, trafficking, and other offenses.[96] If left unaddressed, inequities perpetuate harm and trauma resulting in fatally flawed energy systems that look and operate like a foreclosed past, not restorative future.

[93] Purcell, M., & Brown, J. C. (2005). Against the local trap: scale and the study of environment and development. *Progress in Development Studies, 5*(4), 279–297.

[94] Huber, M. (2022). *Climate change as class war: Building socialism on a warming planet.* Verso.

[95] Llorens, H. (2021). *Making livable worlds: Afro-Puerto Rican women building environmental justice.* University of Washington Press.

[96] Táíwò, O. O. (2022). *Reconsidering reparations.* Oxford University Press.

Try this

Vision, *plan*, *and act*: *Geographic imaginaries can decolonize and transform. What would you prioritize in placemaking or worldmaking? Would your proposal focus on local, regional, national, or international spaces or scales?*

Summary

An energy violence framework coupled with spatial and temporal analysis helps us understand the role of power and how it produces ignorance and injustice throughout the energy sector. New rules often build awkwardly upon earlier regimes and always within budgetary constraints and under political pressure, contributing to the advancement of false climate and energy solutions as politically tenable compromises. Decision-makers ignore expensive lessons of past energy mis-development when assessing the hazards in new sources of energy, resulting in the magnification or creation of sacrifice zones. Transformative change requires spatial and geographic imaginaries in placemaking and worldmaking to reframe the potential of renewable and alternative energy; degrowth planning alongside community-based participatory action research and reparations for past injustice further serve to reduce climate disruption and build a livable world.

Vocabulary

1. amplifying feedback loops
2. blast radius
3. blindspots
4. body burden
5. cluster
6. co-location
7. community-based participatory action research (CBPAR)
8. contested space
9. counter-mapping
10. cradle-to-grave
11. cumulative impacts
12. deathprint
13. direct action
14. displacement
15. downstream
16. ecological debt
17. embodied carbon

18. embodied research
19. embodiment
20. energy governance
21. energy justice
22. energy regime
23. environmental justice
24. fragmentation
25. geographic imaginaries
26. geographic information science (GIS)
27. hotspot
28. hotspot analysis
29. intersectionality
30. investor-owned utilities (IOUs)
31. justice
32. midstream
33. mixed methods research
34. participatory science
35. place-making
36. power lens
37. power over
38. power with
39. praxis
40. productive exclusion
41. prosumer
42. proximity
43. qualitative research methods
44. quantitative research methods
45. racial capitalism
46. regime of measurement
47. remote sensing
48. reparations
49. representation
50. sacrifice zone
51. scale
52. segmentation
53. sink
54. source
55. space
56. spatial analysis
57. spatial distribution

58. spatial planning
59. synergistic effects
60. temporal analysis
61. threat multiplier
62. uneven development
63. upstream
64. vector
65. vested interests

Recommended

Arboleda, M. (2020). *Planetary mine: Territories of extraction under late capitalism*. Verso Books.
Goodall, J. (2023). *The heat will kill you first: Life and death on a scorched planet*. Little, Brown and Company.
Mendez, M. (2020). *Climate change from the streets: How conflict and collaboration strengthen the environmental justice movement*. Yale University.
Monmonier, M. (2018). *How to lie with maps*. University of Chicago Press.
Nair, C. (2022). *Dismantling global white privilege: Equity for a post-western world*. Berrett-Koehler Publishers.
Solomon, B. D., & Calvert, K. E. (Eds.). (2017). *Handbook on the geographies of energy*. Edward Elgar Publishing.
Peluso, N. L., & Watts, M. (Eds.). (2001). *Violent environments*. Cornell University Press.

CHAPTER 2

Coal blooded

Contents

Climate Crisis, Energy Violence
ISBN 978-0-12-819501-7,
https://doi.org/10.1016/B978-0-12-819501-7.00013-8

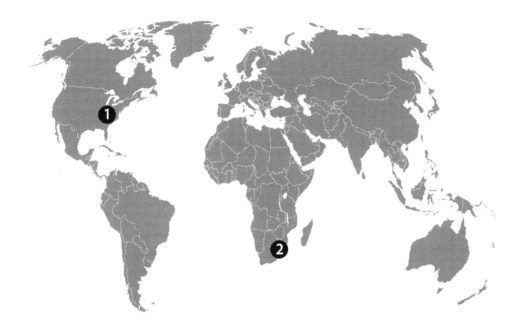

Locations:	1 - Appalachian Region - United States
	2 - KwaZulu-Natal Province - South Africa

Themes:	Spatial control; Structural violence; Physical violence

Subthemes:	Captured state; Legacy pollution; Water stress

Whether you know it or not, industrial pollution is in your blood.[1]

Coal's promise ... and peril

It's hard to imagine how different life would be without the Industrial Revolution and coal-powered electrification and transportation. Thermal coal fueled electricity while metallurgical coal ("coking coal") powered steelmaking. During its heyday (~1850−2010), coal was viewed as cheap and reliable; yet the sale price never reflected its true costs. The coal industry was given large subsidies and was protected from

---

[1] Talpos, S. (2020, February 24). Industrial pollution is in your blood. Is that a form of battery? *Undark*.

regulation, resulting in firms creating and leaving behind dangerous pollution.[2] The soil and rock dislodged by mining as waste, called spoil or overburden, is dumped in valleys and streams. Spoil is a source of **acid mine drainage** (AMD)[3]—the formation and movement of highly acidic water rich in heavy metals from a mining site. Slurry, the residue from cleaning coal, can leach toxic pollutants like arsenic and lead when impounded in ponds or injected into abandoned underground mine shafts.[4]

Coal's harms are not borne fairly. Current standards ignore or mask harms with severe repercussions, including birth defects and cancers.[5] While concentrations vary, all people have in their body harmful contaminants, like heavy metals from coal combustion. Learning from coal can help avoid similar mistakes with other fuel types, including oil and gas (Chapters 5–10). State agencies continue to permit coal projects due to the lobbying power of the industry, but the public has long fought to take away coal companies' **social license to operate** (SLO)— the perception in society that a company or an industry is legitimate and operates in an acceptable fashion. In the case of coal, a majority now feel granting extraction leases or yet more power plant construction is unacceptable. Taking away the social license of fossil fuels was proposed emphatically a decade ago by writer and college professor Bill McKibben when he and his students helped to popularize the goal to keep fossil fuels in the ground.[6]

Coal extraction often relied on tight socio-spatial control in isolated locations where dangerous conditions and repressive tactics, including physical violence, were used to control the workforce. **Physical violence** occurs when a person exerts control over another person through the use of physical force (i.e., hitting, restraining). In turn, these harms have been rendered invisible through commemoration (Chapter 3). Oppression was normalized in the industry—an assumption that the miners' sacrifices were necessary for "progress" is the energy sector's "original sin." As coal use expanded, coal energy exemplified luxury (Fig. 2.1), further disguising harms.

[2] EarthJustice. (2019). Mapping the Coal Ash Contamination.

[3] Holzman, D. C. (2011). Mountaintop Removal Mining: Digging Into Community Health Concerns. *Environmental Health Perspectives*, *119*(11), a476.

[4] Zipper, C. E., & Skousen, J. (2021). Coal's legacy in Appalachia: Lands, waters, and people. *The Extractive Industries and Society*, *8*(4), 100990.

[5] Hendryx, M., O'Donnell, K., & Horn, K. (2008). Lung cancer mortality is elevated in coal-mining areas of Appalachia. *Lung Cancer*, *62*(1), 1–7.

[6] Schifeling, T., & Hoffman, A. J. (2017, December 11). How Bill McKibben's radical idea of fossil-fuel divestment transformed the climate debate. *The Conversation*.

Figure 2.1 Wholesale Coal advertisement, 1887. *(Image credit: S.J. Patterson wholesale coal., ca. 1887. Nov. 11. Photograph. https://www.loc.gov/item/2004667298/.)*

Material outcomes of energy infrastructure and the systems that maintain it demonstrate our larger powerscape.[7] Those with financial capital (i.e., owners, investors)

[7] Leebrick, R. A., & Maples, J. N. (2015). Landscape as arena and spatial narrative in the New River Gorge National River's coal camps: A case study of the Elverton, West Virginia 1914 Strike. *Southeastern Geographer, 55*(4), 474–494.

shaped coalfields to produce cheap energy. Powerscapes morph over time, yet profitable patterns undergird current geographies. The **company town** exemplifies spatial control tied to an industry, as a monopoly owns housing, stores, and even services like sanitation and security.[8] Entrenched coal regions exemplify extreme social and environmental injustice.

Energy violence

Structural violence is systemic inequality, wherefore the (mis)use of power results in injury, death, psychological harm, maldevelopment, or deprivation without accountability. Although each region is unique, power inequalities built into most coal operations multiplied risks for miners. In many coal regions, as in Appalachia, extreme inequities in land ownership created an enduring structural violence that has persisted.[9] For example, high numbers of absentee owners, having been given tax breaks, drained wealth from the region leaving high mortality and morbidity.[10] Firms exploited identity formation among miners created by deep connections among those who lived and worked in tight quarters under trying conditions. With few job options, mine work became a source of intergenerational pride, a way of life (Fig. 2.2).

Uneven power created fertile ground for abuse: rural inhabitants were given pejorative labels like "hill folk" and "redneck."[11] Coalfield communities internalized oppression resulting in caste systems and **labor hierarchy**, or rank by degree of financial security and power (Fig. 2.3). With huge pay gaps between the highest and lowest salaries, coal's hierarchy exemplifies labor inequality.

Coal operators created racial and cultural wedges to impede worker solidarity, splitting the labor market with disparate pay scales for Black versus White laborers. Coal companies sought to weaken labor organizing through policies of **judicious mixture**—fragmentation that was strategically employed. Mines hired roughly 25% foreign, 25% Black, and 50% White workers.[12] This "race wedge" divided the workforce. Mine owners fed competition

[8] Fishback, P. (1992). The Economics of Company Housing: Historical Perspectives from the Coal Fields. *Journal of Law, Economics, and Organization, 8*(2), 346–365.

[9] Guilford, G. (2017, December 30). The 100-year capitalist experiment that keeps Appalachia poor sick and stuck on coal. *Quartz*.

[10] Woolley, S. M., Meacham, S. L., Balmert, L. C., Talbott, E. O., & Buchanich, J. M. (2015). Comparison of mortality disparities in central Appalachian coal-and non—coal-mining counties. *Journal of Occupational and Environmental Medicine, 57*(6), 687–694.

[11] Lewis, R. L. (1993). Appalachian restructuring in historical perspective: coal, culture and social change in West Virginia. *Urban Studies, 30*(2), 299–308.

[12] Bailey, K.R. (1973). A judicious mixture: negroes and immigrants in the West Virginia mines, 1880–1917, *West Virginia History, 34*(2), 141–161.

Figure 2.2 US coal miners. *(Image credit: Library of Congress, Prints & Photographs Division, Farm Security Administration/Office of War Information Black-and-White Negatives.)*

MOST	Position	Worker	Type of Work
↑	owner	generally off-site	investment
-authority -financial security -control over time and others' labor	manager	site administrators	office
	security	officers, armed thugs, shack rouster	office and patrols
↓ LEAST	laborer	immigrants, ethnic minorities, convicts, children	mines
	unpaid laborer	women, children	house

Figure 2.3 Labor hierarchy in company coal towns, 1900s.

among racial groups to encourage peers to see each other as foes or threats and discourage unionization. Nonetheless, tensions boiled over toward the mine bosses due to egregious practices like **cribbing**, the undercounting of mined coal during weigh-in at the end of a shift. A 1912 coal strike demand list illustrates this and other key concerns (Fig. 2.4).

With coal companies' oppressive tactics, miners' support of unions was widespread. In 1933, once unions received state protection, miners joined unions at every mine.

| Operators accept and recognize the union; |
| Miners' rights to free speech and peaceable assembly be restored; |
| Discontinue "black-listing" discharged workers resulting in diminished future work prospects; |
| Compulsory trading at company stores end; |
| 'Cribbing' discontinue; |
| Scales be installed at all mines to weigh coal; |
| Miners be allowed their own check-weighman to check against the company weighman; |
| Two weighmen determine all "docking" penalties. |

Figure 2.4 West Virginia Coal Camp demands, 1912.

Coordinated strikes provided leverage.[13] Unions were successful in improving work conditions, but strikes motivated fear of economic loss among mine owners.[14] Tensions intensified when bosses brought in **strikebreakers**, generally called "scabs" by unions, to undermine work stoppages. US coal operators initially used immigrants as strikebreakers, but soon immigrant miners became integral to US labor uprisings. Mine operators intensified pressure to control workers in response to union organizing. In many coal camps, owners selected candidates most favorably disposed toward their interests and required "their" miners to vote for them.[15]

Vote for the candidates we select or get off the job.

Owners pinned labor organizers with trumped-up criminal charges, forcing them before judges who feared to question. A judge in one such case replied with the statement below when asked why he did not call a Special Grand Jury as permitted under the law:

> "I knew it would be impossible to get a grand jury that was not controlled by the coal operators and mine guards. And I knew that if I started an investigation I would be killed..."

[13] Corbin, D. A. (2015). *Life, Work, and Rebellion in the Coal Fields: The Southern West Virginia Miners, 1880–1922.* West Virginia University Press.

[14] Lewis, R. L. (1993). Appalachian restructuring in historical perspective: coal, culture and social change in West Virginia. *Urban Studies, 30*(2), 299–308.

[15] Lee, H. B. (1969). *Bloodletting in Appalachia: The Story of West Virginia's Four Major Mine Wars and Other Thrilling Incidents Of Its Coal Fields.* West Virginia University Press.

Slow ecological violence

There was no environmental regulation of early coal. To the contrary, the *Power Plant and Industrial Fuel Use Act* made it illegal until as late as 1978 to construct a new power plant in the US without capacity to burn coal.[16]

Even at its cleanest, coal is too dirty.

Fig. 2.5 demonstrates the broad ecological harm from coal. Clean-up following coal pollution is expensive. For example, air pollution contributes to acid rain, harming health as well as lowering property value.[17,18,19]

Fig. 2.6 demonstrates various routes of pollutants into the human body, with potential harm to organs even from trace amounts. Significant exposure to particulate matter can lead to asthma or chronic obstructive pulmonary disease (COPD). Respiratory system damage may increase rates of stroke or cardiac arrest.

At closed and active mines, tons of coal ash are not stored or disposed of safely (Chapter 4). Appalachia, a mountainous area of the eastern US, is one of the world's

Category of Harm	Examples
air pollution	<ul><li>PM (particulate matter)</li><li>SOx (sulfur oxides)</li><li>CO_2 (carbon dioxide)</li><li>NOx (nitrogen oxides)</li><li>smog</li></ul>
water contamination	<ul><li>acid rain</li><li>acid drainage</li><li>heavy metals</li></ul>
contaminated solid waste	<ul><li>coal ash (**Chapter 4**)</li></ul>
soil toxicity[18]	<ul><li>heavy metals</li></ul>
climate change[19]	<ul><li>CO_2 (carbon dioxide) (from coal combustion)</li><li>methane emissions (from coal extraction)</li></ul>

Figure 2.5 Coal harm.

---

[16] Höök, M., & Aleklett, K. (2009). Historical trends in American coal production and a possible future outlook. *International Journal of Coal Geology, 78*(3), 201–216.

[17] Gottlieb, B., & Lockwood, A. (2017). The Life Cycle of Coal and Associated Health Impacts. In *Coal in the 21st Century* (pp. 100–146).

[18] Munawer, M. E. (2018). Human health and environmental impacts of coal combustion and post-combustion wastes. *Journal of Sustainable Mining, 17*(2), 87–96.

[19] Edwards, G. A. (2019). Coal and climate change. *Wiley Interdisciplinary Reviews: Climate Change, 10*(5), e607.

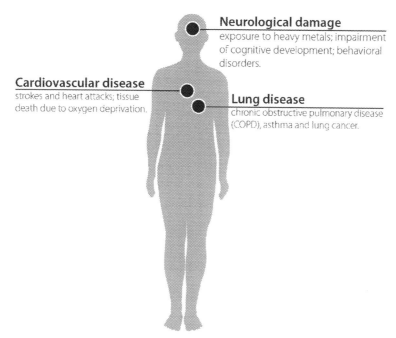

Neurological damage
exposure to heavy metals; impairment of cognitive development; behavioral disorders.

Cardiovascular disease
strokes and heart attacks; tissue death due to oxygen deprivation.

Lung disease
chronic obstructive pulmonary disease (COPD), asthma and lung cancer.

Figure 2.6 Human body and coal chemicals.

main coal-producing regions. Laws are broken more often near poor families.[20] Appalachia significantly lags behind national health indicators.[21] Water throughout large regions is no longer potable. Health impacts from consumption range from persistent diarrhea to birth defects and cancers.[22] Not surprisingly, there's increased risk of depression in Appalachian coal regions,[23] and internationally.[24]

[20] Stretesky, P. B., & Lynch, M. J. (2011). Coal strip mining, mountaintop removal, and the distribution of environmental violations across the United States, 2002–2008. *Landscape Research*, *36*(2), 209–230;

[21] Hendryx, M. (2008). Mortality rates in Appalachian coal mining counties: 24 years behind the nation. *Environmental Justice*, *1*(1), 5–11.

[22] Holzman, D. C. (2011). Mountaintop Removal Mining: Digging Into Community Health Concerns. *Environmental Health Perspectives*, *119*(11), a476.

[23] Hendryx, M., & Innes-Wimsatt, K. A. (2013). Increased risk of depression for people living in coal mining areas of central Appalachia. *Ecopsychology*, *5*(3), 179–187.

[24] Yong, X., Gao, X., Zhang, Z., Ge, H., Sun, X., Ma, X., & Liu, J. (2020). Associations of occupational stress with job burn-out, depression and hypertension in coal miners of Xinjiang, China: a cross-sectional study. *BMJ Open*, *10*(7), e036087.

Disturbance of one's surrounding by mountaintop removal[25] can result in acute feelings of grief, desolation, loss of identity, and powerlessness.[26]

**Mountaintop removal's harms
go so deep and last so long.**[27]

Spatial distribution

Spatial concentration of the most profitable reserves manifested as clustering (Chapter 1). Fig. 2.7 demonstrates clustering in the production of US coalfields, circa the late 1800s.

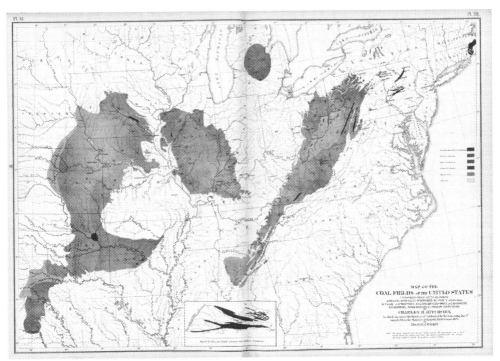

Figure 2.7 Coalfields of the US, 1870. *(Map image credit: United States Census Office. 9th Census, 1870, and Francis Amasa Walker. [New York J. Bien, lith, 1874]. https://www.loc.gov/item/05019329/.)*

[25] Canu, W. H., Jameson, J. P., Steele, E. H., & Denslow, M. (2017). Mountaintop removal coal mining and emergent cases of psychological disorder in Kentucky. *Community Mental Health Journal, 53*(7), 802–810.

[26] T. Kruger, T., Kraus, T., & Kaifie, A. (2022). A changing home: a cross-sectional study of environmental degradation, resettlement and psychological distress in a Western German coal-mining region. *International Journal of Environmental Research and Public Health, 19*(12), 7143.

[27] Britton-Purdy, J. (2016, March 21). The violent remaking of Appalachia. *The Atlantic.*

Social control techniques to maintain a docile workforce coupled with excessive mistreatment in these clusters were central to the success of early coal extraction. **Spatial control**, meaning command over use and access to a location, was an essential tactic of company towns.

They hired police, delivered mail and to a certain extent dispensed justice.[28]

Mine owners employed physical force and surveillance tactics (Fig. 2.8) to keep miners working in abusive conditions. Private security forces attempted to destroy any labor organizing in the mines. Described below, the Coal Mine Wars, some of the worst violence against organized labor in US history, involved machine-gun massacres of miners and their families.[29]

Company town entrances were guarded. Within company towns, a hired "shack rouster" routinely terrorized workers and their families as a means of social control.[30] Postal service workers serving these private jurisdictions were encouraged to destroy

Figure 2.8 Coal company town guard. *(Image credit: Rothstein, Arthur, photographer. Guard at company town. Jefferson County, Alabama. Jefferson County United States Alabama, 1937. Photograph. https://www.loc.gov/item/2017775916/.)*

[28] Boyd, L. W. (1994). The economics of the coal company town: Institutional relationships, monopsony, and distributional conflicts in American coal towns. *The Journal of Economic History, 54*(2), 426–427.

[29] Lee, H. B. (1969). *Bloodletting in Appalachia: The Story of West Virginia's Four Major Mine Wars and Other Thrilling Incidents Of Its Coal Fields.* West Virginia University Press.

[30] Lewis, R. L. (2009). *Black Coal Miners in America: Race, Class and Community Conflict (1780–1980).* University Press of Kentucky.

mail considered subversive—a means of information and communication control. Coal companies owned the houses, stores, and saloons in company towns, creating conditions where miners were wholly dependent for their own precarious survival. Miners were paid in **scrip**—tokens or paper with a monetary value issued as an advance on wages by the company. Scrip could only be used in the company town named, wherein there were usually no other retail establishments, allowing companies to charge exorbitant prices. Payment in scrip eliminated prospects of acquiring generational wealth, resulting in miners being perpetually in debt to their employer.[31] Officially banned as late as 1967, the practice of scrip (Fig. 2.9) exemplified the persistence of coal industry power.[32]

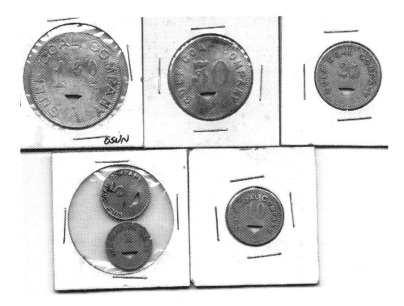

Figure 2.9 Coal company scrip. *(Image credit: Coal town guy at English Wikipedia.)*

Coal sacrifice zones

Coal's sacrifice zones are marginalized and receive toxic burdens and extreme hazards (Chapter 1).[33] Coal communities have been particularly abandoned with **legacy pollution** as responsibility for the release of toxins is persistently unassigned.

---

[31] Fishback, P. V. (1986). Did Coal Miners "Owe Their Souls to the Company Store"? Theory and Evidence from the Early 1900s. *Journal of Economic History*, 1011–1029.

[32] Guilford, G. (2017, December 30). The 100-year capitalist experiment that keeps Appalachia poor sick and stuck on coal. *Quartz*.

[33] Fox, J. (1999). Mountaintop removal in West Virginia: an environmental sacrifice zone. *Organization & Environment, 12*(2), 163–183.

Destruction by multiple mining corporations operating within the same watershed—often with imprecise recordkeeping—deters culpability. Coal regions have few income-generating options until pollution is remediated. Mountaintop removal creates flattened, wasted space—insidiously a preferred topography for security facilities.[34] Prisons have been strategically placed on top of toxic coal mines (Box 2.1).[35] A **spinoff industry** occurs when fossil fuels instigate the creation of new or additional sectors, products or services. This can be more direct, such as petrochemical products being produced alongside coal mines, or less direct, such as prisons forming on an abandoned mine. Spinoff industries are sullied by ties with fossil fuels, whether by pollution, corruption, or other negative influence.

Temporal analysis

The most oppressive mining tactics remain consistent across the globe, from the earliest industrial mines to the present. This includes labor in small spaces where the miner—often a child—is unable to stand or even sit up. Miners pull themselves forward on their elbows. In India, many children have lost their lives extracting coal with hand picks in deep small spaces called "rat hole mines," which are narrow and prone to collapse.[48] While illegal, the Indian government does little to regulate, much less end, these coal mines. These abusive labor practices harken back to the 1800s in the US. Similarly, practices in Canada are marked by abuses, as surveyed by *Boys in the Pits: Child Labour in Coal Mines*.[49]

In 1891, the US Congress passed the first federal regulations to protect miners—a central concern was proper ventilation.[50] The Bureau of Mines founded in 1910 was charged with ameliorating thousands of mining deaths. Fig. 2.10 highlights a timeline showing vast spans of inaction punctuated by a handful of key federal protections.

After the 1977 establishment of the US Mine Safety and Health Administration (MSHA), coal mine fatality rates began to significantly decline. MSHA created rules to reduce the risk of mine collapse. Episodes of fast violence at the job site have indeed dropped, and deaths directly attributed to mine collapse and related workplace fatalities have decreased significantly (Fig. 2.11). MSHA has not been able to address slow violence of coal ash or methane emissions, however.

[34] Schept, J. (2021). Planning prisons and imagining abolition in Appalachia. In *The Routledge International Handbook of Penal Abolition* (pp. 387–398). Routledge.

[35] Perdue, R. T. (2021). Trashing Appalachia: Coal, prisons and whiteness in a region of refuse. *Punishment & Society*, 14624745211011526.

[48] Nath, M. (2018). Violation of human rights: Coal mining areas in jaintia hills, Meghalaya. *Journal of Applied Geochemistry*, 20(3), 361–364.

[49] McIntosh, R. (2000). *Boys in the Pits: Child Labor in Coal Mines*. McGill-Queen's University Press.

[50] Mine Safety and Health Administration. (n.d.). Legislative history of U.S. mine safety and health.

BOX 2.1 Spinoff industry: detention facilities

The US continually adds new facilities to what is already the largest prison system in the world.[36] Super-maximum ("supermax") facilities expanded in the US due to a "tough-on-crime" political climate in the 1990s, during which parole was abolished and sentences lengthened.[37] For-profit correctional facilities are lucrative businesses backed by powerful lobbies.[38] Prisoners, predominantly African-American men, are extracted from troubled urban areas outward to rural facilities as resource "deposits."[39]

At the end of slavery, **convict leasing**, a system of forced penal labor, targeted African-Americans to reinforce post-emancipation racial hierarchy in the US.[40] This racial capitalism benefited southern coal and steel companies and increased competitiveness to capture early national and global markets.[41] Black men were arrested for any behavior deemed inappropriate by white authorities—common "crimes" were vagrancy, changing employers without permission, and loitering.[42] These newly freed African-American "offenders" were then sentenced to long jail terms. Consequently, the US prison population became overwhelmingly Black. Once in the correctional system, Black men were caught up in debt bondage, unable to exit.[43] Convicts were sent to industrial labor sites like dirty, dangerous coal mines. Convict leasing programs reduced budgets at correctional facilities, much like US prison labor programs today.

In the mid-1990s as part of a massive prison-building effort, the Red Onion Supermax State Prison in Wise County, Virginia, was established on top of an old coal mine. Many of the corrections officers arrived at Red Onion after being laid off from jobs in nearby coalfields.[44] Dozens of prisons have been built on or next to mines, landfills, and contaminated sites.[45] Toxic prisons harm inmates and staff. The trauma of the violence in US correctional facilities often leads to mental health problems for prisoners and jailers alike. Prisoner treatment often violates human rights. Many facilities lack potable water and have sewage leaks, polluted air, or deadly heat.[46] Prisoners and correctional officers frequently experience adverse effects with heightened inability to cope with stress, anger, and frustration.[47]

[36] Pellow, D. N. (2017). *What is critical environmental justice?* John Wiley & Sons.

[37] Human Rights Watch. (1999). Red Onion State Prison: Super-Maximum Security Confinement in Virginia.

[38] Cohen, M. (2015). How for-profit prisons have become the biggest lobby no one is talking about. *Washington Post, 28.*

[39] Perdue, R. T. (2018). Linking environmental and criminal injustice: the mining to prison pipeline in central Appalachia. *Environmental Justice, 11*(5), 177–182.

[40] Lewis, R. L. (2009). *Black Coal Miners in America: Race, Class and Community Conflict (1780–1980).* University Press of Kentucky.

[41] Gilmore, R. W. (2022). *Abolition Geography: Essays Toward Liberation.* Verso.

[42] Stoddard, B. (2020). "Slaves of the State": Christianity and Convict Labor in the Postbellum South. *Religions, 11*(12), 651.

[43] Blackmon, D. A. (2009). *Slavery by another name: The re-enslavement of black Americans from the Civil War to World War II.* Anchor.

[44] Solitary: Inside Red Onion.

[45] Greenfield, N. (2018, January 19). The Connection Between Mass Incarceration and Environmental Justice. National Resources Defense Council.

[46] Perdue, R. T. (2018). Linking environmental and criminal injustice: the mining to prison pipeline in central Appalachia. *Environmental Justice, 11*(5), 177–182.

[47] Swanson, K. (2019). Silent killing: The inhumanity of US immigration detention. *Journal of Latin American Geography, 18*(3), 176–187.

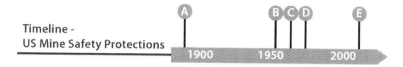

A - First federal statute for mine safety
B - Federal Coal Mine Safety Act of 1952
C - Federal Metal and Nonmetallic Mine Safety Act
D - Federal Coal Mine Health and Safety Act
E - Mine Improvement and New Emergency Response Act

Figure 2.10 A timeline of key US mine safety protections.

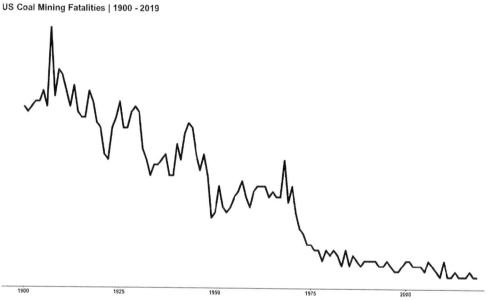

Figure 2.11 US coal mining fatalities (1900–2019). *(Source: Mine Safety and Health Administration (MSHA).)*

As shown in this mining fatalities trend plot, the number of deaths has decreased over time—but there remain challenges to document slow-evolving and poorly diagnosed illnesses like black lung (Chapter 3). Former miners who die earlier than they would have without mining-related exposures may not appear in occupational data.[51] For decades, poorly administered black lung funds fell short of effective and direct benefit to sick

---

[51] Fleming, E. M. & Grulkowski, T. J. (2020, September 3). Mysteries underlying black lung liabilities. Milliman.

and dying miners.[52] Despite a wave of funding in 2022, the future of the black lung payment program remains unstable.

While coal's total costs are still inadequately understood, climate impacts are likely to be the most expensive. By the early twentieth century, the heat-trapping effect of carbon dioxide (CO_2) from coal combustion was connected to the potential of climate change (Fig. 2.12). Little more than a century later, we are acutely experiencing these changes, yet continue to burn coal in spite of its high carbon footprint. The amount of carbon dioxide in the atmosphere in 2022 hit 417.06 parts per million, 50% higher than it was in pre-industrial times.[53]

From its earliest days, the coal industry was aware of dangerous repercussions of methane gas buildup in mining operations. Coalbed methane emissions remain unregulated throughout the world, including in China, which is continually straddled with the highest coalbed emissions[54] and fatal gas explosions in coal mines.[55] Around the globe

March, 1912   **POPULAR MECHANICS**   **341**

The furnaces of the world are now burning about 2,000,000,000 tons of coal a year. When this is burned, uniting with oxygen, it adds about 7,000,000,000 tons of carbon dioxide to the atmosphere yearly. This tends to make the air a more effective blanket for the earth and to raise its temperature. The effect may be considerable in a few centuries.

Figure 2.12 Reference linking coal to climate change, 1912. *(Image credit: Popular Mechanics magazine, March 1912, page 341. Accessed via Wikimedia Commons.)*

[52] Giangreco, L. (2020, August 8). The deadly disease that really scares coal miners. Politico.
[53] Cohen, L. (2023, April 6). Carbon dioxide hits highest sustained rate ever recorded as greenhouse gases creep toward "uncharted levels," NOAA says. CBS News.
[54] Miller, S. M., Michalak, A. M., Detmers, R. G., Hasekamp, O. P., Bruhwiler, L. M., & Schwietzke, S. (2019). China's coal mine methane regulations have not curbed growing emissions. *Nature Communications, 10*(1), 1–8.
[55] Yin, W., Fu, G., Yang, C., Jiang, Z., Zhu, K., & Gao, Y. (2017). Fatal gas explosion accidents on Chinese coal mines and the characteristics of unsafe behaviors: 2000–14. *Safety Science, 92*, 173–179.

coal methane is simply a waste product, rarely used for heating or cooking at mine facilities, powering ventilation systems, or coal drying.[56]

Clean coal is dirty

Attempts to mitigate coal pollution are expensive, with significant tradeoffs and underwhelming results. **Clean coal** is an elusive concept without a standard protocol—the phrase usually refers to technologies capturing carbon emissions from coal combustion. For years companies created a smokescreen by vague advertising promoting the concept of clean coal without intending to move from research to actual projects.[57] Today, carbon capture and storage (CCS) does not represent a mature set of technological solutions—indeed, there is uncertainty as to whether it ever will. Promoted projects in practice operate as overpriced technological experiments that tend to perform poorly to reduce emissions of particulate matter, SOx, NOx, and other regulated pollutants, regardless of their application before construction, during combustion, or post-combustion.[58]

Proposals to create "clean" coal upend earlier promotion of coal as low-cost; in actuality CCS plants are expensive to build and maintain. Carbon capture technologies lower the overall energy efficiency of plants.[59] Retrofitting older plants is costly and increases the power needed for the plant. Conventional CO_2 capture technologies, such as chemical and physical absorption and membrane and cryogenic separation, are all energy-consuming processes. CCS pilots have failed to achieve intended targets for capture, suggesting this expensive technology is unlikely to supercharge greenhouse gas **mitigation**, referring to efforts aimed at reducing and stabilizing the levels of heat-trapping greenhouse gases in the atmosphere. More broadly, mitigation is the action of reducing the severity, seriousness, or painfulness of something. Ironically, CCS will prolong deadly coal combustion, waste financial resources, and divert investments from renewables.[60]

[56] Bibler, C. J., Marshall, J. S., & Pilcher, R. C. (1998). Status of worldwide coal mine methane emissions and use. *International Journal of Coal Geology, 35*(1–4), 283–310; Moore, T. A. (2012). Coalbed methane: a review. *International Journal of Coal Geology, 101,* 36–81.

[57] Weiss, D. J., Kong, N., Schiller, S., & Kougentakis, A. (2008). The clean coal smoke screen. Center for American Progress.

[58] Çelik, P. A., Aksoy, D. Ö., Koca, S. A. B. İ. H. A., Koca, H., & Çabuk, A. (2019). The approach of biodesulfurization for clean coal technologies: a review. *International Journal of Environmental Science and technology, 16*(4), 2115–2132.

[59] Oh, S. Y., Yun, S., & Kim, J. K. (2018). Process integration and design for maximizing energy efficiency of a coal-fired power plant integrated with amine-based CO2 capture process. *Applied Energy, 216,* 311–322.

[60] Downie, C., & Drahos, P. (2017). US institutional pathways to clean coal and shale gas: lessons for China. *Climate Policy, 17*(2), 246–260.

Hundreds of CCS projects have been proposed but only a handful of projects have been constructed. Those tied to coal have been marked by unimpressive results (Fig. 2.13).[61,62]

None of the coal projects that received funding from US Department of Energy (DOE) grant programs are currently operating.[63] Of eight coal projects selected, just one resulted in a completed operating facility. That project, the Petra Nova plant, shut down in 2020 for economic reasons: the CCS technology required an entirely separate gas power plant just to power the scrubber, and the resulting emissions weren't even offset by the Petra Nova technology.[64] Of the other seven DOE coal projects initially chosen, three were withdrawn in early stages as not economically feasible even with subsidies. DOE ended agreements with the four others before construction. Investing in technologies of partial fixes to "clean" coal are dubious at best.[65]

Carbon capture makes little sense as retrofitting for old power plants because the plant will cost more to operate, potentially even doubling operation costs.[66] CSS creates a smokescreen for new coal facilities, wrongly suggesting coal plants are viable in a warming world. Yet, when the attempts are consistently years behind schedule and over budget, this shows CCS to be embarrassingly costly. CCS suggests a type of pathology to be willing to pay such an unnecessarily inflated price to stubbornly cling to coal—without evidence to show likely success in permanently removing sufficient CO_2 from the atmosphere in a timeframe to mitigate run-away climate change. The sad irony

Company (Location)	Estimated Price	Results
Kemper (MS)	$7.5 billion	Too costly to be viable; shut down in 2017[62]
Petra Nova (TX)	$1 billion	Closed in 2020
SaskPower (SK)	$1.5 billion on retrofits of an existing plant	Came online in 2014; by 2021 unable to store at capacity due to technological failure

Figure 2.13 Early coal carbon capture and storage attempts.

61 Hauter, W. (2022, July 31). Carbon capture won't work but it will funnel billions to corporations. Truthout.
62 Fehrenbacher, K. (2017, June 29). Carbon capture suffers a huge setback as Kemper Plant suspends work. *Greentech Media*; Condliffe, J. (2017, June 29). Clean coal's flagship has failed. *Technology Review*.
63 Government Accountability Office. (2021, December 20). Carbon Capture and Storage: Actions Needed to Improve DOE Management of Demonstration Projects.
64 Taft, M. (2021, February 2). The only carbon capture plant in the US just closed. Gismodo.
65 Anchondo, C. (2022, January 10). CCS 'red flag?' World's sole coal project hits snag. *E&E News*.
66 Gearino, D. (2023, July 9). A North Dakota coal plant is being retrofit for carbon capture. It will be a massive test. Fast Company.

is that coal is promoted to help the poor when renewables have been proven to be cheaper, cleaner, and healthier.[67] Additional analysis of CCS in Chapter 8 shows the number of projects currently under development in spite of expensive trials with dismal results.

Illustrative cases

United States

Mining symbolized American industrialization.[68] By the 1880s, coal supplied most US electricity. Despite coal's ascendancy, workers experienced brutal labor conditions. Absentee landowners and mine owners drained wealth from coal regions.[69] Labor abuses were widespread.[70] Coal companies enjoyed a high degree of government protection. On multiple occasions, US presidents used national troops to keep mines open, even as the threat of closure came from an abused workforce demanding protections that should have been granted by existing labor laws.

Weakening unions has been a deliberate tactic of company administrators, as labor organizations seek fairer distribution of benefits. As unions achieved hard-sought gains for better working conditions and higher wages, companies embraced mechanization at every turn as a means to cut costs and avoid future strikes. This tactic is part of a **race-to-the-bottom**, whereby firms seek weak locations with weak social and environmental protections to cut costs and drive down market prices.

US coal unions served an instrumental lobbying role to give voice to mining communities and their concerns,[71] and more broadly were fundamental to labor protections and standards beyond the coal sector. Labor movements aren't homogeneous, and tactics shift over time.[72] Without romanticizing unions or falling prey to industry disparagement of unions, fair treatment of labor organizing is necessary. Subsequent chapters discuss coal, oil, and gas unions—invariably, one pattern remains clear—in fossil fuels, protection of workers' overall well-being is poor.

[67] Ritchie, H. (2020, February 10). What are the safest and cleanest sources of energy. Our World in Data; Masterson, V. (2021). Renewables were the cheapest source of energy in 2020, new report shows. World Economic Forum.

[68] Jevons, W. S. (1866). The Coal Question; An Inquiry Concerning the Progress of the Nation, and the Probable Exhaustion of our Coal-Mines. *Fortnightly*, *6*(34), 505–507.

[69] Goodrich, C. (1925). Nothing but a coal factory. *New Republic*, *44*(563), 91–93; Miller, T. D. (1974). Who owns Appalachia? *The Herald-Dispatch*.

[70] Lee, H. B. (1969). *Bloodletting in Appalachia: The story of West Virginia's four major mine wars and other thrilling incidents of its coal fields*. McClain Printing Co.

[71] Calmatters. (2021, September 23). What's the role of unions in the 21st century?

[72] Abraham, J. (2017). Just transitions for the miners: Labor environmentalism in the Ruhr and Appalachian coalfields. *New Political Science*, *39*(2), 218–240.

Rail extends coal

Coal has been integral to warfare. US coal mining expanded with the Civil War (1861—65) as extraction supported the confederate defense of slavery.[73] Staging of troops and equipment relied upon coal-fired steam locomotives. The capital of the confederacy in Richmond, Virginia, flourished due to coal mining (Box 2.2). Profit from coal led to creation of the Richmond and Danville Railroad, a route in service today.

Try this

Discuss: *Various forms of coal violence overlap. Is there one type of violence (structural, physical, psychological, gender, territorial, etc.) you think is most influential? See* Fig. 1.15 *for definitions of the intersection forms of violence.*

Divide and conquer

A **scarcity mindset** highlights restricted opportunities and uses economic anxiety to create competition in a workforce. Coal unionization brought health and social benefits, but organizing in the coal sector was still laden with a scarcity mindset and racial divide, often intentionally promoted by the industry as a "race wedge."[84] Yet many African American coal miners became ardent supporters of unions and strongly discouraged Black strikebreakers. With dangerous conditions and situations of extreme social control, union leaders fought back using grassroots organizers. Mary "Mother" Jones encouraged African American workers, women, and children to be involved in strikes.[85] Excerpts from her autobiography show the divide and conquer tactics unions faced.[86] Hearing this, Mother Jones decided not to speak, but this exhibits the divisive tactics used to deter organizing for better conditions in the mines.

> Virginia, 1891—We couldn't get a hall to hold a meeting. Everyone was afraid to rent to us. Finally the colored people consented to give us their church for our meeting. Just as we were about to start the colored chairman came to me and said: "Mother, the coal company gave us this ground that the church is on. They have sent word that they will take it from us if we let you speak here."

---

[73] Whisonant, R. C. (2015). *Arming the Confederacy. How Virginia's Minerals Forged the Rebel War Machine.* Springer.

[84] Quoted in Lewis, R. L. (2009). *Black Coal Miners in America: Race, Class and Community Conflict (1780—1980).* University Press of Kentucky.

[85] AFL-CIO (n.d.). Mother Jones.

[86] Jones, M. (1925). *Autobiography of Mother Jones.* CH Kerr.

BOX 2.2 Spinoff industry: railroads

The precursor to the modern-day railroad system was made possible with government subsidies and slave labor. The Chesterfield tramway was the beginning of a rail system that developed along the eastern US.[74] Fig. 2.14 depicts the rail route from Chesterfield mines to Richmond. Trains powered by animals and gravity pulls were replaced by coal-fired steam trains by the 1850s.

Midlothian Mines merged with Tredegar Iron Works, located nearby in Richmond. The Confederacy relied on these firms to manufacture bullets, heavy ordinance, locomotives, and railroad tracks.[75] Women and children performed the dangerous work of making ammunition.

State support for railroads made the region a transportation hub. Enslaved laborers were essential to the operation of early railroads.[76] The growth of the US rail system, as tied to national development, mirrors other locations where coal production thrived. In Europe, coal-fired steam railroads bound together regional outputs.[77] South African rail lines formed east–west corridors still in place today with coastal export facilities tied directly to expansive, interior coal mines.[78] Centrally positioned along the US eastern seaboard, the Port of Virginia bound together outlying coalfields through rail networks to push domestic coal to the international market (Fig. 2.15).[79] Historically, the Port of Virginia also operated as a trafficking lynchpin in the slavery market.[80]

Today, Norfolk Southern, a Fortune 500 company, controls Lambert's Point at the Port of Virginia: one the largest coal export facilities in the Northern Hemisphere. The company operates the core rail route transporting coal from operating coalfields in southwest Virginia to the Port (Fig. 2.16). An estimated 90,000 pounds of coal dust contamination is released along the route yearly.[81] However, the most concentrated harm occurs in majority Black neighborhoods near Lambert's Point.

Lambert's Point generates environmental injustice with coal dust and black soot plaguing surrounding neighborhoods. The proximity of houses and concentration of risk in low-wealth African American neighborhoods is stark (Fig. 2.17). Various other polluting facilities are nearby, including one of the nation's largest shipyards and deepwater ports.

Segregation persists and African American households receive the brunt of localized pollution. Old technologies persist through permits "grandfathered in" (Chapter 3) and, in effect, allow loopholes that permit older facilities to operate free of the restrictions placed on newer facilities. Dust mitigation amounts to merely spraying moisture on coal piles to reduce airborne mobility. To this day, Norfolk Southern obstructs improved environmental measures.[82] Simply covering rail cars would lower the toxic dust in vulnerable neighborhoods.[83]

[74] Midlothian Mines Park. (n.d.). Chesterfield Railroad Timeline: 1824:1851.

[75] Virginia Museum of History and Culture. (n.d.). Why Richmond? Time period 1961–1876.

[76] Black, III, R. C. (1998). *The railroads of the Confederacy*. UNC Press Books.

[77] Verstraete, G. (2010). *Tracking Europe: mobility, diaspora, and the politics of location*. Duke University Press.

[78] Staff. (n.d.). Mpumalanga to Richards Bay Freight Coal Line. *Railway Technology*.

[79] Geiling, N. (2018, March 15). 'This is a matter of life and death': A Virginia community choking on coal dust pleads for help. *Think Progress*.

[80] Eaton, L. (2020, February 14). Retracing the steps of slavery in Hampton Roads. *The Virginian-Pilot*.

[81] Baptiste, N. (2017, October 18). This huge rail company is spewing dust all over a low-income community. *Mother Jones*.

[82] Baptiste, N. (2017, October 18). This huge rail company is spewing dust all over a low-income community. *Mother Jones*.

[83] Byrum, L. (2017, October 19). Why Black communities are resisting coal pollution in Norfolk. New Virginia Majority.

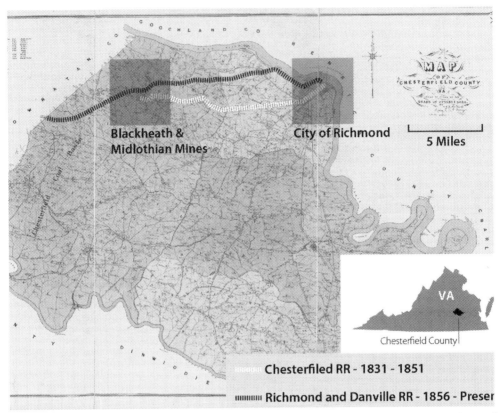

Figure 2.14 The Chesterfield tramway. *(Basemap credit: Map of Chesterfield County, Va, 1888. Library of Congress, Geography and Map Division.)*

Figure 2.15 Rail line feeding Lambert's Point coal export. *(Image credit: Norfolk and Western Ry. Co. Coal Yards, at Lamberts Point, Norfolk, Virginia. 1915. Accessed via Wikimedia Commons.)*

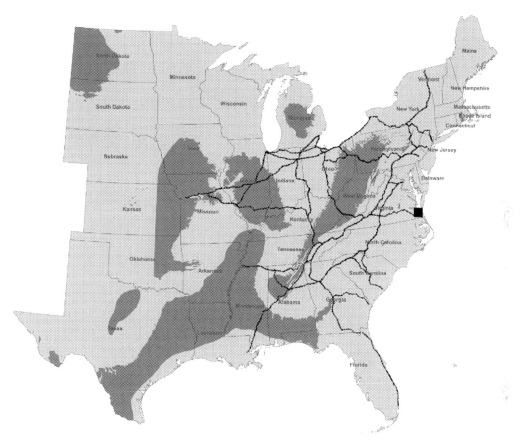

Figure 2.16 Eastern coal railways. *(Sources: Homeland Infrastructure Foundation-Level Data (HIFLD), National Transportation Atlas Database (NTAD), U.S. Census.)*

Labor conflict with coal operators intensified over time. The Coalfield Mine Wars (∼1913—1921) were one of the largest rebellions in American history, second only to the Civil War,[87] and remain the largest labor uprisings in US history. Miners seized

---

[87] Gilmore, N. (2018, September 3). The Buried History of West Virginia's Coal Wars. *Saturday Evening Post.*

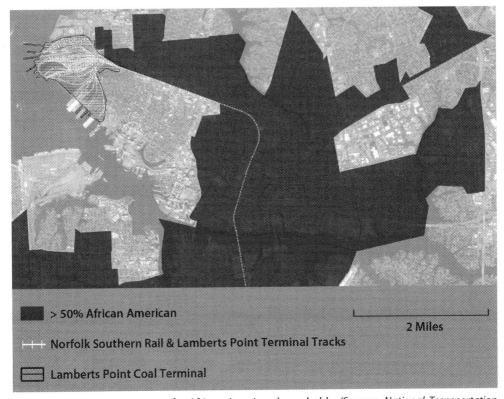

> 50% African American

⊢ ⊢ ⊢ Norfolk Southern Rail & Lamberts Point Terminal Tracks

Lamberts Point Coal Terminal

2 Miles

Figure 2.17 Proximity exposure for African American households. *(Sources: National Transportation Atlas Database (NTAD), U.S. Census, OpenStreetMap terminal extract (Overpass Turbo API), Esri, Maxar, Earthstar Geographics, CNES/Airbus DS, USDA FSA, USGS, Aerogrid, IGN, IGP, and the GIS User Community.)*

and managed 500 square miles.[88] Historical narratives of this period—produced and circulated by those with resources and power—often overlook the roles and perspectives of low-income communities, women, people of color, and immigrants.[89] Fig. 2.18 highlights precursors to, and key events in, the Coalfield Mine Wars. Attention given to these transformational events a century later highlights how they relate to current tensions over

[88] Savage, L. (1990). *Thunder in the Mountains: The West Virginia Mine War 1920-21*. University of Pittsburgh Press.

[89] Bell, S. E., & Braun, Y. A. (2010). Coal, identity, and the gendering of environmental justice activism in central Appalachia. *Gender & Society, 24*(6), 794–813.

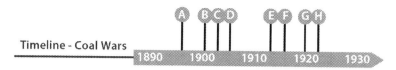

Timeline - Coal Wars

A - Lattimer Massacre
B - 5 Week Stoppage
C - The Great Strike
D - Coal Strikes Hearings
E - Ludlow Massacre and the 10-Day War
F - Colorado Coalfield War
G - Battle of Matewan
H - Battle of Blair Mountain

Figure 2.18 Coalfield Mine Wars timeline.

the loss of jobs in coal regions and press access. Media outlets historically supported company positions and disparaged unions,[90] not unlike recent times.[91]

In 1902—03, the federal government convened hearings on coal strikes after stoppages provoked demands for the nationalization of mines and railroads.[92] The hearings marked a turning point with two outcomes: (1) better wages, hours, and working conditions, and (2) a step toward organized labor obtaining the right to confer with industry and make agreements applicable to all. Nevertheless, conflicts continued. In 1913, approximately 90% of the workforce in the southern Colorado coalfields went on strike. Unionized workers were evicted and families moved into tent colonies, where they controlled entrances to the coal mines to intercept strikebreakers. Ludlow was the largest tent colony, where 1200 striking miners and their families held out for 14 months.[93] Mine owners brought in the Baldwin-Felts Detective Agency to initiate a campaign of harassment using murder, beatings, and the "death special," an armored car that would spray machine-gun fire. With escalating violence and pressure from mine owners, the governor called for reinforcements: **martial law** ensued, suspending habeas corpus and allowing military control over civil functions. Mass incarceration with torture and beating of prisoners followed. Hundreds died when mine guards burnt the Ludlow tent colony to the ground (Fig. 2.19).

When news spread, striking miners from other tent colonies attacked and destroyed mines, fighting with guards and militia across 40 miles for 10 days. The governor called for federal reinforcements, but the strike dragged on for 7 months. Today, the Ludlow

[90] Burt, E. V. (2011). Shocking Atrocities in Colorado: Newspapers' Responses to the Ludlow Massacre. *American Journalism, 28*(3), 61–83.

[91] Andreescu, V., & Schutt, J. E. (2009). Violent Appalachia: the media's role in the creation and perpetuation of an American myth. *JIJIS, 9,* 62.

[92] Grossman, J. (1975). The Coal Strike of 1902. Turning Point in US History. *Monthly Labor Review.*

[93] McGuire, R. (2004). Colorado Coalfield Massacre. *Archaeology, 57*(6), 62–66.

Figure 2.19 Ludlow colony ruins. *(Image credit: Ruins of the Ludlow Colony near Trinidad, Colorado, following an attack by the Colorado National Guard. 1915. Forms part of the George Grantham Bain Collection at the Library of Congress.)*

site is sacrosanct to the labor movement. The Ludlow massacre became a rallying cry for union organizing.[94] It was a cruel massacre: they targeted a family space, so 55 women and children died in the fires or when fleeing the camp, where they were met with a barrage of bullets.

Fig. 2.20 shows the arsenal gathered from another town garrisoned by federal troops. Large amounts of military force were deemed necessary to keep coal workers oppressed, even though their demands were already protected by existing labor laws and constitutional freedoms.

The Coal Mine Wars set the stage for labor reforms, yet also brought about a backlash. Early labor reforms weren't fully solidified before the Battle of Matewan in West Virginia.[95] In response to the United Mine Workers of America (UMWA) organizing in the area, mining companies forced "yellow dog" contracts that prohibited miners from joining unions. When this still did not deter unionization, Baldwin-Felts agents evicted

---

[94]   Walker, M. (2003). The Ludlow Massacre: Class, warfare, and historical memory in southern Colorado. *Historical Archaeology, 37*(3), 66–80.

[95]   Boissoneault, L. (2017, April 25). The coal mining massacre America forgot. *Smithsonian Magazine*.

Figure 2.20 A federally garrisoned town. *(Image credit: West Virginia Mine Wars Museum, Collection of Kenneth King.)*

union miners' families from company housing, generating a shootout and several fatalities.[96] The subsequent murder of Matewan's pro-union Chief of Police, Sid Hatfield, sparked the largest labor uprising in US history. At the Battle of Blair Mountain, approximately 10,000 miners had a 5-day standoff with an army made up of local and state police, state national guard, and federal troops, after they sought to try to prevent the entry of strikebreakers.[97] President Harding brought in thousands of federal army troops and the Air Force dropped bombs in advanced air strikes.[98] One hundred and fifty people died, and their descendants are still fighting historical erasure of their sacrifice (Chapter 3).[99]

Should disruptions in production due to strikes be considered "national emergencies"?[100] On four separate occasions between 1919 and 1921 the US Army indeed intervened in conflicts between miners and coal mine operators.[101] Concerns underlying these historic events persist today. Using state resources to end strikes is not

[96] Archer, W. (1992). From Matewan to Welch: One Man's Thirst for Vengeance. *Appalachian Heritage 20*(2), 9–15.

[97] Nida, B. (2013). Demystifying the hidden hand: Capital and the state at Blair Mountain. *Historical Archaeology, 47*(3), 52–68.

[98] Keeney, C.B. (2018, February 26). The battle of Blair Mountain is still being waged. *The Cultural Landscape Foundation.*

[99] Keeney, C.B. (2021). *The Road to Blair Mountain: Saving a Mine Wars Battlefield from King Coal.* West Virginia University Press.

[100] Warren, E. L. (1951). Thirty-Six Years of "National Emergency" Strikes. *ILR Review, 5*(1), 3–19.

[101] Laurie, C. D. (1991). The United States Army and the Return to Normalcy in Labor Dispute Interventions: The Case of the West Virginia Coal Mine Wars, 1920–1921. *West Virginia History, 50,* 1–24.

uncommon; case in point, the deployment of federal troops in 2016 to quell resistance over the Dakota Access Pipeline (DAPL) at the Standing Rock Indian Reservation. Controlling movement and communications in coal company towns is an earlier rendition of surveillance tactics of Native pipeline organizers (Chapter 7). The use of government resources ties to the long-standing assumption that the state needs to intervene to protect private energy resources, even when corporations mistreat workers and communities.

South Africa's energy injustice

Like Appalachia, coal formed a backbone for inequity in South Africa where coal flourished under the morally bereft system of apartheid.[102] Coal has dominated South African energy since the 1800s, forming the base of its national industrialization.[103] The South African situation exhibits elite capture during domestic operations. Multinational corporations enjoy outsized profits derived from low-wage production, making South African coal highly competitive on export markets. Coal extraction only expanded under South Africa's apartheid, a system of white supremacy and extreme structural violence. During apartheid, the separation of powers between mining corporate conglomerates and the government was virtually indistinguishable.[104] The mining industry, immune to regulation, perpetuated generational harm.

South Africa's coal sector is defined by concentration among five companies. Coal extraction, combustion, and exportation facilities are also spatially concentrated, and a profitable chemical industry has developed alongside coal. The influence of South Africa's chemical industry is felt around the world. Initiated by the government and developed during apartheid, the chemical company Sasol (Box 2.3) is the world's largest coal-to-chemicals producer.

Development of coal, chemicals, and railroads are so intimately bound that it is difficult to tease apart discrete industries and their influence. The landscapes of South Africa's northern region are undoubtedly coal geographies,[108] following earlier patterns in US Appalachia. Established mostly after World War II, South Africa's mining companies

---

[102] Bond, P., & Garcia, A. (2000). *Elite transition: From apartheid to neoliberalism in South Africa*. Pluto Press.

[103] Eskom. (n.d.). Coal Power.; Snyman, C. P., & Botha, W. J. (1993). Coal in South Africa. *Journal of African Earth Sciences, 16*(1–2), 171–180.

[104] Leonard, L. (2018). Mining corporations, democratic meddling, and environmental justice in South Africa. *Social Sciences, 7*(12), 259.

[108] Marais, L., Denoon-Stevens, S., & Cloete, J. (2020). Mining towns and urban sprawl in South Africa. *Land Use Policy, 93*, 103953.

BOX 2.3 Spinoff industry: chemicals

Sasol's "minerals–energy complex" developed from deep ties between industrial conglomerates and the South African government. Its initial objective was to fill demand for explosives and chemicals needed by the mining industry.[105] Over time, Sasol's specialization in coal gasification and petrochemicals intensified and expanded. Production today is prolific across synthetic fuels and diesel, gasoline, kerosene, polymers, and solvents used in a range of products such as adhesives, electronics, fragrances, pharmaceuticals, plastics, and rubber.

South Africa's petrochemical industry was built upon coal gasification.[106] Coal and chemicals prop each other up and concentrate political influence. South Africa's Chemical and Allied Industries Association (CAIA), an industry-led organization, is the custodian of the global Responsible Care initiative launched in 1994. While Responsible Care oversight—essentially a system of planned audits—helps standardize products for export to locations like Europe, industry-run accreditations like this often fall short of **corporate social responsibility** (CSR). The rigor of voluntary self-regulation is simply less stringent than independent, third-party verification.

In South Africa, the chemical industry has secured a role in state science education. CAIA used this influence to oppose climate policy, suggesting it would cause deindustrialization.[107]

housed Black miners in high-density compounds while white mineworkers and families received company houses. Segregated mine housing contributed to a higher rate of Black dispossession.[109] More recently, mining companies privatized worker housing, devolving risks to households.[110] Unsafe housing remains a vexing problem, while miners today are often foreign contract laborers.[111]

From its inception, South African coal extraction has been centralized and deadly. Coal sacrifice zones are situated in vulnerable regions such as KwaZulu-Natal (KZN), part of the homeland of the Zulu Indigenous People. South Africa's coal fields are concentrated in Mpumalanga, where coal extraction and combustion occur in extreme proximity to villagers.[112]

[105] Rafey, W., & Sovacool, B. K. (2011). Competing discourses of energy development: The implications of the Medupi coal-fired power plant in South Africa. *Global Environmental Change, 21*(3), 1141–1151.

[106] Majozi, T., & Veldhuizen, P. (2015). The chemicals industry in South Africa. *American Institute of Chemical Engineers, 46,* 51.

[107] Trollip, H., & Boulle, M. (2017). Challenges associated with implementing climate change mitigation policy in South Africa.

[109] Cloete, J., & Marais, L. (2020). Mine housing in the South African coalfields: the unforeseen consequences of post-apartheid policy. *Housing Studies,* 1–19.

[110] Marais, L. (2018). Housing policy in mining towns: issues of race and risk in South Africa. *International Journal of Housing Policy, 18*(2), 335–345.

[111] Pelders, J., & Nelson, G. (2019). Living conditions of mine workers from eight mines in South Africa. *Development Southern Africa, 36*(3), 265–282.

[112] Olufemi, A. C., Mji, A., & Mukhola, M. S. (2019). Health risks of exposure to air pollutants among students in schools in the vicinities of coal mines. *Energy Exploration & Exploitation, 37*(6), 1638–1656.

Land grabbing in Mpumalanga

South Africa's racial segregation remains widespread. Much of the coal for export and domestic markets is produced by eight mega-mines, seven of them co-located with ethnic regions (Fig. 2.21). Coal industry's **land grabbing** in KZN dispossessed households living in cultural territories.[113] Displacement is particularly devastating for traditional peoples rooted in place who rely on subsistence production.

Community leaders who opposed Tendele Coal Mine were assaulted. The 2020 murder of activist, Fikile Ntshangase (Dedication), drew international scrutiny. She was the deputy chairperson of the Mfolozi Community Environmental Justice Organisation, and a grandmother, who had resolutely fought Tendele's expansion.[114]

- • **Coal Plants**

- ▥▥ **Coal Field Extents**

- ▬ **> 50% Black African Ward**

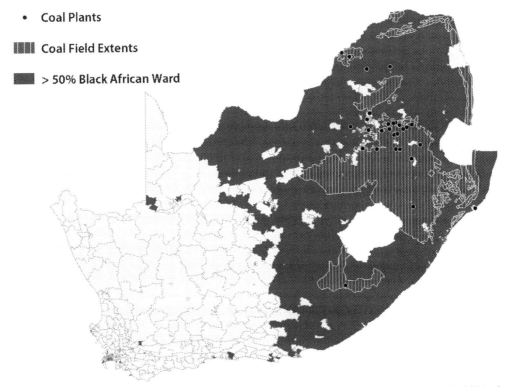

Figure 2.21 Coal plants and race in South Africa. *(Sources: South African Census 2011 - Household Ward Survey, USGS Coal-Bearing Areas in Africa, Global-scale mining polygons (Version 1). PANGAEA, https:// doi.org/10.1594/PANGAEA.910894, Global Coal Plant Tracker.)*

[113] Rolando Mazzuca, C. & Tran, D. (2020, October 27). Somkhele coal mine owned by Tendele, KwaZulu-Natal, South Africa. EJAtlas.
[114] Koko, K. (2020, November 7). How we braved danger to honor Fikile Ntshangase. *Mail & Guardian.*

Violence could have been prevented by the courts. In 2018, impacted residents tried to stop the mine expansion with a lawsuit but the case was thrown out despite lawyers providing evidence that mining operations cause serious environmental harm and have adverse community health effects. Somkhele mine operated without the required national environmental authorization, a land-use authorization, and a waste management license.[115] The government still sides with the company. Tendele admitted to desecrating graves and is working with the state heritage agency to relocate more graves. Destroying burial sites violates the United Nations Declaration on the Rights of Indigenous Peoples (UNDRIP). National protections also exist under the KwaZulu-Natal Heritage Act.

Regional power dynamics are hugely unequal as grassroots resistance faces opaque domestic entities bolstered by slippery but immense foreign backing. Tendele mine was originally financed in part by the World Bank. Currently, the mine is owned by Johannesburg-based Petmin, itself owned by Capital Works Private Equity after delisting from the Johannesburg Stock Exchange in 2017.[116] South Africa is the world's most unequal country, and any financier, business, or trader working in this context shares responsibility for profiteering from violent oppression. While unscrupulous investors prioritize coal exports, people in frontline areas struggle to survive.[117] For example, a network of local women conducted epidemiological studies showing a crisis spurred by mountaintop removal mining.[118] With a documented record of harm, foreign investors enter into fossil fuel geographies like South Africa with knowledge and culpability for these harms.

In 2022, in brief reverse of course, a South African judge blocked expansion of Tendele Mine, sending the case for further review.[119] During the hearing, Tendele admitted flaws in its public participation process and deficiencies in its environmental reports. Yet the company suggested the mines would fail and by mid-2023 national courts had given permission to reinitiate mining at Tendele.[120]

Even getting these cases to court is dangerous for impacted communities, and in South African coal sacrifice zones local women risk their lives to confront companies

[115] Kockott, F. (2018, November 21). Activists lose bid to stop coal mine. Ground Up.
[116] Petmin. (n.d.). About Petmin.
[117] Business and Human Rights Center. (n.d.). Tendele Coal mining (PTY).
[118] Leonard, L. (2019). Traditional leadership, community participation and mining development in South Africa: The case of Fuleni, Saint Lucia, KwaZulu-Natal. *Land Use Policy, 86*, 290–298.
[119] Business and Human Rights Center. (2022, May 5). S. Africa: Court reverses controversial issuing of mining license to Tendele Coal mining in KZN.
[120] Steyn, L. (2023, July 26). Court green lights restart of mining at Tendele anthracite coal mine. News24.

harming their communities.[121] Those on the frontline fear for their lives.[122] Local women documented how mining companies contributed to water scarcity in communities. They reported this to authorities and demanded action, gaining international solidarity.[123] Destruction of water resources by KZN mines makes this arid environment uninhabitable,[124] intensifying conflicts.[125]Like investors, global consumers of South Africa's export coal[126] are responsible too, spurring coal expansion without recognizing its inherent energy violence (Chapter 1).

What is in a name?

Poor administration of coal facilities in South Africa has also not improved with newer power stations like the 4800 MW Medupi Power Station,[127] where Eskom's technical errors have led to a series of delays and concerns about design defects.[128] In the midst of inept project development, greater care was given to naming the facility than to its safe operation as the South African government contracted a group of historians for its name selection.

The original name given to the Medupi Power Station coal plant in 2006 was Project Alpha, referring to the Greek alphabet and system of numerals. To seem as if they were embedded in place, managers switched the name to Project Medupi, meaning "rain that soaks parched lands, giving prosperity"[129] in Sedepi, a Sotho-Tswana language. This is an example of **cultural appropriation**, or adoption of an element of one culture by members of another, especially harmful when members of a dominant group appropriate from

[121] DeCOALinize. (2021). Why we should involve women in project decisions: the case of Lamu Coal Plant.

[122] Bond, P. (2021, October 22). Lessons from the assassination of Fikile Ntshangase: Climate violence, the 'Right to say No!', uncompensated resource extraction, financial profiteering and unpaid ecological debt in South Africa's coal mining belt. Committee for the Abolition of Illegitimate Debt.

[123] GroundWork. (2020, April 8). The Water Crisis in a Time of the COVID-19 Crisis - Women of Somkhele and civil society groups call on government to uphold promises and to provide water, in the time of the COVID-19 threat.

[124] Koko, K. (2020, November 7). Tendele's mine pollutes their air and water, residents' claim. *Mail & Guardian*.

[125] Koko, K. (2020, October 23). Murder of anti-mining activist emboldens KZN community. *Mail & Guardian*.

[126] Bayna, N. (2022, August 15). South African coal exports to Europe surge, shipments to Asia decline. Reuters.

[127] Tshehla, M. (2019, May 1). Kusile and Medupi were destined to fail from the start. Business Live.

[128] Ayemba, D. (2021, May 17). Medupi Power Station in South Africa to become fully operational by end of year. *Construction Review Online*.

[129] Rafey, W., & Sovacool, B. K. (2011). Competing discourses of energy development: The implications of the Medupi coal-fired power plant in South Africa. *Global Environmental Change*, 21(3), 1141–1151.

a minority group. The name Medupi was selected to create a false narrative—now running, the plant emits toxic air emissions and prosperity has not occurred.[130]

The Medupi Project, Eskom's largest single investment, required a $3.75 billion loan from the World Bank—its first to South Africa since the end of apartheid—and $992 million in loans from the African Development Bank and the China Development Bank.[131] The shortsightedness of these international finance institutions (IFIs)—while still profiting from loans—was exposed only a year after the plant came on-line, when the same funders predicted that the plant would not achieve operational profitability. Banks now propose the coal plant will close before its intended date,[132] making it an expansive stranded asset burdening thousands of low-income ratepayers in the midst of a national energy crisis.

In spite of its massive size, the Medupi plant exemplifies government actions that are too little and too late. With an installed capacity of 4.8 GW (GW), Medupi is the world's largest dry-cooled power station. It was supposed to last for 60 years. The plant's debt burden was exacerbated by poor management, sabotage, work strikes, and corruption—all leading to delays and cost overruns.[133] In 2021, after an explosion at the plant due to a hydrogen leak,[134] an investigation revealed major procedural non-compliance, leading to dismissal of nine employees.[135] Mismanagement remains a persisting theme at Eskom plants. There have been a series of protests by Medupi plant workers, including 8,000 workers striking for several weeks in 2015 for better working conditions.[136] In 2022, with anger building from blackouts and dangerous work conditions, striking Medupi workers torched Eskom officials' homes.[137] Eskom's profiteering in one of the most racist places in South Africa[138] reinforces past wrongs, while exposing how international banks profiteer also from energy violence harm.

[130] Bosch, E. (2022, May 21). Watch: Medupi assaults the senses even without crippled Unit 4. *Sunday Times*.

[131] BankTrack. (2021, December 16). Medupi Coal Power Plant Financiers.

[132] Sguazzin, A. (2022, June 2). Funder says Eskom's $8.7 billion power plant won't make money. Bloomberg.

[133] Staff. (2020. May 11). South Africa's Eskom to fix design flaws at Medupi, Kusile coal plants. Reuters.

[134] Parkinson, G. (2021, August 11). World's newest and most expensive plant explodes after hydrogen leak. Renew Economy.

[135] Modise, K. (2022, May 19). Investigation into blast at Medupi Power Plant reveals procedural non-compliance. *Eyewitness News*.

[136] Staff. (2015, May 18). S. Africa's Eskom says most workers resume duty at Medupi after strike. Reuters.

[137] Vecchiatto, P., & Burkhardt, P. (2022, June 29). Workers torch Eskom officials' homes as South Africa cuts power. Bloomberg.

[138] Taylor, T., & Bertrams, N. (2020, April 29). The story of SA's biggest power plant, and its little town. *Mail & Guardian*.

Try this

Research and discuss: *Analyze the Medupi Power Station name. Identify other cases where projects appropriate names from the people they displace.*

Pit-to-port

As in the previous case with Appalachia coal exported at the Port of Virginia, colonial infrastructure in South Africa was a mechanism to reinforce local spatial and social oppressions. Multilateral investors prioritize export terminals as they argue coal is key to Africa's development.[139] The push to export massive quantities of coal creates a hazardscape for local populations in Richards Bay, the biggest coal terminal in Africa (Fig. 2.22).

Figure 2.22 South African hazardscape. *(Image credit: Richards Bay Coal Terminal (RBCT).)*

[139] Urgewald. (2018). A Global Tour.

Prior to becoming an export port in the 1970s, Richards Bay was a coastal town situated on a lagoon.[140] It was the heavy haul trains from the Mpumalanga region that allowed for development of this massive export terminal. Fig. 2.23 depicts Richards Bay's rail network and connected gigantic terminal of South Africa's premier bulk cargo port.

Most railroads in South Africa are run by the company Transnet Freight Rail, formerly known as Spoornet, part of a state-controlled railway administration. Once considered efficient, the deteriorated long train system means small incidents cause major delays with repercussions for electricity supply and price.[141] For example, in 2020, coal piles got wet leading to a significant blackout. In 2021, a series of train derailments drove up prices. In 2022, with supply challenges in spare train parts, copper cable theft, and

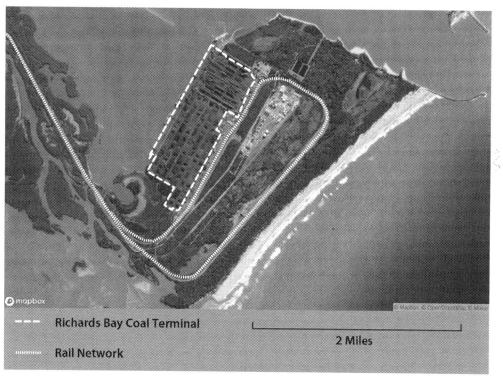

Figure 2.23 Richards Bay rail. *(Map Image credit: © Mapbox, © OpenStreetMap, © Maxar.)*

[140] Nel, E., Hill, T. R., & Goodenough, C. (2004). Global coal demand, South Africa's coal industry and the Richards Bay Coal Terminal. *Geography*, *89*(3), 292–297.

[141] Thackreh, A. (2021, May 5). Transnet warns of possible rail force majeure. Argus Media.

vandalism, coal was delivered to port on trucks, even as this transportation alternative costs four times more, boosts emissions, and clogs roads.[142] Unreliable coal supply has been a growing concern in South Africa, where railroads continually fail to efficiently transport coal to the export terminal (Fig. 2.24).

The state enforces a monopoly with Eskom, the debt-stricken, state-owned utility. South Africa represents a **captured state**, meaning systemic political corruption in which

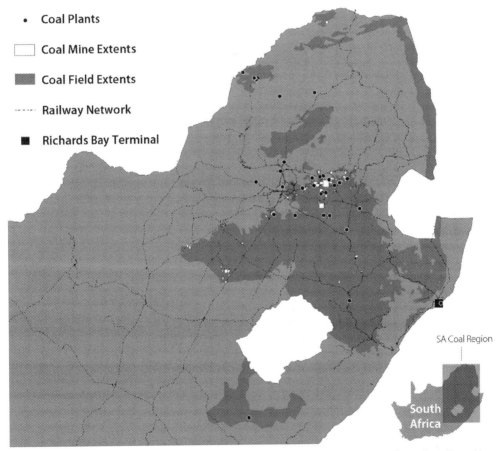

Figure 2.24 South Africa's rail system. *(Sources: USGS Coal-Bearing Areas in Africa, OpenStreetMap railway extract (Overpass Turbo), Global-scale mining polygons (Version 1). PANGAEA, https://doi.org/10.1594/PANGAEA.910894, Global Coal Plant Tracker.)*

[142] Banya, N., & Reid, H. (2022, May 3). South African coal mines turn to trucks as rail services deteriorate. Reuters.

private interests significantly influence state decision-making processes to their own advantage. In 2018, the national treasury bailed out Eskom—yet corruption continued.[143] The former president of South Africa (2009–18) Jacob Zuma served jail time in 2019 for refusing to release information about corruption tied to Eskom tax fraud.[144]

Civil unrest grows with state-controlled companies like Transnet and Eskom calling the shots. Tensions are high after years of **load shedding**—cutting off parts of the supply grid, resulting in perpetual rolling blackouts—that severely impacts the lives of South Africans. **Brownouts**, the sporadic short-term loss of electricity, whether planned or spontaneous, create distrust and disapproval even when they help avoid **blackouts**, which mean a loss of electricity for days or weeks due to factors like grid failure. Eskom has been struggling to provide reliable electricity for years, but load shedding reached a new height in 2023, cutting off power to different areas for 16 out of 32 hours to prevent a full blackout.[145] Power outages to this degree have serious economic and social ramifications. One of the greatest costs is the decrease in access to reliable water, since water and electrical systems are linked in distribution and delivery. Electricity is also essential to sewage treatment and overflow prevention, exacerbated by load shedding, in turn harming people's health and that of marine ecosystems.[146]

Life after coal

Though industrialization was long considered South Africa's greatest strength in the region, much of this domestic industry is being replaced with cheaper imports. High rates of unemployment are rampant, and climate disruption reinforces alternating extremes of heavy rain and searing drought. The weather extremes are pronounced and dangerous: in 2022 a "rain bomb" killed over 400 people.

South Africa's ongoing water crisis is forcing change,[147] with water rationing prominent in urban areas.[148] South Africans urgently need justice-oriented

---

[143] Staff. (2021, June 4). There's still a lot of corruption at Eskom - Expert. eNCA.

[144] Parliamentary Monitoring Group. (2019). Eskom Oversight Report.

[145] Creamer, T. (2023, May 18). Eskom warns of Stage 8 loadshedding this winter but insists risk of blackout is low. *Engineering News*.

[146] du Plessis, A. (2023, January 23). Power cuts in South Africa are playing havoc with the country's water system. The Conversation.

[147] Hedden, S., & Cilliers, J. (2014). Parched prospects-the emerging water crisis in South Africa. *Institute for Security Studies Papers*, 11, 16.

[148] Bischoff-Mattson, Z., Maree, G., Vogel, C., Lynch, A., Olivier, D., & Terblanche, D. (2020). Shape of a water crisis: Practitioner perspectives on urban water scarcity and 'Day Zero' in South Africa. *Water Policy*, 22(2), 193–210.

decarbonization,[149] yet there's sparse international funding for this transition. On the contrary, a boom in intensive mining is occurring with the demands for rare earth elements built into renewable infrastructures. Greening in the wealthy parts feeds mining in South Africa, producing half of the world's palladium and 75% of the world's platinum.[150] Mining for platinum and palladium is dangerous in deep veins.[151] Strikes have occurred at South African metal mines due to poor work conditions. Given this track record, mining contracts for greener equipment and machinery must be accompanied by labor and environmental protections to avoid an **impact paradox**, when an intent to do something positive has negative ramifications.

Talk generous and green, walk stingy and dirty.[152]

South Africa plans to burn coal for many years, while constructing new plants.[153] National leaders remain steadfast to coal despite embracing a goal of carbon neutrality by 2050.[154] With ample potential renewable energy throughout the country, including wave, geothermal, wind, and solar, the promotion of fossil fuels is shortsighted.[155] The phrase "we can't breathe" represents a collective sense of urgency made more pressing by climate change.[156] South Africa is fertile ground for linking social justice with degrowth strategies given its overly dependent industrial and chemical economy.[157] Interconnected strategies for food sovereignty, water rights, and workers' rights converge

[149] Cahill, B. (2020). Just transitions: lesson learned in South Africa and Eastern Europe. Center for Strategic and International Studies.

[150] Besta, N. (2020, October 26). Top five platinum mining companies. *NS Energy*.

[151] Njini, F., & McCorry, J. (2019, October 19). A mile below ground is the dangerous world of precious metals. Bloomberg.

[152] Galvin, M., & Bond, P. (2022, April 22). Durban's latest Rain Bomb kills more than 300 and unveils state climate sloth. International Viewpoint.

[153] Department of Mineral Resources and Energy, Republic of South Africa. (n.d.). Coal Resources.

[154] Lo, J. (2020, September 16). South Africa aims to reach net zero emissions in 2050 - still burning coal. *Climate Home News*.

[155] Huxham, M., Anwar, M., & Nelson, D. (2019). Understanding the impact of a low carbon transition on South Africa. Climate Policy Initiative.

[156] Gerntholtz, L. (2019, September 17). For communities in South Africa climate change is now. *Mail & Guardian*.

[157] Rodríguez-Labajos, B., Yánez, I., Bond, P., Greyl, L., Munguti, S., Ojo, G. U., & Overbeek, W. (2019). Not so natural an alliance? Degrowth and environmental justice movements in the global south. *Ecological Economics, 157*, 175–184.

in Life After Coal (Impilo Ngaphandle Kwamalahle), a campaign to phase out coal and enable a just transition.

South Africa's climate justice movement builds on the historic environmental justice movement that developed in the coal production and exportation sectors due to high toxic burden and exploitative conditions, especially in export areas like South Durban.[158] In addition to grassroots actions and protests, groups have drafted a climate justice charter. Durban's community-based organizations highlight the urgency for change, and bring together labor and civil rights movements to seek justice in access to water, food, and housing.[159] Grassroots initiatives for transformative change exist in labor organizations. The Congress of South African Trade Unions (COSATU) resolved in 2009 that "climate change is one of the greatest threats to our planet and our people." COSATU recognized that "unless the working class and its organizations take up the issue of climate change seriously, all the talk about 'green jobs' will amount to nothing more than an accumulation mechanism for capitalists."[160]

To move beyond coal, both local and global environmental justice and labor movements must connect to a larger climate justice movement first drafted in South Africa in the mid-to-late 2000s. In 2011, the build up to the United Nations Framework Convention on Climate Change (UNFCCC) Conference of Parties (COP) mobilized South African civil society. National trade groups initiated conversation around equity and renewable energy. COSATU's call to action centered on the concept of just transition and sought leadership from organized labor for climate jobs and renewable energy. While grassroots organizations were disappointed by the South African government's failure to advance a sustainable, equitable and just transition, the national climate justice movement that consolidated in Durban was indeed influential in defining international climate justice during the 17th UNFCCC COP and since.

Global attention propelled the National Union of Metalworkers (NUMSA) to develop and publicize their positions on climate change, particularly in the light of the failings of COP17 process, which was seen as demonstrating that the state was not capable of delivering serious climate change mitigation. In 2012, NUMSA adopted two resolutions. The first, Climate Change and Class Struggle, committed to climate justice solutions from below (i.e., the grassroots) as part of a larger struggle for a deep transition to a

[158] Bond, P. (2000). Economic growth, ecological modernization or environmental justice? Conflicting discourses in post-apartheid South Africa. *Capitalism Nature Socialism, 11*(1), 33–61; Scott, D., & Barnett, C. (2009). Something in the air: civic science and contentious environmental politics in post-apartheid South Africa. *Geoforum, 40*(3), 373–382; Leonard, L., & Pelling, M. (2010). Mobilisation and protest: environmental justice in Durban, South Africa. *Local Environment, 15*(2), 137–151.
[159] Bond, P., & Garcia, A. (2000). *Elite transition: From apartheid to neoliberalism in South Africa.* Pluto Press.
[160] Congress of South African Trade Union's Policy Framework on Climate Change (2011). Media statement.

low-carbon economy. The second resolution, "Building a socially owned renewable energy sector in South Africa," involved cooperatives to manufacture renewable technologies locally. In 2018, NUMSA further advanced a platform for socially owned renewable energy with more wind and solar projects to tackle energy poverty.[161] NUMSA aims to meet universal needs, decommodify energy, and provide an equitable dividend to communities and workers while decarbonizing.[162]

Try this

Plan and share: *You are an advisor to the energy minister for South Africa or the US. What timeframe do you propose for the path away from coal? How does your proposal address community and labor concerns?*

China

Coal use has been widespread in China since the eleventh century, when it was used for heating and metallurgy. Later, imperial officials emphasized mining to foster economic development as broader statecraft and coal extraction expanded imperialism into the interior of the country. Most early mines were small and private before the Chinese Communist Party established the People's Republic of China in 1949. Under Maoism, coal mines were restructured, with some becoming local or state-owned mines and some community enterprises. The Ministry of Coal Industry called for "all the people to open coal mines" during the Great Leap Forward (1957–59). Smaller village-owned enterprises were considered successful examples of socialism in that they employed a rural labor force in a diversified local economy.[163] There was peak output in the mid-1990s from small operations. When state-owned operations grew larger, China's leaders shuttered smaller plants.[164] To boost production, the government integrated coal operations, ultimately resulting in more output, but harming communities where plants closed.

Coalbed methane emissions from China's mines, the highest in the world, contribute to the slow violence of climate disruption while causing fast violence with fatal explosions. China is building new coal plants in areas without adequate water resources for cooling operations. Thermal power plants that produce energy from heat using fossil fuels are water-intensive and aggravate existing regional water shortages. **Water stress** occurs

[161] Satgar, V., & Cock, J. (2021). Ecosocialist activism and movements in South Africa. In Engel-Di Mauro, S., Giacomini, T., Isla, A., Löwy, M., & Turner, T. E., Eds. *The Routledge Handbook on Ecosocialism*. L. Brownhill (Ed.). Routledge. 179–188.
[162] Williams, C. (2022, April 27). South Africa's electricity workers can teach us about winning a green new deal. The Real News.
[163] Rui, H. (2005). *Globalization, Transition and Development in China: The Case of the Coal Industry*. Routledge.
[164] Staff. (2020, June 24). China has 250 GW of coal-fired power under development - study. Reuters.

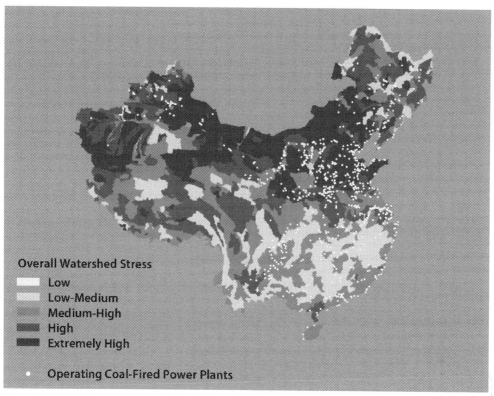

Figure 2.25 China's water stress and coal plants. *(Sources: World Resources Institute (WRI), Global Power Plant Database Version 1.3.0, Aqueduct Global Water Risk, Version 3.)*

when water resources are insufficient for needs. Power plants in water-deficient regions exacerbate water insecurity (Fig. 2.25).[165]

China's coal-fired plants were transferred out of cities because of air pollution. China's coal-fired heating systems generate hazardous emissions and deteriorate winter air quality as well.[166] Elevated coal use for heating during winter increases mortality rates for low-income elderly populations in particular.[167]

[165] Chai, L., Liao, X., Yang, L., & Yan, X. (2018). Assessing life cycle water use and pollution of coal-fired power generation in China using input-output analysis. *Applied Energy, 231,* 951–958.

[166] Fan, M., He, G., & Zhou, M. (2020). The winter choke: coal-fired heating, air pollution, and mortality in China. *Journal of Health Economics,* 71, 102,316.

[167] Lora-Wainwright, A. (2017). *Resigned activism: living with pollution in rural China.* MIT Press.

China consumes half the world's coal yet pledges to become carbon-neutral by 2060.[168] The government is unlikely to meet mitigation goals while building new coal plants.[169] State repression is used to keep plants operating in spite of widespread toxic pollution.[170] Independent media coverage—difficult to find both domestically and internationally—exists but is often suppressed by censorship and even violence. Case in point: In 2023, an environmental watchdog investigating wastewater runoff was brutally beaten in pursuit of evidence.[171]

As in the US, China's path to end coal is protracted, even as losses and damages from ecological breakdown grow exponentially. Products made in China that circle the globe require vast energy resources at a time of growing water crises.[172] If China's past predicts the future, local and regional ethnic and class conflicts will likely intensify with climate disruption, yet slip past international press coverage that could add awareness and pressure to end coal injustice once and for all.

Spatial and temporal synopsis

Space

(1) *Coal sacrifice zones emerge from unfair distribution of costs and benefits.*

(2) *As coal advances so do forms of spatial control over workers and communities, with prominent examples in isolated, surveilled company towns.*

(3) *Spinoff industries in Appalachia, like prisons, exacerbate coal's harms and environmental injustice.*

(4) *Coal intensifies existing crises in its areas of operation; for example, the use of water for coal operations in South Africa and China contributes to local water shortages.*

Time

(1) *Early periods of coal expansion were frequently tied to war (e.g., the US Civil War).*

(2) *A race-to-the-bottom keeps a constant squeeze on workers and undermines environmental regulations, initiating the climate crisis.*

---

[168] Xu Elegant, N. (2020, September 24). China pledges to be carbon neutral – but remains addicted to coal. *Fortune.*

[169] DeVore, C. (2019, December 16). China Goes All-In On Coal While Telling The Rest Of The World To Reduce Emissions. Forbes.

[170] EJ Atlas. (2023, July 31). Xiegou coal mine, Lvliang, Shanxi, China.

[171] Business and Human Rights Resource Center. (2023, April 18). China: Environmental volunteer beaten while investigating alleged pollution by a coal mine in Shanxi Province.

[172] Maizland, L. (2021). China's fight against climate change and environmental degradation. Council on Foreign Relations.

(3) *Coal extraction sacrifice zones like Appalachia and KZN suffer long-term negative consequences, spurring grassroots initiatives for change.*

(4) *Coal's preferential treatment explains its longevity; the false "solution" of "clean" coal highlights the industry's disconnect with a shared climate crisis.*

Summary

Coal has generated socio-ecological contradictions. Supposedly "cheap" coal becomes much more costly—even unaffordable—once restitution and ecological safeguards become "internalized" and we assign true costs to the commodity. The coal industry was buoyed by captured governments, as demonstrated in the US, South Africa, and China, and has been allowed to use physical and structural violence against workers and impacted communities. Coal exacerbates existing crises, such as water scarcity. Firms immorally create false solutions, like CCS, to justify continued coal combustion in the face of the climate crisis.

Vocabulary

1. acid mine drainage
2. blackout
3. brownout
4. captured state
5. carbon capture and storage (CCS)
6. clean coal
7. company town
8. convict leasing
9. corporate social responsibility
10. cribbing
11. cultural appropriation
12. impact paradox
13. judicious mixture
14. labor hierarchy
15. land grabbing
16. legacy pollution
17. load shedding
18. martial law
19. mitigation
20. physical violence
21. race-to-the-bottom
22. race wedge

23. scarcity mindset
24. scrip
25. shack rouster
26. social license to operate (SLO)
27. spatial control
28. spinoff industry
29. strikebreakers
30. structural violence
31. water stress

Recommended

Bond, P., & Garcia, A. (2000). *Elite transition: From apartheid to neoliberalism in South Africa*. Pluto Press.

Lee, H. B. (1969). *Bloodletting in Appalachia: The story of West Virginia's four major mine wars and other thrilling incidents of its coal fields*. Morgantown: West Virginia University Press.

Lewis, R. L. (2009). *Black coal miners in America: Race, class and community conflict (1780-1980)*. University Press of Kentucky.

Malm, A. (2016). *Fossil capital: The rise of steam power and the roots of global warming*. Verso Books.

Titler, G. J. (1972). *Hell in Harlan*. BJW Printers.

Wu, S. X. (2020). *Empires of coal*. Stanford University Press.

CHAPTER 3

Black lung

Contents

Climate Crisis, Energy Violence
ISBN 978-0-12-819501-7,
https://doi.org/10.1016/B978-0-12-819501-7.00009-6

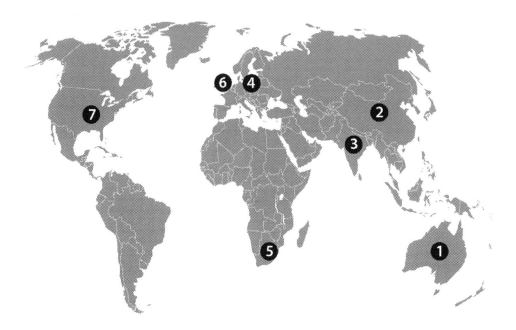

Locations: 1 - Australia
2 - China
3 - India
4 - Poland
5 - South Africa
6 - United Kingdom
7 - United States

Themes: Bureaucratic violence; Health disparity; Erasure

Subthemes: Bankruptcy; Commemoration; Lobbying

They litigate you until you're dead.

Untold narratives

Coal industry necropolitics—politics involving the power to decide who lives and who dies—is not unique to the US. Many romanticize the plight of miners making true costs, including early death, as part of an accepted narrative. The unjust treatment of mining communities remains largely unexplored, blanketed by industry's pervasive and persistent narrative coupling coal to national progress. Laborers are treated as disposable when they develop occupational diseases like black lung. If there's commemoration, miners who died on the job are frequently treated as martyrs without mention that safeguards to prevent harm were often insufficient.

Tourism sites almost universally treat coal-powered industrialization and mining infrastructure positively in spite of patterns of disease, trauma, and displacement.[1] Honest depictions of coal would require accurate **commemoration**—the collective process of remembrance and preserving memories of important events and deeds in sites or ceremonies. Personal coal tragedies found in material culture inside the homes of former miners are seldom part of public telling; if included, coal commemoration sites could be living places of healing.[2]

Grassroots or folk sites of memorializing often contrast with industrial tourism focused on technologies and structures of mines, quarries, and company towns. Struggles to protect labor rights have been poorly acknowledged. Commemoration "from below" in coal country shares overlooked community stories and the cultural landscape. There's muted public memorialization of labor union victories that led to better working conditions. A **union**—be it a trade or labor union—is an organization of workers intent on maintaining or improving the conditions of their employment, such as obtaining better wages or safer working conditions. The continued power of industry is one reason for the lack of prominent pro-labor sites.[3] Yet communities have recorded their own labor histories in music, storytelling, and independent publishing. Contemporary versions of this are found in Praxis 3. Amidst structural violence, hard-working communities invested in places, gaining a degree of self-sufficiency. Miner families across Appalachia's coal country typically raised and butchered their own livestock, hunted, tended gardens, and canned produce. Although land ownership was concentrated in corporate holds or owned by absentee families, Appalachians built and maintained tight communities. Helping out "holler to holler" remains a tradition in a region accustomed to state neglect.

Tourism can offset some loss in economically depressed coal regions. Overly nostalgic sites ignore harms from mining, even reifying loss of life tied to corporate negligence. For example, the Upper Big Branch Memorial in West Virginia commemorates miners killed in an explosion (Fig. 3.1) without mentioning how the company shirked responsibility. The site could have created awareness about necessary reforms. Instead, funded by coal

[1] Bora, I., & Voiculescu, M. (2021). Resettlement, intergenerational memory, place attachment, and place identity in Roşia Jiu coal mine-Gorj County, Romania. *Journal of Rural Studies, 86*, 578—586.

[2] Rohse, M., Day, R., & Llewellyn, D. (2020). Toward an emotional energy geography: Attending to emotions and affects in a former coal mining community in South Wales, UK. *Geoforum, 110*, 136—146.

[3] Leebrick, R. A., & Maples, J. N. (2015). Landscape as arena and spatial narrative in the New River Gorge National River's coal camps: A case study of the Elverton, West Virginia 1914 strike. *Southeastern Geographer, 55*(4), 474—494.

Figure 3.1 Upper Big Branch Memorial. *(Image credit: Upper Big Branch Memorial, West Virginia Explorer.)*

and coal-related industry, this memorial portrays miners as martyrs while avoiding naming the violence that killed them.

> The roadside memorial plaza [...] stands as a solemn reminder of the human cost that West Virginians have so dearly paid to power this great nation.[4]

Coal tourism should educate visitors. Sites can provide a valuable lens for developing place knowledge and cautionary narratives to avoid repeating mistakes. It's important to record the history of **black lung**—coal workers' pneumoconiosis (CWP). Black lung is socially produced and preventable, yet miners are still dying from it.[5] This loss of life could have been prevented. Rates had leveled off but, in 2012, black lung disease began to trend upwards again in US coal areas. By 2018, Appalachia miners were experiencing black lung disease at rates not seen for 25 years.[6] In Kentucky, a new commemoration is the first to honor those who died of black lung[7] where 200 lives have been lost to the disease from just one community. More typically, romanticized commemorations miss active opportunites to challenge the abuse of power that has allowed black lung resurgence.

[4] Upper Big Branch Miners Memorial. (n.d.). Remember April 5, 2010.

[5] Sainato, M. (2020, January, 24). 'It's really tragic.' why are coal miners still dying from black lung disease? *The Guardian.*

[6] Berkes, H. (2018, July 19). Black Lung rate hits 25-Year high in Appalachian coal mining states. *NPR.*

[7] Boles, S. (2019, October 13). *New Kentucky memorial honors miners who died of Black Lung.* Ohio Valley Resource.

Figure 3.2 Campaigning on false hope. *(Image credit: Donald Trump Rally, Wilkes-Barre, PA. October 10, 2016 - Matt Smith, Shutterstock (ID 1427778647).)*

False promises

Black lung emerged out of structural violence. Politicians often promise to improve conditions in coal country, but history suggests the unmitigated byproduct of this industry is just more sacrifice zones and abandoned workers. As policymakers romanticize coal mining as a quintessential American occupation, actual worker protections remain inadequate. Through sloganeering about domestic jobs and energy independence, politicians exploit blue collar worker identification while avoiding proactive protection from industry abuses. For example, when running for office, former US President Donald Trump exploited the dream of a coal revival (Fig. 3.2). During his administration, however, he did little to support miners, even those sick with black lung.[8]

Black lung protections came out of US coal miners' mobilization for reforms starting in the 1960s. The work of coal unions to defend the health of miners brought national attention to occupational hazards. The tragic nature of black lung pierced walls of silence

---

[8] Usenick, H. N. (2019). The heart of Appalachia is on life support: How West Virginia can revive its economy in a post-coal era. *Uniiversity of Pittsburgh Law Review, 81,* 851.

around mining deaths, yet it took until 1973 to pass the Black Lung Benefits Act.[9] Accordingly, disease rates dropped over the next decade but gains were short-lived. Industry intensified lobbying with the goal to avoid contributing funds to black lung programs.

Coal's decline, while creating ecological benefits, brings additional coal community harm. Company bankruptcies mean near-abandonment of former miners, some with life-threatening illnesses. **Bankruptcy** is a legal process through which people, corporations, or other entities who cannot repay debts seek relief. Some bankruptcies are intentional abandonment of responsibility. Bankruptcy settlements in the coal industry often dismantle agreements with unions and release corporations from their obligations to retirees and widows.[10]

As black lung funding and programs were cut, disease and death resurged.[11] Yet policy makers ignored trending data. The lack of urgent remediation demonstrated a state beholden to industry, a willingness to abandon workers. To make matters worse, some mining companies knowingly undermine protections granted by law. Evasive methods include **dust sampling fraud**, acts to avoid coal dust legislation and avoid penalization—for example, tampering with monitoring devices to hide dust level overages.[12] Widespread and persistent gaming of monitoring efforts has been documented.[13]

Captured state

Lobbying is lawfully attempting to influence the actions, policies, or decisions of government officials. Fossil fuel companies spend billions of dollars each year lobbying for policies that will streamline costs to increase their profit. For example, a common tactic among fossil fuel companies is **tax avoidance**: the minimization or avoidance of taxes including shifting profits to offshore tax havens.[14] The sector also prolifically recoups **subsidies**—these are sums of money granted by the government or a public body to assist an industry or business.[15]

[9] Smith, B. E. (1981). Black lung: The social production of disease. *International Journal of Health Services*, *11*(3), 343–359.
[10] Moritz-Rabson, D. (2019, October 30). Eleven coal companies have filed for bankruptcy since Trump took office. *Newsweek*.
[11] Smith, B. E. (2020). *Digging our own graves*. Haymarket Books.
[12] Oppegard, T. (1997). Coal dust sampling fraud. *Appalachian Heritage*, *25*(2), 24. U.S. Attorney's Office, Western District of Virginia (2020, August 12). Mine owner and foreman sentenced to prison for dust sampling fraud.
[13] Detrow, S. (2012, July 10). "An addiction to cheat" on coal mine dust inspections. StateImpact Pennsylvania.
[14] Market Forces. (n.d). Do you pay more tax than the big fossil fuel companies?
[15] Mann, R. F. (2018). The tax treatment of coal. *Tax Law and the environment: A multidisciplinary and worldwide perspective*, 59.

Regulatory capture occurs when a political entity, policymaker, or managing agency is co-opted to serve the commercial, ideological, or political interests of a regulated industry.[16] For example, stringent coal dust control proposals aimed to reduce black lung repeatedly faced industry headwinds. New dust rules established in 2014 then inappropriately targeted dust in aggregate, in effect creating a loophole for hazardous silica to persist. This regulatory failure led to preventable fatalities of coal miners.

Crystalline silica is often a core component of coal mine dust—it's a known human carcinogen that directly contributes to black lung disease and lung cancer.[17] Silicosis, a type of potentially fatal lung disease officially called pulmonary fibrosis, comes from breathing in silica. In the 1990s, black lung disease levels rose in the US when mining operations targeted ever narrower seams laden with quartz, in turn creating excessive silica dust.[18] This dust type is roughly 20 times more damaging to the lung as other coal dust.[19] Surface miners, particularly drillers, blasters, and thin seam and high wall miners, remained acutely at risk,[20] yet regulators ignored silica. Strategies to reduce disease prevalence by targeting high-occurrence regions and occupations are known and adoptable.[21] Research identifying conditions contributing to excessive disease rates could have been used to target policies and programs. For example, higher rates of black lung are associated with working in a small mine (<50 employees) and low coal seam height (<43 inches). Other influential factors include the age of the miner and years in the occupation.[22]

Disease prevention methods in coal mining have been woefully insufficient. As coal operations go bankrupt, workforce liabilities are foisted on taxpayers. The Black Lung Disability Trust Fund (Fig. 3.3), providing insurance and a living stipend for miners, is projected to balloon with debt. Funds were severely mismanaged and public liability for coal's private gain grows.

[16] Carter, R. M., & Morgan, R. K. (2018). Regulatory trust and failure—a case study of coal seam gas in New South Wales, Australia. *Journal of Environmental Planning and Management, 61*(10), 1789—1804.

[17] Trechera, P., Moreno, T., Córdoba, P., Moreno, N., Zhuang, X., Li, B., ... & Querol, X. (2021). Comprehensive evaluation of potential coal mine dust emissions in an open-pit coal mine in Northwest China. *International Journal of Coal Geology, 235*, 103677.

[18] Casey, J.P. (2020, May 28). Black lung is back: Why more is needed to fight dust. Mining Technology.

[19] Bodenhamer, A., & Shriver, T. E. (2020). Environmental health advocacy and industry obstruction: The case of black lung disease. *Rural Sociology.*

[20] Doney, B. C., Blackley, D., Hale, J. M., Halldin, C., Kurth, L., Syamlal, G., & Laney, A. S. (2020). Respirable coal mine dust at surface mines, United States, 1982—2017. *American Journal of Industrial Medicine, 63*(3), 232—239.

[21] Mazurek, J. M., Wood, J., Blackley, D. J., & Weissman, D. N. (2018). Coal workers' pneumoconiosis attributable years of potential life lost to life expectancy and potential life lost before age 65 years United States, 1999—2016. *Morbidity and Mortality Weekly Report, 67*(30), 819.

[22] Hall, N. B., Blackley, D. J., Halldin, C. N., & Laney, A. S. (2019). Current review of pneumoconiosis among US coal miners. *Current Environmental Health Reports,* 1—11.

Black Lung Disability Trust Fund - Simulated Outstanding Debt Moderate Case Assumptions, 2018 - 2050

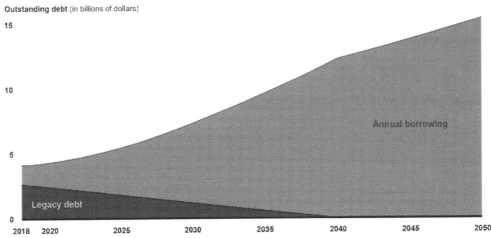

Figure 3.3 Black lung fund debt projections. *(Credit: Black Lung Benefits Program Report, 2018, US Government Accountability Office [GAO-18-351]. Chart data based on US Department of Labor and Treasury, the Energy Information Administration and the Office of Management and Budget.)*

In 2022, the US Congress passed the Inflation Reduction Act (IRA), which included a permanent extension of the Black Lung Excise Tax. Yet fundamental flaws in the program were left unaddressed—primarily, miners are denied benefits because their breathing is not deemed sufficiently impaired.[23] The fact that miners with lung disease are stopped from receiving support is an example of **bureaucratic violence**—this form of violence occurs when bureaucrats justify following ill-fitting or incomplete rules, doing little to go above or beyond standard protocol to prevent catastrophe. Civil servants inundated with detailed codes lose perspective. A bureaucratic dead zone of the imagination occurs as regulators fail to think critically about their work and its implications.[24] A numb imagination is in direct contrast with the geographic and spatial imaginaries needed for regenerative worldmaking (Conclusion). Bureaucrats uphold violent rules even if mandates are broken; they become apathetic about changing norms even as the results remain unsatisfactory or harmful.

[23] Farrish, J. (2020, March 8). Miners hope for Black lung benefits at Boone Memorial event. *The Register-Herald.*

[24] Graeber, D. (2012). Dead zones of the imagination: On violence, bureaucracy, and interpretive labor: The Malinowski Memorial Lecture, 2006. *HAU: Journal of Ethnographic Theory, 2*(2), 105—128.

Energy violence

No other industry has black lung rates as high as coal mining.[25] Black lung disease is a slow and painful death—an incremental suffocation over time. Symptoms include coughing up black mucus, very severe shortness of breath, and low blood oxygen levels, which puts stress on other organs, such as the heart and brain. While miners understand their occupation is risky, many miners feel trapped by the fact that mining is the only regional work option and they can't afford to relocate.[26] Once diagnosed, miners often feel pressure to continue mining.

> **We were young men, just kids.**
> **We killed ourselves here.**[27]

Regulatory violence occurs when policymakers and permitting agencies control a select number of criteria toxins while ignoring co-pollutants, accompanying risk, and underlying vulnerabilities.[28] Time is often a critical dimension—the tactic of grandfathering is a prime example (Fig. 3.4). To be "grandfathered in" (i.e., to have acquired rights) means to be exempt based on a provision in which an old rule continues to apply to some existing situations, while a new rule will apply to all future cases.

Tactics	Description
<u>loophole</u>	using an ambiguity or inadequacy in a law to circumvent its purpose
<u>regulatory rollback</u>	'backsliding' of standards, often following lawsuits from industry
<u>grandfathering</u>	allowing existing facilities to continue to use old equipment or standards

Figure 3.4 Key regulatory violence tactics.

Grandfathering exempts older facilities where executives argue new regulations would require costly upgrades or retrofits to existing sites. Since new rules are already phased in so that industry has time to adapt, grandfathering outlives its useful purpose, and exemptions are extended, allowing a company to avoid responsibility for reducing dangerous pollutants

[25] Potera, C. (2019). Black lung disease resurges in Appalachian coal miners. *AJN The American Journal of Nursing, 119*(4), 14.

[26] Almberg, K. S., Friedman, L. S., Rose, C. S., Go, L. H., & Cohen, R. A. (2020). Progression of coal workers' pneumoconiosis absent further exposure. *Occupational and Environmental Medicine, 77*(11), 748−751.

[27] A miner quoted in PBS. (2019, January 22). Coal's deadly dust. *Frontline.*

[28] Earle, C. (2016). Survey says-an argument for more frontloaded FERC public use provider determinations as a means of streamlining the commission's regulatory role over interstate natural gas pipeline operators. *William & Mary Environmental Law and Policy Review, 41*, 711.

for years or decades—as an example of racial violence, polluting plants in majority-minority neighborhoods are allowed continual emissions surpassing safe levels.[29]

Try this

Do a news search on a particular location with coal energy. Look for examples of grandfathering, loopholes, or other regulatory violence in the coal sector.

Mine owners capitalize on uneven development, exploiting poverty and marginalization (Chapter 1)—in particular **health disparity** (Fig. 3.5), where health differences are closely linked with social, economic, and environmental disadvantages, particularly acute in mining regions. Like many struggling rural communities, the Appalachian medical system lags behind national health care standards.[30] Geographic location significantly contributes to an individual's ability to achieve or maintain good health. Coal communities face various forms of exclusion, including fewer medical facilities in low-wealth areas where rates of illness are high.

To avoid responsibility, industry often blames those being harmed. For example, the West Virginia's Manufacturing Association (WVMA) released a 2019 statement[31] arguing that West Virginians drink so little water and eat so little fish compared to the rest of the country, it would be unfair to the polluting industry in the state to hold them to the same standards as thirstier, fish-eating locations.[32] In the same statement, WVMA further suggested that people who have higher body weight can handle more contamination.

Poverty
Racism and other forms of bigotry
Compound environmental threats (i.e., hotspots)
Inadequate access to health care
Educational inequalities

Figure 3.5 Key factors in health disparity.

---

[29] Patterson, J., Fink, K., Wasserman, K., Starbuck, A., Sartor, A., Hatcher, J., & Fleming, J. (2012). Coal blooded: Putting profits before people. National Association for the Advancement of Colored People.

[30] Hendryx, M. (2008). Mortality rates in Appalachian coal mining counties: 24 years behind the nation. *Environmental Justice*, 1(1), 5—11.

[31] West Virginia Manufacturing Association. (2019, March 11). WVMA statement on human health criteria development.

[32] O'Neil, L. (2019, March 13). They say if we are fatter we can accept more pollutants. *Welcome to hell world*.

Myth of the free market

Throughout the twentieth century, coal was inaccurately promoted as a cheap source of energy due in large part to public subsidies. Subsidies create bias toward particular fossil fuels, even when other fuels are socially, ecologically, and economically superior. In effect, subsidies operate to reinforce slow violence. In addition to direct and indirect subsidies, the public bears unaccounted **externalized costs** that remain absent from market prices. Health care expenditures and environmental remediation are just two examples.[33] Budget expenditures propagating fossil fuels are funds not available for other essential needs. For example, the US has spent 10 times more on fossil fuel subsidies than on educational expenses in recent years.[34] Struggling schools have become a feature of the US education system, which is rife with disparity and consistently fails low-income children.

Subsidies relieve market pressure on inefficient and dirty operations, slowing the transition to renewable energy.[35] From 1950 to 2010, the federal government provided the coal industry with more than $70 billion in tax breaks and subsidies. Further, coal companies benefit from tariffs on foreign coal and indirect aid to steel smelters, railroads, and industries that burn coal.

Myth of the benevolent state

Coupled with coal's manipulation of the "free market," a fundamental tension lies at the core of the sector between (1) maximizing the value of a company's estate when bankruptcy occurs, and (2) regulatory structures that mandate parties internalize costs of their behavior.[36] A demonstrable case of this tension lies again in the administration of federal black lung funds (Fig. 3.6). Even as "Part B" cases filed before 1973 differ in payout rates from "Part C" cases in the years since 1973, taken together they point to the historically prodigious burden left by coal. This burden is not borne by the industry alone, or even equally. Increasingly, rate payers and tax payers are left covering the fund's debt burden.

---

[33] Tschofen, P., Azevedo, I. L., & Muller, N. Z. (2019). Fine particulate matter damages and value added in the US economy. *Proceedings of the National Academy of Sciences, 116*(40), 19857–19862.

[34] Ellmoor, J. (2019, June 15). United States spends 10 times more on fossil fuels subsidies than education. *Forbes*.

[35] Kotchen, M. J. (2021). The producer benefits of implicit fossil fuel subsidies in the United States. *Proceedings of the National Academy of Sciences, 118*(14).

[36] Macey, J., & Salovaara, J. (2019). Bankruptcy as bailout: Coal company insolvency and the erosion of federal law. *Stanford Law Review, 71*, 879–962.

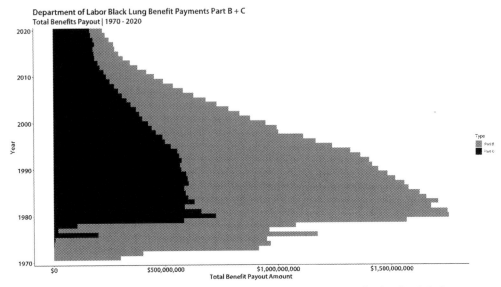

Figure 3.6 Black lung fund payouts. *(Source: US Department of Labor Statistics.)*

While fighting against regulations to protect workers, coal companies have sucessfully exploited the regulated system to recover costs, meaning taxpayers pay for decades. Underscoring the sector's cynicism, an infamous coal baron—who himself contracted black lung—applied for federal benefits to treat his own black lung disease, after opposing more stringent coal dust regulations for years.[37]

Malicious Murray

An example of the abusive burden Murray Energy put on the public, the Murray Energy company, whose owner and CEO was influential climate denier Robert Murray (1940—2020), went bankrupt twice.[38] Months after seeking his second bankruptcy and leaving taxpayers to pay millions,[39] Murray began a new family business under new leadership, instantly creating the largest privately owned US coal operation

---

[37] Volcovici, V. (2020, August 1). Coal baron Murray seeks U.S. benefits to treat his black lung disease—report. Reuters.

[38] Downie, C. (2017). Fighting for King Coal's crown: Business actors in the US coal and utility industries. *Global Environmental Politics*, *17*(1), 21—39.

[39] Friedman, L. (2019, December 17). A coal baron funded climate denialism as his company spiraled into bankruptcy. *New York Times*.

(described in more detail later).[40] Murray's story shows how powerful corporations create public costs.[41] Fig. 3.7 provides a timeline of Murray Energy's maneuvering.

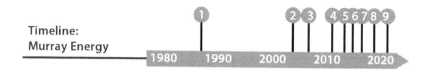

Timeline:
Murray Energy

1 - Murray Energy founded
2 - Crandall Canyon Mine collapse
3 - $1.85 million penalty for safety violations at Crandall Canyon Mine
4 - Murray sues the EPA
5 - Murray sues publisher for slander
6 - Murray Energy sues NYT and HBO for libel
7 - Confidential memo to Trump Administration -
    blueprint of 16 detailed requests
8 - Murray Energy files for bankruptcy protection
9 - Murray Energy emerges out of bankruptcy under a new company

Figure 3.7 Murray Energy timeline.

In August 2007, Murray Energy became a household name as its Crandall mine in Utah dramatically collapsed, leaving miners trapped in spite of week's long rescue efforts broadcast moment by moment on national television. The collapse entailed a 69-acre section of the mine, registered regionally as a 3.9 earthquake and ultimately entombed six miners underground.[42] To underscore the seriousness of the collapse, criminal charges against individual coal executives were lodged in 2012—a rare occurrence.[43] Violations contributed to the collapse, including excessive use of retreat mining that carved out "too many voids," creating the structural instability.[44] The Mining Safety and Health Administration (MSHA) determined the mine was destined to fail because the company made

[40] Sheppard, K. (2002, October 9). CREW files FEC complaint over coal company's coerced campaign donations. *Mother Jones*; Banerjee, N. (2012, August 29). Ohio miners say they were forced to attend Romney rally. *Los Angeles Times*; Danno, S. M. (2013). Murray Energy corporation v. McCarthy. *Public Land and Resources Law Review*, (7), 16.

[41] Frazier, R. (2019, November 21). After Murray Energy bankruptcy, what's the future of coal? WESA FM.

[42] Associated Press. (2008, August 6). Year after mine collapse, many failures clear. NBC News.

[43] Rolston, J. S. (2010). Risky business: neoliberalism and workplace safety in Wyoming coal mines. *Human Organization*, 331–342; CNN (2008, July 24). Feds blame mine operators for fatal collapse; Associated Press. (2008, August 6). Year after mine collapse, many failures clear. NBC News.

[44] Rolston, J. S. (2010). Risky business: Neoliberalism and workplace safety in Wyoming coal mines. *Human Organization*, 331–342.

critical miscalculations and didn't report early warning signs. MSHA itself was found at fault for lax oversight and for their haphazard rescue, during which three people died.[45]

Years later, when the federal government began to increase safeguards in the coal sector, Murray called this a "war on coal."[46] He won an initial lawsuit to weaken EPA standards,[47] followed by reversal in higher courts. Murray assaulted freedom of the press. He sued media outlets who criticized him for defamation and slander[48] and sought to lift protections on journalists that would make them reveal sources. Luckily, he was unsuccessful in this endeavor, which would have significantly eroded independence in news coverage.

Murray's reach would achieve new levels following the 2016 election of Donald Trump as US president. In 2018 Murray wrote a confidential memo to the Trump Administration with 16 industry-friendly declarations, since revealed to be a blueprint for the administration.[49] Andrew Wheeler, a former Murray executive, became the EPA Director and rolled back coal regulations.

As Murray's companies were facing bankruptcy in 2019, Murray funded climate denialism. Because Murray did not cover benefits owed to staff when he claimed bankruptcy, US taxpayers were left paying Murray's workers.[50] Murray Energy was reborn as a new company, where Murray became chairman of the board.[51] The new company, American Consolidated Natural Resources (ACNR), acquired mines in five states.[52] Restructuring transferred $8 billion of Murray's company debt and legacy liabilities to the state to be borne by taxpayers. It also allowed ACNR to access financing, even as they were laying off workers.[53]

Murray Energy shows how current coal industry safeguards are insufficient. There's a gaping chasm between legal compliance with health and safety obligations and effective

[45] Dreger, D. S., Ford, S. R., & Walter, W. R. (2008). Source analysis of the Crandall Canyon, Utah, mine collapse. *Science, 321*(5886), 217–217.

[46] Nagle, J. C. (2017). The war on coal. *LSU Journal of Energy Law and Resources, 5*(1), 6.

[47] Tollefson, J. (2019). Air pollution science under siege at US environment agency. *Nature, 568*(7750), 15–17; Danno, S. M. (2013). Murray Energy corporation v. McCarthy. *Public Land and Resources Law Review*, (7), 16; Taylor, P. B. (2017). Murray Energy corporation v. Administrator of environmental protection agency. *Public Land & Resources Law Review*, (8), 12.

[48] Feuer, A. (2017, June 21). Case tests limits of law protecting journalists' sources. *New York Times*; Supreme Court of Ohio; Spayd, L. (2017, May 10). A rare libel suit against the Times. *New York Times*.

[49] Friedman, L. (2018, January 9). How a coal baron's wish list became President Trump's to-do list. *New York Times*.

[50] Dow, J. (2019, December 18). Bob Murray paid for science denial instead of his coal workers' wages as the company went bankrupt. Electrek.

[51] Staff. (2020, December 31). Murray Energy becomes ACNR; founder dies. *The Times Leader*.

[52] Raby, J. (2020, September 17). Coal giant Murray Energy out of bankruptcy under new name. AP News.

[53] Staff. (2021, June 23). Prep fire at Monongalia County coal operation. WV Metro News.

workplace protection.[54] Ironically, Robert Murray died in 2020 of lung disease associated with work in the mines earlier in his career.

Industry silencing

As discussed in the example of Robert Murray, free press can come under attack if they critique coal bosses. Independent views can be hard to come by in coal regions. For decades, **do-it-yourself (DIY) publishing** (Box 3.1) provided essential access to

BOX 3.1 Independent media

Coal regions, areas where mining provides a significant portion of jobs and revenue, have unique politics. One of the challenges in coal regions is access to information. Oppression relies on controlling what people know about alternatives. The Appalachian labor movement fights to be heard. *So Much to be Angry About* is a robust collection from the Appalachian Movement Press, including classics like *Up! Up!* written by Don West for the Kentucky Workers Alliance.

```
SOLIDARITY FOREVER
When the union's inspiration
Through the workers blood shall run
There can be no power greater
Anywhere beneath the sun.
Yet what force on earth is weaker
Than the feeble strength of one?
But the union makes us strong.[55]
```

The largest US union covering miners, the United Mine Workers of America (UMWA), worked for decades for black lung benefits and other essential protections.[56] Nationally and in local chapters, the UMWA has been fighting for legislation to protect workers caught in bankruptcy proceedings and a comprehensive fund to stimulate employment in communities experiencing job loss.[57] Where coal unions were permitted to function, worker protections improved.[58] In contrast, union busting paved the way for regulatory violence as captured politicians skirted their oversight role. One of the groups most identified as working to change this political and economic inequality is the UMWA. In spite of its efforts, decades of hostile treatment by self-interested firms have dampened the union's power.

(Continued)

[54] Kirkwood, S., & Kenafacke, G. (2017, September 7). Black lung and compliance: The great divide. Mining Risk Review 2017. Willis Towers Watson.

[55] Songwriter Don West In Slifer, S. (2021). *So much to be angry about: Appalachian movement press and radical DIY publishing, 1969–1979.* West Virginia University Press.

[56] Patterson, B. (2019, February 20). West Virginia Coal Miners rally for Black Lung legislation. *Ohio Valley Resource.*

[57] UMWA (United Mine Workers of America). (n.d.). Current Legislation.

[58] Smith, B.E. (2020). *Digging our own graves: Coal miners and the struggle over Black Lung Disease,* Updated Edition. Chicago: Haymarket Books.

> ## BOX 3.1 Independent media—cont'd
>
> Harlan County became known as "Bloody Harlan" following violence against UMWA from 1937 to 1941. Workers who joined the union were immediately discharged, as they were forced to sign a contract binding them to not join any mine labor organization. These contracts violated provisions of the National Labor Relations Act passed in 1935 to protect the rights of employees to organize and bargain collectively. Yet, in the decades since, the coal industry has been unrelenting in attempts to weaken unions and organized labor as these are shown to improve labor conditions.

information censored by mainstream media. DIY is a method of building, modifying, or repairing things by oneself without the direct aid of professionals or experts. DIY economies are core to Appalachian self-reliance and survival (Chapter 2). DIY publishing has been central to recording regional experiences of structural and physical violence.

Community and labor organizing can bring incomparable joy,[59] but the stress and workload can be unrelenting. The power and resource differentials between company owners like Robert Murray, who have connections at the highest levels of government, and workers like miners who struggle to leverage their power in spite of their contributions to regional economies, are significant. Since worker benefits cut into profit margins, firms lobby against these necessary changes. Due to the widespread existence of captured politicians, policy makers lend additional power to industry, reinforcing harms from sacrifice zones and threatening the workforce.

Try this

Considering specific types of energy violence in the coal sector; how might governmental reforms and independent oversight of the energy sector encourage more transparency and equity? What challenges remain?

Spatial distribution

Both regionally and globally, coal's costs and benefits tend to be poorly distributed across space.[60] In select industrialized regions, coal combustion continues to expand. While significantly concentrated in Asia, the US and Germany continue to both produce and consume coal (Fig. 3.8).

The phaseout of coal exemplifies uneven development patterns (Fig. 3.9).[61] There's a double standard: cut dirty fossil fuels in wealthy locations, while maintaining and even

---

[59] Brown, A. M. (2019). *Pleasure activism: The politics of feeling good.* AK Press.

[60] Ghosh, A. (2018). *The great derangement: Climate change and the unthinkable.* Penguin UK.

[61] Fouquet, R. (2016). Historical energy transitions: Speed, prices and system transformation. *Energy Research & Social Science, 22,* 7—12.

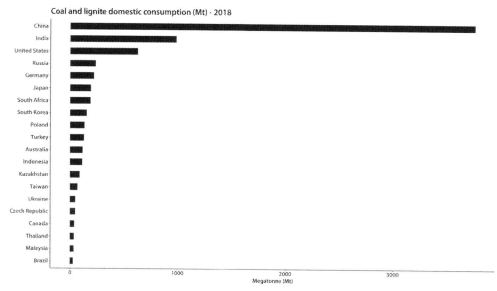

Figure 3.8 Top coal consumers. *(Source: Enerdata.)*

- **Emerging Coal Plants** *
   * 8% of nearly 13,000 plants within the Global Energy Monitor dataset categorized as announced, under construction, permitted or pre-permitted.

Figure 3.9 Emerging coal plants, 2020. *(Source: Global Energy Monitor, Global Coal Plant Tracker, accessed 7/1/20.)*

intensifying coal and polluting fuels in low-income areas. This division in the global energy sector is consistent with larger inequities between rich and poor.

International banks (Chapter 2) have historically promoted coal as essential for the poor. Coal projects justified to address poverty benefit absentee landowners and corporate executives, while life-threatening pollution in "host" communities creates sacrifice zones. Degraded sites experience harm long after beneficial heat or light from energy abates.[62]

Try this

Fig. 3.9 *demonstrates new coal plants, providing a visual representation of uneven development. What do the new areas of coal development tell us about global economics and politics?*

Temporal analysis

The resurgence of black lung disease is slow violence. From the time of disease onset, several courses of action are possible.[63] Even if a miner has no further exposure, the disease can still progress (Fig. 3.10).

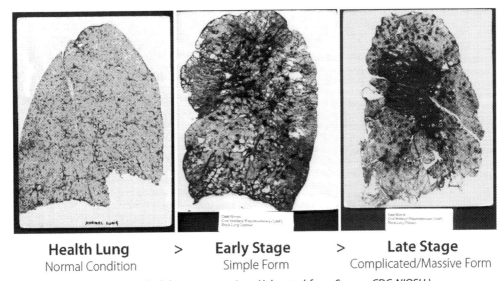

| **Health Lung** | > | **Early Stage** | > | **Late Stage** |
| Normal Condition | | Simple Form | | Complicated/Massive Form |

Figure 3.10 Black lung progression. *(Adapted from Source: CDC-NIOSH.)*

---

[62] Maldonado, J. K. (2017). Transforming coastal louisiana into an energy sacrifice zone. *ExtrACTION: Impacts, engagements, and alternative futures*, 108.

[63] Almberg, K. S., Friedman, L. S., Rose, C. S., Go, L. H. T., & Cohen, R. A. (2020). Progression of coal workers' pneumoconiosis absent further exposure. *Occupational & Environmental Medicine*, 77(11), 748–751.

BOX 3.2 Spin-off industry: Black lung lawyers

Black lung lawyers are an example of **derivative exploitation**, the economic offshoots of harm which are often described or even tallied economically as "progress" though they potentially compound problems. Miners are dependent on lawyers to access any benefit. At the same time, the lawyers view this work as simply an occupation, similar to those focused on asbestos litigation. There are not enough affordable lawyers to attend to the number of sick miners with black lung claims. Cases are hard to win, in part because research has been inadequate.[65]

Since policy makers have not adequately implemented protections to prevent disease or to treat illness once it appears, sick miners have had to fight for medical care through the courts.[64] A profitable spin-off industry from coal is black lung lawyers (Box 3.2).

Illustrative cases

Internationally, black lung educational and medical programs have been established to positive effect, demonstrating stabilization or even declining rates of disease.[66] Over time however, state agencies reduced oversight and attention, allowing opportunities for industry to lobby to reduce or avoid protections. These delays and setbacks have contributed anew to growth of the disease even after causes and prevention tactics were established. The following international coal locations uniquely illustrate how worker's health has been undercut by states sponsorship of coal interests.

United Kingdom

For the first half of the 20th century, the UK medical community lacked a clear under-standing of the health risks from mining. British doctors assumed coal-dust actually *pro-tected* workers from tuberculosis and other lung diseases. Nationalization of the coal industry, passage of pro-labor legislation, and the inclusion of pneumoconiosis (black lung) into Workmen's Compensation in 1943 all helped shift power toward labor. Growth of the coal union and increase in social medicine led to re-examination of dis-ease. Unlike the US, the UK experienced an explosion of interest in lung diseases

[64] Hamby, C. (2014). Black lung claims by 1100 coal miners may have been wrongly denied. *The center for public integrity*; Hamby, C. (2013). As experts recognize new form of black lung, coal industry follows familiar pattern of denial. *The center for public integrity*.

[65] Petsonk, E. L., Rose, C., & Cohen, R. (2013). Coal mine dust lung disease. New lessons from an old exposure. *American Journal of Respiratory and Critical Care Medicine, 187*(11), 1178–1185.

[66] Marek, K., & Lebecki, K. (1999). Occurrence and prevention of coal miners' pneumoconiosis in Poland. *American Journal of Industrial Medicine, 36*(6), 610–617.

affecting miners during the 1950s due to concern about air pollution.[67] Nevertheless, UK government officials seemingly worked at cross purposes to keep coal production up despite clear knowledge of horrendous mine conditions. Labor parties and unions were left to worry about impacts to employment, even as miners often underreported lung conditions and returned to work after diagnosis. Justifications to underestimate the extent of disease emerged across the sector. Less miners were diagnosed because companies sought to avoid payment for detailed lung scans.

**Workers were incidental;
people were sacrificed,
sometimes knowingly, sometimes not[68]**

Today, sick miners face a precarious future. Phasing out coal quickly without support for workers may be weaponized as class warfare.[69] The UK moved the farthest away from coal considering its historical dominance.[70] UK mines have largely shut down, resulting in the lowest levels of coal combustion in 250 years.[71]

Red–green alliances (i.e., alliances between socialist and green parties) benefit both workers and environmentalists. There's intentional work in the UK to maintain coal communities, recognizing families have been exploited across centuries.[72] Blue–green alliances, cooperation between labor organizations and environmental groups, are discussed in Chapter 9.

China

China's black lung rates are likely underestimated.[73] Study of the disease within China has been delayed and inadequately covered.[74] A challenge with identifying black lung in China is the grouping of this illness into a general category of occupational lung

---

[67] Staff. (2016, January 11). Miners' lung disease: 'Thousands may have Pneumoconiosis.' BBC.

[68] Rosner, D. (2008). Arthur McIvor and Ronald Johnston, Miners' lung: A history of dust disease in British coal mining, Studies in Labor History. *Medical History, 52*(4), 535–537.

[69] Wishart, R. (2019). Class capacities and climate politics: Coal and conflict in the United States energy policy-planning network. *Energy Research & Social Science, 48*, 151–165.

[70] Evans, S., & Pearce, R. (2018). How UK transformed its electricity supply in just a decade. *Carbon Brief.*

[71] McKie, R. (2020, August 9). Is this the end for coal 'king coal' in Britain? *The Guardian.*

[72] Johnstone, P., & Hielscher, S. (2017). Phasing out coal, sustaining coal communities? Living with technological decline in sustainability pathways. *The Extractive Industries and Society, 4*(3), 457–461.

[73] Mo, J., Wang, L., Au, W., & Su, M. (2014). Prevalence of coal workers' pneumoconiosis in China: A systematic analysis of 2001–2011 studies. *International Journal of Hygiene and Environmental Health, 217*(1), 46–51.

[74] Perret, J. L., Plush, B., Lachapelle, P., Hinks, T. S., Walter, C., Clarke, P., ... & Stewart, A. (2017). Coal mine dust lung disease in the modern era. *Respirology, 22*(4), 662–670.

disease.[75] There are high death rates from coal in China, though it is not known how much is from black lung.[76] Chinese pneumoconiosis cases could reach epidemic levels as the time for coal workers to contract black lung seems to be shortening. In spite of this impending risk, direct intervention in dust exposure reduction remains largely absent.[77] Like other international coal geographies, targeted interventions in China's coal regions could indeed reduce fatalities.

India

Accurate data on black lung in India are lacking.[78] The state spreads disinformation throughout its sectors, including that of coal tourism.[79] Typically, this tourism commemorates coal by focusing on technical feats, such as the deepest mine,[80] without paying sufficient attention to risks.

India's coal realities are much murkier than its spin. Indians overall breathe some of the most toxic air in the world.[81] Low-income workers, including minors, have no protections from coal dust laden with heavy metals.[82] Children in the coal industry experience some of the greatest harm.[83]

Australia

Black lung experts have examined the resurgence of the disease in Australia,[84] resulting in recommendations to reduce exposures near vulnerable population centers, such as

[75] Wang, X. R., & Christiani, D. C. (2003). Occupational lung disease in China. *International Journal of Occupational and Environmental Health*, 9(4), 320–325.

[76] Buchanan, S., Burt, E., & Orris, P. (2014). Beyond black lung: Scientific evidence of health effects from coal use in electricity generation. *Journal of Public Health Policy*, 35(3), 266–277.

[77] Graber, J. M. (2018). Application of the Delphi method to reduce disability and mortality from coal mine dust lung disease in China; a new approach to an old problem. *Occupational and Environmental Medicine*, 75(9), 615–616.

[78] Kaul, R. (2019, October 9). Miners at a higher risk of lung disease with no cure for most: Experts. *Hindustan Times*.

[79] TNN. (2016, December 20). First time in India, coal mine tourism open for visitors; Western coalfields join hands with Maharashtra. *The Times of India*.

[80] Singh, R. S., & Ghosh, P. (2019). Potential of mining tourism: A study of select coal mines of Paschim Bardhaman District, West Bengal. *Indian Journal of Landscape Systems and Ecological Studies*, 42(1), 101–114.

[81] Greenpeace. (2020, January). Airpocalypse IV: Assessment of Air Pollution in Indian Cities and National Ambient Air Quality Monitoring Program.

[82] Rout, T. K., Masto, R. E., Padhy, P. K., Ram, L. C., George, J., & Joshi, G. (2015). Heavy metals in dusts from commercial and residential areas of Jharia coal mining town. *Environmental Earth Sciences*, 73(1), 347–359.

[83] Lydersen, K. (2011, May 25). From U.S. to India, the Casualties of Coal. *In These Times*.

[84] Parliament of Australia. (2016). Fifth interim report: Black lung 'It has buggered my life.'

daycare centers, schools, hospitals, and senior centers.[85] Black lung in workers is most pronounced in Queensland,[86] where the majority of Australia's higher quality metallurgical coal is produced.

Historically, strong political interests during coal privatization in Australia led to denial of worker benefits.[87] Black lung programs were catastrophic failures[88] and deliberately underfunded.[89] After re-emergence of the disease, research identified the cost of inaction.[90] Showing regulatory capture,[91] inspectors are too close with industry.[92] To complicate matters, coal dust protections continue to be undercut.[93]

Through state propagation of regional mega projects, public finances have been sunk into intensive yet expansive coal infrastructures. A boom drove up local costs, people went into debt, followed by price collapse.[94] The capture of Australian politics has further delayed transition,[95] and Australia's energy sector workforce itself has become increasingly migratory. As workers move from project to project, local economies are left with extraction and contamination. Australian-based companies often rely on a **fly in, fly out (FIFO) workforce**, a method of employing people in remote areas by flying them temporarily to the work site instead of relocating them on a more permanent basis.

Touring mines, mining tourists

In the US, South Africa, India, Australia, and other locations marked by a long history of mining and tourism, coal tourism too often becomes romanticized and commemorated in superficial ways. This also occurs with historic silver and gold mines, similarly built on brutal labor conditions and toxic releases.[96] Profiting from mis-education about mining serves as an example of derivative exploitation. Mine tours usually limit discussions to immediate

[85] The Thoracic Society of Australia and New Zealand (n.d.). Advocacy: Coal workers pneumoconiosis.
[86] Perret, J. L., Plush, B., Lachapelle, P., Hinks, T. S., Walter, C., Clarke, P., ... & Stewart, A. (2017). Coal mine dust lung disease in the modern era. *Respirology, 22*(4), 662–670.
[87] Australian Institute of Health and Safety. (2016, August 25). First case of black lung disease in an open cut mine.
[88] Mellor, L., & Riga, R. (2017, May 29). Black lung inquiry finds 'catastrophic failure.' ABC.
[89] O'brien, C., & McKillop, C. (2017, March 22). Black lung detection in Queensland "deliberately underfunded, under-resourced." ABC News.
[90] McCall, C. (2017). The cost of complacency—black lung in Australia. *The Lancet, 390*(10096), 727–729; Australian Mining (n.d.). Black lung archives.
[91] Queensland Parliament. (2017). Black Lung, White Lies: Coal Workers' Pneumoconiosis Select Committee, Inquiry Report.
[92] Rose, C. (2017). Resurgence of black lung in the U.S. and Australia: lessons from medical surveillance.
[93] Atkin, M. (2018, June 7). Coal miner's death after silicosis diagnosis a warning on dangerous dust levels. ABC.
[94] Ritter, D. (2018). *The Coal Truth*. UWA Publishing.
[95] Krien, A. (2017). The long goodbye: Coal, coral and Australia's climate deadlock. *Quarterly Essay*.
[96] Pretes, M. (2002). Touring mines and mining tourists. *Annals of Tourism Research, 29*(2), 439–456.

surroundings—quarries and ponds, mining roads, exposed rock profiles, fossils, pieces of extraction machinery, and other physical and technical objects. Simultaneously, space is limited for any substantial coverage of negative repercussions, such as political corruption, occupational diseases like black lung, public health costs from acid rain or from asthma near power plants, and toxic legacies. Comprehensive education that incorporates tradeoffs and admits industrial harms would be a step forward toward sustainable alternatives.

Particularly in Australia, the situation has reached a crossroads: yet more coal or a turn towards tourism? There is growing concern over how climate change threatens Australia's once-thriving nature tourism industry, comprising charming koalas and popular barrier reefs. For many years, however, Australians were unwilling to admit the obvious—that the Great Barrier Reef is in grave danger—or acknowledge the causes. The most obvious source of the coal industry's power is the tale told about the sector's significance, which in reality pales in comparison with the tourism industry. Invariably these alleged benefits are exaggerated—the value the coal industry brings to the public in the form of revenue is prone to hyperbole. Although Australia is the top coal exporter, the industry employs fewer overall workers than do the arts and recreation sectors. In fact, it's the greater benefits of marine tourism revenue, resulting from approximately two million people visiting the Reef annually, and the marine tourism job sector, which employs approximately 60,000 workers, that are put at risk by climate disruption.[97] As reported by the Australian government, the Reef's health is critically important to the stability and value of the Reef tourism industry. The greatest threat to the Reef continues to be climate change, followed by, and associated with, coastal development and land-based run-off.[98]

In India, the push for coal tourism is state driven. The Indian government has embraced coal mining tourism as part of a broader initiative to showcase how operations in coal mines have "minimal impact on the environment."[99] Tourists can "travel to the depths of underground mines and observe the operations taking place at the open-cast mine …."[100] If done carefully, it could bring extra revenue to coal regions and demonstrate phases of reclamation following operations.[101] Mining regions are often well-connected to the outer world by rail, used by trains or reworked as biking or riding trails. In rural India, ecotourism could provide better jobs (i.e., less physically and ecologically

[97] Great Barrier Reef Marine Park Authority, Australian Government. (2019). Outlook Report 2019.
[98] Great Barrier Reef Marine Park Authority, Australian Government. (2022, August 24). Tourism on the Great Barrier Reef.
[99] TNN. (2016, September 13). Coal tourism likely at mining spots. *The Times of India*.
[100] TNN. (2016, December 20). First time in India, coal mine tourism open for visitors; Western coalfields join hands with Maharashtra. *The Times of India*.
[101] Singh, R. S., & Ghosh, P. (2021). Geotourism potential of coal mines: An appraisal of Sonepur-Bazari open cast project, India. *International Journal of Geoheritage and Parks, 9*(2), 172–181.

risky, more sustainable) than new coal mines,[102] while having the added benefit of showing "lessons learned." When coal tourism in any region fails to highlight the important histories of workers as well as environmental education on fundamental topics such as pollution and climate change, it's a lost opportunity.

Educational tourism is also emerging in South Africa, where lessons from its history of coal abound. Heritage tourism, based on the use of abandoned mine facilities, is a feature of development projects in Johannesburg (Gold Reef City), Kimberley, and Pilgrim's Rest.[103] As in the US, India, and the UK, communities seek tourism alongside the demise of the coal sector.[104] A rural example in South Africa focuses on participatory education with coal-impacted communities, a contrast to the historically inadequate involvement of community leaders in the coal sector.[105]

While international tourism revenues are expected to decline as a result of climate change,[106] the number of domestic travelers engaging their local area and surrounding region can be expected to rise. Done well, tourism can develop forward and backward linkages to support local communities. For example, the company True Pigments is cleaning and restoring Sunday Creek in Ohio using technology that turns pollution from historic coal mining into vibrant pigments for use in paint and other products. True Pigments advances community benefit and creates jobs while making the water cleaner.[107] Grassroots-focused networks provide opportunities for international tourism, such as education tours addressing core themes like sustainability, equity, gender, clean water, and healthcare. These initiatives are bittersweet as they build from tragedy and hardship.

How should Appalachia commemorate coal?

Early US mining practices were particularly repugnant as slaveholders, knowing the risk of death, purchased life insurance on the Black laborers they forced into the mines. The earliest commercial coal emerged around 1735 in the eastern US; Blackheath Mines in central Virginia stood as the biggest and most dangerous regional mine operation. To this point, insurers began to revoke the practice of black mine worker policies taken by slaveholders.[108] Explosions were

[102] Katsineris, S. (2014). India: Forests and tigers versus coal mines. *Guardian (Sydney)*, (1662), 6.
[103] Binns, T., & Nel, E. (2002). Tourism as a local development strategy in South Africa. *Geographical Journal*, 168(3), 235−247.
[104] Binns, T., & Nel, E. (2003). The village in a game park: local response to the demise of coal mining in KwaZulu-Natal, South Africa. *Economic Geography*, 79(1), 41−66.
[105] Leonard, L. (2019). Traditional leadership, community participation and mining development in South Africa: The case of Fuleni, Saint Lucia, KwaZulu-Natal. *Land Use Policy*, 86, 290−298.
[106] Anser, M. K., Yousaf, Z., Awan, U., Nassani, A. A., Qazi Abro, M. M., & Zaman, K. (2020). Identifying the carbon emissions damage to international tourism: turn a blind eye. *Sustainability*, 12(5), 1937.
[107] True Pigments. (n.d.). Mission and Impact.
[108] McCartney, M.W. (1989). Historic overview of the Midlothian coal mining historical tract Chesterfield County, Virginia.

increasingly common at Blackheath and other deep mines of the time due to the buildup of explosive gases, particularly methane. Experienced overseers knew how to ventilate the gases, but the situation was made more dangerous by the necessary use of headlamps which, at the time, featured wicks and open flames. Predictable disasters were labeled as "accidents."

```
Dreadful Accident
Adams Sentinel
March 25, 1839
We learn from the Washington Globe that on Saturday night an explosion
took place in Heth's Pit [Blackheath Mine], (a coal mine situated about 12
miles from Richmond, Va., in the county of Chesterfield,) by which it is said
that sixty-three negroes have been killed or buried alive. The shaft is 800
feet deep — deeper, probably than any other in the United States — and as the
falling in of earth has been considerable, there is no probability that any of
the persons below, if now alive, can be extricated [...] The shaft and engine
are but little injured.[109]
```

Blackheath Mines eventually shut down because of continual dangerous fires and explosions.[110] The mine's current-day commemoration (Fig. 3.11) focuses on the owners as military officers and the quality of coal, not mine operations or its workers. Local renditions for official signage haven't corrected the record in spite of calls from local descendants to do so.[111]

African American erasure

In 1882, a large explosion killed dozens of workers at central Virginia's Midlothian Mines,[112] one of the nation's first industrial coal fields (Fig. 3.12).[113] Those who died and the majority of the work crew were mainly children and African Americans.[114] In the weeks following this tragedy, and during the pain and hardship experienced by critically burned miners, media coverage focused on the timely payment to the owners of the

---

[109] United States Mine Rescue Association. (n.d.). Mines disasters in the United States: Black heath colliery explosion. https://usminedisasters.miningquiz.com/.

[110] Lewis, R. L. (2009). *Black coal miners in America: Race, class and community conflict (1780–1980)*. University Press of Kentucky.

[111] Stewart, I.A. (2022, June 20). Researchers work to honor enslaved coal miners of Chesterfield. VPM News.

[112] Burtchett, B.I. (1983). A history of the village of Midlothian, Virginia, emphasizing the period 1835–1935. Unpublished Master's thesis, University of Richmond.

[113] Frantel, N.C. (2008). *Chesterfield County Virginia uncovered: The records of death and slave insurance records for the coal mining industry 1810–1895*. Heritage Books.

[114] Llovio, L. (2014, January 17). Slavery figured in the history of Midlothian Mines. *The Richmond Dispatch*.

Figure 3.11 Black Heath commemoration. *(Image credit: Mary Finley-Brook, Author.)*

indentured workers (who were "rented" after the end of slavery) instead of the suffering of the injured and dead. Today's historical markers at the site of the tragedy focus on rescue missions.[115] The root causes of this suffering and the fates of the people most harmed are absent. African American **erasure**, a form of silencing, occurs as information is excised from dominant narratives and sometimes lost altogether.

---

[115] The Historical Marker Database. (2016). Mid-Lothian Mines and Railroads.

Figure 3.12 Midlothian methane massacre. *(Image credit: VCU Libraries Commons.)*

An accurate rendition would point to the risks of open-flame lighting in mines, well known at the time. The only hint on the commemorative signage at the Midlothian Mines site is the word "violent" describing the methane explosion: "a violent methane explosion trapped 32 in the Grove Shaft." Deadly mining operations continued after the Midlothian disaster—owners continued to send crews of workers with flammable head-lamps deep underground, giving evidence that workers were viewed as expendable.[116]

Today, the property where the Midlothian Grove Shaft mine was located has been donated to the county and serves as a park. Historical markers celebrate the importance of coal and the related railroad infrastructure. This presents an embellished history and ignores that these early coal operations were dependent on Black labor, whether free or enslaved, and that mine owners partnered with Tredegar Iron Mills in Richmond to make weapons for Confederate troops to defend slavery (Chapter 2).[117] The historical signage here lacks mention of

[116] Zallen, J. B. (2014). *American lucifers: Makers and masters of the means of light, 1750–1900.* Doctoral dissertation, Harvard University.

[117] Lewis, R. L. (2009). *Black coal miners in America: Race, class and community conflict (1780–1980).* University Press of Kentucky.

Midlothian coal as the basis for building Confederate cannons. Ignoring ecological harms and erasing the histories of marginalized ethnic and racial groups (i.e., "**whitewashing**"), as exemplified in these historical markers, creates a limited understanding of the past (Fig. 3.13). The continued normalization of past and present racial violence in the region overshadows emerging local narratives about Black coal miners and their descendents.

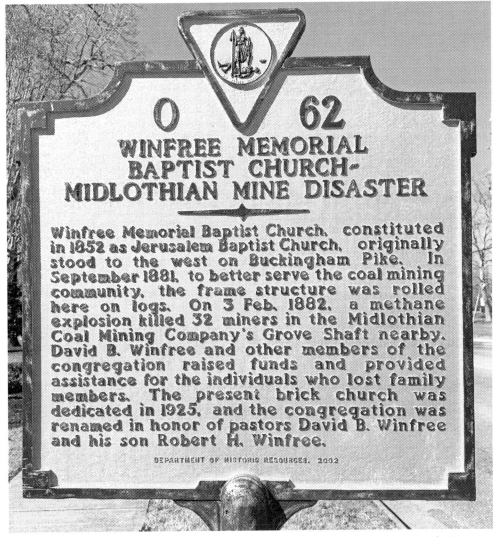

Figure 3.13 Whitewashed commemoration. *(Image credit: Mary Finley-Brook, Author.)*

Memorializing Blair Mountain

The armed uprising at Blair Mountain in 1921 was America's greatest labor uprising and the largest armed insurrection since the Civil War. The battle was part of a wave of revolutions and worker uprisings in the wake of World War I (i.e., in Russia, Hungary, Ireland, Mexico).

> **Coal Mine Wars were the most potent challenge to the industrial power structure in US history.**

Today, the struggle continues following a decade of legal actions between 2009 and 2018 that involved listing and delisting the Blair Mountain site in official historical designations.[118] Tensions resurfaced in the context of expanding mountaintop coal mining as companies sought to mine Blair Mountain yet again, potentially desecrating one of the foremost historical sites of the US labor movement. Large, absentee conglomerates own most of the battlefield's nearly 1700 acres on which they have obtained extraction permits. In 2018, an industry lawsuit sought to destroy and erase the site altogether.[119]

The Mine Wars Museum in Matawan, West Virginia, spotlights buried histories of the conflict. Relatives of miners suggest the Blair Mountain story has never been accurately told.

> **This is the first time that our people are in charge of our own history.**[120]

In 2021, The Battle of Blair Mountain Centennial events sought to amend the record with participatory events, retracing history, and retelling stories previously lost or downplayed.[121]

---

[118] Miskin, K. (2018). Blair mountain battlefield back on national register of historic places. *West Virginia Gazette*.

[119] Soodlater, R. (2018, January 31). In the battle for Blair Mountain, coal is threatening to bury labor history. *The Progressive*.

[120] West Virginia Mine Wars Museum (n.d.). Welcome.

[121] Keeney, C.B. (2021). *The road to Blair mountain: Saving a mine wars battlefield from king coal*. West Virginia University Press.

Revenue from tourism now surpasses that from coal in various former mining areas. Tourism could help pay for clean-up from prior exploitation. For example, the annual West Virginia Big Coal River "Tour de Coal" for kayakers raises river restoration funds.[122] In such commemorative events, a fundamental goal should be to provide accurate education about mining. However, coal tourism is often lumped in with other forms of recreational tourism and must compete to attract visitors.[123] Alternatively, sustainable tourism operators may support local protests.[124]

Try this

How might educational tourism help avoid future harms? Who should make decisions about memorialization at contested sites like Blair Mountain (i.e., local residents, descendants of people killed, coal companies, historians, etc.)?

Spatial and temporal synopsis

Space

(1) *Coal's phaseout and the distribution of coal's costs and benefits is spatially uneven, with harms like black lung borne disproportionately by low-income communities.*

(2) *Black lung rates vary depending on the geological area. Disease and fatalities could have been reduced if known prevention strategies were adopted in these regions.*

Time

(1) *Black lung saw a resurgence because people stopped paying attention over time. Coal tourism could teach us about past mistakes and injustices to prevent harms like this.*

(2) *Black lung progresses in a person over time, even after exposure ends.*

(3) *Industry lobbying and regulatory capture have allowed gaps between the start of disease and when miners receive necessary benefits, while company bankruptcies leave expensive remediation to taxpayers.*

[122] Coal River Group. (n.d.). Tour de Coal.

[123] Armis, R., & Kanegae, H. (2021). Regional competitiveness of a post-mining city in tourism: Ombilin coal mining heritage of Sawahlunto, Indonesia. *Regional Science Policy & Practice*, *13*(6), 1888–1910.

[124] Hales, R., & Larkin, I. (2018). Successful action in the public sphere: the case of a sustainable tourism-led community protest against coal scam gas mining in Australia. *Journal of Sustainable Tourism*, *26*(6), 927–941.

Summary

Government agencies step lightly around the powerful coal industry, often generating a national energy culture that lacks accountability. This has resulted in a preventable yet un-addressed prevalence of black lung, a long-ignored public health emergency rooted in regulatory violence based on historical unevenness. Black lung exposes coal's slow physical violence as unfortunate workers become victims to a socially created disease—a clear example of necropolitics. Resurgence spans globally and is likely to be underreported. As lessons from black lung escape media attention, we risk repeating errors with other energy types. Accurate commemoration of coal country is important to prevent mistakes.

Vocabulary

1. bankruptcy
2. black lung
3. bureaucratic violence
4. commemoration
5. derivative exploitation
6. do-it-yourself (DIY) publishing
7. dust sampling fraud
8. erasure
9. externalized costs
10. fly in, fly out worker (FIFO)
11. grandfathering
12. health disparity
13. lobbying
14. loophole
15. regulatory capture
16. regulatory rollback
17. regulatory violence
18. subsidies
19. tax avoidance
20. union
21. whitewashing

Recommended

Hamby, C. (2020). *Soul full of coal dust: A fight for breath and justice in Appalachia*. Little, Brown & Co.
Smith, B. E. (2020). *Digging our own graves: Coal miners and the struggle over black lung disease*. Haymarket Books.

CHAPTER 4

King Coal's ashes

Contents

Climate Crisis, Energy Violence
ISBN 978-0-12-819501-7,
https://doi.org/10.1016/B978-0-12-819501-7.00010-2

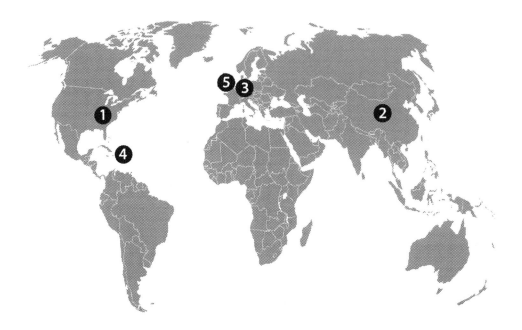

Locations: 1 - Appalachian Region - United States
2 - China
3 - Germany
4 - Puerto Rico
5 - United Kingdom

Themes: Legal violence; Mobility; Opposition

Subthemes: Fugitive emissions; Internal colonization

Power plants are lifeblood and bane.

Is King Coal finally dethroned?

The novel *King Coal* by Upton Sinclair describes the unchallenged power of coal companies more than a century ago.[1] Damages from coal combustion have been evident for centuries. The earliest environmental protections in Europe were later rolled back—a common pattern seen in the energy sector. In 1306, nobles across England traveled to London to participate in Parliament and were struck by the noxious smell from widespread burning of a sooty black rock. After public demonstrations against stinky coal, the king banned its use.[2] Nonetheless, by the late 1500s, with population growth and forest depletion threatening supply of firewood, coal gained traction anew. With time, London then developed some of the most contaminated urban air recorded,[3] demonstrated in episodes like the great smog of 1952.[4] With increased production and use, fatalities likewise trended upwards. In 1966, an infrastructure failure caused slurry to cascade downhill killing 144 people, mostly children in school.[5] An official inquiry following this tragedy, known as the Aberfan Disaster, concluded that the catastrophe was avoidable.

Marred by deadly air pollution over generations, British politicians finally pushed for coal's termination, resulting in an 84% decrease in coal combustion across the UK from 2014 to 2019.[6] The UK's shift is noteworthy given coal's long-standing sway. In the 1950s, formation of the European Coal and Steel Community, the precursor to the European Union (EU), was driven by industry to advance coal and steel growth across borders.[7] Coal fed the development of railroads and industrialization generally—core antecedents for European regionalization. It was a dramatic reversal when the EU moved away from coal, even subsidizing renewable energy transition in recalcitrant locations.[8]

---

[1] Sinclair, U. (1917). *King Coal: A novel.* Macmillans.

[2] Freese, B. (2016). *Coal: A human history.* Basic Books.

[3] Brimblecombe, P. (1977). London air pollution, 1500–1900. *Atmospheric Environment (1967), 11*(12), 1157–1162.

[4] Polivka, B. J. (2018). The great London smog of 1952. *AJN The American Journal of Nursing, 118*(4), 57–61.

[5] Blakemore, E. (2020, December 14). How the 1966 Aberfan Mine Disaster became Elizabeth II's biggest regret. History.com.

[6] Mathis, C. F. (2018). King Coal rules: Accepting or refusing coal dependency in Victorian Britain. *Revue Française de Civilisation Britannique. French Journal of British Studies, 23*(XXIII-3).

[7] Mason, H. L. (2013). *The European coal and steel community: experiment in supranationalism.* Springer; Bebr, G. (1953). The European Coal and Steel Community: A political and legal innovation. *Yale LJ, 63*, 1.

[8] World Bank. (2021). World Bank and European Commission to support Poland to transition out of coal.

Globally, coal use decline began around 2013, even as waves of resurgence continue regionally.[9] World leaders committed to coal's phase out found reason to backtrack as Russia invaded Ukraine in 2022. Yet the overwhelming cost in human health is clear.[10] Data shows emissions from coal-powered plants cause at least 9,500—12,100 excess deaths each year within the EU's former 28 member states (pre-Brexit).[11] These deaths are not allocated evenly, with 1,800—2,260 in Germany, 1,270—1,670 in the UK, 1,470—1,840 in Poland, and 2,800—3,600 in Romania, Bulgaria and Greece. Emissions from Europe's coal plants contribute to excess mortality outside the EU as well.

As a matter of course, the coal industry accepts harm: actions to reduce public health costs or safeguard the environment are perceived as unnecessary impediments to profit.[12] One such example of coal entrenchment was Poland, where leaders continually undermined impending change.[13] Poland sought to delay Europe's transition away from the sector and vetoed EU climate policies to protect its domestic coal industry.[14] The power bloc fighting transition in Poland included coal unions: in 2020, Polish unions finally agreed to coal phaseout by the late year of 2049.

Coal remains deeply embedded in political institutions and coal firms seek to maintain relevance. Fig. 4.1 depicts early twentieth century fuel switching pressures from coal to oil. Today, the energy transition has become a much more complicated process with a proliferation of greener fuel types and technologies. Markets have begun shifting away from coal for electricity, although coal for steel and coal ash for reuse remain strong. Countries with coal but no oil or fossil gas have pushed technologies for hybrid or alternative products, such as China's coal-based synthetic natural gas (SNG) and South Africa's coal gasification to produce syngas. Deep structural transformation is necessary to address power inequalities and exploitation built into King Coal.

[9] Marques, L. (2020). The regression to Coal. In *Capitalism and environmental collapse* (pp. 157—167). Springer.

[10] Kravchenko, J., & Ruhl, L. S. (2021). Coal combustion residuals and health. In *Practical applications of medical geology* (pp. 429—474). Springer.

[11] Kushta, J., Paisi, N., Van Der Gon, H. D., & Lelieveld, J. (2021). Disease burden and excess mortality from coal-fired power plant emissions in Europe. *Environmental Research Letters, 16*(4), 045010.

[12] Seow, V. (2022). *Carbon technocracy: Energy regimes in modern East Asia.* University of Chicago Press.

[13] Vasev, N. (2017). Governing energy while neglecting health—The case of Poland. *Health Policy, 121*(11), 1147—1153.

[14] Brauers, H., & Oei, P. Y. (2020). The political economy of coal in Poland: Drivers and barriers for a shift away from fossil fuels. *Energy Policy, 144*, 111621.

OLD KING COAL'S CROWN IN DANGER.

Figure 4.1 King Coal. *(Pughe, J. S. , Artist. Old king coal's crown in danger / J.S. Pughe. , 1902. N.Y.: J. Ottmann Lith. Co., Puck Bldg. Photograph. https://www.loc.gov/item/2010652154/.)*

Figure 4.2 Ende Gelände. *(Image credit: Ende Gelände. Accessed via Wikimedia Commons.)*

Coal has lost its SLO (Chapter 2), as a majority around the world now support coal's reduction.[15] Instead of merely asking for change (i.e., the majority of petitions, rallies, and marches), organizers use disruptive protests to draw attention and leverage power. For example, Ende Gelände (German for "here and no further") protesters target large-scale coal operators via direct actions, such as mass civil disobedience to pause operations (Fig. 4.2).[16]

---

[15] Roberts, D. (2017, November 21). New global survey reveals that everyone loves green energy—Especially the Chinese. And everyone hates coal. Vox.

[16] Sander, H. (2017). Ende Gelände: Anti-Kohle-Proteste in Deutschland. Forschungsjournal Soziale Bewegungen, 30(1), 26—36; Buckland, K. (2017). Ende Gelände: Desobediencia en el Antropoceno. *Ecología Política, 53,* 94—98.

Ende Gelände emerged from anti-coal and anti-nuclear movements, the degrowth movement, climate camps, forest occupations, and other grassroots initiatives.[17] Ende Gelände uses large-scale, high-density tactics to increase pressure, to create a level of safety when facing police, and deter arrests due to the high number of participants.

As seen throughout history, intensification of public and private surveillance and control tactics tried to limit protesters' influence. States criminalize environmental protest by creating harsher sentences and bigger fines. Some governments repress key activists, who report restrictions to travel or communication.[18] Meanwhile media coverage at outlets like *Fox News* (US)[19] and *Daily Mail* (Australia) tends toward character assassination.[20] In spite of challenges, climate justice movements are emerging around the world in response to the lack of action from industry-captured politicians. People feel the need to engage in civil disobedience (Chapter 1), a form of nonviolent resistance.[21] Actions are versatile and imaginative, similar to the Suffrage, labor, temperance, and anti-slavery movements. A century ago, the Suffrage movement used public actions, street speaking, demonstrations, and hunger strikes. Civil rights activists relied on direct action to create necessary change. A modern movement to end coal similarly defends civil and human rights.

Energy violence

The majority of coal ash waste is fly ash, a light material that is highly mobile when airborne. Coal companies commonly profit off reuse of their hazardous waste laden with arsenic, chromium, cadmium, mercury, radioactive isotopes, and other dangerous pollutants[22]—a form of **crisis capitalism** wherein exploitative investments leverage disruptions and emergencies. Ash reuse allows for profitable sales by the coal industry after facing pressure to reduce large quantities of haphazardly stored coal waste—tons of ash were commonly held in large retaining ponds near thermal plants in prior decades.

---

[17] Temper, L. (2019). Radical climate politics: from Ogoniland to Ende Gelände. In *Routledge Handbook of Radical Politics* (pp. 97–106). Routledge; Swift, A. (2020, October 2). Ende Gelände 2020. *The Ecologist.*

[18] Human Rights Law Center (2022, June 23). Civil society groups warn against police overreach in NSW climate defenders raid.

[19] Haring, B. (2019, September 21). Fox news host Tucker Carlson on climate strikes: "Adults hoping to exploit children for political purposes." Deadline.

[20] Airs, K. (2022, June 28). Exclusive: How the secret and VERY glamorous past of Australia's most infamous climate pest explains why she LOVES being the center of attention—After 'paraysing Sydney by chaining herself to a car with a bike lock.' Daily Mail.

[21] Malm, A. (2021). *How to blow up a pipeline*. Verso Books.

[22] Zierold, K. M., Sears, C. G., Myers, J. V., Brock, G. N., Zhang, C. H., & Sears, L. (2022). Exposure to coal ash and depression in children aged 6–14 years old. *Environmental Research*, 114005.

Coal contamination **mobility**, or capacity for movement, occurs with wind and water. The movement of pollution from coal waste is documented through the traceable contaminants it contains. Lightweight ash transports easily into the homes and bodies of frontline communities. Pollutants also spread while coal ash is wet, causing slow violence through leaching and fast violence when dikes rupture and other structures fail. Coal ash leaks toxins to drinking water, surface water, and groundwater. A significant amount of coal ash remains wet in holding ponds that can leach. Ash ponds are often unlined and located near water bodies.[23] As facilities age, structures often cannot withstand extreme weather.[24] For example, a Tennessee Valley Authority (TVA) dam breach near Kingston in 2008 spread coal slurry across 300 acres.[25] The botched cleanup efforts lead to the deaths of dozens of workers after they were denied necessary protective gear. Their families are still fighting for compensation.

Children are especially vulnerable to the release of heavy metals from impounded coal ash. Compared with adults, children are more susceptible to the effects of pollution as they have higher rates of respiration, are located closer to the ground, and have increased hand-to-mouth behaviors.[26] Exposure to heavy metals like mercury from coal ash are of particular concern.[27] Children living in proximity to coal-fired power plants have poorer neurobehavioral outcomes compared to children not living in close proximity to power plants.[28] Some heavy metal exposure, as is the case with lead poisoning, can cause irreversible, permanent harm to babies and children.

Your household trash is handled more consistently.

Regulatory violence occurs when coal ash is exempted from being listed as hazardous despite its contaminants documented as harmful to human health. This was exemplified a decade ago by the US EPA after coal companies lobbied against the "hazardous" label so they would be able to widely market coal ash to deflect disposal costs.[29] Moving ash

23 Earthjustice. (2019, November 6). Mapping the coal ash contamination.
24 Harkness, J. S., Sulkin, B., & Vengosh, A. (2016). Evidence for coal ash ponds leaking in the southeastern United States. *Environmental Science & Technology, 50*(12), 6583–6592.
25 Bourne, Jr., J. K. (2019, February 19). Coal's other dark side: Toxic ash that poison water and people. National Geographic.
26 Zierold, K. M., Sears, C. G., Myers, J. V., Brock, G. N., Zhang, C. H., & Sears, L. (2022). Exposure to coal ash and depression in children aged 6–14 years old. *Environmental Research*, 114005.
27 Ku, P., Tsui, M. T. K., Liu, S., Corson, K. B., Williams, A. S., Monteverde, M. R., … & Rublee, P. A. (2021). Examination of mercury contamination from a recent coal ash spill into the Dan River, North Carolina, United States. *Ecotoxicology and Environmental Safety, 208*, 111469.
28 Sears, C. G., & Zierold, K. M. (2017). Health of children living near coal ash. *Global Pediatric Health, 4*, 2333794X17720330.
29 Morris, E. K. (2017). No one likes an ash hole: Advocating for a management scheme that prioritizes beneficial utilization of coal ash in the United States and Georgia through domestic and international comparisons. *Georgia Journal of International & Comparative Law, 46*, 789.

without warning labels adds to occupational risk, particularly if safety equipment may not be given to, or worn by, those directly handling ash materials.

No one likes an ash hole

Production of millions of tons of **coal combustion residuals** (CCRs), commonly known as coal ash, creates a global health crisis.[30] Coal ash is increasingly considered a resource for recovering rare earth elements.[31] CCR can remain loose (unencapsulated) or bind into products such as drywall, concrete, or bricks (encapsulated).

With coal ash reuse, toxicants, haze, dust, and aerosol pollution require attention. Metals and metalloids in ash can be many times more concentrated than in the parent coal.[32] In the US, China, India, and many other countries, coal ash is not legally considered a hazardous waste, thus there is limited regulation or transparency.[33] To reduce storage and generate additional profit, coal ash is sold to be used as a soil additive or in construction material, among a range of uses. Coal ash markets create incentives for the coal industry to continue the ecological and health risks of coal combustion, even as other electricity sources are less expensive and cleaner. Since the 1970s, approximately 1.5 billion tons of coal ash in the US have been put to **beneficial reuse**: state-sanctioned sale of industry by-products, in this case toxic. Based on industry bias, this is also called "beneficiation."[34] After lobbying by coal industry groups, coal ash is not listed as a hazardous waste by the EPA, further enacting harm on workers and the public. More recent legislation prioritizing local workers in ash pond clean up and closure seems positive,[35] but because ash is indeed toxic yet not labeled as hazardous, often workers use inadequate precautions for safe handling.

[30] Hall, K. (2019). Uncontainable threat: the nation's coal ash ponds. *Emory Law Journal, 69*(1); Munawer, M. E. (2018). Human health and environmental impacts of coal combustion and post-combustion wastes. *Journal of Sustainable Mining, 17*(2), 87—96.

[31] Wang, Z., Dai, S., Zou, J., French, D., & Graham, I. T. (2019). Rare earth elements and yttrium in coal ash from the Luzhou power plant in Sichuan, Southwest China: Concentration, characterization and optimized extraction. *International Journal of Coal Geology, 203,* 1—14.

[32] Verma, C., Madan, S., & Hussain, A. (2016). Heavy metal contamination of groundwater due to fly ash disposal of coal-fired thermal power plant, Parichha, Jhansi, India. *Cogent Engineering, 3*(1), 1179243.

[33] Kamal, N. M., Beddu, S., Syamsir, A., Mohammad, D., Itam, Z., Hamid, Z. A. A., & Manan, T. S. A. (2019). Immobilization of heavy metals for building materials in the construction industry—An overview. *Materials Today: Proceedings, 17,* 787—791.

[34] Han, J., Yu, D., Wang, Q., Yu, N., Wu, J., Liu, Y., ... & Pan, H. (2022). Beneficiation of coal ash from ash silos of six Chinese power plants and its risk assessment of hazardous elements for land application. *Process Safety and Environmental Protection, 160,* 641—649.

[35] Code of Virginia. § 10.1—1402.03. Closure of certain coal combustion residuals units.

As global coal combustion declines in aggregate, interest in harvesting ponded ash increases.[36] Profits should cover harms to impacted areas from storage ponds and from exposures during the harvest, transportation, and processing of ash. Whether ash is left in ponds, landfilled, or reused, global cases show toxic heavy metals persist. The need for improved monitoring and enforcement via federal policy only occurred after a series of preventable disasters. For over 40 years, coal ash in the US has had an unclear designation as a "special waste."[37] In the Solid Waste Disposal Act Amendments of 1980, US Congress exempted fossil fuel combustion waste from being listed explicitly as hazardous. One key argument was that it was produced and stored in large volumes, ironically underscoring yet normalizing coal's prodigious waste.

Spatial distribution

If not kept wet, coal ash tends to become airborne, spreading contamination. Dust from open sources is termed "fugitive" because it is not discharged in a confined flow. **Fugitive emissions** do not pass through vents, stacks, or other intentional openings.[38] With fugitive emissions, as with legacy pollution, it can be difficult to assign responsibility.[39] Crystalline silica exposure during handling and transport—since fly ash contains silica—can cause silicosis if inhaled (Chapter 3).

Coal ash contaminants—and their cumulative health impact on individuals and communities—escape state regulation as **non-point source pollution**, or pollution outside of a site regulated as a "point source." **Point source pollution** occurs from any single identifiable source from which pollutants are discharged, such as a pipe, ditch, or smokestack. Under US federal rules, coal ash is not regulated as a single source pollutant.[40] Because ash reuse is largely unregulated, there is nothing to prevent a "coal ash corridor" as was the case in Morrisville, North Carolina. Here ash from a Duke Energy steam plant was sold as material fill for embankments and dumped

---

[36] Innocenti, G., Benkeser, D. J., Dase, J. E., Wirth, X., Sievers, C., & Kurtis, K. E. (2021). Beneficiation of ponded coal ash through chemi-mechanical grinding. *Fuel*, 299, 120892.

[37] Environmental Protection Agency. (2022, June 22). Special wastes.

[38] Mueller, S. F., Mallard, J. W., Mao, Q., & Shaw, S. L. (2015). Emission factors for fugitive dust from bulldozers working on a coal pile. *Journal of the Air & Waste Management Association*, 65(1), 27–40.

[39] Madrigano, J., Osorio, J. C., Bautista, E., Chavez, R., Chaisson, C. F., Meza, E., ... & Chari, R. (2018). Fugitive chemicals and environmental justice: a model for environmental monitoring following climate-related disasters. *Environmental Justice*, 11(3), 95–100.

[40] Crowder, J. (2018). Notice to SCOTUS: Coal ash should be a point source discharge under the clean water act. *Vermont Journal of Environmental Law*, 19, 89.

throughout the local region. Many locations lacked linings to protect groundwater,[41] creating a thyroid cancer cluster impacting young people.[42]

Once ash is harvested and re-deposited, leaching rates depend on many factors, including the acidity or pH level of the substrate material and the size of particles. In the US, ash is added to a series of products without labels identifying its inclusion due to beneficial reuse promoted by the EPA.[43] This program bastardized the purpose behind industrial symbiosis: an ecological relationship where the waste products from one operation become a source material for another as part of a **circular economy**.[44] Circular economy is a model of production and consumption that involves sharing, reusing, refurbishing, and recycling existing materials and products as long as possible. It decouples economic activity from the consumption of finite resources but must avoid toxic inputs that will persist.

There are dozens of commercial ash marketing firms. This "green" solution is viewed favorably in the construction and engineering fields, often suggesting that products with fly ash are not only cheaper but structurally better or have improved performance when compared to virgin materials. Coal ash reuse in concrete is common internationally. Cement containing fly ash is generally cheaper, and replacing a portion of the concrete mix with fly ash is argued to increase strength and permeability.[45] Environmental controls for coal ash reuse in cement are simply nonexistent. China, the greatest producer of cement,[46] further adds wastes like steel slag in addition to coal ash.

While there is obvious value from funneling coal's toxic by-product materials into positive uses,[47] policy makers need to take care not to promote a damaging industry. For example, drywall from coal ash can introduce mercury into the environment without safety guidance to prevent circulation.[48] Fly ash is used as a replacement for natural

[41] Suggs, M. (2019, March 31). Mooresville's 'coal ash corridor' is largest concentration in the state. *Mooresville Tribune*.

[42] Ortiz, E. (2020, January 4). Teen's cancer uncovers a mystery in one North Carolina town: why here? NBC News.

[43] Environmental Protection Agency. (2021). Coal ash reuse.

[44] Chertow, M.R., & Lombardi, D.R. (2005). Quantifying economic and environmental benefits of co-located firms. *Environmental Science & Technology*, *39*(17), 6535–6541.

[45] Rafieizonooz, M., Khankhaje, E., & Rezania, S. (2022). Assessment of environmental and chemical properties of coal ashes including fly ash and bottom ash, and coal ash concrete. *Journal of Building Engineering*, *49*, 104040.

[46] Nidheesh, P. V., & Kumar, M. S. (2019). An overview of environmental sustainability in cement and steel production. *Journal of Cleaner Production*, *231*, 856–871.

[47] Das, S. K., Mishra, S., Das, D., Mustakim, S. M., Kaze, C. R., & Parhi, P. K. (2021). Characterization and utilization of coal ash for synthesis of building materials. In *Clean coal technologies* (pp. 487–509). Springer, Cham.

[48] Healthy Building Network (2020). Selecting the wrong drywall could introduce mercury into the environment.

gypsum in drywall, also referred to as sheetrock, a pervasive interior wall construction material.[49] Disposal of ash-laden drywall after its use, such as in construction and demolition debris, can contribute to high arsenic levels in landfills[50]—exposure to arsenic has been linked to various cancers, including bladder, kidney, liver, and prostate cancers. Demolition piles from extreme weather, like hurricanes, tornadoes, and floods, often include large quantities of drywall without special handling to avoid dangerous runoff or emissions.[51] Where facilities exist, drywall may be recycled,[52] extending coal ash risks into new cycles of use.

Another major coal ash use is as an additive for soil. Coal ash is argued to help nutrient supplementation since it contains macronutrients (K, Ca, Mg, S, P) and micronutrients (Fe, Cu, Zn, Mn, Ni, Mo, B) essential to plants. However, long-term applications could result in an undesirable imbalance of nutrient supply, heavy metals pollution, food toxicity, or other safety risks. Pollutants in ash could potentially harm soil microbes beneficial for farming.[53] Additional research in a range of soil types (i.e., acidic, basic) and conditions (i.e., wet, dry) would provide a clearer picture of repercussions.

An underlying goal of a circular economy is to decouple economic activity from the consumption of finite resources. The circular economy should support the transition to renewable energy, not only because of the benefits of resources that can be regenerated as opposed to those that are exhaustive, but to avoid the harms of extraction and toxin-laden reuse exemplified by coal (Chapters 2 and 3).

Temporal analysis

Chemicals from coal will persist for centuries. Examining the full **life cycle** of energy types—life cycle analysis (LCA), also termed "cradle-to-grave"—is a helpful tool to compare energy types by offering a more complete picture of both benefits and harms from inception through disposal. The lack of a holistic assessment of coal—including its waste **afterlife**—leads to underestimation of true costs. For example, the afterlife of coal is lengthy, with heavy metal pollution and climate change. These damages continue for decades after coal's brief use for heat and energy.

[49] Ndukwe, I., & Yuan, Q. (2016). Drywall (gyproc plasterboard) recycling and reuse as a compost-bulking agent in Canada and North America: A review. *Recycling, 1*(3), 311–320.

[50] Zhang, J., Kim, H., Dubey, B., & Townsend, T. (2017). Arsenic leaching and speciation in C&D debris landfills and the relationship with gypsum drywall content. *Waste Management, 59*, 324–329.

[51] Environmental Protection Agency. (2019). Planning for natural disaster debris.

[52] Environmental Protection Agency. (2014). Best management practices to prevent and control hydrogen sulfide and reduced sulfur compound emissions at landfills that dispose of gypsum drywall.

[53] Nayak, A. K., Raja, R., Rao, K. S., Shukla, A. K., Mohanty, S., Shahid, M., ... & Lal, B. (2015). Effect of fly ash application on soil microbial response and heavy metal accumulation in soil and rice plant. *Ecotoxicology and Environmental Safety, 114*, 257–262.

To shirk responsibility, coal companies exploit legal and jurisdictional constraints through legal loopholes. Impacted families go to court with damning evidence of death or illness linked to corporate dishonesty and negligence, yet they bounce from trial to trial without fair or just resolution. Corporate lawyers fine-tune legal ploys that permit firms to exploit regulatory weaknesses or gaps. Even clearly demonstrative legal cases, often involving whistleblowers with a strong evidentiary record of intentional misdeeds and fraud, drag on decade after decade. **Protracted lawfare** means to delay and pervert the legal process. US coal ash lawsuits have been drawn out and are difficult to win without a coherent regulatory framework at the federal level. However, beginning in 2018, coal-fired electric utilities were finally compelled to publicly report groundwater monitoring data for the first time following the passage of the federal Coal Combustion Residuals Rule.[54] Even under this rule, hundreds of sites have been shown to leach harmful toxins;[55] and more than 100 are out of compliance.[56]

Illustrative cases

Litigation has become necessary to force remediation and seek compensation for US families suffering from death and disease.[57] While billions of profits were amassed from US coal regions, it remains rare to see "clean closure" of storage coal ash sites to create jobs and protect communities.[58] Examples of risky coal ash mobility and reuse in the UK, India, and Australia are less studied, but existing evidence and concerns are presented below.

United States

Congress passed the Resource Conservation and Recovery Act (RCRA) in 1976 to set national goals for protecting human health and the environment from solid and hazardous waste disposal. In 1978, the EPA considered treating coal ash as hazardous waste under RCRA, but deferred judgment. When Congress finally acted in 1980, following aggressive lobbying from industry, it was to *exempt* CCRs from classification as hazardous waste.

---

[54] Environmental Protection Agency. (2022). Disposal of coal combustion residuals from electric utilities rulemakings.

[55] Earthjustice. (2019, November 6). Mapping the coal ash contamination.

[56] Earthjustice. (2022, June 14). 101 coal plants with toxic sludge threatening our drinking water.

[57] Knoblauch, J. A. (2021, August 10). Court victories signal hope for communities threatened by coal ash.

[58] Earthjustice. (2021). Cleaning up coal ash for good.

Coal utilities would remain free from regulations and costs associated with management of hazardous CCR until 2014, in spite of massive quantities of coal ash across hundreds of sites. After standards were put in place, its taken years to bring sites into compliance; the process is still incomplete as numerous sites stand uncompliant, loaded with cancer-causing chemicals.[59]

Under the Trump Administration (2017–21), the coal lobby pushed for the backsliding of federal standards to obtain more time to clean up dangerous coal ash ponds. While there has been a stronger role for the EPA under the Biden Administration since 2021, companies are not being held accountable to clean up as soon as possible.

The four cases below implicating the US exemplify how those with less political and economic power experience greater harm. In the case of the Kingston coal ash disaster, while reformed federal management of coal ash was a positive policy outcome, its huge cost was counted not only in dollars but in dozens of unnecessary fatalities.[60]

Deadly secrets in Kingston

The energy industry's ties to the military are deep and wide. As one example, Tennessee Valley Authority's (TVA) coal plant site in Kingston, Tennessee—west of Knoxville— became cross-contaminated with nuclear waste from the nearby Oak Ridge nuclear facility, a national security complex instrumental in building atomic bombs.[61] Early radiation "management" involved diluting low-radioactivity waste for release and storing higher concentration waste until natural decay would occur.[62] Leakage from buried and stored wastes meant cancer rates increased in proximate local populations.[63] Downstream from Oak Ridge in Kingston sits TVA's nine-unit, coal-fired power plant: site of one of the worst industrial spills in US history. In 2008, aging retaining walls of the coal plant's dikes ruptured, releasing huge amounts of toxic slurry (Fig. 4.3).

Following the spill, Greenpeace requested a criminal investigation, arguing the disaster was preventable. Warning signs prior to the breach demonstrated a need to act (Fig. 4.4).[64,65]

[59] Earthjustice. (2022, June 14). 101 coal plants with toxic sludge threatening our drinking water.

[60] Sturgis, S. (2022, September 12). EPA sued for failing to regulate hundreds of coal ash dumps. *The Louisiana Weekly*.

[61] Gaffney, A. (2020, December 15). A legacy of contamination. Grist.

[62] Browder, F.N. (1989). Radioactive waste management at Oak Ridge National Laboratory. Office of Science and Technical Information.

[63] Mangano, J. J. (1994). Cancer mortality near Oak Ridge, Tennessee. *International Journal of Health Services, 24*(3), 521–533.

[64] White, C. (2008, December 24). Ash spill: Two prior breaches at retention site. *Knoxville News Sentinel*.

[65] Ashtracker. (2019, July 19). Kingston fossil plant.

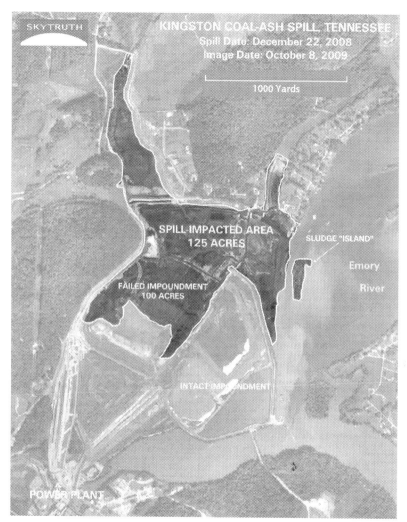

Figure 4.3 Kingston coal plant. *(Map image credit: Skytruth.)*

Concerns about structural stability at the site were documented by the 1980s.[64]
Notable breaches at the site in 2003 and 2006.[65]
Coal ash spills caused ecological harm like fish kills in other locations.

Figure 4.4 Warning signals.

TVA subcontracted the global Jacobs Engineering firm for cleanup management. By late 2009, cleanup workers were reporting illnesses following acute and prolonged exposure. Reports would later emerge showing worker deception, denial of protective gear, threatened firings, safety test tampering, and other mistreatment.[66]

Early on in the remediation process, all operations were suddenly halted. As a background fact, regulators allowed remediation to include ash mixing with surrounding areas as long as TVA promised to monitor all activity.[67] Reasons for the operation shutdown were not publicly available. A hidden history of nuclear ties did not surface until 2020: radioactive waste from the Oak Ridge nuclear facility was found to have commingled at the Kingston Coal Plant site via the shared Clinch River waterway, yet had been first unknown, then kept tightly secret in the cleanup process. With time, the confluence of nuclear waste and toxic coal ash contamination would publicly emerge. Between 2010 and 2017 half of the plant's 33 groundwater monitoring wells demonstrated pollution above federal advisory levels for cobalt, selenium, manganese, arsenic, sulfate, beryllium, lead, and molybdenum.[68] Wells with unsafe levels were located by water bodies, meaning contamination could migrate into streams or rivers.

More than 40 former workers have died of illnesses including brain cancer, lung cancer, and leukemia in the TVA plant debacle.[69] Hundreds more are now ill. In 2013, ex-workers and their families sued in District Court.[70] The lawsuit was dismissed and later overturned on appeal in the Sixth Circuit Court. In 2018, a federal jury ruled in favor of the workers. In 2019, there was initiation of Phase 2 of the trial, wherein individuals could seek compensation for loss and damages. These proceedings were held up by Jacobs Engineering as it sought immunity, which the judge then denied in 2022.[71] While TVA itself has escaped direct legal responsibility, in 2023 Jacobs Engineering reached an undisclosed settlement with 200 workers after the decade-long litigation.[72]

This case underscores the need for the EPA to re-examine its coal ash rules.[73] This was vividly demonstrated as workers were denied protective gear during intense exposures.

[66] Satterfield, J. (2019, April 4). TVA contract workers say they are being exposed to toxic coal ash and flue gas. *Knox News*.
[67] Environmental Protection Agency. (2016). Kingston coal ash release site.
[68] Ashtracker. (2019, July 19. Kingston fossil plant.
[69] Bourne, Jr., J. K. (2019, February 19). Coal's other dark side: toxic ash that can poison water and people. *National Geographic*.
[70] Satterfield, J. (2019, February 5). Kingston coal ash case: From spill to sickness to lawsuits. *Knoxville News-Sentinel*.
[71] Grzincic, B. (2022, May 19). TVA contractor can't ditch $3 bln suit by coal-ash cleanup workers—6th circ. Reuters.
[72] Loller, T. (2023, May 23). Contractor says it has settled lawsuit with sick and dying coal ash workers. Associated Press.
[73] Environmental Protection Agency. (2019). Coal ash rule.

Environmental racism influenced disposal of waste from the Kingston site in 2010. Forty thousand shipments of waste[74] were hauled by trucks and trains 300 miles to a landfill by the predominantly African American town of Uniontown, Alabama (Fig. 4.5).[75]

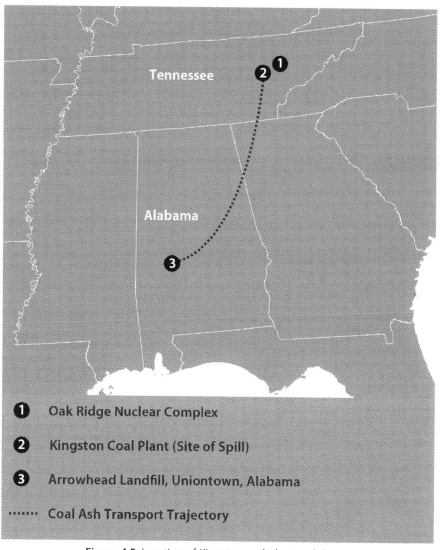

Figure 4.5 Location of Kingston coal plant and dump.

[74] Gaffney, A. (2020, December 15). A legacy of Contamination. Grist.
[75] Bienkowski, B. (2016, January 14). Spotlight hits coal ash impact on poor and minority communities. *Environmental Health News.*

TVA transported toxic materials to the Arrowhead Landfill with approval from the Alabama state government.[76] This example of disparate dumping in Black communities in the US South isn't isolated.[77] In fact, in 2021 national reports showed that Uniontown was still being flooded with out-of-state trash.[78]

Ash mobility in Chesapeake

Coal related activities in Chesapeake, Virginia, underscore a long-standing weakness in EPA protection measures: the difficulty to prove and attribute direct harm from mobile pollutants. In 2022, Earthjustice sued the EPA over uncontained coal ash in the Chesapeake case—one of nearly 300 sites in 38 states that constituted the legal action.[79]

Over 6 decades the Chesapeake Energy Center (1953–2014) generated power (Fig. 4.6), creating approximately 3.4 million tons of coal ash. Ash was stored in unlined pits at the site until 1984, when an impoundment was built by Dominion Energy (formerly Virginia Electric and Power Company—VEPCO).[80]

Dominion Energy sold 1.5 million tons of coal ash to the Battlefield Golf Club to develop a new 217-acre golf course.[81] As trucks transferred ash from 2002 to 2007, it blew into nearby residential areas, but no entity was found culpable for damages. Federal rules address point source pollution—yet ash in transport is not regulated as a single source, even in this case that featured a fixed, direct route from the point of combustion to the disposal area (Fig. 4.7).

Complicating matters, groundwater at, and near, the Battlefield Golf Course (Fig. 4.8) has become contaminated with toxic arsenic, beryllium, cobalt, lead, lithium, selenium, and radium.[82] Water bodies adjacent to the Battlefield Golf Course and Chesapeake Energy Center flow to the Chesapeake Bay. Eastern Virginia experiences hurricanes, heavy storms, and frequent flooding. Risks of spreading legacy contamination increase with **extreme weather**, meaning a time and location in which weather, climate, or environmental conditions—such as temperature, precipitation, drought, or

[76] Engelman-Lado, M., Bustos, C., Leslie-Bole, H., & Leung, P. (2020, April 13). Environmental Injustice in Uniontown, Alabama, decades after the Civil Rights Act of 1964: It's time for action. American Bar Association.
[77] Bullard, R. D. (2008). *Dumping in Dixie: Race, Class, and Environmental Quality*. Westview Press.
[78] Pillion, D. (2022, January 13). Millions of pounds of garbage from other states again flooding rural Alabama. Al.com.
[79] Earthjustice. (2022, August 25). EPA faces lawsuit for exempting half a billion tons of toxic coal ash from health protections.
[80] Zullo, R. (2017, March 23). Federal Judge finds arsenic from Dominion coal ash violated clean water act. *Richmond Times-Dispatch*.
[81] Applegate, A. (2014, December 20). Environmentalists threaten Chesapeake coal-ash suit. *The Virginian-Pilot*.
[82] Earthjustice. (2019). Mapping the coal ash contamination.

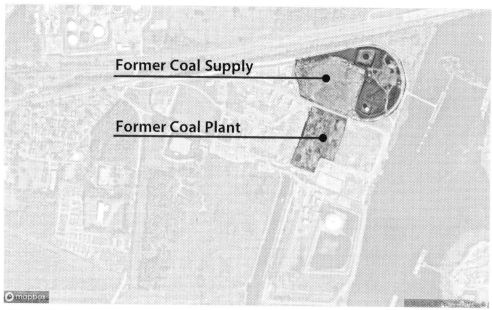

Figure 4.6 Chesapeake Energy Center. *(Map image credit: © Mapbox, © OpenStreetMap, © Maxar.)*

flooding—occurs as an outlier near or exceeding an end of a range of historical measurements.[83]

Dominion Energy knew fly ash's risks and kept them secret for 7 years.[84]

Dominion Energy escaped responsibility for the coal ash harms at Chesapeake Energy Center and the Battlefield Golf Course after withholding of information and deliberate misinformation. In 2009, 383 nearby residents sued the golf course owners, developers, contractors, and others for $1 billion in compensation for distress.[85] That same year, 62 nearby homeowners sued Dominion Energy for $1.25 billion for contaminating local groundwater.[86] In 2014, golf course owners sued VEPCO (Dominion) for $40 billion for knowingly lying about the safety of coal ash.[87] In 2015, a partner company in the

[83] Herring, D. (2020, October 29). What is an "extreme event"? Is there evidence global warming has caused or contributed to any particular extreme event? National Oceanic and Atmospheric Administration.

[84] McCabe, R. (2009, May 3). Dominion kept a 7-year secret on fly ash's environmental risks. *The Virginian-Pilot*.

[85] Rostami, M. (2012, February 22). Chesapeake fly suit against Dominion refiled. *The Virginian-Pilot*.

[86] Perks, R. (2009, August 27). Golf course or coal ash landfill? Natural Resources Defense Council.

[87] Himmel, C. (2014, December 11). Utility accused of lying to build toxic golf course. Courthouse News.

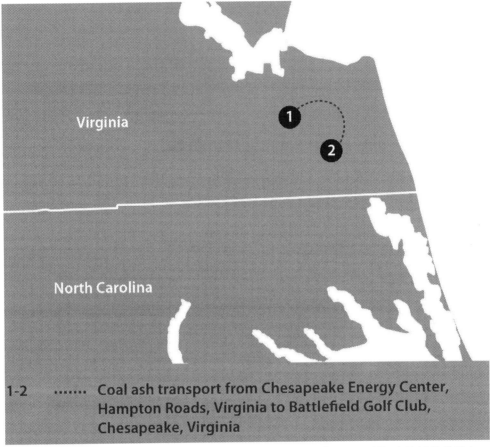

1-2 ······· **Coal ash transport from Chesapeake Energy Center, Hampton Roads, Virginia to Battlefield Golf Club, Chesapeake, Virginia**

Figure 4.7 Transport of coal ash between Chesapeake Energy Center and Battlefield Golf Club.

venture—CPM Virginia, LLC (CPM)—filed a complaint against VEPCO (Dominion) for breach of contract, fraud, misrepresentation, estoppel, nuisance, breach of warranties, negligence, and interference with prospective business advantage. CPM alleged that Dominion insisted the compounds were "safe as dirt" despite having "multiple reports indicating a high potential for environmental contamination due to inadequate reagent binders and the proximity of household water wells, drinking water aquifers and ecologically sensitive waterways."[88] There have been several other unsuccessful lawsuits, including from the owner of CPM, who developed cancer following ash exposure.

---

[88] Ibid.

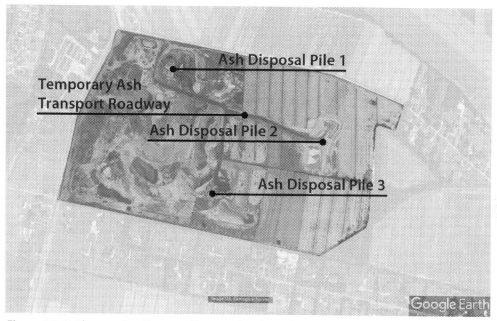

Figure 4.8 Battleground golf course. *(Sources: Google Earth, United States Geological Survey (USGS).)*

Multiple cases were tossed out, claiming the plaintiffs "lacked standing" because of narrow rules defining eligibility for damage claims.

While Chesapeake Energy Center stopped burning coal in 2014, then demolished in 2016, Virginia's Department of Environmental Quality (DEQ) continued to support Dominion's stance that the ash was safe.[89] DEQ continued to allow release of water from ash ponds into state waters until 2016 based on the notion that "dilution is a solution"—a common euphemism characterizing this dubious environmental management practice. Also in 2016, DEQ permitted a "mixing zone" located within Virginia's James River whereby toxic chemical levels were allowed to exceed safe limits as long as dilution occurs as an end result.[90]

In 2017, a federal judge ruled Dominion's coal ash violated the Clean Water Act.[91] Federal coal ash rules had finally been developed at this point, and the site was treated as

---

[89] McCabe, R. (2010, October 3). Former worker: Agency's OK to use fly ash is 'unconscionable.' *The Virginian-Pilot*.

[90] Staff. (2016, February 26). Virginia allows Dominion to exceed toxic limits for James River dumping. NBC 12.

[91] Zullo, R. (2017, March 23). Federal judge finds arsenic from Dominion coal ash violated clean water act. *Richmond Times-Dispatch*.

an inactive surface impoundment. In 2017 and 2018, 9,000 tons of ash were sold for reuse. However, persistent coal ash at the Chesapeake facility remains problematic. While the 2022 Earthjustice national lawsuit against the EPA was successful,[92] and Dominion Energy is now required to move an additional three million tons of coal ash at the Chesapeake facility, the proximate communities will be exposed further as Dominion secures its permit for removal and transport.[93] While cleanup action of Dominion's coal ash is indeed court-ordered, exactly which route, method and final destination remains unknown to a vulnerable public.

Notably, Chesapeake Power Station is located in a majority-minority and low-income environmental justice community that is heavily burdened with toxic exposures from military and industrial sites.[94] Groundwater monitoring reports near the storage ponds show arsenic, beryllium, cobalt, lead, lithium, molybdenum, selenium, and radium in excess of safe levels in existing laws.[95] A demonstrable case of toxic hotspot occurring across time (Chapter 1), the developer of the Keystone XL Pipeline, TC Energy (formerly Transcanada), sought approval in 2023 from the Federal Energy Regulatory Commission (FERC) for a pipeline that would require fossil gas infrastructure atop this contaminated coal site. This vulnerable community now faces additional harm after 60 years of coal contamination.

A sacrifice zone - Guayama, Puerto Rico (US)

Puerto Rican coal ash contamination is an ongoing public health crisis marked by the longstanding colonial relationship between the US mainland and the island of Puerto Rico.[96] The AES Corporation, a Fortune 500 company traded on the New York Stock Exchange (NYSE), constructed a large, $800 million coal plant in Guayama—a region to the south of Puerto Rico where significant Afro-Caribbean and Indigenous descendant populations reside.[97]

[92] Earthjustice. (2022, August 25). EPA faces lawsuit for exempting half a billion tons of toxic coal ash from health protections.

[93] Hafner, K (2023, June 17). After years of legal battles, Dominion energy to remove 3 million tons of coal ash at former Chesapeake plant.

[94] Environmental Protection Agency. (2022, August 3). Power plants and neighboring communities.

[95] Golder Associates. (2022, January 31). 2021 CCR Annual groundwater monitoring and corrective action report. Chesapeake Energy Center Bottom Ash Pond.

[96] Hill, B. E. (2020). Bad policy, disastrous consequences: Coal-fired power in Puerto Rico. *The Environmental Law Reporter, 50*, 10017.

[97] Santiago, R. & González Cuascut, Y. (2016, November 28). Coal ash protesters arrested in Puerto Rico. *Clean Energy*.

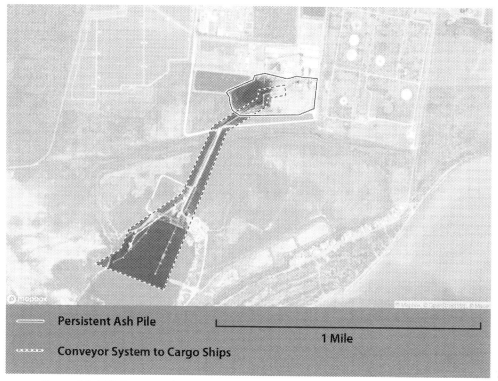

Figure 4.9 AES Guayama. *(Map image credit: © Mapbox, © OpenStreetMap, © Maxar.)*

Since its 2002 inception, the 454 MW plant in Guayama has burned an average rate of 250 tons of coal per hour, producing 300,000 tons of coal ash annually.[98] In the initial agreement, AES promised to ship the ash out of Puerto Rico, and the transboundary contamination that followed over the next decades crossed hundreds of miles.

In its agreement for the Guayama site (Fig. 4.9 above), AES committed to exporting ash if it could not find a "beneficial use" on the island of Puerto Rico. Accordingly, AES used cargo ships to transport ash to the Dominican Republic, offloading at Samaná Bay.[99] Local populations were told the ash was safe and could be used in construction or household needs. Residents used ash on walkways and floors, causing exposures resulting in

[98] Garrabrants, A., Kosson, D., DeLapp, R., & Kariher, P. (2012). Leaching behavior of "AGREMAX" collected from a coal-fired power plant in Puerto Rico. Environmental Protection Agency.

[99] Alonso, O. (2016, March 2). Something happened in Arroyo Barril. Centro de Periodismo Investigativo.

miscarriages and birth defects. Of the 1,290 pregnancies registered during the year 2005 in Samaná, more than 30% ended in miscarriage.[100] The government ultimately recommended women avoid pregnancy altogether. Coal ash dumping in the Dominican Republic ended within a year, yet the consequences lingered. In 2009, a lawsuit was filed on the part of several parents who lost children or fetuses. Before going to trial in 2016, AES settled. Dominican officials accepted $2 million dollars within a $6 million "side agreement" with AES that stipulated: (1) exoneration from future charges, (2) the Dominican state would be liable for claims against AES related to the coal ash, and (3) Dominican citizens would incur AES legal expenses with corporate lawyers paid $200–500 per hour.[101] This side agreement exemplifies legal violence.

Since its inception, the AES plant treated the ash as safe and distributed it throughout the south of the island via thousands of truck trips, often to unlined dumps. Over two million tons of ash were disposed of across 33 sites;[102] AES further sold ash for reuse in construction materials. An EPA study in 2012 documented leaching from disposal sites as coal ash was spread carelessly.[103] In some scenarios, coal ash was deposited within a few meters of public water wells, irrigation canals, streams, farms, and wetlands. Large quantities of ash were used "without appropriate engineering controls or where placement on the land has apparently far exceeded those necessary for the engineering use." Lightweight ash flying off of trucks was found to be yet another means of mobility and dispersal of toxic contaminants.

They promise jobs, and bring ashes.[104]

With growing pressure to get rid of its ash, AES marketed the ash waste product under the brand name Agremax. The product was a mix of ash with water used to create a concrete-like material marketed for applications in roads and construction, among other uses. Puerto Rico banned ash dumping in 2017 after dozens of individual local governments prohibited ash dumping in their communities. AES's ash mound grew (Fig. 4.10),

[100] Alonso, O. (2018, December 20). Arroyo Barril: coal ash and death remain 15 years later. Centro de Periodismo Investigativo.

[101] Feeley, J. & Chediak, M. (2016, April 4). Power Company AES settles claims that it killed or deformed babies with dumped coal ash. Bloomberg.

[102] Feliciano, I. & Green, Z. (2018, April 18). Coal ash raising concerns of public health risks in Puerto Rico. PBS Newshour.

[103] Garrabrants, A., Kosson, D., DeLapp, R., & Kariher, P. (2012). Leaching behavior of "AGREMAX" collected from a coal-fired power plant in Puerto Rico. Environmental Protection Agency.

[104] Alfonso, O. (2016, 8 de marzo). They promised jobs … and brought ashes. Centro de Periodismo Investigativo.

Figure 4.10 Coal ash pile in Guayama, Puerto Rico.[105] *(Image credit: Leandro Fabrizi Rios, Center for Investigative Journalism.)*

reaching the height of a 12-story building by the time Hurricane Maria hit the area in the fall of 2017.

AES' ash exposures have continued well beyond Puerto Rico and the Dominican Republic.[106] AES transported coal ash by all means available—truck, train, and barge. In 2019, AES sought to relieve their growing Puerto Rican coal ash problem by exporting ash even to the mainland US. Companies haul long distances to facilities that have low "tipping fees," meaning the amount to dump per ton. AES' dump site selection created public outrage over disparity of harms. For example, one site in Osceola, Florida, is marked by a large Puerto Rican community with an influx of people fleeing the primary impacts of Hurricane Maria.[107] These waste flows to the cheapest disposal site expressed complex patterns of differentiated dumping (Fig. 4.11), exemplified when corporations export waste to accessible markets with the lowest fees (i.e., dumping/tipping fees, import taxes).[108]

[105] Alfonso, O. (2018, March 14). Toxics from AES ash are contaminating groundwater. Centro de Periodismo Investigativo.

[106] Hill, B. E. (2020). Bad policy, disastrous consequences: Coal-fired power in Puerto Rico. *The Environmental Law Reporter, 50*, 10017.

[107] Calma, J. (2019, May 16). Puerto Rico got rid of its coal ash pits now the company responsible is moving them to Florida. Grist.

[108] Winslow, P. (2022, September 20). How Puerto Rico's banned coal ash winds up in rural Georgia. Energy News Network.

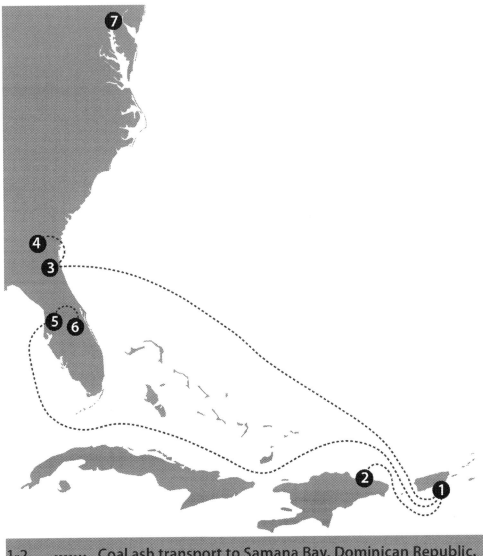

1-2   ······· Coal ash transport to Samana Bay, Dominican Republic.
1-3-4 ······· Coal ash transport to Jacksonville, Florida with landfill
                deposit at Chesser Island, Florida.
1-5-6 ······· Coal ash transport to Tampa, Florida with landfill
                deposit at Saint Cloud, Florida.
  7   EPA, Washington, D.C.

Figure 4.11 Ash transports from AES's Guayama coal plant.

In recent years, AES has quietly shipped waste to yet further locations with detrimental results. In 2021, a boat carrying AES's Agremax capsized off the US coast of Georgia. The cargo ship had minimum protection against damage or outflow of cargo. Ash was carried in a large open hull as permitted by the US Coast Guard specifically because coal ash isn't regulated as a hazardous substance.[109] AES did not clean up the spill and was not held accountable for the pollution dispersed.

Guayama = ground zero

AES's irresponsible treatment of ash has caused imminent and substantial endangerment to proximate human health, not to mention the health of those impacted at far distances from the plant itself. Ruth Santiago, a lawyer confronting the public health crisis from AES's coal ash for more than a decade,[110] testified in a 2021 congressional subcommittee on the high health costs to the local community surrounding the Guayama plant after 20 years of coal combustion. Santiago cited harms recorded in an epidemiological study by a team of researchers at University of Puerto Rico:[111] there are elevated rates of respiratory disease, cardiovascular disease, pediatric asthma, and spontaneous abortions, among other indicators of contamination.[112] Bioaccumulation where contamination concentrates in the food chain is a concern for local inhabitants given the use of wild sources of food from surrounding marine estuaries.[113]

A 2021 letter from the EPA documents a large number of serious regulatory concerns.[114] The same year, AES refused to testify to a House Committee on Natural Resources[115] yet sought a taxpayer bailout.[116] After decades of poisoning families, AES

[109] Boles, S. (2021, June 4). Ship carrying coal ash from Puerto Rico to Georgia spills 'very nasty stuff' off the coast of Jax. WJCT News.

[110] Lloréns, H. (2017, May 5). The making of a community activist. *Sapiens*.

[111] Santiago, R. (2021, June 30). Written testimony for hearing to examine the impacts and retirement of the coal-fired power plant in Guayama, Puerto Rico, Subcommittee on Oversight and Investigation, House Committee on Natural Resources, Congress of the United States.

[112] Santiago, R. (2012). Imminent and substantial endangerment to human health and the environment from use of coal ash as fill material at construction sites in Puerto Rico: A case study. *Procedia-Social and Behavioral Sciences*, *37*, 389—396.

[113] Martinez—Colon, M., Alegria, H., Huber, A., Kubra-Gul, H., & Kurt-Karakus, P. (2021). Bio-accumulation and biomagnification of potential toxic elements (PTEs): An Avicennia germinans—Uca rapax trophic transfer story from Jobos Bay, Puerto Rico. *Ecological Indicators*, *121*, 107038.

[114] Environmental Protection Agency (EPA). (2022, January 11). EPA letter to AES-Puerto Rico.

[115] Earthjustice. (2021, June 29). Press release: AES Puerto Rico and Luma energy refuse to testify at congressional hearing investigating AES coal plant.

[116] Earthjustice. (2022, January 31). Press release: Earthjustice response to AES Puerto Rico Seeking tax payer bailout.

argued operating costs were too high as a result of environmental regulation.[117] AES's toxic ash has not only spread to the Southern Coast Aquifer, the drinking water source for a large population on the island, but has been transported by truck, train, and barge, dispersing more risk.[118]

The status of Puerto Rico as an "unoccupied US territory" has led to disproportionate pollution on the island, as well as inadequate healthcare.[119] **Internal colonization**, the domination of one group by another within a country,[120] is demonstrated in Guayama through environmental injustice.[121] The impacted community demanded AES clean up the ash[122] and depart from Puerto Rico.[123] EPA's documentation and warning letters have proven insufficient to the scale of the problem—community members with the local environmental organization Comité Diálogo Ambiental have begun to map sites of ash contamination and they seek their own locally controlled renewable energy solutions (Conclusion). Yet, in a continuation of energy colonialism, AES and other US companies like New Fortress Energy (NFE) only push ever more fossil fuels transported from the US mainland (Chapter 9).

Try this

Research and publicize: *If you are from an impacted community in Puerto Rico, what would be your message to the world? How would you communicate it? What frontline communities near you experience energy violence? See Praxis 2 to publish an EJ Atlas entry, article or zine if you have reliable knowledge of injustice.*

---

[117] Alfonso, O. (2022, January 26). AES gives government of Puerto Rico an ultimatum for economic bailout. Centro del Periodismo Investigativo.

[118] Lloréns, H. (2020). Puerto Rico's coal-ash material publics and the summer 2019 Boricua uprising. Society + Space.

[119] Bonilla-Soto, L. A. (2019). United States congressmen support the legalization of environmental health injustice in Puerto Rico. International Journal of Public Health, 64(1), 59–66.

[120] Maldonado-Torres, N. (2016). Colonialism, neocolonial, internal colonialism, the postcolonial, coloniality, and decoloniality. In Critical terms in Caribbean and Latin American thought (pp. 67–78). Palgrave Macmillan.

[121] Lloréns, H. (2016). In Puerto Rico, environmental injustice and racism inflame protests over coal ash. The Conversation, 8.

[122] de Onís, C. M., Lloréns, H., Colón Pérez, M., & García-Lloréns, K. (2021). La justicia ambiental es para ti y para mí/Environmental Justice Is for You and Me.

[123] de Onís, C. M., Lloréns, H., & Santiago, R. (2020). ¡Ustedes tienen que limpiar las cenizas e irse de Puerto Rico para siempre!: la lucha por la justicia ambiental, climática y energética como trasfondo del verano de Revolución Boricua 2019. Editora Educación Emergente.

India

The Indian government speaks of a commitment to renewables while still expanding its domestic coal sector.[124] Particulate matter spreads to nearby communities from drilling and mining,[125] and there is dangerous air pollution from numerous coal power plants.[126]

India's coal has a high ash content.[127] Chemical washes have been developed to reduce the ash content of Indian coal,[128] but reduction below a certain level is difficult. India has a massive coal ash disposal stream, producing three times more coal ash than municipal waste annually. This waste stream flows beyond national borders. Ships carrying coal ash from India have capsized—with dozens of wrecks reported in the past years releasing hazardous waste.[129]

Indian ash storage ponds lack adequate design standards.[130] With more than 70 mishaps in the last decade,[131] the situation appears to be getting worse.[132] These toxic incidents at coal sites release contaminants with potential to harm all organs of the human body.[133] Heavy metals were found in storage ponds near the plants, which in turn are leaching into groundwater.[134]

[124] Roy, B., & Schaffartzik, A. (2021). Talk renewables, walk coal: The paradox of India's energy transition. *Ecological Economics, 180,* 106871.
[125] Sahu, S. P., & Patra, A. K. (2020). Development and assessment of multiple regression and neural network models for prediction of respirable PM in the vicinity of a surface coal mine in India. *Arabian Journal of Geosciences, 13*(17), 1–16.
[126] Kopas, J., York, E., Jin, X., Harish, S. P., Kennedy, R., Shen, S. V., & Urpelainen, J. (2020). Environmental justice in India: Incidence of Air Pollution from Coal-Fired Power Plants. *Ecological Economics, 176,* 106711.
[127] Morris, E. K. (2017). No one likes an ash hole: Advocating for a management scheme that prioritizes beneficial utilization of coal ash in the United States and Georgia through domestic and international comparisons. *Georgia Journal of International & Comparative Law, 46,* 789.
[128] Dash, P. S., Kumar, S. S., Banerjee, P. K., & Ganguly, S. (2013). Chemical leaching of high-ash Indian coals for production of low-ash clean coal. *Mineral Processing and Extractive Metallurgy Review, 34*(4), 223–239.
[129] Acharya, N. (2020, October 23). Fly ash in India: A free movement of toxicity to Bangladesh. *Mongabay.*
[130] Shah, D. (2021, September 15). Why India needs coal-ash pond design standards ASAP. The Wire.
[131] Shah, D., & Narayan, S. (2020). Coal ash in India—A compendium of disasters, environmental and health risks. Healthy Energy Initiative/Community Environmental Monitoring.
[132] Roy, E. (2022, July 23). 76 mishaps in 10 years. Coal ash, India's big under radar danger notes study. *The Indian Express.*
[133] Goswami, G., Purohit, B., & Arora, J. (2016). Fly ash waste management: A congenial approach. *International Journal of Science Technology and Management, 5*(11): 42–52.
[134] Verma, C., Madan, S., & Hussain, A. (2016). Heavy metal contamination of groundwater due to fly ash disposal of coal-fired thermal power plant, Parichha, Jhansi, India. *Cogent Engineering, 3*(1), 1179243.

India considered this hazardous waste more as a marketing opportunity than as a public health threat. The ash crisis created momentum to send the ash off-site for construction purposes.[135] The World Bank promoted reuse of coal ash in India to make bricks and reduce GHGs as a solution to utilize waste and save topsoil.[136] International agencies added incentives by buying carbon credits from producers of the bricks. Indian brick-making operations rely heavily on child labor, creating situations described outright as slavery due in large part to labor conditions that violate India's Minimum Wage Act (1948) and Bonded Labor Act (1976).

India's coal sector faces a crisis beyond ash alone.[137] Sporadic **opposition**—meaning organized resistance or dissent expressed in action or argument—is hampering coal operations across the country.[138] For example, workers with Coal India were forced to strike to pressure for employment conditions that were promised at the time of hiring.[139] Opposition from Indigenous Peoples have led to large protests.[140] More recently, protests have spread to new urban centers spurring anti-coal flash mobs and criticism of coal from You-Tubers.[141]

China

While each country has responded to the harms of coal ash in unique ways, a common thread is invariably to downplay dangerous risks to vulnerable workers and communities[142] in spite of generating concerns for human health.[143] The Chinese government began acknowledging heavy pollution in 2013, yet its core stance was that pollution is an

[135] Rathore, V. (2020). Toward a greener construction, one fly ash brick at a time.

[136] World Bank. (2012). Fly ash bricks reduce emissions.

[137] Pande Lavakare, J. (2020, June 30). Swelling Indian National Opposition as Modi plans to expand coal mining. *Health Policy Watch*.

[138] Global Energy Monitor. (2021) Opposition to coal in India.

[139] Staff. (2022, April 13). Coal mines struggle to expand as protests slow work in India. Mint.

[140] Staff. (2021, October 21). Coal mine plans spark huge protest from India's tribal people. Survival International.

[141] Staff. (2022, August 23). Flash mobs and YouTubers boost anti-coal mining protests in India. *Deccan Herald*.

[142] Yang, Y., Huang, Q., & Wang, Q. (2012). Ignoring emissions of Hg from coal ash and desulfurized gypsum will lead to ineffective mercury control in coal-fired power plants in China. *Environmental Science and Technology, 46*(6), 3058–3059.

[143] He, Y., Luo, Q., & Hu, H. (2012). Situation analysis and countermeasures of China's fly ash pollution prevention and control. *Procedia Environmental Sciences, 16*, 690–696.

inevitable by-product of rapid economic growth.[144] China's development has resulted in hundreds of "cancer villages."[145]

After awareness of the coal ash problem, actual harms were ignored.[146] Mercury poisoning is a long-term health problem well-understood around the world and yet mercury content of fly ash from coal-fired power plants was poorly evaluated in China until recently.[147] According to a study by scientists in China and the US, Chinese coal ash has been identified as potentially too radioactive for reuse in construction,[148] yet utilization is widespread.[149] Due to the large quantities of coal ash produced annually, costs of disposal and containment have been a motivating factor in reuse markets. Coal ash was in demand as a construction material to support the rapid growth of cities.

In upstream extraction, coal mining's contribution to ethnic conflict in Mongolia has been long documented.[150] In the region, Chinese coal workers and communities have been ignored and silenced.[151] China remains the leading producer and consumer of coal, and as such has global influence on coal use. To date, China continues to reject binding international commitments for the diminishment of coal, citing economic development as its primary priority.[152]

Australia

Boosted by state subsidies,[153] Australia became the world's largest coal exporter.[154] Exports of coal couple with ash reuse products designed for importers like China. Australian

[144] Hong, X. (2021). Changes in environmental profiles of cancer villages: A case study of Aizheng Cun in China. *The Professional Geographer, 73*(2), 200–212.

[145] Liu, L. (2010). Made in China: cancer villages. *Environment: Science and Policy for Sustainable Development, 52*(2), 8–21.

[146] Si. M. (2010, March 10). Coal ash cloud looms large over China. *Inside Climate News*; Gang, Z. (2010, September 16). The devastating effects of coal ash pollution in China. *The Guardian*.

[147] Chen, Q., Chen, L., Li, J., Guo, Y., Wang, Y., Wei, W., … & Yang, Y. (2022). Increasing mercury risk of fly ash generated from coal-fired power plants in China. *Journal of Hazardous Materials, 429*, 128296.

[148] Staff. (2017). Chinese coal ash is too radioactive for use. *Asian Scientist*.

[149] Luo, Y., Wu, Y., Ma, S., Zheng, S., Zhang, Y., & Chu, P. K. (2021). Utilization of coal fly ash in China: a mini-review on challenges and future directions. *Environmental Science and Pollution Research, 28*(15), 18727–18740.

[150] Liu, L., Liu, J., & Zhang, Z. (2014). Environmental justice and sustainability impact assessment: In search of solutions to ethnic conflicts caused by coal mining in Inner Mongolia, China. *Sustainability, 6*(12), 8756–8774.

[151] Audin, J. (2020, April 28). The coal transition in Datong: An ethnographic perspective. *Made In China Journal*.

[152] Associated Press. (2022, April 25). China promotes coal in setback for efforts to cut emissions. *NPR*.

[153] Slattery, C. H. (2019). 'Fossil Fueling the Apocalypse': Australian Coal Subsidies and the Agreement on Subsidies and Countervailing Measures. *World Trade Review, 18*(1), 109–132.

[154] Geoscience Australia. Government of Australia. (n.d.). Coal.

firms repackage coal ash for cement, as well as concrete, mine filler, and agricultural soil amendment.[155] Like ash reuse illustrated in Puerto Rico and other cases, Australian's domestic ash management is not immune from a toxic legacy failing vulnerable communities.[156]

Open pit mining is prevalent and the country has 4 of the top 10 biggest coal mines in the world.[157] Acute mining damages in Australia are severe. Land conflicts proliferate, and Aboriginal populations' traditional rights are repeatedly violated.[158] Internal colonization represents the continuation of economic and structural oppression and exemplifies territorial violence. Another attempted mitigating tactic of corporations is financial support for Aboriginal candidates, which is often perceived as "blood money."[159]

Opposition to coal harm in Australia is growing. A primary target is the foreign-owned Adani Group, launched domestically in 2014. This vast corporation headquartered in India known for its ecological damage continually faces crises of public perception.[160] The Australian campaign against Adani has been creative in its methods of protest against the huge Carmichael coal and rail project in Queensland. Savvy social media raises awareness—this growing anti-coal movement is not centralized and comprises thousands of individuals and community groups across Australia whose commitment and versatility have been difficult for Adani to counter.[161] Bold on-the-ground actions coupled with digital platforms innovatively and effectively extend coal protest (Fig. 4.12). Growing grassroots opposition against the Carmichael mine demonstrates a shifting tide toward fossil-free energy.

Global opposition creates a dark stain on the reputation of banks supporting the Carmichael project. In 2022, the State Bank of India committed "green" bonds to Carmichael, bringing additional condemnation for use of these "toxic bonds."[162] Some financiers attempt to weather public opposition to coal with forms of greenwashing about new technologies for coal.

[155] Park, J. H., Ji, S. W., Shin, H. Y., Jo, H., & Ahn, J. W. (2019). Recycling of coal ash and related environmental issues in Australia. *Resources Recycling, 28*(4), 15—22.
[156] Earthjustice. (2019). Unearthing Australia's toxic coal ash legacy.
[157] Murray, J. (2020, February 12). What are the five biggest coal mines in fossil-fuel reliant Australia? *NS Energy*.
[158] Davidson, H. (2019, August 4). Australia to be sued over mining projects 'unmerciful' destruction of Indigenous land. *The Guardian*.
[159] Giannacopoulus, M. (2019, February 15). Blood money: Why mining giants are backing an Indigenous Voice to Parliament. ABC.
[160] Casey, J. (2020, October 11). Adani launches new mining brand. *World Coal*.
[161] Hall, M. (2020, July 20). Anadi's Charmichael coal mine controversy explained. *Mining Technology*.
[162] Sum of Us. (2022, May 22). Bank bonds initiative: five key takeaways from the launch webinar. Banktrack.

Figure 4.12 Blockade Adani. *(Image credit: Julian Meehan. Accessed via Flickr.)*

Pressure to shift away from coal comes from investment risks. Market shifts away from fossil fuels create **stranded assets**, meaning assets that lose value before the end of their planned lifespan. Assets can become stranded due to climate policy and action (Chapter 9); they can also become stranded due to environmental factors like water scarcity. **Water-stranded assets** occur as financiers fail to consider ecological constraints or climate change, and thus site projects in unsuitable zones prone to water risk.[163] Today, the Carmichael mine project is understood as a stranded asset case study; if not by activist pressure in the short term, then aquifer depletion in the longer term.[164] Forty insurers are now among the more than 100 that have sworn to steer clear of the Adani mine or abstain from coal in general.[165] Australian coal companies are now desperate enough to try to self-insure through various schemes.[166] In 2018, an editorial in the *Financial Times*, an influential business paper, called steps to end coal insurance a "welcome and logical development."[167]

[163] Swiss Federal Office for the Environment. (2022). High and dry: How water issues are stranding assets.
[164] CDP (n.d.). Carmichael coal mine.
[165] Market Forces (2022). The Adani list.
[166] Bosshard, P. (2021). Self-insuring coal: A desperate ploy by an industry without a future. Insure Our Future.
[167] Insure Our Future. (2018, January 9). World's leading business paper calls for an end to coal insurance.

Even as financial shifts threaten coal, they are not occurring quickly enough to protect public health. With urgency and agency, blockades, occupation, and protest at strategic points (Fig. 4.13) halt or delay momentum or impede access, discussed further in Chapter 7. Coal blockages directly create delays and cost coal companies money, in turn reminding investors that coal's SOL is no longer valid, diminishing potential future returns.

Figure 4.13 Equipment blockade in Australia. *(Image credit: Blockade Australia, November 16, 2021. Accessed via Wikimedia Commons.)*

The organizing network Blockade Australia intensified its actions in 2022, as did efforts to repress participants, bringing condemnation of state security actions violating the civil rights of blockade participants and concern for whistleblowers.[168] Releasing an industry-captured state, like Australia, is still a challenge—yet it is important to reinvigorate democracy.[169] As governments continue to defend coal, they place their own citizens at further

---

[168] Civicus. (2022, August 29). Arrest of climate protesters, increasing anti-protest laws, and continued prosecution of whistleblowers in Australia.

[169] Ritter, D. (2018). *The Coal Truth: The fight to stop Adani, defeat the big polluters and reclaim our democracy.* UWA Publishing.

risk from climate and litigation. In 2022, the Australian government was found culpable in a lawsuit for violating the human rights of Indigenous Torres Strait Islanders by failing to adequately protect them from the severe impacts of climate change. The successful plaintiffs used the July 2022 UN recognition of the human right to a safe and healthy environment as part of the justification for their case.[170] As courts around the world continue to receive dozens of climate cases charging states and corporations for causing expensive harm, the economics of climate loss and damage will accelerate the long overdue end of coal use for electricity, making way for less expensive and cleaner options to flourish.

Try this

Engage and envision: Take a role as a stakeholder in the Australian Adani coal dispute:[171]
- *Mine owners, operators*
- *Native groups and allied land right advocates*
- *Workers and job seekers*
- *Government (national, local)*
- *Environmental and activist organizations*
- *UN and multilateral organizations*
- *Global banks and investors*

What is the position of the group you represent? How and why do these stakeholders' perspectives differ? What could be common ground (i.e., shared goals) between these groups?

Spatial and temporal synopsis

Space

(1) *Coal ash mobility causes fugitive pollution, and ash management results are uneven, differentiated and unfair.*

(2) *Energy colonialism, including racialized internal colonization, concentrates harm and risk, forming hazardscapes in poor areas.*

(3) *The extended and mobile harm from coal ash exemplifies the need for full life cycle analysis of energy options, extending in space as well as time.*

Time

(1) *Companies use protracted lawfare to shirk responsibility for coal waste's harms.*

(2) *The costs of coal ash clean up and public health impacts continue for years and decades following the end of coal production and use.*

---

[170] Surma, K. (2022, October 1). In pivotal climate case, UN panel says Australia violated islanders human rights. Inside Climate News.

[171] Albeck-Ripka, L. (2020, June 11). Mining firm plans to destroy Indigenous Australian sites, despite outcry. *New York Times*.

(3) *"Beneficial" reuse of coal ash is a form of disaster capitalism, whereby profits to coal firms prolong the harms of coal use and its waste.*

Summary

Typical accounting systems do not cover coal's life cycle, including the lingering impacts from coal ash. With full life cycle analysis, coal becomes much more costly—even unaffordable—once restitution and ecological safeguards are internalized into prices. Mismanagement of coal ash has intensified harms—disparities are widespread with unfair distribution of damages, especially in colonized territories like Puerto Rico. Because ash was not labeled as hazardous and there are large quantities requiring disposal, it was designated for beneficial reuse, entering regional and global supply chains without safeguards and further prolonging harm. Responding social movements take away coal's social license to operate, putting their bodies on the line to pressure industry-captured agencies and politicians to action. As financial risks grow, insurers have no choice but to move away from coal. If markets do not force closures first, companies will hold water-stranded assets.

Vocabulary

1. afterlife
2. beneficial reuse
3. circular economy
4. crisis capitalism
5. coal combustion residuals (CCR)
6. extreme weather
7. fugitive emissions
8. internal colonization
9. life cycle
10. mobility
11. non-point source pollution
12. opposition
13. point-source pollution
14. protracted lawfare
15. stranded assets
16. water-stranded assets
17. water stress

Recommended

de Onis, C. M. (2021). *Energy islands: Metaphors of power, extractivism, and justice in Puerto Rico*. University of California Press.
Lahiri-Dutt, K. (2016). *The coal nation: Histories, ecologies and politics of coal in India*. Routledge.
Ritter, D. (2018). *The Coal Truth: The fight to stop Adani, defeat the big polluters and reclaim our democracy*. UWA Publishing.
Seow, V. (2022). *Carbon technocracy: Energy Regimes in Modern East Asia*. University of Chicago Press.
Sinclair, U. (1917). *King Coal: A novel*. Macmillan.

CHAPTER 5

Drilled

Contents

Climate Crisis, Energy Violence
ISBN 978-0-12-819501-7,
https://doi.org/10.1016/B978-0-12-819501-7.00002-3

Locations: 1 - Alberta Province - Canada
2 - Saudi Arabia
3 - United States

Themes: Economic violence; Negligence; Spill

Subthemes: Captive ratepayers; Return on investment; Safety

The way to make money is when blood is running in the streets.[1]

Oil's promise … and peril

Hard to imagine today, oil was once an unwanted byproduct. However, it became an essential building block of imperial capital-intensive expansion.[2] As shown below, spills and explosions are often avoidable, yet firms prioritize shareholder payouts over safe, reliable and economically sound energy systems. Inadequate regulatory oversight means that risk has been normalized. Case in point: risks of explosion are uneven and could be

---

[1] New World Encyclopedia. (2022, August 3). John D. Rockefeller.
[2] Editors. (2020, August 24). First American Oil Well. *American Oil and Gas Historical Society*.

reduced by focusing on areas of greatest negligence. **Negligence**, covered in detail below, is the failure to use reasonable care; at large oil sites, it often results in damage, injury or death.

Oil has been a corrupting force in many countries. Politicians bend and rework rules for the oil and gas industry, and companies take advantage of preferential treatment to increase profits and consolidate power. The monopoly power of John D. Rockefeller's company Standard Oil in the US inspired the creation of **antitrust measures**—legislation preventing or controlling monopolies. Even as the Standard Oil monopoly was eventually dismantled, the sector continued to receive advantages allowing powerful families and companies to reconsolidate again. Antitrust laws form a necessary basis for energy democracy,[3] yet have proven short to fully curtail the oil industry's abuses of power.

A century ago, electric utilities were allowed to become monopolies through a process known as the **regulatory compact**: the idea was that these monopolistic investor-owned utilities (IOUs) with centralized ownership, generation, and distribution would build just one set of transmission and distribution infrastructure and thus protect the public interest. Yet, once utilities secured competitive advantage as natural monopolies, quality of service suffered and customers were forced to overpay for big uncompetitive projects with high startup costs.[4]

IOUs operate with a **guaranteed return on investments**, meaning that whatever happens to the market, companies will always get back what they invest in exchange for a reliable energy supply. Yet, the regulatory compact removes community choices about types of energy,[5] and who benefits or loses when risks threaten reliability. **Captive ratepayers**, customers without a choice in utility provider because of a monopoly in service provision and contracts, pay for expensive infrastructure through utility bills.

Return on investment (ROI) is used to evaluate the efficiency of an investment or to compare efficiencies across different investments. ROI examines net income and investment as a performance measure over a period of time. ROI has become a core metric for **market-based management**, a business ideology promoting continuous improvement of competitive position[6]: the driving philosophy of Koch Industries, a multinational corporation entrenched in fossil fuels.

The excessive power of oil companies and energy utilities invariably produces negative consequences for consumers. IOUs took advantage of the regulatory compact to "overbuild" expensive, profit-driven infrastructure, stealing funds from ratepayers and

[3] Farrell, J. (2021, August 16). The Role of Antitrust Law in Creating Energy Democracy. Clean Technica.

[4] Del Fiacco, J. (2020, July 9). Is energy still a 'natural monopoly'? Institute for Local Self-Reliance.

[5] Baker, S. (2021). *Revolutionary Power: An Activist's Guide to the Energy Transition*. Island Press.

[6] Koch, C.G. (2015). *Good Profit*. Crown Business.

taxpayers, meaning there is less finance available for other needs including education and health care. In light of the patterns of abuse of the regulatory compact, grassroots initiatives aim to distribute energy supply so that utilities are more affordable and more responsive to community needs. The oil industry has consistently sought to maintain monopoly control by obstructing distributed energy production that allows consumers to shape decisions.

Energy violence

Entitled through monopoly, structural violence is a foundational method of oil's powerful grasp on markets. Over time, Standard Oil gained complete control of US production. Maneuvering around state laws, Rockefeller's lawyers connived a form of partnership to completely centralize holdings in the Standard Oil Trust in 1882. More than 20 years later, Ida Tarbell, a leading muckraker in her day, issued the first significant public attack on Rockefeller in 1904,

They never played fair. That ruined their greatness.[7]

Oil markets have many externalized costs: negative social and ecological impacts generated by producers but not paid by them as damages remain outside the price of a good or service. Externalizing costs means companies have lower expenses and higher profits, yet society pays. For example, people pay more to clean up pollution and for healthcare as a result of sulfur oxides (SOx) emissions from oil combustion, even as SOx could be scrubbed if companies added pollution controls. Climate change and the loss of ecosystem services are other examples of externalities from oil's pricing structure containing harm for generations.[8] By accepting externalities as a matter of course, they are normalized and in turn reinforce sacrificial thinking.

A lack of power is common in hotspots of pollution—loci of oil's many externalities. As seen in Fig. 5.1,[9] toxic pollutants are numerous and severe in crude oil. The oil industry thrives on slow structural violence, always looking to avoid enforceable regulatory controls. Oil companies lobby to impede social and ecological safeguards to make profit.

[7] Tarbell, I. M. (2010). *The History of the Standard Oil Company*. Cosimo, Inc.

[8] Allred, B. W., Smith, W. K., Twidwell, D., Haggerty, J. H., Running, S. W., Naugle, D. E., & Fuhlendort, S. D. (2015). Ecosystem services lost to oil and gas in North America. *Science, 348*, 401–402.

[9] Mendoza-Cano, O., Trujillo, X., Huerta, M., Ríos-Silva, M., Lugo-Radillo, A., Bricio-Barrios, J. A., … & Murillo-Zamora, E. (2023). Assessing the relationship between energy-related methane emissions and the burden of cardiovascular diseases: a cross-sectional study of 73 countries. *Scientific Reports, 13*(1), 13515.

Category	Examples	Health Concerns
Volatile organic compounds (VOCs)	Benzene, toluene, ethylbenzene, and xylenes (BTEX)	Many VOCs have toxic, mutagenic and/or **carcinogenic** (meaning it has potential to cause cancer) properties.
Polycyclic aromatic hydrocarbons (PAHs)	Air pollution from petroleum when it is incompletely combusted	Many PAHs have toxic, mutagenic and/or carcinogenic properties.
Gases	methane and carbon dioxide	Cardiovascular disease[9]
Heavy metals	Hg, V, Cu, Pb, Zn, Cd, Fe, Mn, Ni	Biomagnification in the food chain Damage to organs, especially brain damage for babies and children

Figure 5.1 Common oil pollutants.

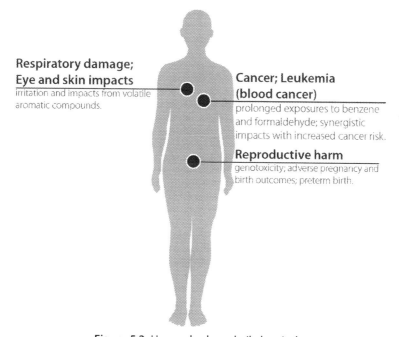

Figure 5.2 Human body and oil chemicals.

Bodily exposure pathways include inhalation, direct eye or skin contact, and ingestion or eating petroleum-contaminated food (Fig. 5.2 above). Long-term exposure to oil pollution has been documented to create a range of conditions reducing quality of life including dermatitis, visual and auditory hallucinations, and/or gastrointestinal disorders. Frontline oil production exposures in Ecuador raised the incidence of childhood leukemia and caused pregnant women to miscarry at increased rates (Chapter 6).

Category of Failure	Examples
Ignorance	• Inadequate training, knowledge and experience to provide safety guidance
Cost-cutting Safety Measures	• Maintenance or loss prevention programs are inadequate • Expenditures on safety measures are insufficient
Negligence	• Failure to conduct comprehensive and timely safety inspections • Unethical or unprofessional conduct (i.e., 'culture of deviance,' coverups) • Inadequate consideration of data, including extreme weather and climate
Upheaval and Unrest	• Political conflict or upheaval • Terrorist activities; sabotage • Labor unrest • Vandalism
Denialism	• Disputes scientific evidence of existential climate breakdown

Figure 5.3 Causes of oil sector safety failures.

Due to cost savings, there are a number of fast violence risks with oil. A key lesson from fatal pipeline accidents is that the public needs to be more aware of risks.[10] Negligence, as discussed in Fig. 5.3 above, can result in damage, injury, or death. Oil sector safety risks include: (1) the aging of pipeline infrastructure, (2) business mergers and splits causing an ever-shifting complex of parent companies and subsidiaries with a propensity for shirking responsibility, and (3) lax safety protocols and inadequate prevention mechanisms during recessions or periods of state dysfunction, like war.

Underlying risks intensify due to cost-saving measures, including automation. Industry-wide movement toward mechanization and remote management reduces the number of jobs and causes delays during emergencies as a result of non-local control staff. Fig. 5.4 demonstrates a system used to collect data from remote terminal units (RTUs) located across oil infrastructure. Federal safety review of Supervisory Control and Data Acquisition (SCADA) revealed inadequate controller training, controller fatigue, and ineffective leak detection systems.[11]

A rather rudimentary method oil companies use to reduce the loss of life from accidents on remotely managed pipelines is to educate the public on how to respond in the case of an emergency. Educating people with visible markers showing where risks are located (Fig. 5.5) may help minimize damage. However, infrastructure often blends

---

[10] Biezma, M. V., Andrés, M. A., Agudo, D., & Briz, E. (2020). Most fatal oil & gas pipeline accidents through history: A lessons learned approach. *Engineering failure analysis*, *110*, 104446.

[11] National Transportation Safety Board (2005). Supervisory Control and Data (SCADA) in Liquid Pipelines. Safety Study.

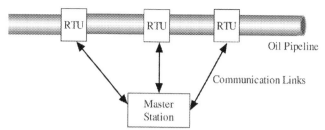

Figure 5.4 A pipeline's remote control system. *(Image credit: Improving Security of Oil Pipeline SCADA Systems Using Service-Oriented Architectures. Subramanian, Nary, Springer Nature, 2008.)*

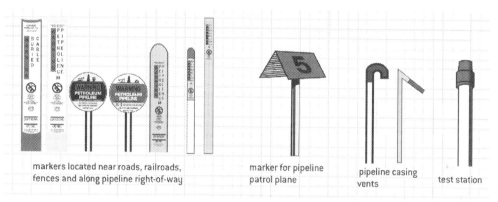

Figure 5.5 Pipeline warning markers. *(Image credit: Landowner's Guide to Pipelines, Third Edition, 2016. Pipeline Safety Trust.)*

into the background, meaning it becomes part of the landscape and is taken for granted. When recognizing a **spill**, an unpermitted release of polluting materials, the burden of responsibility is placed on the public. People who know how leaks and spills look and smell are expected to notify personnel to mitigate harm before a larger accident, fire, or explosion occurs (Fig. 5.6). Multi-million-dollar companies put the onus on the public to stay alert and to report spills. A 2020 Colonial Pipeline spill in North Carolina was found by teenagers recreating in a rural area. Often time is critical: the family of a Kentucky woman killed in a separate 2019 pipeline explosion is suing the pipeline company

Sensory Input	Potential Indicators of a Leak
sight	• liquid pools • discolored or abnormally dry soil/vegetation • continuous bubbling in wet or flooded areas • an oily sheen on water surfaces • vaporous fog • blowing dirt around a pipeline area • fire coming from the ground or appearing to burn aboveground • dead or discolored plants in an otherwise healthy area of vegetation or frozen ground during warm weather • exposed pipeline following a natural disaster such as flood or earthquake
sound	• volume can range from a quiet hissing to a loud roar depending on the size of the leak
smell	• a gaseous odor may accompany pipeline leaks

Figure 5.6 How to recognize a pipeline release.[12]

in part due to its failure to identify and correct hazardous conditions in time to prevent loss of life.[13]

Methods to remediate petroleum-contaminated soil, such as incineration, landfill, leaching, chemical oxidation, and microbiological treatment, vary in efficacy and cost. Contaminants in petroleum are hazardous. When released to soil, these pollutants can decrease microbes and deplete fertility for a long time.[14] Damages cluster along infrastructure, such as the Keystone XL Pipeline, a crude oil pipeline with a long series of spills.

TC Energy's oil spills

The owner of Keystone XL, Transcanada—now called TC Energy—is responsible for dozens of spills on this and other systems. Two Keystone XL leaks in the pipeline right-of-way (Fig. 5.7) released nearly 12,000 barrels of oil. Faulty materials and pipeline construction errors caused a 2017 accident in South Dakota, while defects in pipe manufacturing caused a 2019 North Dakota accident.[15]

---

[12] Pipeline and Hazardous Material Safety Administration. (2017). Pipeline leak recognition and what to do.

[13] Kobin, B. (2019, September 10). Family of women killed in Kentucky pipeline explosion sues pipeline owner. *Courier Journal*.

[14] Sui, X., Wang, X., Li, Y., & Ji, H. (2021). Remediation of Petroleum-Contaminated Soils with Microbial and Microbial Combined Methods: Advances, Mechanisms, and Challenges. *Sustainability*, 13(16), 9267.

[15] LeFebvre, B. (2021, August 23). Severe oil leaks worsened Keystone XL's spill record, GAO finds. Politico.

Figure 5.7 Two large Keystone pipeline spills. *(Image credit: Pipeline Safety: Information on Keystone Accidents and DOT Oversight. GAO-21-588. NTSB (top); PHMSA/TC Energy (bottom).)*

Large oil companies like TC Energy own thousands of miles of pipes with clear patterns of leakage, yet do minimal maintenance and upkeep. Shared concerns about water unites anti-pipeline resistance across space forming solidarity even if infrastructure is not physically linked.[16] Grassroots opposition movements have grown because lawsuits seeking reparation from oil damages are slow (Chapter 6) and rarely fully hold oil companies accountable for spill damages. In the end, all pipelines leak (Fig. 5.8).

Activists around the world have been inspired by a decade of Keystone XL protests (2011—2021).[17] More than 1,000 people were arrested at sit-ins protesting Keystone XL, launching anti-pipeline activism into the mainstream seeking to take away the oil industry's SOL. Keystone XL protesters fought for a decade to halt the project—grassroots organizing came up against the entrenched power of TransCanada (TC Energy), one of the biggest North American energy companies.

Keystone XL crystallized intense public opposition to tar sands oil and set the stage for opposition to the Dakota Access Pipeline (DAPL). In 2014, DAPL's route was moved

Figure 5.8 Tar sands blockade. *(Image credit: Keystone XL Protest - Tar Sands Blockade, Laura Borealis, 2012. Accessed via Flickr.)*

[16] Finley—Brook, M., Williams, T. L., Caron-Sheppard, J. A., & Jaromin, M. K. (2018). Critical energy justice in US natural gas infrastructuring. *Energy Research & Social Science, 41,* 176—190.

[17] Henn. J. (2021, December 31). Here is how we defeated the Keystone XL Pipeline. *Sierra Club; Magazine;* Malm, A. (2021). *How to blow up a pipeline.* Verso.

close to Indigenous territories intentionally to avoid a prior route considered too risky to the city of Bismark's water.[18] DAPL broke treaty rights and intentionally sacrificed Indigenous lands, people, and water. "Water is Life" became a unifying slogan for an international movement with Indigenous People, doctors, veterans, and hundreds of allies standing ground against DAPL construction at Standing Rock. Water Protectors are still in prison as a result of these protests (Chapter 6).

In 2022, an Illinois court vacated DAPL expansion, remanding state regulators for poorly evidenced finding of public need. Regulators were also found to have abused their discretion when finding "irrelevant" the objectors' evidence that the pipeline's operator had been fined for safety and environmental violations.[19] Some companies are repeat offenders, knowing they will go unpunished or that fines will be low enough to easily absorb.

Midstream transportation of oil and clusters of industrial development scattered along their routes imposes risk along pipelines. TC Energy's 2700-mile Keystone pipeline exemplifies injustice and structural violence at both ends. At the upstream end, tar sands oil is some of the most damaging energy, both ecologically and socially. At the downstream end is a toxic hotspot in Port Arthur, Texas, home of several major chemical plants and export terminals in the Gulf.

The Biden Administration canceled the Keystone XL permit in 2021. While this stands as a victory for the anti-pipeline movement, new fossil fuel buildout emerges under different guises. TC Energy is now emphasizing growth in liquefied natural gas (LNG) exports from western Canada and eastern US (Chapter 10).

Colonial spills

Entrenched power continues undaunted in existing systems as with the Koch Brothers, who financed efforts to promote oil sector and other deregulation for years, and are the largest investors in Colonial Pipeline.[20] The 2000-mile Colonial Pipeline transports petroleum liquids—a crack in a previously installed repair sleeve lead to the largest US terrestrial oil spill in August, 2020.[21] This was not Colonial's first big release. In 2016, Colonial reported a gasoline spill first recognized through satellite imagery showing discoloration from pollution near the pipeline, demonstrating an important role for remote sensing (Chapter 1) as long-distance detection.[22] This particular damage (Fig. 5.9) was detected first by private sector satellite imagery company Planet Labs PBC.

[18] Smithsonian Institute. (2018). Treaties still matter: the Dakota Access Pipeline.

[19] Sanicola, L. (2022, January 12). Illinois court vacates approval of Dakota Access pipeline capacity expansion. Reuters.

[20] Freitas, Jr., G., Blas, J., & Wethe, D. (2021, May 14). Colonial Pipeline has been a lucrative cash cow for many years. Bloomberg.

[21] Soraghan, M. (2022, July 25). N.C. pipeline caused largest U.S. gasoline spill, records say. E&E News.

[22] Planet. (2017, May 17). The anatomy of a pipeline accident: the Colonial Pipeline spill. Medium.

Impacted Pond upstream of Peel Creek, tributary of the Cahaba River.

Figure 5.9 Colonial Pipeline's "leaks". *(Image credit: Overflight Imagery, dated 09/18/2016, US Environmental Protection Agency (EPA).)*

As part of its operations, Colonial Pipeline provides dividends to the well-endowed Koch Industries.[23] Colonial Pipeline saw a net income of $421 million in 2019, a gain of nearly 32 cents for every dollar of revenue.[24] Yet Colonial demonstrates a pattern of releases from sites that are not sufficiently maintained or monitored.

Damages from its 2020 spill cost $105 million, as stated in Colonial's incident report. According to the EPA, Colonial should revamp monitoring and use a better leak

[23] Helman, C. (2016, November 01). Lessons from the Colonial outage, let the Koch Bros. build more. Forbes.

[24] Freitas, Jr., G., Blas, J., & Wethe, D. (2021, May 14). Colonial Pipeline has been a lucrative cash cow for many years. Bloomberg.

detection system.[25] As a major player in the oil sector, Colonial influences national standards despite their terrible record.[26] In its 2020 spill first found by teenagers riding off-road ATVs in Huntersville, North Carolina , the state had to resort to suing Colonial for nearly $5 million for not providing data, not adequately remediating, and not maintaining sampling for toxic substances including VOCs and lead. Part of the suit established a consent decree in which the company was ordered to maintain a series of data gathering, remediation work, and monitoring protocol. Fig. 5.10 shows the concentric pattern of 281 testing and removal wells in a 12-acre vicinity of the spill.

Figure 5.10 Colonial Pipeline leak assessment. *(Image credit: Workers examing testing and removal well map, Colonial Pipeline, Hutersville, 2021. David Boraks, WFAE - Charlotte's NPR News Source.)*

[25] Marusak, J. (2021, July 12). Colonial Pipeline could face enormous daily fine after massive North Carolina fuel leak, feds say. *The Times News*.

[26] Rees, J. H. (1988). Colonial Pipeline v. Brown: Georgia Constructs a State Constitutional Limit on Punitive Damages. *Am. J. Trial Advoc.*, *12*, 511; Smith Jr., A. B. (1993). Colonial pipeline Enoree River oil spill: A case history. *International Oil Spill Conference*. 1: 165−168.

Offshore oil spills

Offshore oil incidents can be identified with remote sensors. Fig. 5.11 shows synthetic aperture radar (SAR) technology over the coast of California. Remote sensing technologies are limited by their temporal and spatial resolutions—spill and leak detection is only as good as current technology and monitoring protocols allow.[27] Current regulatory practices reflect over-reliance on cheaper distance monitoring without addressing gaps in enforcement or piecemeal regulatory approaches.

Once offshore spills occur, remediation often involves chemical dispersants, creating new risks of economic disruption, biodiversity costs, and negative health impacts. Chemical composition of dispersants is considered a trade secret.[28] Studies of occupational hazards from cleanup are ongoing; some workers continue to experience negative health impacts, such as in the US Gulf after the *Deepwater Horizon* disaster.[29] Yet Gulf Coast spills are tiny compared to global hotspots like the Niger Delta or Ecuador's Oriente (Chapter 6).

Newport Beach, CA

— Oil Slick Extent - October 1st - 4th, 2021

Figure 5.11 Remote detection of offshore spill. *(Source: Copernicus Sentinel, Sentinel-1 SAR instrument, 2021.)*

[27] Taxin, A., & Melley, B. (2021, October 20). Coast Guard had earlier notice about California oil spill. Associated Press.
[28] Lichtveld, M., Sherchan, S., Gam, K. B., Kwok, R. K., Mundorf, C., Shankar, A., & Soares, L. (2016). The Deepwater Horizon oil spill through the lens of human health and the ecosystem. *Current Environmental Health Reports, 3*(4), 370–378.
[29] National Institute of Environmental Health Sciences. (n.d.). GuLF STUDY.

Deepwater disaster

British Petroleum's *Deepwater Horizon* disaster of 2010 began with an explosion and subsequent sinking of the massive rig killing 11 people, leading to the release of nearly five million barrels of oil into the Gulf of Mexico. Fig. 5.12 shows the relative size of recent US offshore spills, including the *Deepwater Horizon* spill.

Long-term consequences of spills are poorly understood.[30] GuLFSTUDY is one of the first health studies to gather detailed data over time: approximately 150,000 people from around the world participated in Deepwater cleanup. GuLFSTUDY's 25,000 participants completed questionnaires on stress, anxiety, depression, and post-traumatic stress disorder (PTSD).[31] Follow-up studies examined experiences during the mitigation,

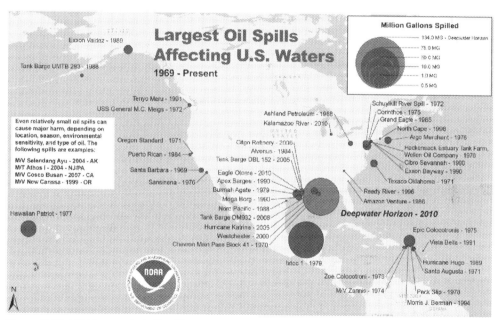

Figure 5.12 NOAA recorded oil spills, 1969–present. *(Source: US Department of Commerce, National Oceanic and Atmospheric Administration (NOAA).)*

[30] Croisant, S. A., Lin, Y. L., Shearer, J. J., Prochaska, J., Phillips–Savoy, A., Gee, J., ... & Elferink, C. (2017). The Gulf coast health alliance: health risks related to the Macondo spill (GC-HARMS) study: self-reported health effects. *International Journal of Environmental Research and Public Health*, 14(11), 1328.

[31] Hu, M. D., Lawrence, K. G., Gall, M., Emrich, C. T., Bodkin, M. R., Jackson, W. B., ... & Sandler, D. P. (2021). Natural hazards and mental health among US Gulf Coast residents. *Journal of Exposure Science & Environmental Epidemiology*, 1–10.

including smelling chemicals, contact with oil, heat stress, or dead animal recovery.[32] People who spend more time at a cleanup or those that have exposure to flaring or burning or direct contact with dispersants may experience increased harm.

Studies found stress emerges from disruption to local industries (e.g., fishing, oil and gas, and tourism), uncertainty about health effects, and previous experiences with disasters. Contact with oil and experiencing financial harm were two key factors that led to depression following the spill,[33] with long-term implications for vulnerable groups such as children.[34]

Lessons from the BP disaster highlight the need for better prevention and improved response. Ineffective use of Corexit, a type of chemical dispersant, is an example of "crisis capitalism" (Chapter 4). The EPA approved using dispersants Corexit 9527A and 9500A below the surface of the water for the first time following the *Deepwater Horizon* explosion (Box 5.1).[35]

Spatial distribution

In the 1890s, operators in the town of Los Angeles (then a population of 50,000) began drilling some of the most productive oil fields in history. By 1930, California was producing nearly a quarter of the world's oil and its population had grown significantly. In neighborhoods like Signal Hill (Fig. 5.13), residences became the understory to a skyline of oil wells.

There are 68 named oil fields within the 450-square mile area of the Los Angeles Basin. Oil rigs defined a grid-type layout of planned new development. For decades, families resided with oil. Ten of these fields were proven giants, containing more than a billion barrels of oil. There were no meaningful setbacks historically in Signal Hill, shown today in Fig. 5.14. **Setbacks** are the absolute minimum distance that must be maintained between any energy facility, such as a drilling or producing well, a pipeline or a gas plant, and a dwelling or public facility.

---

[32] Gam, K. B., Engel, L. S., Kwok, R. K., Curry, M. D., Stewart, P. A., Stenzel, M. R., ... & Sandler, D. P. (2018). Association between Deepwater Horizon oil spill response and cleanup work experiences and lung function. *Environment International, 121*, 695—702.

[33] Kwok, R. K., McGrath, J. A., Lowe, S. R., Engel, L. S., Jackson second, W. B., Curry, M. D., ... & Sandler, D. P. (2017). Mental health indicators associated with oil spill response and clean-up: cross-sectional analysis of the GuLF STUDY cohort. *The Lancet Public Health, 2*(12), e560—e567.

[34] Beedasy, J., Petkova, E. P., Lackner, S., & Sury, J. (2020). Gulf Coast parents speak: children's health in the aftermath of the Deepwater Horizon oil spill. *Environmental Hazards*, 1—16.

[35] Maldonado, J.K. (2019). *Seeking Justice in an Energy Sacrifice Zone*. Routledge.

BOX 5.1 Spinoff industry: chemical dispersants

Risk spurred expansion of chemical industries to clean up spills. Experimental dispersants consisting of unproven substances are often used to save time and money, potentially making the ecological and economic situation worse. Dispersants do not remove oil—they work with wind and waves to accelerate dispersal by allowing the oil to mix with water.

BP chose to spray dispersants across the water's surface and inject them directly at the well site, applying 1.84 million gallons by boat and airplane. BP was empowered to pick Corexit under the Toxic Substances Control Act, which allows companies to release unproven substances into the air or water—in most cases, they don't even have to disclose what is in the product. Following the use of Corexit in the Gulf, EPA studies found that of 18 dispersants approved for use with oil spills, 12 are more effective on southern Louisiana crude oil than Corexit, and the toxicity of those 12 was either comparable to Corexit or less toxic.

Given harmful implications for food chains and regional economies, these practices are an example of regulatory violence due to the absence of safeguards and protective measures.[36] Another example of regulatory violence is the maze of offshore pipelines in the Gulf.[37] It is hard to assign responsibility for harm after infrastructure is abandoned, leaving the costs to taxpayers after profits have accrued in the private sector. Public opinion has shifted such that most people in the US now oppose expansion of offshore production, in part due to high ecological and social costs following oil spills.[38]

Figure 5.13 Signal Hill, California. *(Image credit: Signal Hill Oil Field. Aerograph Co, 1923. US Library of Congress Prints & Photographs Division (digital ID pan.6a17401).)*

US oil setbacks typically range between 200–1000 feet.[39] Dense neighborhoods around infrastructure demonstrate bias toward property value over risks to human life.

[36] Sriram, K., Lin, G. X., Jefferson, A. M., Goldsmith, W. T., Jackson, M., McKinney, W., ... & Castranova, V. (2011). Neurotoxicity following acute inhalation exposure to the oil dispersant COREXIT EC9500A. *Journal of Toxicology and Environmental Health, Part A, 74*(21), 1405–1418.

[37] Sneath, S. (2021, August 5). Louisiana needs sand to rebuild its coast. Old oil and gas pipelines are blocking the way. Washington Post.

[38] Jones, B. (2018, January 30). More Americans oppose than favor increased offshore drilling. Pew Research.

[39] Ericson, S. J., Kaffine, D. T., & Maniloff, P. (2020). Costs of increasing oil and gas setbacks are initially modest but rise sharply. *Energy Policy, 146*, 111749.

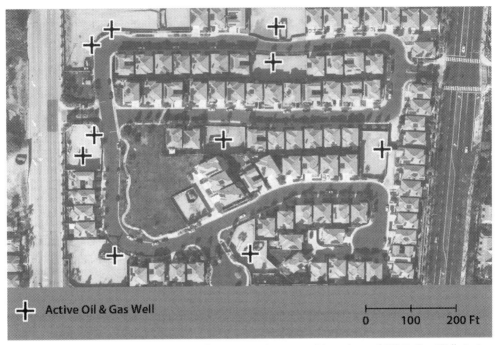

Figure 5.14 Proximity to oil in Signal Hill. *(Sources: Counts and Locations of Oil & Gas Wells in Los Angeles County (April 2019), Division of Oil, Gas and Geothermal Resources; Esri, Maxar, Earthstar Geographics, CNES/Airbus DS, USDA FSA, USGS, Aerogrid, IGN, IGP, and the GIS User Community.)*

Urban operators enter where they can at lowest costs. California, often considered a leader in climate and environmental policy, as late as 2022 finally passed a new state law phasing out new oil wells over the next 2 decades[40]—an inordinately long timeframe given the urgency of the climate crisis and the high health costs for burdened neighborhoods.

Redlining and racism

Redlining occurs when services are withheld from potential customers who reside in neighborhoods classified as a financial risk. Historically, it has been proven a racialized practice with harmful generational outcomes. **Racism** is prejudice, discrimination, or antagonism directed against a person or people on the basis of skin color and hatred toward people of color. **EJScreen**, an EPA screening tool used to identify race and

[40] Witt, E. (2022, March 3). The End of Oil Drilling in L.A. *The New Yorker.*

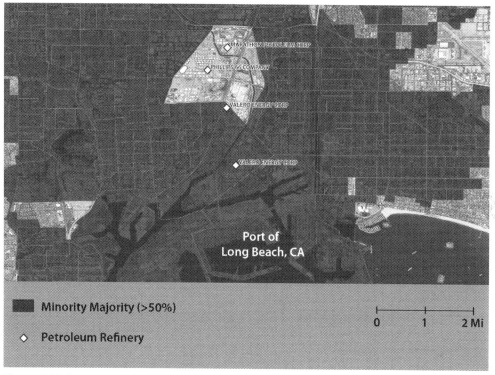

Figure 5.15 Environmental justice—refinery proximity. *(Sources: US Census Bureau American Community Survey (ACS 5-Year Survey 2020); US Energy Information Administration (EIA); Esri, Maxar, Earthstar Geographics, CNES/Airbus DS, USDA FSA, USGS, Aerogrid, IGN, IGP, and the GIS User Community.)*

poverty in relation to pollution,[41] demonstrates grossly unfair spatial patterns in the Los Angeles Basin (Fig. 5.15).[42]

Southern California was the epicenter of a booming oil industry from 1892 to the 1930s. Los Angeles rapidly grew from a town into the nation's fifth largest city. By the end of the oil rush, Los Angeles had also become one of the most racially segregated US cities. As petroleum fueled Los Angeles' massive sprawl, toxins such as benzene, toluene, xylene, and formaldehyde seeped into land, water, and air, harming Black

[41] Environmental Protection Agency. (2022, April 1). EJSsreen: Environmental Justice Screening and Mapping Tool.

[42] Cumming, D. G. (2018). Black gold, white power: Mapping oil, real estate, and racial segregation in the Los Angeles Basin, 1900—1939. *Engaging Science, Technology, and Society, 4,* 85—110.

and Latino residents and compounding the economic inequalities of redlining with an environmental toxicity waged daily against their bodies.[43] Impacts of this disinvestment in majority-minority areas of Los Angeles continue; for example, lack of tree canopy is one characteristic of underserved neighborhoods.

Oil infrastructuring

Networks are systems of interconnected people or things. Oil infrastructure networks are material expressions of power, though material elements remain largely hidden once in place, carrying oil underground across vast distances. The invisibility becomes a convenient fact for companies—especially when control over bottlenecks, potential blockages, or other forms of socio-spatial control can be wielded for economic or political gain. **Bottlenecks** form at narrow passageways that can physically impede or slow flow. Bottlenecks in transportation are a constraint, and can become a chokepoint. Strategic control points of supply networks influence oil prices. Oil companies consolidate political and economic power to reduce friction—such as bottlenecks—in their operations. For example, Koch Industries' Pine Bend Refinery located at the outskirts of Minneapolis, Minnesota is a strategic site where highly valuable refined products are made, requiring overt control to avoid becoming a bottleneck. Geographic placement of the refinery was strategic—imported Canadian crude is "sour" meaning it contains high amounts of sulfur requiring processing in special facilities small in number,[44] thus controlling the flow of several essential products to manufacturing sectors. The Pine Bend Refinery was able to maintain this spatial advantage and industrial domination through its grandfathered status, allowing avoidance of Clean Air Act regulations.

Koch Industries operates an expansive network, often referred to as the "Kochtopus": control emanates from the top, spreading outwards through tentacle-like arms. Until recently, Koch Industries was a hidden giant. Millions use products that Koch refines and distributes like gasoline, jet fuel, and chemicals, without ever seeing the Koch name attached.[45] An example of corporate strategy lies in the political role of the Koch family, which has been able to assert power behind the scenes. Unlike publicly traded companies where more transparency is necessary, the private Koch Industries leveraged its opacity in a vast industrial and political network[46] that has grown ever

---

[43] Cumming, D. G. (2018). Black gold, white power: Mapping oil, real estate, and racial segregation in the Los Angeles Basin, 1900–1939. *Engaging Science, Technology, and Society, 4*, 85–110.

[44] Leonard, C. (2019, September 26). Yes, America is rigged. Here is what I learned from reporting on Koch Industries. *Time.*

[45] Helman, C. (2021, May 14). Cyber-ransom of $5m 'Nothing' to Colonial Pipeline, Which Has Paid Hundreds of Millions in Dividends to Billionaire Koch Family. Forbes.

[46] Shaw, A., Meyer, T., & Barker, K. (2014, February 14). How dark money flows through the Koch Network. Propublica.

more complicated with a maze of limited liability companies (LLCs).[47] These companies operate across vast fields of influence, such as digital media firms.[48] One favorite Koch target audience is university students, who they seek to influence and recruit to help leverage their power.

The impact the Koch brothers have had in shaping acceptance of deregulation cannot be overstated. **Deregulation** is the intentional reduction or elimination of government power or repeal of rules. The Koch brothers have spent millions of dollars on organizations fighting climate mitigation policy.[49] Their market-based, anti-state intervention ideology helped set the stage for Trumpism and a major wave of environmental deregulation (Chapters 3 and 8).[50]

While the tenets behind market-based management are ideological for the Kochs,[51] the results are decidedly structural and physical violence. Structural violence appears repeatedly in the advocacy work of brothers Charles and David Koch to dismantle workers unions. For example, **union busting** occurred at the Pine Bend Oil Refinery in the mid-1970s.[52] Union busting can involve a range of activities undertaken to disrupt or prevent the formation of trade unions and their attempts to grow their membership. In this instance, deliveries across a union picket line weakened the strike. Dating back to the coal wars (Chapter 3), the deterioration of union bargaining power is a through line in the erosion of rights of coal workers. Legal and illegal methods were deployed at Pine Bend. As the picket line was breached by both coercion and physical violence, state power failed in its prime protective function. An important event in labor's fraught relationship with oil's industrial power, the Pine Bend episode also points to the importance of access to information. As Koch Industries operates as private companies, its finances are not readily available for independent analysis. Koch Industries use this to their advantage much like the Saudi company Aramco discussed below. The Koch brothers have been relatively secretive about how they earn their money, further deploying **shadow stock**: contracts paid out to upper management that confer the benefits of owning company stock without public disclosure. In reality, 80% of Koch Industries has been owned by the two brothers alone. This is due to their elaborate **self dealing**, or selling downstream

[47] Staff. (2019, August 29). The Kochtopus's Garden. *Economist*.

[48] Lennon, W. (2018, October 9). An introduction to the Koch digital media network. Open Secrets.

[49] Geary, J. (2019). The Dark Money of Climate Change. *ESSAI*, *17*(1), 17.

[50] Pulido, L., Bruno, T., Faiver-Serna, C., & Galentine, C. (2019). Environmental deregulation, spectacular racism, and white nationalism in the Trump era. *Annals of the American Association of Geographers*, *109*(2), 520–532.

[51] Koch, C. G. (2007). *The science of success: How market-based management built the world's largest private company*. John Wiley & Sons.

[52] Leonard, C. (2020). *Kochland: The secret history of Koch Industries and corporate power in America*. Simon & Schuster.

products to entities that are forms of the same (i.e., subsidiaries, partners), a typical characteristic of shell companies.

Preceding the Kochs by decades, John D. Rockefeller amassed a fortune using a similar network of secret trusts (or shell companies) to build an unrivaled monopoly. Rockefeller's spatial strategy included sweetheart deals made with railroads to control supplies. Like the Kochtopus, Fig. 5.16 depicts Rockefeller's Standard Oil as an octopus with tentacles around the steel, copper, and shipping industries—as well as political seats of power.

Antitrust laws tried to break up the concentration of economic and political power in the petroleum industry. Consolidation returned: Standard Oil and other companies that were split up realigned as ExxonMobil. The Federal Trade Commission brought the "Exxon case" in the early 1970s, during which it issued a complaint against the eight leading companies in the petroleum industry, alleging that the companies monopolized the domestic refinery industry. These strong companies and their captured politicians were not dissuaded.

Policymakers created rules that would reinforce earlier advantages. Oil shocks in the 1970s fed a scarcity mindset that has remained strong in the decades since. Any social or ecological threat that might increase the cost of oil, including events such as war or storm damage, reverberates in fear about the price of gas increasing, yet oil companies raise the price at the pumps even higher—a practice known as **price gouging**. This is another example of crisis capitalism discussed in Chapter 4.

Oil company sway over government remains excessive and is a fundamental driver of pervasive climate denial hampering political institutions. Government oversight

Figure 5.16 Rockefeller's spatial strategy. *(Image credit: Illus. in: Puck, v.56, no.1436(1904 September 7), centerfold. US Library of Congress Prints & Photographs Division (Item ID 2001695241).)*

continues to be weakened, and politicians remain unwilling to institute necessary reform measures even in the face of windfall profits. In 2022, a year-long Congressional Oversight Committee documented multiple companies operating to increase emissions while they publicly championed lower emissions. Oil industry malfeasance is nothing new, but the stakes and depth of its deceptive practices are still alarming. Compounding its deceit, the oil industry has been fully aware of its culpability of climate risk across 4 decades of BAU.

Oil companies weaponize their climate pledges to further secure the industry's license to operate.[53]

Temporal analysis

US imperialism laid the foundation of a global oil economy. US companies strangled domestic controls when and where they entered, creating oil fields as resource colonies. Oil violence in Nigeria and Ecuador (Chapter 6)—abandonment following a boom—reflects global trends.

Consolidation in the earliest phases of global oil has implications today. The big companies known as the Seven Sisters (Royal Dutch Shell, British Petroleum, Gulf, Exxon, Mobil, Texaco, and Chevron) held concessions obtaining exclusive right to search for, produce, and distribute oil. In exchange, firms gave shares of the profit in royalties to the nations in which they operated. This arrangement gave oil majors critical property rights throughout global locations, including Venezuela and Middle Eastern nations. By 1950, the Seven Sisters possessed over 98% of petroleum production market share. Fig. 5.17 depicts the later phases in the global oil market. Power to control prices today rests in large part with the Organization of Petroleum Exporting Countries (OPEC).

Oil-producing nations grew increasingly critical of their relative share of oil receipts from the Seven Sisters. In 1960, Saudi Arabia, Iraq, Iran, Kuwait, and Venezuela formed OPEC to defend the interests of the member states.[54] The growth in importance of OPEC wasn't fully clear until the 1970s. OPEC continued to emerge from a loose confederacy into a strong cartel, forcing the Seven Sisters to sign 50—50 profit-sharing agreements.[55]

---

[53] Marsh, R. (2022, December 9). Big Oil has engaged in a long-running climate disinformation campaign while raking in profits, lawmakers find. CNN.

[54] Wood, A. D., Mason, C. F., & Finnoff, D. (2016). OPEC, the Seven Sisters, and oil market dominance: An evolutionary game theory and agent-based modeling approach. *Journal of Economic Behavior & Organization, 132*, 66—78.

[55] Noguera, J. (2017). The Seven Sisters versus OPEC: Solving the mystery of the petroleum market structure. *Energy Economics, 64*, 298—305.

Phase	Dates	Description
1	1859-1947	Relatively stable prices based on the one-base price system controlled by an oil cartel known as the Seven Sisters
2	1947-1971	Transition to the two-base price system with an increase in oil supplies from the Middle East and the rise of OPEC
3	1971-1985	End of international concession agreements in OPEC countries; trend toward nationalization of deposits
4	1985-2005	Transition to stock market-based oil trade, rising volatility and the development of risk management mechanisms
5	2005-2020	Financial instruments decided global oil trade, leading to unsustainable expansion
6	2020-present	Intensifying evidence of climate disruption; peak oil; divestment; stranded assets

Figure 5.17 Phases of the oil market.

The 1973 oil embargo was an important turning point—concerns about energy independence and energy security entered the domestic politics of targeted countries with significant geopolitical ramifications. The public became more susceptible to energy industry manipulation due to the perceptions of insecurity—a vulnerability that oil companies used to their advantage.

OPEC continued to grow in strength and expand throughout the phase structure of the oil market. Fig. 5.18 shows the location of members, though the story is more complex than regional distribution alone. OPEC exhibits internal complexity, with some members having more weight than others, even though its network is often perceived as a monolith.

OPEC's shifting power is reflective of a changing world.[56] Spatial analysis highlights potential chokepoints as petroleum moves along maritime routes (Fig. 5.19). The Strait of Hormuz and the Strait of Malacca are important sites of oil transit.[57] A blockage of the Strait of Malacca would serve as a bottleneck in the supply of oil to East Asia—this was a central inspiration for China's Belt and Road Initiative (BRI) (Chapter 8) to diversify energy supply. China is the largest trade partner for Saudi Arabia, the greatest oil exporter in the Middle East.[58]

[56] Chatzky, A., & Siripurapu, A. (2020, April 9). OPEC in a Changing World. Council on Foreign Relations.
[57] Barden, J. (2019, June 20). The Strait of Hormuz is the world's most important transit chokepoint. Energy Information Administration.
[58] Abbhi, A. (2015, May 23). Yemen: A Battle for Energy Supremacy. E-International Relations.

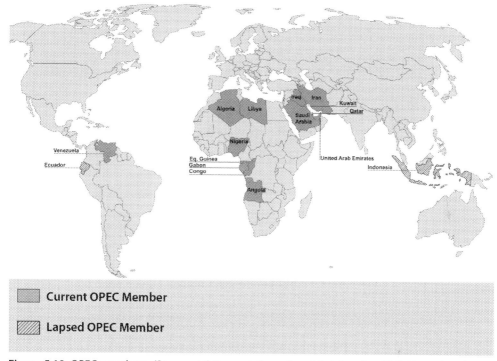

Figure 5.18 OPEC members. *(Sources: US Energy Information Administration (EIA); Natural Earth; Esri, Maxar, Earthstar Geographics, CNES/Airbus DS, USDA FSA, USGS, Aerogrid, IGN, IGP, and the GIS User Community.)*

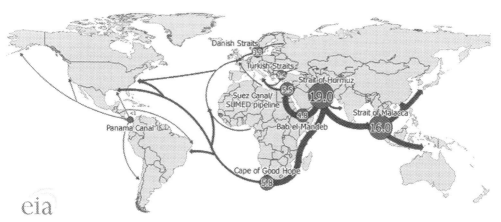

Figure 5.19 Oil flows in 2016. *(Map image credit: US Energy Information Administration (EIA), 2016.)*

Regaining our senses

In aggregate, from the years 2015–2020,[59] oil stood as the dominant fossil fuel, but today it is anything but secure. **Peak oil** occurs when the maximum rate of extraction of petroleum is reached, after which it is expected to enter terminal decline. The sharpness of decline is dependent on multiple political and economic factors, including climate policies and fuel subsidies. In 2020, the Rockefeller Foundation announced full divestment of its portfolio from fossil fuels—the last 3%—was imminent.[60] **Divestment** involves selling shares of stocks, bonds, or investment funds. This is both ironic and significant as the foundation's wealth itself stems from Rockefeller's helm of Standard Oil; the announcement indicates fossil fuels are losing their historical edge.

Oil's history is replete with examples of corporate maneuvering to create political and economic environments favorable to its pursuit of profit. While its injustices are widespread and costly, its most harmful impact is climate disinformation. Even though lawsuits increasingly target individual companies for their specific acts of climate denial, the true costs of decades of delay and inaction have only begun to be quantified. In 2021, Oxford Sustainable Finance estimated the cost of delay at $150 billion dollars per year.[61] The costs that fossil fuel companies owe for climate reparations alone are hundreds of billions (Conclusion).

Returning to the timeline of phases of the oil industry in Fig. 5.17 (located above), the current sixth stage, a phase of oil divestment, has too long been delayed.[62] Decades prior, oil companies became aware of the reality that much of the planet's remaining petroleum resources will need to remain underground. Instead of scaling back, they withheld information from the public. Two of the most aggressive propagators of climate denialism are ExxonMobil and the Koch Brothers.[63] ExxonMobil faces dozens of legal cases for scientific dishonesty—for decades, the company suggested global warming was unfounded, even while executives planned for climatic shifts they knew to be imminent.[64] Fig. 5.20 highlights ExxonMobil's pernicious denial across decades.[65]

---

[59] International Energy Agency. (2020). Global Energy Review. Report Extract: Oil.

[60] Rockefeller Brothers Fund. (n.d.). Fossil Fuel Divestment. The Rockefeller Foundation (2020, December 18). The Rockefeller Foundation commits to divesting from fossil fuels.

[61] Oxford Sustainable Finance Group. (2021, November 21). US $150 billion per year: Tallying the costs of delayed climate action.

[62] Leavitt, K. (2021, August 9). A 'death knell' for fossil fuels: What the UN's alarming new climate report means for Alberta. *Toronto Star*.

[63] Mayer, J. (2017). *Dark money: The hidden history of the billionaires behind the rise of the radical right*. Anchor Books; Leonard, C. (2019). *Kochland: The secret history of Koch industries and corporate power in America*. Simon and Schuster.

[64] Carter, L. (2021, June 30). Inside Exxon's playbook: How America's biggest oil company continues to oppose action on climate change. Unearthed.

[65] Jennings, K., Grandoni, D., & Rust, S. (2015, October 23). How Exxon went from leader to skeptic on climate change research. *Los Angeles Times*.

1977- 78 - James Black of Exxon's Products Research Division presents to management and creates an internal briefing paper informing company leaders "...there is general scientific agreement that the most likely manner in which mankind is influencing the global climate is through carbon dioxide release from the burning of fossil fuels."

1989 - Exxon and other fossil fuel companies create the Global Climate Coalition (GCC).[68]

1997 - Exxon CEO Lee Raymond tells the 15th World Petroleum Congress, "Currently, the scientific evidence is inconclusive as to whether human activities are having a significant effect on the global climate."

Ongoing - Exxon misleads;[69] see the #ExxonKnew campaign[70] and greenwashing of ineffective pollution abatement strategies like CCS

Figure 5.20 Examples of influential Exxon climate denialism.[67–69]

In dozens of lawsuits against oil companies, evidence of subterfuge continues to emerge.[66]

ExxonMobil, like other corporations and influencers, demonstrates the usefulness of mapping power (Praxis 1) to elucidate and connect power brokers in the energy sector. Power maps are useful analytical methods and visual tools to show influence and identify nexus of power in order to prioritize social change actions. As an illustration of the type of data relevant to elucidating influential ties in a power map, see the Corporate Accountability Project's entry on key decision-makers connected to ExxonMobil.[70]

Divestment

Divestiture from stigmatized industries over time has included tobacco, child labor, weapons, or firearms production. The most famous divestment movement targeted

[66] Supran, G., & Oreskes, N. (2021). Rhetoric and frame analysis of ExxonMobil's climate change communications. *One Earth*, 4(5), 696–719.
[67] Hasemyer, D., & Cushman, Jr., J. H. (2015, October 22). Exxon sowed doubt about climate science for decades by stressing uncertainty. Inside Climate News.
[68] ExxonMobil (n.d.). Climate change.
[69] #ExxonKnew (n.d.). Exxon knew about climate change a half a century ago. They deceived the public, misled their shareholders, and robbed humanity of a generation's time to reverse climate change.
[70] Lil Sis. (2022). ExxonMobil. Corporate Accountability Project.

South Africa during Apartheid.[71] Starting around 2011, the fossil fuel divestment movements targeting fossil fuels, which historically had tended to be localized around particular institutions like a retirement pension fund or stocks or bonds owned by a church or a school, sought to catalyze multi-scale fossil-free energy transition.[72]

While fossil fuel divestment had an early strong push from academia, the movement is led today by faith and philanthropic organizations.[73] The divestment movement has been mainly focused on North America and Europe and spread to Australia. The European Investment Bank was the first to phase out all fossil fuels[74]—with institutions like this making change, the finance sector norms begin to shift. More than 40 trillion dollars are scheduled for divestment globally—many of these committed divestments are bundled into investment strategies designed to avoid abrupt change, but prolong actual, measured energy transition.

Global movements also aim to constrain supplies to force transition away from oil.[75] One international example is the **fossil fuel non-proliferation treaty**[76] advanced by Pacific Island countries and others since global policy at the UN and in other institutions is not effectively mitigating harm from oil and gas.[77]

Climate litigation

No government is adequately protecting nor preparing future generations for the climate crisis. In the face of this political inaction,[78] youth climate lawsuits are proliferating and will continue to grow.[79] Early youth climate cases are shown in Fig. 5.21—both their number and geographic scope have continued to expand. Of particular note, the original Juliana v United States case brought by Our Children's Trust now represents more than 100 young people across multiple US states.

While litigation grows, Many cases have languished in pre-trial marked by industrial and state agency stalling tactics. Taken together, more than 1,600 US climate cases now

[71] Mangat, R., Dalby, S., & Paterson, M. (2018). Divestment discourse: war, justice, morality and money. *Environmental Politics*, 27(2), 187—208.

[72] Ayling, J., & Gunningham, N. (2017). Non-state governance and climate policy: the fossil fuel divestment movement. *Climate Policy*, 17(2), 131—149.

[73] Healy, N., & Barry, J. (2017). Politicizing energy justice and energy system transitions: Fossil fuel divestment and a "just transition". *Energy Policy*, 108, 451—459.

[74] Editorial Board. (2019, November 17). Growing demand for fossil fuels is a wake up call. *Financial Times*.

[75] Piggot, G. (2018). The influence of social movements on policies that constrain fossil fuel supply. *Climate Policy*, 18(7), 942—954.

[76] Newell, P., & Simms, A. (2019). Toward a fossil fuel non-proliferation treaty. *Climate Policy*, 1—12.

[77] Fossil Fuel Treaty. (n.d.). The Fossil Fuel Non-proliferation Treaty.

[78] See, for example, Our Children's Trust. (n.d.). Major Court Orders and Filings.

[79] Blumm, M. C., & Wood, M. C. (2017). No ordinary lawsuit: climate change, due process, and the public trust doctrine. *American University Law Review*, 67, 1.

Year	Country	Cases
2012	Uganda	Mbabazi and Others v. The Attorney General and National Environmental Management Authority
2015	US	Juliana v. United States
2016	Pakistan	Ali v. Federation of Pakistan
2017	India	Pandey v. India
2018	Colombia	Future Generations v. Ministry of the Environment (Colombia) and Others
2018	Canada	ENVironnement JEUnesse v. Canada
2020	Ecuador	Herrera Carrion et al. v. Ministry of the Environment et al.
2021	UK	Plan B Earth and Others v. Prime Minister
2021	Germany	Otis Hoffman, et al. v. State of Mecklenburg-Vorpommern; Tristan Runge, et al. v. State of Saxony; Leonie Frank, et. al v. State of Saarland; Luca Salis, et al. v. State of Sachsen-Anhalt
2021	Australia	Nature Conservation Council v. New South Wales (NSW) Nature Conservation Council of NSW v. Minister for Water, Property and Housing; Environmental Justice Australia (EJA) v. Australia

Figure 5.21 Early youth climate lawsuits.

exist targeting different jurisdictions and governance levels from local to regional to national as well as more than 700 international cases.[80] In these cases, children and young people constitute more than one quarter of all plaintiffs in rights-based strategic climate litigation cases filed globally up to 2021.[81] The resistance of governments to address these complaints is a reflection of failing social contracts. Children and youth are unfairly burdened, their fundamental human rights violated; litigation now stands as a global tactic holding governments accountable.[82]

[80] Sabin Center for Climate Change Law, Columbia Law School. (n.d.). Global Climate Change Litigation.
[81] Donger, E. (2022). Children and youth in strategic climate litigation: Advancing rights through legal argument and legal mobilization. *Transnational Environmental Law, 11*(2), 263–289.
[82] Parker, L., Mestre, J., Jodoin, S., & Wewerinke-Singh, M. (2022). When the kids put climate change on trial: youth-focused rights-based climate litigation around the world. *Journal of Human Rights and the Environment, 13*(1), 64–89.

Exercise

Research and communicate: examine international climate lawsuits—how could these change the stakes? Why are they not receiving more public attention? How could you and others publicize these efforts?

Illustrative cases

As Peak Oil consolidates, countries like Saudi Arabia become more resistant to energy transition, determined to extract oil until the last drop. With increasingly acute climate change and political conflict, countries of the Middle East appear to shelve proactive intervention in simmering regional distress. For example, experts predict water scarcity will severely limit extractive practices that rely so heavily on its availability. Tensions continue to build in the Middle Eastern region where collaborative actions often fall short. Even in less politically charged areas, like Canada, expansion of some of the most ecologically harmful oil in unconventional tar sands deposits builds social tension (Chapter 6). What does the expansion of extreme energy in spite of its burgeoning high costs and political and environmental hazards say about market-driven growth?

Saudi Arabia

The Saudi Kingdom is intimately tied to oil and the rise of OPEC. Nearly 100% of Saudi Arabia's electricity is generated from fossil fuels. Today Aramco is the world's most profitable oil company by far (Fig. 5.22)—yet when the Saudi Arabian Oil

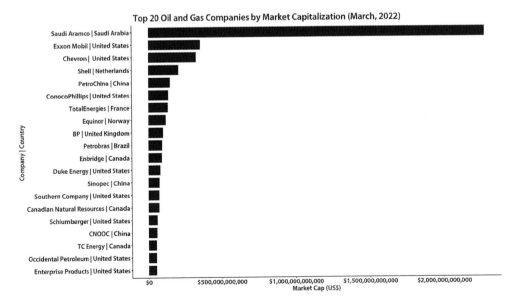

Figure 5.22 Aramco, top oil and gas company, 2021. *(Source: Companies Market Cap (https://companiesmarketcap.com/). Accessed March 23, 2022.)*

Company (Saudi Aramco) formed in 1933, no Saudi people had experience in the oil industry.[83]

Dating back to its original 1933 inception through a concession agreement between the Saudi Arabian government and the Standard Oil Company of California, Saudi Aramco has been ascendant. In the years from 1970 to 2019, Saudi Aramco faced very limited public scrutiny. In 1975, Aramco built the Trans-Arabian Pipeline—at the time it was the world's largest oil pipeline system. With the Trans-Arabian Pipeline development came a myriad of oil and gas infrastructure including gas gathering, treating, processing, and transmitting systems to provide fuel and feedstocks (Fig. 5.23).

Intimately tied to the Saudi Arabian government, Saudi Aramco operates with impunity. Today, the Saudi Arabian government has a nearly 100% stake in Aramco, increasing fourfold since 1973. Aramco's finances remain hidden in spite of near-complete nationalized status. In 2019, Aramco did sell three billion shares on the public market, but this only amounts to a 1.5% stake. At the time, Aramco was producing more than 13 million barrels per day—five times that of ExxonMobil's daily production average.

Through infrastructure consolidation, Saudi Arabia created valuable efficiencies—less mass lifted per unit of oil produced, and less energy used for fluid separation, handling, treatment, and reinjection. In turn, these efficiencies aided a domestic network of power generation, petrochemical processing, refining industries, desalination, and cement and steel facilities.

Aramco has ambitions to expand globally as an energy and chemical company. In 1989, Aramco internationalized in a joint venture with Texaco in the US. By 2017, the Saudis became the sole owner of North America's largest single-site crude oil refinery at Port Arthur, Texas (Chapter 1). US operations do not achieve the same efficiencies as domestic Saudi infrastructure. Compared to US operations, Aramco in Saudi Arabia shows lowered emissions, lower flaring rates, fewer leaks, and lower water use.[84]

Aramco describes its relationship with oil like a sensual romantic encounter[85]—soothing any environmental and climate inconveniences through advanced technology hype.[86] While an improvement, Saudi feats do not deserve to be romanticized—prolific flaring and leaks remain. Saudi royals seek to keep subsidizing production, making no sacrifices to their high-carbon economy.[87] Their plan for net-

---

[83] Delventhal, S. (2021, October 13). What is Saudi Aramco? Investopedia.

[84] Aramco. (2020). The Master Gas System - fueling a nation; Aramco. (n.d.). Our history: driven by the curiosity to explore.

[85] Saudi Green Initiative. (n.d.). Championing climate action at home and abroad.

[86] Aramco. (2021). Cracking Crudes Tantalizing Secrets.

[87] EL Aassar, M. (2021, October 26). Saudi Arabia will be talking 'green' but making no sacrifices on oil at COP26. *Fortune.*

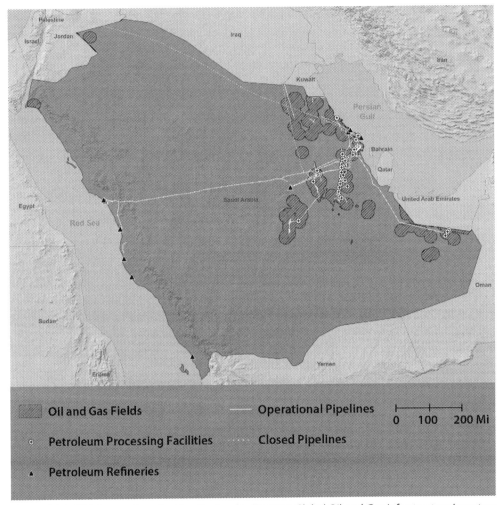

Figure 5.23 Oil infrastructure, Arabian Peninsula. *(Sources: Global Oil and Gas Infrastructure Inventory (GOGI), Natural Earth.)*

zero carbon emissions by 2060 is not sufficient for necessary global mitigation goals. Saudi climate denialism is tied to its national objective to be the last man standing in the oil industry.[88] Saudi royals seek to produce oil to the last drop.[89] To aid their aims, historic control over information has left the Saudi Arabian public largely compliant and

[88]   Krane, J. (2021). The Bottom of the Barrel: Saudi Aramco and Global Climate Action. Baker Institute.

[89]   Aramco. (2021, October 26). Aramco expands focus on emerging sectors at Future Investment Initiative.

complicit, looking askance at anthropogenic drivers of climate change (i.e., human activity like the combustion of fossil fuels).[90]

The climate crisis itself will very likely dampen Saudi Arabia's oil extraction ambitions. Water scarcity will force change in Saudi Arabia, an arid place that lacks permanent water-bodies. Saudi Arabia relies on an unsustainable and inefficient use of oil resources to operate their 30 desalination plants to supply potable water—more than 50% of Saudi's energy is used to desalinize water.[91] While oil income has largely offset desalination costs for now, climate change itself portends even higher costs.[92] Current practices may prove to be little more than temporary stopgaps: the fact remains that wastewater reuse and desalination are compensatory measures.

Conflict and oil

Oil and political strife often go hand in hand, especially in the Middle East. Fig. 5.24 depicts sites of oil conflict on the Arabian Peninsula, where tensions spill across borders. For

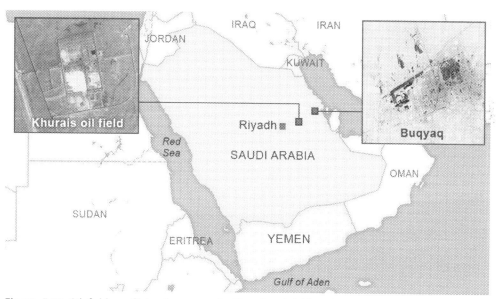

Figure 5.24 Oil field conflicts. *(Image credit: Khurais oil field and Buqyaq, Saudi Arabia. Voice of American English News (VOA), 2019.)*

---

[90] EL Aassar, M. (2021, October 26). Saudi Arabia will be talking 'green' but making no sacrifices on oil at COP26. *Fortune.*

[91] Rambo, K. A., Warsinger, D. M., Shanbhogue, S. J., & Ghoniem, A. F. (2017). Water-energy nexus in Saudi Arabia. *Energy Procedia, 105*, 3837–3843.

[92] Tarawneh, Q. Y., & Chowdhury, S. (2018). Trends of climate change in Saudi Arabia: Implications on water resources. *Climate, 6*(1), 8.

example, attacks have been made on Aramco facilities by a group originating in Yemen and supported by OPEC member Iran known as the Houthis.[93] These ballistic missile and drone attacks were in response to aggressive Saudi involvement in Yemen.[94]

Buqyaq (Fig. 5.24 above) is Saudi Arabia's largest global oil processing facility. The 2019 drone strike on the facility created the single largest daily oil supply disruption in history.[95] There are differing accounts on who was potentially responsible—Houthis in Yemen claimed responsibility for the strikes.

Abandoned oil

Since the early days of the Yemen civil war in 2015, the oil tanker *FSO Safer* has been stranded off the coast of Yemen, south of the Ras Isa Terminal as situated in Fig. 5. 25. In

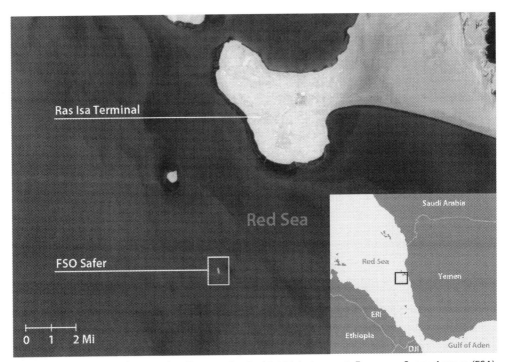

Figure 5.25 Location of *FSO Safer*. (Sources: Sentinel-2 L2A Instrument, European Space Agency (ESA), Natural Earth.)

[93] Bapat, N. (2019, September 17). The larger implications of the oil attacks in Saudi Arabia. *Political Violence at a Glance.*

[94] Cohen, A. (2021, March 8). Iranian-backed Houthis strike Saudi oil facility. Forbes.

[95] Verrastro, F. A. (2019, September 18). Attack on Saudi Oil infrastructure: we may have dodged a bullet, at least for now. Center for Strategic and International Studies.

2022, the UN unveiled a plan to transfer the tanker's oil to another vessel that would require crowdfunding of $80 million dollars.[96] Total cleanup costs have been estimated at $20 billion if the transfer were not completed in time in order to avoid the complete failure of a disastrous oil spill. In 2023, the UN purchased the transfer boat,[97] and the "timebomb" ship was defused.[98]

The ship, stranded in Yemen's war zone for nearly a decade, was literally a ticking time bomb with the potential to release into open water more than a million barrels of oil.[99] If *FSO Safer* dumped this cargo, an ecological disaster could have breached Yemen's north-western coastline, causing 10 million people to lose access to drinking water, destroying the Red Sea's coral reef, mangroves, and turtle nesting sites. Fig. 5.26 maps simulations of

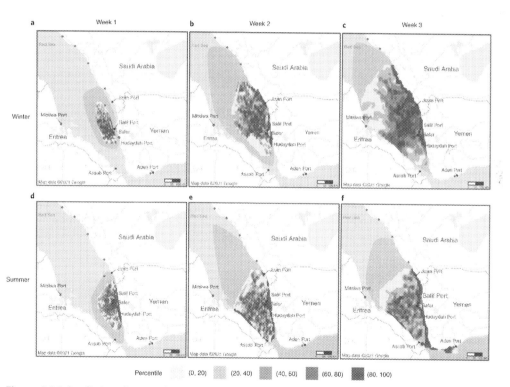

Figure 5.26 Predicting disaster. *(Source: Simulated surface oil concentration. Huynh, B.Q., Kwong, L.H., Kiang, M.V. et al. Public health impacts of an imminent Red Sea oil spill. Nat Sustain 4, 1084—1091 (2021).)*

[96] United Nations. (2022, May 11). Yemen: $33 million pledged to address decaying oil tanker threat.
[97] Exarheas, A. (2023, March 16). UN Buys Very Large Crude Carrier for FSO Safer Operation. *Rigzone.*
[98] Peachey, P. (2023, June 28). Tugs arrive to prepare for 'time-bomb' tanker replacement. *TradeWinds.*
[99] Caesar, E. (2021, October 11). The ship that became a bomb. *New Yorker Magazine.*

a potential spill spread over a first 3 week period based on historical trends across seasonal weather patterns, both winter and summer.[100]

This was a successful recovery. Left to its own, The *FSO Safer* could have severely disrupted traffic into a critical Yemeni port for months on end, cutting off humanitarian aid to Yemen upon which 20 million people now rely on to survive. Nonetheless, the *FSO Safer* debacle represents a long history of oil infrastructure abandonment, plaguing current and future generations who will no longer benefit from the use of fossil fuels. Future populations will live with constrained resources, transition burdens throughout the energy sector, continual disaster, and pervasive oil pollution remediation.

Oil companies are banking on abandonment in order to avoid responsibility for billions of dollars for cleanup. Investment in lobbying against environmental regulation—a core practice of regulatory violence—normalizes abandonment. Oil companies are poised to follow the path of coal companies to take advantage of settlements and eventual bankruptcy claims (Chapter 2). Regulatory reform to mitigate overall damages is urgent; moreover, creating bonds and other investments for oil companies to set aside monies to cover the costs of damages during the same periods when they are earning profit. Without intervention, the lessons of coal will go unheeded, and harmful patterns of public burden will be unbound through succeeding fuels of oil and gas.

Canadian tar sands

Oil extracted from tar sand regions deserves unique attention as an "extreme" energy due to the prodigal means necessary to derive marketable oil products from bitumen—a thick, sticky, black oil—that vexingly clings to sands and clay in both open-pit and in-situ mines. Once extracted, diluents are employed to thin product for pipeline transportation to refineries; these diluents are considered **proprietary information** (an industry trade secret). Much like the chemicals used in fracking (Chapter 9), diluents are known to contain benzene, a human carcinogen, and other harmful pollutants.[101]

Controversy has arisen with the development of tar sands,[102] considered the most destructive oil type and practice due to its inherent environmental racism, water pollution, energy debt, and high GHG emissions.[103] The remoteness of the tar sands means

---

[100] Huynh, B. Q., Kwong, L. H., Kiang, M. V., Chin, E. T., Mohareb, A. M., Jumaan, A. O., ... & Rehkopf, D. H. (2021). Public health impacts of an imminent Red Sea oil spill. *Nature Sustainability*, 4(12), 1084–1091.

[101] Song, L. A. (2012, June 26). Dilbit Primer: How it's different from conventional coal. Inside Climate News.

[102] Kelly, E. N., Short, J. W., Schindler, D. W., Hodson, P. V., Ma, M., Kwan, A. K., & Fortin, B. L. (2009). Oil sands development contributes polycyclic aromatic compounds to the Athabasca River and its tributaries. *Proceedings of the National Academy of Sciences*, 106(52), 22346–22351.

[103] Leahy, S. (2019, April 11). This is the world's most destructive oil operation - and it's growing. National Geographic.

Alberta Tar Sands
(North of Fort McMurray)

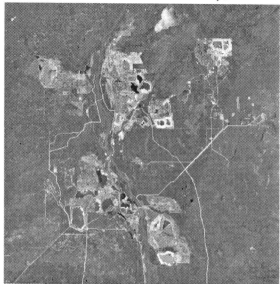

Manhattan
New York

10 Miles

Figure 5.27 Tar sands hazardscape. *(Source: Esri, Maxar, Earthstar Geographics, CNES/Airbus DS, USDA FSA, USGS, Aerogrid, IGN, IGP, and the GIS User Community.)*

damage remains largely invisible to national and international consumers; all the while, ecological and climate implications of expansive sites like Fort McMurray (Fig. 5.27 above) are wide reaching.[104] As with the broader oil industry, tar sands oil extraction has been kept alive by state handouts. Companies increasingly expect taxpayer funds to keep this energy-intensive oil product alive in an era of decarbonization pressure. When LCA methods are used to evaluate tar sands, its ecological and societal debt weigh heavy on its assumed benefits.

Try this

Research and reflect*: Tar sands oil is one of the most extreme forms of energy. Compare the temporary moment of energy use with tar sands oil's life cycle impacts, including the destruction of ancient cultures and landscapes.*

---

[104] Parlee, B. L. (2015). Avoiding the resource curse: Indigenous communities and Canada's oil sands. *World Development, 74,* 425−436.

Bomb trains, bomb trucks

New forms of oil transportation continue to be developed as mobility from producing regions to consumer markets remains a critical cost to the industry, especially with high opposition to new pipeline construction. With adaptations designed to increase mobility, new forms of fast violence manifest themselves on unsuspecting populations. Trains in the US and Canada move volatile fossil fuels by rail through dense residential areas without many being aware.[105] **Bomb trains**—locomotives carrying volatile oil or gas—create vast areas of intermittent risk with homes, schools, and public spaces located directly in the blast radius. Precautions are largely ignored, and there is a general lack of security around our rail systems due to industry greed and regulatory failure.[106] For example, in 2013 fires from a runaway oil train destroyed a downtown area of Lac-Megantic, Quebec, Canada, killing 47 people and burning for 3 days (Fig. 5.28). The

Figure 5.28 Bomb train derailment. *(Image credit: Picture taken from a Sûreté du Québec helicopter of Lac-Mégantic, the day of the derailment. Sûreté du Québec, 6 July, 2013. Accessed via Wikimedia Commons.)*

---

[105] Jackson, E. (2021, June 30). Updated National Energy and Petrochemical Map. Fractracker Alliance.
[106] Mikulka. J. (2019). *Bomb Trains: How Industry Greed and Regulatory Failure Put the Public At Risk.*

resulting damage has been described as a corporate crime scene. An investigation uncovered 18 factors that contributed to the disaster. Many of these factors were the result of cost-cutting measures, though they ultimately led to a hugely expensive disaster.

The Montreal, Maine and Atlantic (MMA) railroad, which operated the derailed train, retained just a small amount of insurance and avoided responsibility by declaring bankruptcy; in the end, even its owner escaped criminal charges and easily settled just one civil suit.[107] In place of serious repercussions, a low-level employee who worked the night of the accident and who had set the brakes in accordance with company policy was arrested at his house, where he and his son were thrown to the ground in what a Canadian journalist suggested was a politically motivated stunt to deflect attention from those truly responsible.[108]

A global hedge fund later bought MMA, placing trains on the same tracks, leading to yet more explosions and spills in the intervening years. After the fatal disaster in Lac-Megantic, the oil and rail industry increased (not decreased) the amount of oil moving by rail—all without additional safeguards such as mechanical devices placed on tracks to prevent future runaway trains. To this day, trains often operate with just a handful of onboard workers—a pervasive cost-saving measure laden with dire consequences.

The crude we are moving today is different from 10 years ago.

Bakken oil and some types of crude have high vapor pressures, containing more dissolved gas—a known increased risk within the industry. These "extra ingredients" make these oil types more profitable. Stabilization methods that would remove fossil gas liquids from the crude oil mix—and sell each separately—are not widely available. Since there were no regulations to force stabilization, the oil industry takes advantage of their ability to ship the volatile mix in spite of the danger.

Oil companies divert attention from gas capture, flaring, or other stabilization methods prior to transport; instead, finger-pointing is directed at braking systems and tank cars. After deadly explosions, the rail and oil industries use public relations staff to reassure the public that safety remains their top priority. Mainstream media mirrors these sentiments uncritically, in spite of all evidence to the contrary.

The US Pipeline and Hazardous Materials Safety Administration (PHMSA), whose task is to oversee oil transportation, is heavily influenced by industry. For example, the American Petroleum Institute (API), a powerful oil and gas trade association, offers safety and testing guidelines (i.e., "industry standards"), and regulators often allow these initial frames to influence final rulemaking—such as those related to gas concentrations in train shipments. In other instances, API has weighed in to suggest that industry standards need further deliberation—an effective method to spur doubt among policymakers to delay or avoid regulation altogether.

[107] *CBC News.* (2018). 'Jury made the right decision,' says former MMA chairman of Lac-Mégantic verdict.

[108] Mikulka. J. (2019). *Bomb Trains: How Industry Greed and Regulatory Failure Put the Public At Risk.*

Conflicts of interest

Former regulators who have left their regulatory positions and can now speak openly suggest the need to **follow-the-money**, a research strategy to track policy repercussions from donations, lobbying, and other forms of influence.[109] In the case of insufficient regulations and ongoing lack of enforcement of proposed reforms to stop "bomb trains," one can "follow a money trail" between policymakers in Washington, DC, back to powerful lobbying arms of the API.

Relationships between politicians and oil companies remain all too close when there is conflict of interest in politicians' personal investments or revolving door tendencies between policymaking positions and industry roles. A **revolving door** occurs as private sector employees are often hired into government positions, or former government regulators move to industry positions. Large consulting firms often have oil and gas or utility companies as major clients. Less obvious influences emerge serving together on directive boards, commissions, and other decision-making bodies.[110] An extensive network of professionals defending the oil industry contributes to its hegemonic control.[111]

> ### A "revolving door" culture between government and industry is threatening climate progress.[112]

Insufficient safeguards for transport and the use of dangerous fuels is an international problem extending from oil to its myriad of derivatives. Where there are fatal accidents, practices fit current rules,[113] suggesting regulatory violence. Investigations of fatal accidents often point to a combination of events that lead to an explosion, including but not limited to equipment failure, human error, and above average heat or cold. Not surprisingly, research finds that situations are deadlier where detection equipment is not used and oversight is lax.

The oil sector demonstrates power over space and mobility. While regulatory change becomes increasingly needed, industry resistance is both pervasive and effective. Examples of oil-related disasters and climate denial show that we do not have regulatory mechanisms in place to safeguard society, individuals, and the planet itself from the oil industry. Regulatory gaps are pervasive, difficult to fill, but urgently need policy intervention. Youth and student led advocacy movements such as Campus Climate Network (previously Fossil Free Research) demand their educational institutions call out and refuse

[109] Center for Responsive Politics. (1994). Follow the Money Handbook.
[110] Grasso, M. (2019). Oily politics: A critical assessment of the oil and gas industry's contribution to climate change. *Energy Research & Social Science, 50*, 106–115.
[111] Maddow, R. (2019). *Blowout: Corrupted democracy, rogue state Russia, and the richest, most destructive industry on earth.* Crown Publishing Group.
[112] Meyer, C. (2022, October 26). How oil and gas lobbyists build 'very close relationships' with politicians and governments. *The Narwhal.*
[113] Wood, M. H. (n.d.). Lessons learned from LPG/LNG. European Union.

research funds and influence linked to fossil fuel industries.[114] While the transition of the Rockefeller Foundation and others is groundbreaking, politicians and policymakers continue to overrely on oil corporations to inform and influence regulation to industry's advantage. As finite limits are surpassed, global heating and climate disruption continues to increase in frequency and intensity, world markets are not immune. For all its wealth generation within multinational corporations and national companies over the past century, oil cannot buy a livable planet.

Spatial and temporal synopsis

Space

(1) *The regulatory compact and guaranteed rates of return allow IOUs to consolidate power and overbuild infrastructure, holding ratepayers captive.*

(2) *Oil infrastructure networks are largely hidden, making it easier for companies to use socio-spatial control tactics, like control over transport bottlenecks.*

(3) *Oil infrastructuring is often spatially unfair, compounding discriminatory practices like redlining.*

(4) *Spills and explosions are frequently tied to negligence; the highest risks are uneven and often concentrated.*

Time

(1) *Oil is becoming an insecure industry—the period after peak oil parallels the economic decline in coal towns (Chapters 2 and 3); progressive water scarcity is among the deciding factors for peak oil.*

(2) *Following antitrust measures that weakened companies' power, companies re-emerged with new consolidation, pushing for deregulation and prolonging oil's longevity.*

(3) *Movements against oil use tactics like climate litigation (e.g., youth climate lawsuits), divestment, and constraining supplies (e.g., fossil fuel non-proliferation treaty).*

Summary

The violence of oil is multifaceted, and physical, structural, regulatory, and territorial violence is common. Oil operations are linked to power, leading to abuse of the regulatory compact and constant attempts to overreach into statecraft. Infrastructuring is spatially uneven and unfair, with negligence producing avoidable violence with spills and explosions; this violence also leads to spinoff industries, like the chemical dispersants used for spill cleanup. Denialism continues with industry greenwashing even as the urgency of the climate

---

[114] Fossil Fuel Research (n.d.). No more fossil fuel research money.

crisis quickens. Global movements seek to quell oil with litigation, divestment, and supply constrictions. The increasing insecurity of industry, marked by abandonment after peak oil and climate risks like water stress, further threatens oil's future.

Vocabulary

1. antitrust rules
2. bomb trains
3. bottleneck
4. captive ratepayer
5. carcinogenic
6. deregulation
7. divestment
8. follow-the-money
9. fossil fuel non-proliferation treaty
10. EJScreen
11. guaranteed return-on-investment
12. market-based management
13. negligence
14. network
15. peak oil
16. price gouging
17. proprietary information
18. racism
19. redlining
20. regulatory compact
21. return on investment (ROI)
22. self dealing
23. setbacks
24. shadow stock
25. spill
26. union-busting

Recommended

Battler, L. (2015). *Endangered hydrocarbons*. BookThug.
Huber, M. T. (2013). *Lifeblood: Oil, freedom, and the forces of capital*. University of Minnesota Press.
Juhasz, A. (2008). *The tyranny of oil*. Harper.
Leonard, C. (2019). *Kochland: The secret history of Koch industries and corporate power in America*. Simon & Schuster.
Malm, A. (2021). *White skin, black fuel: On the danger of fossil fascism*. Verso Books.

CHAPTER 6

Crude

Contents

Climate Crisis, Energy Violence
ISBN 978-0-12-819501-7,
https://doi.org/10.1016/B978-0-12-819501-7.00006-0

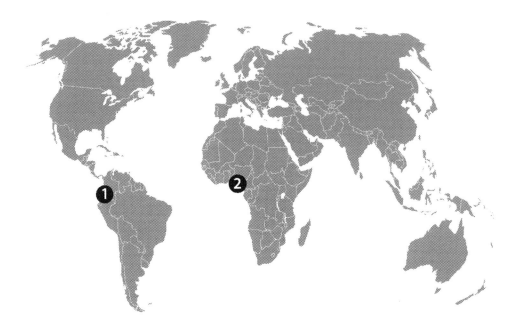

Locations: 1 - Lago Agrio Canton - Equador
              2 - Niger Delta Region - Nigeria

Themes: Decolonization; Indigenous rights; Racial violence

Subthemes: Boom and bust; Flare; Petrostate

Different energy boom, same mistakes.

Geographies of neglect

It is nearly always more expensive to clean up pollution than to prevent harm. Failing this lesson, polluters continue to skirt responsibility for damages, building sacrifice zones as inevitable corollaries to profit. Among numerous violent oil sites, two stand out—Ecuador and Nigeria—with criminal destruction of homelands and decades of delayed justice.

A **petrostate** is a country whose economy is heavily dependent on oil or fossil gas. Both oil cases experienced "Boom then Bust"—a fundamental cycle of modern fossil fuel economies. With the "bust" phase come the realities of **abandonment**—the severing of critical resources of sustenance. To make matters worse, oil hotspots tend to crowd out dynamic, varied economies, narrowing national wealth strictly to fossil fuel extraction, polluting environments, and corrupting institutions.[1] In Ecuador and Nigeria this has meant elites captured fortunes while plaguing their people with failing governments. International institutions largely choose neutrality over action.[2] When legal intervention has finally occurred, ethnic territories are deep in crisis, poisoned by flagrant practices.[3]

International arbitration of the oil industry's harm has produced meager results. For Nigeria, just four farmers and their communities received $16 million in 2022 from Shell.[4] Ecuador's never-ending, unresolved lawsuits exemplify corporate impunity.[5] Encouragingly, however, global climate litigation seems to be entering a new phase marked by two strategic themes—one looking to the past, the other to the future. The first involves suits based on material remediation of deadly harms; and the second, increasingly prevalent struggles of youth to protect their rights to a livable future.

Over decades, the oil industry has avoided regulation. Flaring, explosions, spills, leaks, and ruptures are intrinsic to upstream oil extraction hotspots with lax oversight. Oil is kept alive by political maneuvering and economic subsidies, even after its harms are calculated and alternatives have been found. The oil industry uses its political power, state subsidies, and secured markets to retain its hegemony against cleaner, cheaper alternatives. Even while investors and insurers are increasingly turning away from oil,[6] greenwashed operations and oil derivatives prop up this fossil fuel, a vexing obstacle to urgent energy transformations (Conclusion).

Energy violence

Closely associated with oil hotspots, protracted lawfare (Chapter 4) occurs as claims from injured frontline populations remain unresolved decade upon decade—in the case of Ecuador and Texaco, nearly 30 years after the start of exposures.[7] This is an example of **legal violence**: manipulation of legal and legislative weaknesses or loopholes that either results in or has a high likelihood of resulting in injury, harm, or death.

[1] Porter, D., & Watts, M. (2017). Righting the resource curse: Institutional politics and state capabilities in Edo state, Nigeria. *The Journal of Development Studies*, 53(2), 249–263.

[2] Donnelly-Saalfield, J. (2009). Irreparable Harms: How the Devastating Effects of Oil Extraction in Nigeria Have Not Been Remedied by Nigerian Courts, the African Commission, or US Courts. *Hastings W.-Nw. J. Envt'l L. & Pol'y*, 15, 371.

[3] Lindén, O., & Pålsson, J. (2013). Oil contamination in Ogoniland, Niger delta. *Ambio*, 42(6), 685–701.

[4] Staff. (2022, December 23). Shell to pay $16 m to Nigerian farmers over oil damage. BBC.

[5] Orellana Lopez, A. (2019, March 27). Chevron versus Ecuador: international arbitration and corporate impunity. Open Democracy.

[6] Insure Our Future (2022). 2022 Scorecard on Insurance, Fossil Fuels and the Climate Emergency.

[7] Joseph, S. (2012). Protracted lawfare: The tale of Chevron Texaco in the Amazon. *Journal of Human Rights and the Environment* 3(1): 70–91. (p. 70).

Legal violence extends and follows centuries of structural violence—colonization has put Nigeria and Ecuador in weakened positions at the entry of foreign oil companies and, after brutal corporate exploitation, were left in yet greater insecurity stripped of state protections made worse by foreign debt and austerity measures. Hampered by the overhang of state debt burdens, countries throughout the global South are unable to staunch exploitative, outward resources flows. Resource transfers—most often to the global North—should at minimum cover basic domestic infrastructure projects, yet for most Nigerians and many Ecuadorians energy poverty marked by unreliable and unaffordable electricity is a daily reality. **Debt treadmills**, whether state or personal circumstance, are unending cycles of payment for debt service alone, forfeiting capital that could be freed for meaningful goods and services.

Additional cycles of exploitation coincide with oil extraction, including debt traps laid by international market **neoliberalism**—a combination of deregulation, privatization, and structural adjustment.[8] After earlier colonization, new phases of exploitation took hold in global oil hotspots, many concurrent with **neocolonialism**[9] leaving countries including Nigeria with foreign debts now skyrocketing beyond control.[10] Today's creditors are the World Bank, International Monetary Fund, Afrexim, African Development Bank, and bilateral lenders, including China, Japan, and India.

Similar to Nigeria, Ecuador's debt burden has reached crippling heights. In 2022, Ecuador's assets were frozen due to an unresolved dispute with Anglo-French oil company Perenco on the charge of unlawfully ending a production-sharing agreement.[11] This crisis came to a head as Ecuador simply could no longer service its debt, forcing restructure with China, the country's largest financial partner.[12] True to pattern, the global South's oil hotspots endure colonization, then neocolonialism, eroding statehood while intensifying vulnerabilites upon citizenry.

In international relations amongst creditors and debtors, terminology is telling: "assistance" often implies a relationship of superiority, while "aid" from one location becomes a paternalistic "band-aid"—insufficient to heal glaring wealth disparity and reliant on continual resource exploitation. In the case of Ecuador, energy transition investments from Chinese firms in effect reinforce patterns of oppression established long ago by 16th century Spanish colonialism. While multifacted in its contemporary usage and meanings, **decolonization** points to the undoing of colonialism, giving way to cultural, psychological, and economic freedom, enabling all groups to practice self-determination

[8] Harvey, D. (2007). *A brief history of neoliberalism.* Oxford University Press.

[9] Nkrumah, K. (1965). Neo-Colonialism, the last stage of imperialism. Thomas Nelson & Sons, Ltd.

[10] Tunji, S. (2022, September 20). Nigeria's debt hits N42.84tn amidst revenue crisis. Punch.

[11] Valencia, A. (2022, August 9). Ecuador says debt payments not affected by Luxembourg asset freeze. Reuters.

[12] Staff. (2022, September 20). Ecuador reaches deal with China to restructure the debt. Reuters.

over land, culture, political and economic systems[13]. By allowing today's patterns of predatory lending reinforced by international finance institutions (IFIs), decolonization remains an unrealized project with energy transformation stymied. The past repeats itself regardless of energy type, delivering not freedom but yet more structural violence marked by racial and geographic inequities. After centuries of exploitation, it remains unclear if countries like Nigeria will ever fully decolonize.[14]

Spatial distribution

Fig. 6.1 demonstrates the pattern and concentration of global petroleum fields, denoting both Nigerian and Ecuadorian hotspots. While pollution is found across all major oil geographies (Chapter 5), the level of contamination is often higher in countries with economic constraints and political disorder.

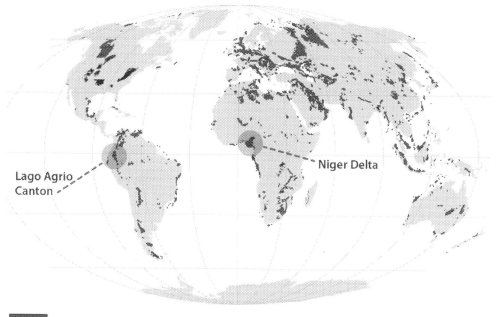

Global Petroleum Fields

Figure 6.1 Global petroleum fields. *(Sources: Global Oil and Gas Infrastructure (GOGI) Inventory, Natural Earth.)*

[13] Heckenberg, R. and O'Dowd, M.F. (2022, June 22). Explainer: what is decolonisation? *The Conversation.*

[14] Ityonzughul, T. T., Gbamwuan, A., Igba, D. M., & Olajire, O. (2020). Beyond colonial entrapment: profiling the economic challenges of nation building in post-colonial Nigeria. *Journal of Nation-building & Policy Studies, 4*(2), 83—101.

Oil is a finite and spatially uneven resource. Deposits are not necessarily located where there is greatest need for energy, meaning that transportation adds costs and risks when compared to locally produced power. Extraction costs are often borne in areas located far away from where the oil is enjoyed—demonstrating oil colonialism.[15]

Firms invest in long-term partnerships around oil aimed to assure future access and dissuade competitors. The role of corrupt oil elites in the domestic political sphere is a major impediment to public well-being and trust in many petrostates.[16]Adding violence to distrust, state agencies, their militaries and paramilitary insurgents all vie for power over oil resources resulting in "blood oil".

When compared to the relative stability of earlier phases of oil development (Chapter 5), contemporary oil markets are complex and dynamic. Even with diversification from traditional investments, like those of the World Bank and Western neoliberal economies, politically and economically marginalized nations and regions continually experience a mismatch of global finance to the well-being of their local economies. Structural violence continues—for example, Shell Oil of Nigeria is an offshoot of Royal Dutch Shell, a British multinational company headquartered in London.

The reach of the state and of dominant multilateral aid agencies reinforces oil domination and colonial relationships with spatial reach above and below the Earth's surface, requiring attention to vertical space. Analyzing oil in Nigeria and Ecuador demonstrates the extent of loss and damage across regions and vertically therein. Energy studies marked by spatial analysis with a **topographical approach** draw attention to the forms and features of land surfaces and vertical and horizontal spaces of power, highlighting terrain, relief, and pollution pathways (Fig. 6.2) with vertical and horizontal expressions.

Air pollution	• Gas flaring; pollution from oil refineries
Water pollution	• Oil spills during releasing contaminants to groundwater, surface water and drinking water
Land contamination	• Disposal of liquid refinery effluents containing phenols, cyanides, sulfides and more • Disposal of drill cuttings mixed with drilling muds • Oil spills

Figure 6.2 Examples of pathways for effluent release.

[15] Watts, M. (2008). Blood oil: The anatomy of a petro-insurgency in the Niger delta. *Focaal, 2008*(52), 18—38.

[16] Lyall, A. (2018). A moral economy of oil: Corruption narratives and oil elites in Ecuador. *Culture, Theory and Critique, 59*(4), 380—399.

The assignment of blocks and leases leverages the potential of oil extraction over any ecological value. A topographical approach contextualizes oil extraction—concession blocks, oil pads, pipeline corridors, and refining facilities—to illuminate intersecting ecological and social processes occurring throughout both vertical and horizontal space. **Horizontal space** is any area below, at, or above the Earth's surface, generally running parallel to or covering the Earth's surface (i.e., a property lot, a city block). **Vertical space** is identified by its location above or below the surface of the Earth (Fig. 6.3). Oil commodifies horizontal and vertical realms.

Privileged use of space and resources is abundantly clear in and around oil boom-towns. A topographical approach helps clarify the full extent of spatial patterns of expansion, such as herringbone-patterned forest clearance across the Amazon resulting from the creation of oil entry roads. Even though oil entry roads facilitate additional deforestation, Ecuador permits oil and gas extraction in national parks.[17] This creates a pressing need for protected areas for uncontacted Indigenous Peoples. In 2007, the Ecuadorian government established by presidential decree a 7,580 square kilometer **untouchable zone**—a geographic area restricted from oil and gas extraction or logging, designed to protect the Tagaeri and Taromenane: two Indigenous groups living in voluntary isolation within their historical territories.

The social construction of a region is integral to the slow yet effective movement towards energy transition. **Regions** are zones, territories, or areas that share ecological and cultural characteristics. Regional solutions originating within West Africa and the Amazon are frequently overlooked based on the dominance of international

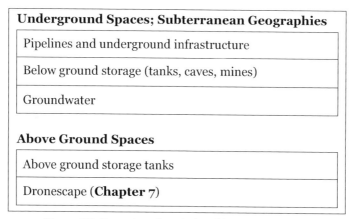

Figure 6.3 Examples of vertical and horizontal spaces.

[17] Finer, M., Jenkins, C. N., Pimm, S. L., Keane, B., & Ross, C. (2008). Oil and gas projects in the western Amazon: threats to wilderness, biodiversity, and indigenous peoples. *PloS one, 3*(8), e2932.

organizations like multilateral banks and aid agencies. For example, Indigenous Federations in the Amazon have proposed a bioregional approach focused on the sacred headwaters of the Amazon.[18] A **bioregion**—a region whose limits are naturally defined by topographic and biological features (such as ecosystems)—is an essential resource for Native communities, yet is overlooked when planning oil concessions.

Temporal analysis

Is there a way to temper the **boom** (a period of rapid economic growth or expansion) and **bust** (a period of downturn or failure)? There are clear temporal components of a **boomtown**—a town experiencing a business and population boom. The idea of a boomtown may often become oversimplified based on a romanticized notion of resource booms.[19] There are some general patterns that are recognizable. First and often, a "preboom" is marked by propaganda and boosterism for jobs and new economic ventures. **Boosterism**, meaning excessive promotion to privilege a place or sector toward a particular agenda, feeds lust for a boom and thus contributes to an inevitable bust due to hasty, narrowly focused economic and environmental planning. Boosters tend to ignore constraints and negatives, rather relying on rhetoric and misinformation. Boosterism often leads to negatives, such as traffic congestion and water pollution or overuse. A bust economy following a boom is often protracted, marked by unresolved harms from too rapid growth. In both cases analyzed below, overreliance on oil had dire consequences.

Illustrative cases

Nigeria

The violence of oil in Nigeria, Africa's most populous nation and largest oil producer, is well documented. Royal Dutch Shell entered Nigeria in 1938,[20] and most oil extraction has occurred since the 1950s. Yet, after decades of extraction more than half of Nigeria's population is caught in debilitating poverty,[21] coupled with extreme flooding in the face of impending climate impacts.[22]

---

[18] Amazon Sacred Headwaters Technical Team. (2021). Bioregional Plan 2030.

[19] Romich, E., Civittolo, D., & Bowen, N. (2017). Characteristics of a boomtown. The Ohio State University.

[20] Watts, M. (2008). Imperial Oil: The Anatomy of a Nigerian Oil Insurgency. *Erdkunde*, 27–39.

[21] Watts, M. (2004). Oil as money: the devil's excrement and the spectacle of black gold. *Reading Economic Geography*, 205–19.

[22] Maishman, E. (2022, October 16). Nigeria floods: 'Overwhelming' disaster leaves more than 600 people dead. BBC.

Since oil is the mainstay of Nigeria's economy, price changes bring economy-wide repercussions, leading to hyper-dependency on oil markets. Much of Nigeria's recent economic diversification has narrowly focused on fossil gas production, a corollary of oil extraction, increasing dependency on fossil fuels. Paradoxically, Nigeria imports nearly all refined petroleum products, a display of structural violence.

Throughout global oil hotspots, **corruption** is often prolific: dishonesty and fraud by those in power, typically involving bribery. Systematic abuses in public finance, rampant in the 1970s under military regimes, have continued under civilian regimes to the present.[23] Tactics include inflated government contracts, over-invoicing of imports, illegal foreign exchange deals, smuggling, and fictitious payrolls.[24] Corruption leads to regulatory capture (Chapter 3) in Niger Delta, with harm to cultural and natural resources (Fig. 6.4).

As patterns of environmental racism in the delta have been normalized, ethnic minorities face genocidal practices justified for cheap oil.[25] Oil pollution made particular locations simply inhabitable, creating a public health crisis (Fig. 6.5).[26] Nigeria's oil disasters are all underscored by historical oppression similar to Ecuador (discussed below). From the decimation of local, artisanal fishing to ever-expanding sovereign debt, Nigeria's delta is a calamity and tragedy.[27]

As is typical of global oil hotspots, benefits accrue to foreign companies and the ruling **elites**—individuals who hold a disproportionate amount of wealth, privilege, and political power. Costs, on the other hand, are distributed unequally where extraction occurs. Approximately 250 ethnic groups reside in this risk geography. Nigeria's population on average is young, with more than 40% of the population age 14 or younger.

If we protest, they send soldiers.[28]
We have graduates going hungry, without jobs.
And they bring people from Lagos to work here.[29]

Niger Delta violence isn't limited to just slow violence of air and water pollution. Throughout the 1990s, tensions arose between the native Ogoni people of the Niger

[23] Watts, M. J. (2019). State as Illusion? A Commentary on Moral Economies of Corruption. *Comparative Studies of South Asia, Africa and the Middle East, 39*(3), 551–558.

[24] Orjinmo, N. (2022, October 23). Nigeria's stolen oil, the military and a man named Government. BBC.

[25] Westra, L. (1998). Development and Environmental Racism: The Case of Ken Saro-Wiwa and the Ogoni. *Race, Gender & Class*, 152–162.

[26] Lindén, O., & Pålsson, J. (2013). Oil contamination in Ogoniland, Niger delta. *Ambio, 42*(6), 685–701.

[27] Amotsuka, P. O. (2021). Declining Artisanal Fishing in the Niger Delta Region: A Baseline Survey of Finima Community of Rivers State, Nigeria. *African Journal of General Agriculture, 6*(3).

[28] Eghare W.O. Ojhogar, quoted in Amnesty International. (2006). Nigeria: Oil, poverty and Violence.

[29] Heide, I. (2018, March 27). Decades of injustice, oil and violence in the Niger delta. Amnesty International.

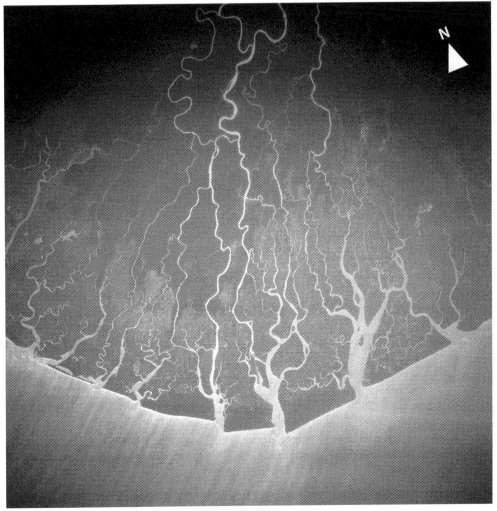

Figure 6.4 Niger delta. *(Map image credit: NASA, Photo #:STS61C-42-72, January, 1986.)*

Delta and the oil company Shell. In 1993, the Movement for the Survival of the Ogoni People (MOSOP) organized large protests against Shell and the government, often occupying oil facilities.

MOSOP was an ethnic movement and an effective regional approach to demand power. Because of MOSOP's success in motivating grassroots involvement, the Nigerian government raided Ogoni villages and arrested several protest leaders. Nine of the

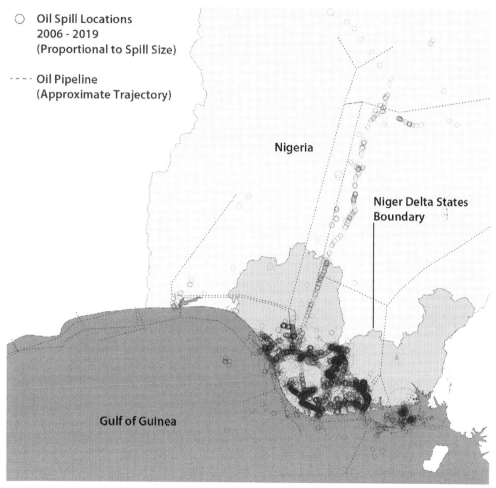

Figure 6.5 Nigeria's polluted delta. *(Sources: National Oil Spill Detection and Response Agency (NOS-DRA); Global Oil and Gas Infrastructure (GOGI) Inventory.)*

arrested—Ken Saro-Wiwa being the most prominent—were later executed with widespread condemnation from human rights organizations (Dedication).[30] Throughout the early 1990s, Nigerian military and police forces attacked Ogoni villages, beating, raping, killing, and arresting residents.[31] The assassination of the **Ogoni 9** by their government exhibits fast violence with seeming impunity. For a few years after these murders, Nigeria

[30] Idowu, A. A. (1999). Human rights, environmental degradation and oil multinational companies in Nigeria: The Ogoniland episode. *Netherlands Quarterly of Human Rights, 17*(2), 161–184.

[31] Llewellyn, A. (2022, August 2). After 20 years, Indonesian ExxonMobil accusers eye day in court. Al Jazeera.

was suspended from the Commonwealth of Nations.[32] This was a low sanction for these homicides.[33] The punishment may not be enough to deter similar future behavior.[34]

Shell tried to avoid culpability for the murder of the Ogoni 9. In 2003, Shell Nigeria acknowledged "we sometimes feed conflict by the way we award contracts, gain access to land, and deal with community representatives" and that it intended to improve its contracting practices. In 2009, Shell offered to settle the Ken Saro-Wiwa case with $15.5 million—while denying any wrongdoing and calling the settlement a humanitarian gesture. The settlement came days before the start of a trial expected to reveal Shell's activities in the Niger Delta.

Human rights abuses and ecological disasters will never be fully remedied in court. Ogoni-9's family members continue to demand justice and remain in litigation. Petitioners allege that Shell aided and abetted atrocities by providing the Nigerian forces with food, transportation and compensation, as well as by allowing the Nigerian military to use respondents' property as a staging ground for attacks.[35] Shell "got away with" criminal behavior.[36]

The Niger Delta experiences the equivalent of an Exxon Valdez spill every year.[37]

For years, Shell disputed the alleged volume of oil spilled in the Niger Delta—the burden was rather placed on impacted communities and allies in the nonprofit sector to prove harm. Loss and damage remains widespread, yet securing reparations remains an exceptual event. Niger Delta residents continue to live with the toxic devastation. Fig. 6.6 shows Eric Dooh, a local community representative in Oganiland, with his hand covered in oil that coated nearby waterways, damage resulting from a 2004 Shell pipeline leak.

In 2012, 15,000 members of the Bodo community in the Niger Delta filed a lawsuit against Royal Dutch Shell.[38] They sought compensation for two oil spills in 2008 and 2009 for losses to their health, livelihoods, and land. Bodo plaintiffs argued that the old, poorly maintained pipelines caused spills and Shell did not respond adequately to spill

---

[32] Holzer, B. (2007). Framing the corporation: Royal Dutch/Shell and human rights woes in Nigeria. *Journal of Consumer Policy, 30*(3), 281–301.

[33] Konne, B. R. (2014). Inadequate monitoring and enforcement in the Nigerian oil industry: the case of Shell and Ogoniland. *Cornell Int'l LJ, 47*, 181.

[34] Pilkington, E. (2009). Shell pays out $15.5 m over Saro-Wiwa killing. *The Guardian*.

[35] Supreme Court of the United States. (2012). Kiobel et al. versus Royal Dutch Petroleum Company et al.

[36] Gilblom, K. (2018, September 28). Shell tries to come clean on its dirty past in Nigeria. Bloomberg.

[37] Watts, M. J. (2021). Hyper-Extractivism and the Global Oil Assemblage: Visible and Invisible Networks in Frontier Spaces. In *Our Extractive Age* (pp. 205–248). Routledge. Pp. 208.

[38] Staff. (2012, March 23). Shell lawsuit (re oil spills and Bodo community in Nigeria). Business and Human Rights Resource Center.

Figure 6.6 Niger Delta sacrifice zone. *(Image credit: Environmental Degradation in Nigeria, Ucheke, May, 2019. Accessed voa Wikimedia Commons.)*

alerts. To avoid cleanup costs, Shell argued its Nigerian subsidiary, the Shell Petroleum Development Company, was liable.

After years of pressure, Royal Dutch Shell finally accepted responsibility for the spills. Shell agreed to pay a settlement and provide remediation assistance—yet the actual price of damage to Ogoniland is much higher. Although just a small part of the total contaminated area, cleanup in Bodo remains slow and uncertain, with inadequate settlement monies.[39,40] While future settlements hold the potential to alter norms of corporate impunity,[41] settlements thus far have done little to deter negligence (Chapter 5), as spills

[39] Mbachu, D. (2020, July 1). The Toxic Legacy of 60 Years of Abundant Oil. Bloomberg Law.

[40] Amnesty International. (2020). On trial: Shell in Nigeria. Legal Actions Against the Oil Multinational.

[41] Princewill, N., & Shveda, K. (2022, May 25). Shell escaped liability for oil spills in Nigeria for years. Then four farmers took them to court - and won. CNN.

continue to be reported including at the Shell facility in Bodo.[42] Across the Niger Delta, hundreds of spills continue to occur year in, year out.

Wasteful flaring

Where there's oil, there'll be gas. If gas builds up within an extraction system it can result in explosions. Flaring—the venting and burning of gas—is the cheapest "remedy." A staggering amount of fossil gas from Shell's operations in Nigeria is flared as a "waste" product. **Flares** are open flames used to release pressure and burn off gas. Flaring gas releases hazardous air pollutants even as flaring reduces GHG emissions.[43] There is a trade-off. Each year, about 150 billion cubic meters (bcm) of fossil gas are flared, emitting 400 million tons of CO_2 emissions and other pollutants. Historically, Russia, Iraq, Iran, the US, Algeria, Venezuela, and Nigeria have remained the top-seven gas-flaring countries year over year. These seven countries produce 40% of the world's oil each year, but account for roughly 65% of global gas flaring.[44]

Despite Nigeria's 1984 ban on gas flaring, a lack of political will and effective regulation allows the practice to continue largely unabated.[45] If Nigeria had flared less and utilized the associated gas from crude oil exploitation from its 1958 inception through to 2004, the country could have earned an estimated $32 billion profit (Fig. 6.7).[46] Gas diverted from flaring could have annually provided Nigerians billions of kilowatt-hours of electricity, in a nation where a high percentage of the population lives without electricity.

In 2005, impacted parties from the Niger Delta Iwherekan community in the Gbemre v. Shell Petroleum lawsuit argued that Shell was flagrantly negligent in its flaring practices over decades. The court ruled in favor of the plaintiffs, agreeing that their right to a clean environment was violated. The judge ordered an immediate cessation to flaring coupled with revised policy to address human rights obligations.[47]

The Nigerian state remains an ineffective regulator and enables the oil industry's worst practices. For example, in 2018, the Nigerian President signed Flare Gas Regulations into law that focused on ownership and marketing issues, leaving communities

[42] Ibunge, B. (2022, September 3). Bodo community laments four oil spills in 1 month from faulty Shell facilities. *This Day Live.*

[43] Ajugwo, A. (2013). Negative Effects of Gas Flaring: The Nigerian Experience. Journal of Environmental Pollution and Human Health. 1(1) 6–8.

[44] World Bank. (2021). Global Gas Flaring Tracker Project.

[45] May, J. R., & Dayo, T. (2019). Dignity and Environmental Justice in Nigeria: The Case of Gbemre v. Shell. *Widener Law Review*, 25, 269.

[46] Amaechi, C. F., & Emejulu, M. J. (2021). Cost–Benefit Analysis of Associated Gas Flaring in the Niger–Delta Area of Nigeria (a case study of 1958–2004). *Journal of Applied Sciences and Environmental Management*, 25(3), 363–369.

[47] Federal High Court of Nigeria. (2005). Gbemre v Shell Petroleum Development Company of Nigeria.

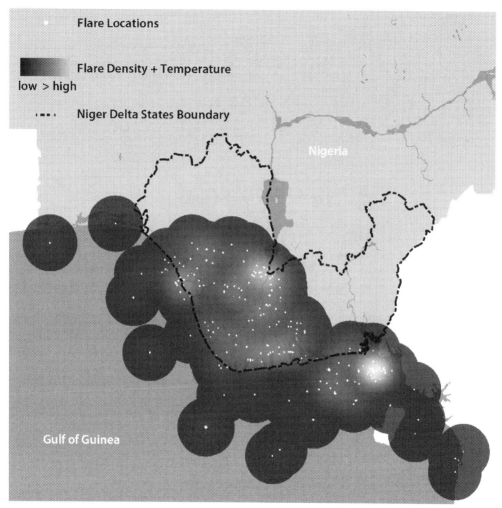

Figure 6.7 Flare intensity. *(Source: Global Survey of Natural Gas Flaring from Visible Infrared Imaging Radiometer Suite (2018-Present), accessed via SkyTruth.)*

where flaring persists without legal tools for remediation. As a signatory to the World Bank's Global Gas Flaring Reduction (GGFR) Partnership, Nigeria is pledging to end global flaring by 2030. Based on its record, skepticism remains that Nigeria will be able to effectively comply—agencies would need to monitor and track via remote sensing and report consistently to a transparent data platform. Across the Niger Delta, flare stacks (Fig. 6.8) are placed not only at the largest facilities, but along extractive water canals in hazardous proximity to villages.

Boom!

Armed factions make Nigeria a tinderbox of potential conflict. **Militancy**, which is the use of confrontational or violent methods in support of a cause, can be a response to oppression. The oil sector became a battleground as guerrilla groups sought to lash out

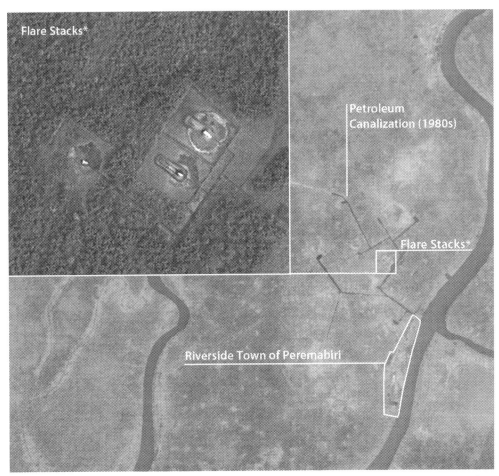

Figure 6.8 Extraction by water canals. *(Sources: Google Earth, 2021 Maxar Technologies, CNES/Airbus; Copernicus Sentinel data 2019, processed by ESA via Sentinel Hub.)*

at state entities and oil companies.[48] Sabotage involves intentional destruction of infrastructure or equipment and theft of oil.[49] Diverging positions exist within Nigeria and internationally about oil militancy: some are concerned that extreme responses will be

[48] Onuoha, F. C. (2008). Oil pipeline sabotage in Nigeria: Dimensions, actors and implications for national security. *African Security Studies, 17*(3), 99–115.

[49] Odalonu, B. H., & Eronmhonsele, J. (2015). The irony of amnesty program: incessant oil theft and illegal bunkering in the Niger Delta region of Nigeria. *International Journal of Humanities and Social Science Invention, 4*(8), 09–18.

used to justify ever more violence, while others argue the people of Nigeria have experienced too much deadly exploitation and must do whatever it takes to gain liberation.[50]

The Movement for the Emancipation of the Niger Delta (MEND), one of the largest militant groups in the region,[51] criticized ecological degradation from oil. In lieu of regulatory protections from the state to stop foreign exploitation, groups such as MEND take extreme actions, including kidnap-for-ransom of oil workers.

Armed assaults staged on oil production sites, outright pipeline destruction, and oil theft for black market profit were all tactics of MEND, whose leaders received amnesty in 2009. Oil militancy again surged in 2016 through a network of actors known as the Niger Delta Avengers (NDA).[52] Key demands from the NDA included site remediation and transfer of a percent of oil ownership to the people of the region.[53] Today, violence continues to play a role in shaping leadership structures across the Delta.[54] While onlookers not directly impacted by oil pollution may condemn militant organizing, they often lack viable solutions, watching passively for decades as governments, corporations, and international agencies allow human and environmental rights violations to persist unabated. In the case of Nigeria's Niger Delta, extreme energy poverty and energy insecurity compound vulnerabilities to violence.

In Nigeria, where safeguards and education are insufficient, explosions and fires occur with regularity at work sites,[55] although accidents are not fully reported in media or scholarly literature.[56] Accidents are often rural, where pipelines are exposed and unprotected and responses are haphazard. For example, in 2008 the Ijegun Pipeline explosion killed more than 100 after a bulldozer ruptured a pipeline near a school in which dozens of children died. During the chaos of the explosion, police and emergency personnel were seen fleeing instead of assisting school children.

The most-deadly Nigerian incident was the Jesse explosion in 1998: more than 1,000 people died and the fire burned for 5 days.[57] The national emergency response team did

[50] Watts, M. (2009). The rule of oil: Petro-politics and the anatomy of an insurgency. *Journal of African Development*, *11*(2), 27–56.

[51] Duffold, C. (2010, October 4). Who are Nigeria's Mend oil militants? BBC News.

[52] Onuoha, F. C. (2016). The resurgence of militancy in Nigeria's oil-rich Niger Delta and the dangers of militarisation. *Al Jazeera Center for Studies Report*, 8.

[53] Ejeh, O. P. A. W. A. (2017). The Niger Delta Avengers (NDA) War Against The Nigeria State & Multinational Oil Companies. *Journal of Humanities and Social Sciences*, *22*(3): 54–63.

[54] Ebiede, T. M., & Nyiayaana, K. (2022). How violence shapes contentious traditional leadership in Nigeria's Niger Delta. *Violence: An International Journal*, 26330024221090542.

[55] Staff. (2020, July 8). Explosion at Nigeria Oil Facility Kills Seven. Agence France-Presse.

[56] Carlson, L. C., Rogers, T. T., Kamara, T. B., Rybarczyk, M. M., Leow, J. J., Kirsch, T. D., & Kushner, A. L. (2015). Petroleum pipeline explosions in sub-Saharan Africa: A comprehensive systematic review of the academic and lay literature. *Burns*, *41*(3), 497–501.

[57] Akande, S. (2018, February 6). Nigeria's worst fire outbreak killed 1098 people in Delta state. *Pulse*.

not contain the fire and foreign assistance was necessary. The cause of the incident remains contested: some claim the fatal episode occurred following a lack of necessary maintenance. The company suggested oil thieves caused the disaster. The high death toll can be attributed in part to mistrust. For example, severely burned victims reportedly fled clinics for fear of being blamed for causing the fire. Nigerians lament the silences surrounding the event, leaving future responses ill-prepared faced with oil infrastructure failure and explosion.[58]

Most readers of this book, if caught in an accidental explosion, would feel free to go to hospital for life-saving treatment, without fear of being wrongly accused of sabotage. Such security should not be taken for granted. Near the site of the 1998 Jesse explosion, 2020 saw yet another tragedy known as the Oviri explosion which claimed the lives of 250 children and adults as they flocked to the site of a damaged gasoline pipeline, canisters in hand to scoop up fuel for roadside sale. Reports following the incident suggest two metal vessels clanged together and created a spark leading to the large explosion.[59] Oil theft via siphoning off oil or petrol from pipelines is common in Nigeria.[60] This highly dangerous practice is a reminder of how people living in sacrifice zones embody greater violence from the oil sector. Insecurity shapes perceptions of risk; people tolerate higher risk when they lack energy security, political security, and economic security.

Instead of addressing problems, Nigerian oil operations shifted to different spaces. Over the past decade, in part due to the high rate of sabotage of oil pipelines on land, there have been more rigs built offshore, further aligning domestic extraction to international export (Fig. 6.9). The intensification of oil is closely followed by gas exploitation. In 2022, the Nigerian President announced the country had commenced a "Decade of Gas,"[61] without recognition of any of the tradeoffs or drawbacks from fossil gas (Chapter 8). As with oil, corruption remains widespread with gas, suggesting Nigeria is unlikely to move smoothly to a diversified energy sector which can both meet the needs of its growing population and achieve its stated net-zero objectives.[62]

Shine on

Energy poverty in Nigeria persists in spite of exported oil, and still-available oil, gas, and solar resources. The majority of Nigerians resort to burning biomass (i.e., wood, manure, industrial or agricultural waste) for electricity—a cruel irony worsened by indoor air pollution, particularly harmful to the health of women and girls.

[58] Ibid.
[59] Staff. (2000, July 12). Nigerian pipeline explosion kills 250. *Oil and Gas Journal*.
[60] Okoli, A. C., & Orinya, S. (2013). Oil pipeline vandalism and Nigeria's national security. *Global Journal of Human Social Science, 13*(5), 67–75.
[61] Anyaogu, I. (2021, June 10). 'Decade of Gas': Nigeria moves from slogan to execution. *Business Day*.
[62] Climate Action Tracker. (2022). Nigeria.

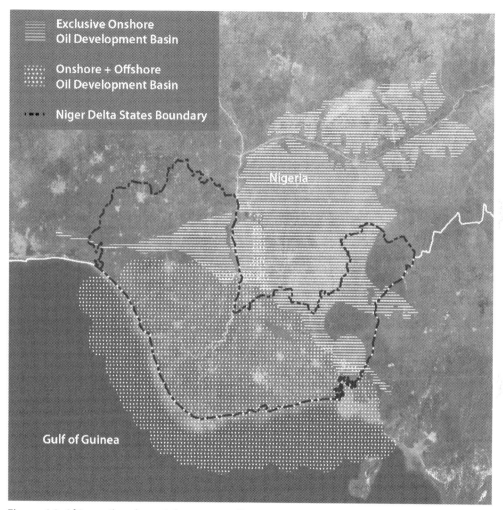

Figure 6.9 African oil and gas infrastructure. *(Sources: Global Oil and Gas Infrastructure (GOGI) Inventory; Esri, Maxar, Earthstar Geographics, CNES/Airbus DS, USDA FSA, USGS, Aerogrid, ING, IGP, and the GIS User Community.)*

Nigerians would benefit from locally produced clean renewable energy, yet infrastructure and knowledge-sharing have been tenuous. Nigeria has excellent potential for renewables, including wind, wave, geothermal, and solar energy.[63] Nigerian solar

[63] Adeyanju, G. C., Osobajo, O. A., Otitoju, A., & Ajide, O. (2020). Exploring the potentials, barriers and options for support in the Nigeria renewable energy industry. *Discover Sustainability*, 1(1), 1—14.

energy is not only its most affordable source,[64] but increasing water scarcity across the continent positions this renewable as a better utility-scale option than past investments in hydroelectric dams.[65]

In spite of the continent's renewables potential, state and donor subsidies continue to bolster Africa's fossil infrastructure. In particular, foreign and domestic investment in gas and particularly the liquefaction trains on Bonny Island immediately offshore the Niger Delta, is risky because it contributes to climate disruption coupled with sea level rise. In 2022 Nigeria LNG declared a *"force majeure"* due to flooding.[66] This legal safeguard is often exercised by fossil fuel industries to avoid liability, declaring disruptive events as beyond foreseeable control.

Unlike unending cycles of fossil fuel development, effective energy transformation should address enduring energy poverty, particularly acute in Nigeria. Reparations can be targeted and more effective than aggregated, low-interest loans laden with private sector constraints (Conclusion). Power can be owned and managed by the communities who use it. Small-scale, distributed, off-grid electrical systems, particularly effective when applied to more rural locations, is one of multiple applications.[67]

Energy transformation shifts power literally to the community level, avoiding the insertion of renewable energy sources into oppressive hierarchies typical of narrow energy transition projects. To illustrate this challenge, Shell's 2022 Renewable Energy division purchased Nigerian solar company Daystar.[68] As a subsidiary of Shell, Daystar runs the risk of extending exploitative practices common in its parent company if policies are not intentionally designed. Who governs renewable energy and how distribution occurs can make the difference between mere subsistence or security and prosperity in a local territory.[69]

The Nigerian Energy Compact committed to electrify five million homes using solar technologies. Jobs in renewables could double as a result, with solar employment

---

[64] Agbo, E. P., Edet, C. O., Magu, T. O., Njok, A. O., Ekpo, C. M., & Louis, H. (2021). Solar energy: A panacea for the electricity generation crisis in Nigeria. *Heliyon*, 7(5), e07016.

[65] International Renewable Energy Agency (n.d.). West African Clean Energy Corridor.

[66] Clowes, W. (2022, October 17). Nigerian LNG declares force majeure after flooding. Bloomberg.

[67] Adeyemi-Kayode, T. M., Misra, S., & Damaševičius, R. (2021). Impact analysis of renewable energy based generation in West Africa—a case study of Nigeria. Problemy Ekorozwoju, 16(1).

[68] Onu, E., & Osae-Brown, A. (2022, September 8). Shell acquires Nigerian solar firm in first Africa power buy. Bloomberg.

[69] Power For All. (2022). Powering Jobs Census 2022: Focus on Nigeria.

surpassing employment in Nigeria's fossil fuel sector.[70] Yet success remains uncertain within market-based models generated from banks and corporations.[71]

In Nigeria, renewable energies are largely driven by the private sector. Projects have a mixed record of accomplishment and have earned limited support at the local level. Structural inequalities throughout the energy sector impede "just" energy transitions, those which advance social justice while providing affordable and sustainable electricity.[72] For more holistic transformations, programs need to target local benefits, including training and access to technical and financial resources, for low-wealth Nigerians living in energy poverty and facing food and water crises.[73]

Microgrids and solar equipment that lower costs of electricity are becoming commonplace, but there are still access and usage gaps.[74] Greater effort is required to develop renewables in challenging contexts—one-size-fits-all approaches cannot merely be inserted across localities. Locally driven energy transformation requires active participation across all project phases from earliest planning to project completion, and can include regional manufacturing of renewables equipment and materials, further spurring local employment.

Try this

Research and advocate*: What would energy transformation look like for Nigeria? If you were an advisor for a sustainable energy policy or for a community-based project, what would you promote? Why and how?*

Ecuador

How did Ecuador succumb to such high levels of energy debt given its oil wealth? Ecuador's oil history long predates its incorporation into OPEC in 1974 (Chapter 5). As in Nigeria, an oil boom in Ecuador involving unscrupulous foreign firms led to a public health emergency.[75] Upon their exit, remaining state firms continued unsustainable extraction. Most recently, the Ecuadorian state has embraced renewables with Chinese financing, largely in the hydroelectric sector. Large dams are ineffective in times of water crisis, creating yet more energy insecurity.

[70] Onu, E. (2022, September 30). Jobs in clean energy in Nigeria to double by 2023, reports says. Bloomberg.

[71] Muhammad, R. (2022, May 30). How solar is driving Nigeria's post-pandemic green recovery. *The Guardian Ng*.

[72] Edomah, N. (2021). The governance of energy transition: lessons from the Nigerian electricity sector. *Energy, Sustainability and Society, 11*(1), 1—12.

[73] Casey, J. P. (2022, February 16). Electrifying Nigeria: could solar power one million households?.

[74] International Renewable Energy Agency. (2018, March 8). Empowering women in Nigeria with solar.

[75] Vásquez, P. I. (2014). *Oil Sparks in the Amazon: Local Conflicts, Indigenous Populations, and Natural Resources*. University of Georgia Press.

Modern exploration started in 1878,[76] growing exponentially in the mid- to late 20th century with the introduction of an insatiable US and global market. While Shell Oil received the first oil concession in the Amazonian region in 1937, expansion skyrocketed as Texaco-Gulf started extraction c. 1967. Fig. 6.10 depicts the evidence of the oil boom in its later stages around Nueva Loja, a town founded in the 1960s as a Texaco basecamp. Deforestation frequently occurs with oil infrastructure including roads—visible as the mainly vertical and horizontal network pattern interrupting forested land. After companies make initial road cuts, smaller informal roads and forest clearings expand outward.

The Amazon, viewed as an extraction frontier by oil firms, is highly contested. The state owns the resources under the soil in Ecuador like much of Latin America.[77] The government's oil concessions overlap with the ancestral lands of Indigenous groups; some of this is titled, meaning tribes have formal rights under law.

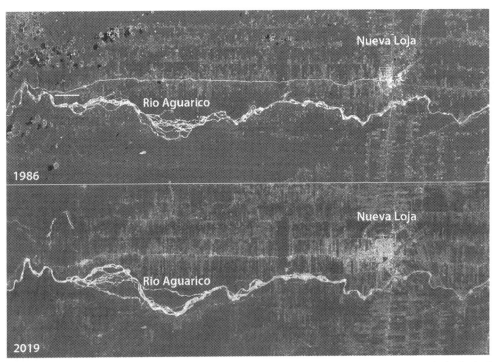

Figure 6.10 Infrastructure expansion and deforestation around Nueva Loja. *(Sources: U.S. Geological Survey (USGS), Landsat L5-TM, L7-ETM.)*

[76] Peláez-Samaniego, M. R., Garcia-Perez, M., Cortez, L. A. B., Oscullo, J., & Olmedo, G. (2007). Energy sector in Ecuador: Current status. *Energy Policy, 35*(8), 4177–4189.

[77] Haley, S. (2004). Institutional assets for negotiating the terms of development: Indigenous collective action and oil in Ecuador and Alaska. *Economic Development and Cultural Change, 53*(1), 191–213.

Spatial organization of the oil industry in Ecuador demonstrates a focus on crude export. Whether foreign-dominated or state-run, little domestic processing occurs, leaving Ecuadorians with an unrefined commodity of lower value. With historical predilection for crude export coupled with nearly nonexistent refining, an often crippling reliance on global markets emerged over time.[78] Notably, Koch Industries in the US strategically and obsessively hedged against this type of vulnerability by exerting spatial control over each and every component of its oil production, processing, and transport network (Chapter 5). In contrast, Ecuador features little more than clusters of extraction hotspots tightly wound to coastal export facilities—spatial evidence of energy colonialism (Fig. 6.11).

Exceedingly leaky pipelines cross the Amazon. Hundreds of unlined waste pits and decades of dumping and spilling produced an environmental disaster. After billions of dollars of oil damages, conservative estimates of fatalities suggest 1,400 could die as a

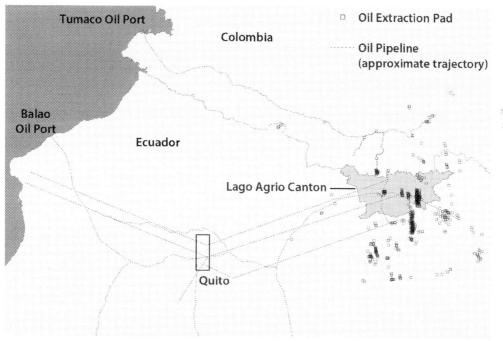

Figure 6.11 Ecuador's oil infrastructure. *(Sources: Global Oil and Gas Infrastructure (GOGI) Inventory; oil pad features extracted via OpenStreetMap (Overpass Turbo API).)*

---

[78] Ponce-Jara, M. A., Castro, M., Pelaez-Samaniego, M. R., Espinoza-Abad, J. L., & Ruiz, E. (2018). Electricity sector in Ecuador: An overview of the 2007—2017 decade. *Energy Policy, 113*, 513—522.

result of the contamination. Casualties could exceed 10 times that depending on the speed and effectiveness of remediation.[79] Assessments must consider the underreporting of regions with insufficient medical treatment.[80] Cancer rates exhibit geographical differences due to patterns of exposures.[81] Increases in miscarriages are tell-tale markers of toxic energy history.[82]

Oil created a health emergency among Ecuador's most vulnerable populations.[83] Thousands of impacted residents in five Amazonian tribes initiated a lawsuit against Chevron-Texaco in 1993. Legally assigning responsibility proved vexing. Hand-overs between Texaco, PetroEcuador, and later Chevron-Texaco complicated matters. Chevron acquired Texaco in 2001; the subsequent 2003 lawsuit meant a much larger, defiant defendant in Chevron-Texaco, who refused remediation and settlement when found guilty of dumping toxic waste water and spilling crude. The court ordered Chevron-Texaco to pay $18 billion designated for impacted communities in Lago Agrio, situated in oil blocks (Fig. 6.12).

If Chevron-Texaco had paid its court-ordered fines, a precedent for oil majors would have been established for gross negligence, setting the stage for more responsible practices. Instead, the Lago Agrio tragedy demonstrates how corporate oil firms skirt accountability.[84] Utilizing an arbitration process known as the Investor—State Dispute Settlement mechanism, nested within the Ecuador—United States Bilateral Investment Treaty,[85] Chevron-Texaco was able to leverage a convoluted, slow-paced international arbitration process seemingly tailor-made for corporate impunity.[86] Chevron-Texaco

[79] Kimerling, J. (2016). Habitat at Human Rights: Indigenous Huaorani in the Amazon Rainforest, Oil, and Ome Yasuni. *Vermont Law Review* 44: 445—524.

[80] Kelsh, M. A., Morimoto, L., & Lau, E. (2009). Cancer mortality and oil production in the Amazon Region of Ecuador, 1990—2005. *International Archives of Occupational and Environmental Health, 82*(3), 381.

[81] Hurtig, A. K., & San Sebastián, M. (2002). Geographical differences in cancer incidence in the Amazon basin of Ecuador in relation to residence near oil fields. *International Journal of Epidemiology, 31*(5), 1021—1027.

[82] San Sebastián, M., Armstrong, B., & Stephens, C. (2002). Outcomes of pregnancy among women living in the proximity of oil fields in the Amazon basin of Ecuador. *International Journal of Occupational and Environmental Health, 8*(4), 312—319.

[83] San Sebastián, M., & Hurtig, A. K. (2005). Oil development and health in the Amazon basin of Ecuador: the popular epidemiology process. *Social Science & Medicine, 60*(4), 799—807.

[84] Lu, F., Valdivia, G., & Silva, N. L. (2017). Oil, Revolution, and Indigenous Citizenship in Ecuador. London: Palgrave.

[85] Pellegrini, L., Arsel, M., Orta-Martínez, M., & Mena, C. F. (2020). International investment agreements, human rights, and environmental justice: The Texaco/Chevron case from the Ecuadorian Amazon. Journal of International Economic Law, 23(2), 455—468.

[86] Langewiesche, W. (2007, April 3). Jungle Law. *Vanity Fair*.

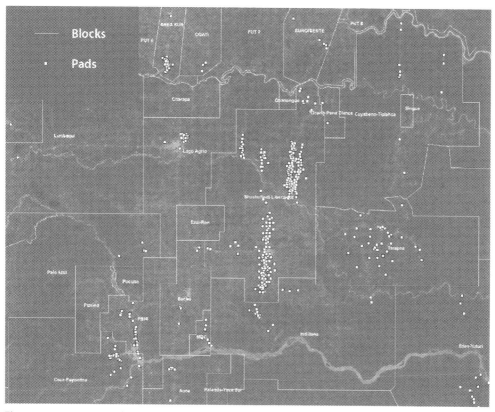

Figure 6.12 Lago Agrio block and pad. *(Sources: The Amazon Cooperation Treaty Organization (OTCA); oil pad features extracted via OpenStreetMap (Overpass Turbo API); Esri, Maxar, Earthstar Geographics, CNES/Airbus DS, USDA FSA, USGS, Aerogrid, ING, IGP, and the GIS User Community.)*

largely escaped responsibility through decades of legal cases.[87] For Lago Agrio, dangerous exposures have destroyed lives, while those in seats of power remain distant. Fig. 6.13 highlights important events; this is not an exhaustive list as lawsuits have been spun off to multiple countries.[88]

Protracted litigation opens the door to further harms, while closing it on justice delivered.[89] US courts focused on the missteps of the Ecuadorian government. Extreme pollution threatens customary lifestyles, causing displacement and cultural erosion.

[87] Kimberling, J. (2005). Indigenous peoples and the oil frontier in Amazonia: The case of Ecuador, ChevronTexaco, and Aguinda v. Texaco. *NYU Journal of International Law and Politics, 38*, 413.

[88] Keefe, P. R. (2012). Reversal of Fortune: The Lago Agrio Litigation. *Stanford Journal of Complex Litigation, 1*, 199.

[89] Kimerling, J. (2012). Lessons from the Chevron Ecuador Litigation: The Proposed Intervenors' perspective. *Stanford Journal of Complex Litigation, 1*, 241.

1 - Texaco enters Lago Agrio, Ecuador
2 - Inception of lawsuits against Texaco for environmental damages
3 - Chevron acquisition of Texaco
4 - Inception of lawsuit against ChevronTexaco within United States
5 - 18 billion Ecuadorean court judgement against ChevronTexaco
6 - Complaint against CEO of Chevron submitted to ICC
7 - Judicial harrassment, suspension, disbarment and incarceration of
    Steven Donziger, continuing into 2022

Figure 6.13 Lago Agrio timeline.

Psychological violence occurs most strongly in impacted communities, particularly when police, military, or paramilitary troops protect oil company interests at the expense of local groups.[90]

Harm extends out to professionals who serve oil-impacted communities, such as lawyer Steve Donzinger, who experienced excessive and punitive meddling and censorship.[91] Donziger's situation is one of the most extreme cases of legal assault by an energy company. Donziger was effective in bringing charges against ChevronTexaco[92] and sought criminal contempt charges against executives[93]; after ChevronTexaco was found guilty in Ecuador, the firm brought charges against him in the US that led to his disbarment and frozen bank accounts.[94]

[90] Finley—Brook, M., & Jordan, O. (2019). Energy Transition and Uneven Development. *Routledge Handbook of Latin American Development*. Routledge Press. pp. 446—457.
[91] Pellegrini, L., Arsel, M., Orta-Martínez, M., & Mena, C. F. (2020). International investment agreements, human rights, and environmental justice: The Texaco/Chevron case from the Ecuadorian Amazon. *Journal of International Economic Law, 23*(2), 455—468.
[92] Donziger, S. R. (2004). Rainforest Chernobyl: Litigating indigenous rights and the environment in Latin America. *Human Rights Brief, 11*(2), 1—4.
[93] Ofrias, L., & Roecker, G. (2019). Organized criminals, human rights defenders, and oil companies: Weaponization of the RICO Act across jurisdictional borders. *Focaal, 2019*(85), 37—50.
[94] Mella, R. A. (2016). The Enforcement of Foreign Judgments in the United States: The Chevron Corp v. Donziger Case. *NYU Journal of International Law and Politics, 49*, 635.

Why not just pay for damages in Ecuador? Instead hundreds of lawyers across 60 firms were hired to track Donziger's movements.[95] False allegations stemmed from a controversial witness coached and paid by ChevronTexaco, and a specific judge with ties to industry.[96] Donziger appealed his conviction yet spent 2 years under house arrest, something unheard of previously.[97] The US Department of Justice (DOJ) further refused jurisdiction, leading to a final US Supreme Court denial of Donziger's appeal.[98]

Try this

Research and engage: *Assess Texaco-Chevron's legal violence toward lawyer Steve Donzinger. Write an opinion piece for a newspaper addressing the extent that corporations should be able to attack professionals, such as lawyers or investigative journalists.*

Figure 6.14 Flaring in the Amazon. *(Image credit: Shoreline, Napo River, South of Lago Agrio, Vince Smith, 2010. Accessed via Flickr.)*

[95] Lerner, S. (2020, January 29). How the environmental lawyer who won a massive judgment against Chevron lost everything. The Intercept.

[96] Savage, K. and Westervelt, A. (2020, November 6). Trial of Attorney who fought Chevron's pollution in Ecuador begins Monday. Drilled News.

[97] Mindock, C. (2022, September 21). Chevron adversary Steven Donziger appeals conviction to the US Supreme Court. Reuters.

[98] Raymond, N. (2023, March 27). U.S. Supreme Court lets Chevron foe Donziger's contempt conviction stand. Reuters.

Figure 6.15 Spill in Lago Agrio. *(Image credit: Depollution, Lago Agrio, Julien Gomba, 2007. Accessed via Flickr.)*

Left behind in litigation's wake, the evidence of Lago Agrio's contamination can be detected, juxtaposed against the Amazonian rain forest. Looking close at Fig. 6.14 above, an oil and gas flare first appears hidden, yet they are exceedingly toxic and now have become a local, persistent feature. Both Figs. 6.15 above and 6.16 demonstrate the flagrant spills and abandonment that followed sloppy extraction practices as oil wastes create public health emergencies.[99] Crisscrossing the forest, oil's haphazard maze of infrastructure creates a regional hazardscape (Fig. 6.17).

[99] Raftopoulos, M. (2017). Contemporary debates on socio-environmental conflicts, extractivism and human rights in Latin America. *The International Journal of Human Rights* 21(4): 387–404.

Figure 6.16 Amazonian sacrifice zone. *(Image credit: Texaco's signature, Lago Agrio, Julien Gomba, 2007. Accessed via Flickr.)*

In this compromised terrain, the Ecuadorian state has used the military against its own people to defend the root cause of environmental degradation—oil extraction.[100] Armed security protect private drilling sites; state troops work for companies in exchange for food, vehicles and fuel.[101] The military routinely charges firms for transportation and other services. As self-financing activities are required to cover gaps in state budgetary allocations, the Ecuadorian military is "for hire" in the service of oil.[102]

[100] Finley—Brook, M. (2019). Extreme Energy Injustice and the Expansion of Capital. In *Organized Violence and the Expansion of Capital*. University of Regina Press. pp. 23—47.

[101] Vásquez, P. I. 2014. *Oil Sparks in the Amazon: Local Conflicts, Indigenous Populations, and Natural Resources*. Athens: University of Georgia Press.

[102] Jaskoski, M. (2013). *Military Politics and Democracy in the Andes*. The John Hopkins University Press.

Figure 6.17 Hazardscape. *(Image credit: Ecuador's drain, Lago Agrio, Julien Gomba, 2007. Accessed via Flickr.)*

Insecurity

To disrupt the longstanding pattern of US oil imperialism, Ecuador decided to seek national control with a new Constitution in 2008 making the government responsible for energy, seeking a programme of hydroelectric dam buildout with surplus to export energy.[103] The debt burden skyrocketed. While the US grip loosened, China took advantage with devastating consequences.[104] China has earned a reputation for **debt trap diplomacy**,[105] an international financial relationship where a creditor country or

---

[103] Finley–Brook, M. (2019). Extreme Energy Injustice and the Expansion of Capital. In *Organized Violence and the Expansion of Capital*. University of Regina Press. pp. 23—47.

[104] Freeman, C. P. (2017). Dam diplomacy? China's new neighborhood policy and Chinese dam-building companies. *Water International, 42*(2), 187—206; Warner, J. F., van Dijk, J. H., & Hidalgo, J. P. (2017). Old wine in new bottles: The adaptive capacity of the hydraulic mission in Ecuador. *Water Alternatives, 10*(2), 332—340.

[105] Brautigam, D. (2020). A critical look at Chinese 'debt-trap diplomacy': The rise of a meme. *Area Development and Policy, 5*(1), 1—14.

institution extends debt to a borrowing nation to increase the lender's political leverage. The use of this phrase has been contested, although the inequality of these financial relationships is clear. Chinese neoimperialism has unique characteristics,[106] but overall implications for trade partners repeat patterns from US imperialism and Spanish colonialism.

The need for power sharing under clear parameters transcends geography.

As one example with direct ties to oil, the Coca Codo Sinclair dam created harmful results distant from its proposed intent.[107] The largest energy project in Ecuador sought to shift unsustainable oil development, even relocating a pipeline to make way for the dam. Ecuadorian President Correa's broader goal was to end imperialism and dependence on conventional development funding sources.[108] Instead, the Coca Codo Sinclair project extended structural violence between China and Ecuador through inequitable trade terms between these two creditor and debtor nations.

Chinese development interventions blur boundaries between private sector expansion and state geopolitics.[109] Coca Codo Sinclair dam construction loans from the Chinese Export-Import Bank were $1.7 billion, with an interest rate of 7% over 15 years. Ecuador paid China with oil.[110] In 2018, this meant China received 80% of Ecuador's oil production. Lack of competitors bidding for contracts and requirements tied to specific operators or builders reinforced Chinese advantage.[111] Dam benefits leaked out of Ecuador.[112] As much as 75% of the construction materials had Chinese origin and a Chinese consortium held majority ownership.[113]

[106] Harvey, D. (2005). Neoliberalism 'with Chinese characteristics'. In *A brief history of neoliberalism*. Oxford University Press.

[107] Finley—Brook, M. (2021). Latin American hydropower sacrifice zones. In *Routledge Energy Democracy Handbook*, Feldpausch-Parker, A., Ed.

[108] Ellis, R. E. (2018). Ecuador's Leveraging of China to Pursue an Alternative Political and Development Path. *Journal of Indo-Pacific Affairs* 1,1: 79—104.

[109] Gerlak, A. K., Saguier, M., Mills-Novoa, M., Fearnside, P. M., & Albrecht, T. R. (2020). Dams, Chinese investments, and EIAs: A race to the bottom in South America? *Ambio*, 49(1), 156—164.

[110] Aidoo, R., Martin, P., Ye, M., & Quiroga, D. (2017). Footprints of the Dragon: China's Oil Diplomacy and Its Impacts on Sustainable Development Policy in Ecuador and Ghana. *International Development Policy*, 8(8.1).

[111] Siciliano, G., Del Bene, D., Scheidel, A., Liu, J., & Urban, F. (2019). Environmental justice and Chinese dam-building in the global South. *Current Opinion in Environmental Sustainability*, 37, 20—27.

[112] Ellis, R. E. (2018). Ecuador's Leveraging of China to Pursue an Alternative Political and Development Path. *Journal of Indo-Pacific Affairs* 1,1: 79—104.

[113] Garzón, P., & Castro, D. (2018). China—Ecuador Relations and the Development of the Hydro Sector. *Building*, 24: 37—57.

In 2020, Ecuador's tallest waterfall disappeared from its location on the Coca River.[114] The stream feeding the San Rafael waterfall failed in what was seemingly a sinkhole caused by hydro infrastructure construction.[115] The collapse of the hydro project spilled oil, poisoning communities.[116] The oil spill was not effectively remediated.[117] This particular catastrophe occurred against a backdrop of meager state transparency about oil spills generally, and impunity in cases that are heard.[118]

The Coca Codo Sinclair project is one of a series of new dams expected to generate more electricity than Ecuador's domestic need. The goal has always been to export power. To finance the dam projects, Ecuador initiated oil extraction from the previously protected Yasuní National Park,[119] a biodiverse protected area and Indigenous territory. Yasuní was the focus of a decade-long international campaign to pay to leave oil in the ground.[120] The proposal held promise for mitigating climate change; however, necessary finance fell short.[121] Oil companies built access roads into the primary forests of the Yasuní region. Old-growth Amazon rainforest is increasingly rare. A major portion of Ecuador's untapped oil reserves is under the Yasuní National Park in the Amazon.

Conflicts and tensions have continued between Native populations in Ecuador and the government. In 2022, 18 days of strikes involved thousands of Indigenous people blocking roads.[122] Protests were quelled only after the promise of permitting reform.[123] Further along in 2023, a successful public vote for the protection of remaining areas in Yasuní was taken—the Ecuadorian people chose to leave the oil in the ground.[124]

[114] Rivera-Parra, J. L., Vizcarra, C., Mora, K., Mayorga, H., & Dueñas, J. C. (2020). Spatial distribution of oil spills in the north eastern Ecuadorian Amazon: A comprehensive review of possible threats. *Biological Conservation*, *252*, 108820.

[115] NASA Earth Observatory. (2020, March 28). The disappearance of Ecuador's tallest waterfall.

[116] Krøijer, S. (2019). In the spirit of oil: Unintended flows and leaky lives in Northeastern Ecuador. In *Indigenous Life Projects and Extractivism* (pp. 95—118). Palgrave Macmillan.

[117] Alvaro, M. (2020, May). Dam implicated in waterfall collapse and oil spill. EcoAmericas.

[118] Velez Zuazo, A. E. & Romo, V. (2022, May 23). Stained by Oil: A history of spills and impunity in Peru, Colombia, Ecuador and Bolivia.

[119] Garzón, P. (2018). Implicaciones de la relación entre China y América Latina. Una mirada al caso ecuatoriano. *Ecología Política*, (56), 80—88.

[120] Finer, M., Moncel, R., & Jenkins, C. N. (2010). Leaving the oil under the Amazon: Ecuador's Yasuní-ITT Initiative. *Biotropica*, *42*(1), 63—66.

[121] Kingsbury, D. V., Kramarz, T., & Jacques, K. (2019). Populism or Petrostate?: The Afterlives of Ecuador's Yasuní-ITT Initiative. *Society & Natural Resources*, *32*(5), 530—547.

[122] Stand Earth. (2022, July 1). Solidarity statement: Indigenous Movement in Ecuador ends protests after achieving major policy shifts on day 18 of the strike.

[123] Palma, J. (2022, October 12). Ecuador's indigenous groups accept oil and mining freeze, but await new law. Dialogo Chino.

[124] Watts. J. (2023, August 24). The message from Ecuador is clear: people will vote to keep oil in the ground. *The Guardian*.

Try this

Debate: What are the appropriate roles of the US and China in the energy sector in Africa and Latin America? What type of relationship or intervention, if any, would you recommend? Why?

Granting rights to nature

Some jurisdictions are adopting laws granting rights to nature as a means for protection. In 2008, Ecuador was the first country to adopt **rights of nature** (RoN) in their constitution, followed by Bolivia in 2010, New Zealand in 2014, and Chile in 2022. In Ecuador, nature rights emerged from Indigenous cosmovision, or worldview, that understands Mother Earth (Pachamama) as sacred. In the Quechua language, **sumak kawsay** evokes a thriving collective based on the fullness of community life—this cosmovision is very roughly translated as **buen vivir** in Spanish. Native worldviews like this lack division between people and environment. Thus buen vivir requires more than introducing alternative views into an existing legal framework[125]—a primary objective is decolonization from state authorites steeped in capitalistic modes of extraction and growth.

Ecuador exhibits contradictions in populism as a petrostate.[126] Constitution reforms from 2008 could have translated into an energy transformation yet have only delivered energy transition. Ecuador's RoN remains more rhetoric than reality.[127] Nature's rights were used to block a foreign mine threatening local water; however, other cases have not been successful. Oil operations are often argued to be of pressing national interest, narrowing consideration of alternatives. Given the overlap between oil concessions and Indigenous territories, impacted communities have no choice but to engage with the state, put into a defensive position seeking basic respect for human, territorial, and natural rights.

Movements of alternative self-governance are strong in the Amazon and have a history of resistance to polluting industries. An Amazonian system of governance exists in the Confederation of Indigenous Nationalities of Ecuador (CONAIE), and in the regional Confederation of Nationalities of Ecuadorian Amazon (CONFENIAE) and the Coordinator of Indigenous Organizations of the Amazon River Basin (COICA). These confederations generate viable alternatives to the heavy hand of the state. For

[125] Akchurin, M. (2015). Constructing the rights of nature: Constitutional reform, mobilization, and environmental protection in Ecuador. *Law & Social Inquiry, 40*(4), 937—968.
[126] Valladares, C., & Boelens, R. (2017). Extractivism and the rights of nature: governmentality, 'convenient communities' and epistemic pacts in Ecuador. *Environmental Politics, 26*(6), 1015—1034.
[127] Kotzé, L. J., & Calzadilla, P. V. (2017). Somewhere between rhetoric and reality: environmental constitutionalism and the rights of nature in Ecuador. *Transnational Environmental Law, 6*(3), 401—433.

example, Amazon Sacred Headwaters organized by Indigenous federations exhibits bioregional planning based in radical, direct democracy.[128]

The community of Sarayaku, located southward of Lago Agrio, professes a vision of a Living Forest. "Power to the Protectors" with Amazon Watch has assisted Sarayaku with solarization in order to communicate globally.[129] Sarayaku's Indigenous Kichwa people, threatened by new oil expansion, received equipment and training to monitor and denounce territorial violence. The Sarayaku have also been subject of a landmark ruling in Sarayaku vs Ecuador at the Inter-American Court of Human Rights (IACHR).[130] This favorable ruling upheld Indigenous rights after an oil company had encroached on their ancestral lands.[131] There have been more than 1,000 spills in the Ecuadorian Amazon—spills which are considered an acceptable cost of business for the oil industry, yet a deep environmental wound for the Amazon's Indigenous communities.

Against the backdrop of oil's harm, Ecuador's transition to renewable sources has been slow with one exception: large dams, fraught with problems of their own making.[132] Initiatives like electric solar boats for transportation on the Amazon river[133] struggle to obtain funding. Opportunities for energy transformation in Ecuador remain largely unexplored, although wind, solar, biomass, and other options do indeed exist. A distributed supply of microgrids based on a mix of renewable energy sources (Conclusion) can support regional and local self-governance drawing from Amazonian cultural and governance traditions. Indigenous struggles for sovereignty find complementary modes in global collectives for strength and solidarity in the face of repression, who use counter-mapping as one strategy to point alternative paths forward.[134]

Both Nigeria and Ecuador require political and structural change. Energy transition (Introduction) in isolation brings limited benefit when it assumes an additive approach for expanding renewable energy sources alongside fossil fuels. Energy transformation addresses structural inequality, making deeper changes in not only energy types but also in governance. A key difference between limited energy transition and more

---

[128] Amazon Sacred Headwaters (n.d.). Bioregional plan 2030.

[129] Mazabanda, C. (2019, July 11). Power to the protectors in Ecuador. Amazon Watch.

[130] Koenig, K. (2022). Sarayaku's Kawsak Sacha is what the world needs now!. Amazon Watch.

[131] Palma, J. (2022, October 12). Ecuador's indigenous groups accept oil and mining freeze, but await new law. Dialogo Chino.

[132] Norouzi, N., & Fani, M. (2022). Post-Covid-19 Energy Transition Strategies: Even Reaching 100% Renewable in Ecuador by 2055 is not Enough to Face Climate Change Issue. *Iranian (Iranica) Journal of Energy & Environment*, *13*(1), 1–9.

[133] J. Ordoñez, J. Espinoza, J. Jara-Alvear and F. Guamán (2015. Electric-solar boats: an option for sustainable river transportation in the Ecuadorian Amazon, *Collection of open conferences in research transport*, 333.

[134] Orangotango (n.d.). Untangling the Strategies of Capital: Toward a Critical Atlas of Ecuador.

comprehensive energy transformation is equitable decision making procedures that can deter the rise of authoritarianism as temperatures rise and climate disruption intensifies.

If the status quo intensification of fossil fuels in petrostates like Ecuador and Nigeria continues unabated, so too our oil addiction- as a collective pathology.[135] After a full career researching Nigeria and other oil economies, geographer Michael Watts concludes our oil addiction is a social (and systemic) issue. It does not belong only to individual consumers; it is built into hydrocarbon capitalism itself, protected by corrupt politicians and captured state agencies. As climate disruption reminds us all through its consequences—particularly pernicious in global oil hotspots—what and when will be the breakup with large-scale oil consumption?

Try this

Research and reflect: *What might be achieved if rights of nature were upheld? What limitations might continue even with formally recognized rights of nature, even those added to the constitution, such as in Ecuador?*

Spatial and temporal synopsis

Space

(1) *Internal and external colonization and structural violence concentrate in ethnic territories with valuable resources; decolonization is inhibited by the geographical and racial injustice of oil.*

(2) *Remote siting of toxic sites means that customers that do not live near waste pits, flares, or other sacrifice zones might not know or care about harm.*

(3) *Identifying vertical and horizontal spaces in a topographical approach draws attention to hidden infrastructure and broader damages from oil.*

Time

(1) *Oil is an insecure industry demonstrating booms and busts: following peak oil, there will be abandonment.*

(2) *In formerly colonized southern petrostates like Nigeria and Ecuador, the harms of colonization extend with new cycles of exploitation coinciding with neoliberalism and neocolonialsm.*

(3) *The mismanagement of oil (e.g., unaccounted spills and flaring) accumulates, expanding harms and resource theft from exploited communities.*

(4) *Companies weaponize legal violence like protracted lawfare even as communities fight back to end oil's harms.*

[135] Watts, M. (2021). There Will be Blood: Oil Curse, Fossil Dependency and Petro-addiction. *New Formations, 103*(103), 10–42.

Summary

Protracted lawfare allows international companies who cause oil pollution to avoid costly repercussions, abandoning heavily polluted areas without reparations. The petrostates of Nigeria and Ecuador are also responsible and continue to extract with wasteful methods even after international firms have departed. Both countries have been left in situations of insecurity and indebtedness. Corrupt leaders extract additional resources, reinforcing oppression of ethnic minorities and extending sacrifice zones. Structural violence threatens local initiatives that embrace renewable energy, instead advancing yet more expensive and expansive fossil gas infrastructure.

Vocabulary

1. abandonment
2. bioregion
3. boom
4. boomtown
5. boosterism
6. buen vivir
7. bust
8. corruption
9. debt trap diplomacy
10. debt treadmill
11. decolonization
12. elites
13. flares
14. horizontal space
15. legal violence
16. militancy
17. neoimperialism
18. neoliberalism
19. Ogoni-9
20. petrostate
21. region
22. rights of nature (RoN)
23. sumak kawsay
24. topographical approach
25. untouchable zone
26. vertical space

Recommended

Books

Kashi, E., & Watts, M. (2008). *Curse of the black gold*. New York: Powerhouse.
Shapiro, J., & McNeish, J. A. (2021). *Our extractive age: Expressions of violence and resistance*. Taylor & Francis.

Video

Cullen, D. (2023). Hell in Niger delta. *Spelling Mistakes Cost Lives*.

CHAPTER 7

Rigged

Contents

Climate Crisis, Energy Violence
ISBN 978-0-12-819501-7,
https://doi.org/10.1016/B978-0-12-819501-7.00004-7

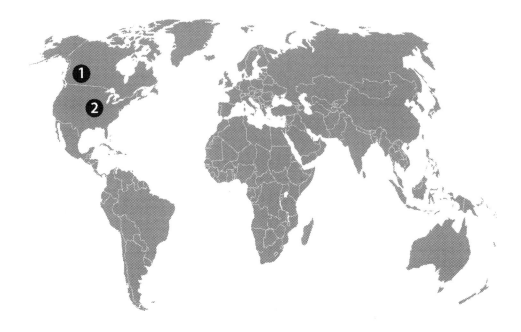

Locations: 1 - Alberta Province - Canada
2 - United States

Themes: Environmental racism; Surveillance; Territorial violence

Subthemes: Indigenous Peoples; Insecurity; Militancy

Pollution is colonialism.[1]

Power, privilege, and prejudice

Environmental racism is found both in practice and policy resulting in the slow poisoning of people of color.[2] Racism is prejudice, discrimination, or antagonism directed against a person or people on the basis of skin color, and hatred toward people of color. Historically, oil companies have taken advantage of racial inequality. Effective environmental justice (Introduction) requires meaningful involvement in ecological laws,

[1] Liboiron, M. (2021). *Pollution Is Colonialism*. Duke University Press.
[2] Turner, R. (2016). The slow poisoning of black bodies: A lesson in environmental racism and hidden violence. *Meridians*, *15*(1), 189–204.

policies, and decisions leading to fair, equitable, and non-discriminatory treatment during resource governance so that no group of people has to bear a disproportionate share of negative consequences resulting from policies, decisions, and actions. While some reforms have been instituted, environmental regulatory agencies continue to abide and enforce policy that fails to deliver meaningful justice for impacted communities—often Indigenous and of color. In practice, state environmental regulatory agencies frequently reinforce ethnocentric, northern, and Western ideals.

Race is a central factor in climate injustice at a global scale with dire and fatal consequences.[3] Whitewashed environmental justice doesn't attend to anti-Black racism and is silent on reparations designed to achieve equity. Broadly, environmental policy and practice lacks context from histories of oppression due to **global white privilege**, the dangerous, disproven belief that lighter-skinned people constitute a superior race,[4] with material and physical implications. In a prominent US example, the governmental response cast Indigenous water defenders as a problem to be contained and silenced in order to construct the Dakota Access Pipeline (DAPL). As DAPL planners sought to divert water impacts to a majority white location onto Native communities, water defenders put their own bodies on the line to protect their land and fundamental human rights.[5]

The energy system reflects power.

DAPL's upstream extraction source in Alberta, Canada is expansive tar sands fields, a prime example of extreme energy marked by limited environmental controls and containment resulting in polluted waterway impacts to customary subsistence from fishing, hunting, and gathering.[6] In aggregate, GHGs from tar sands are exceedingly high, while the net energy gain of its oil products is limited.

Closely linked to justice, **security** entails freedom from danger or threat. Tar sands oil extraction threatens Indigenous peoples directly in two ways. Climate breakdown itself becomes genocidal for Native populations in Canada's Arctic, where warming depletes ice critical to hunting and food storage. Contrarily, state actors transform Indigenous subjects from special stakeholders into a "security" risk, thus enabling extraordinary means of control even as resources are stolen from their historic territories.

[3] Gonzalez, C. (2020). Climate change, race, and migration. *Journal of Law and Political Economy, 1*(1).

[4] Nair, C. (2022). *Dismantling Global White Privilege: Equity for a Post-Western World*. Berrett-Koehler Publishers.

[5] Thorbecke, C. (2016, November 3). Why a previously proposed route for the Dakota Access Pipeline was rejected. ABC News.

[6] Rosa, L., Davis, K. F., Rulli, M. C., & D'Odorico, P. (2017). Environmental consequences of oil production from oil sands. *Earth's Future, 5*(2), 158–170.

Sioux scholar Nick Estes reminds that in an expected process of truth and then reconciliation,

"We haven't even gotten to the truth part."

Pipelining of Indigenous territories demonstrates the need to confront the colonial foundations of institutions and networks that inconspicuously yet powerfully reproduce inequity and disinformation. Nearly every country in the world recognized Indigenous Peoples rights to **Free, Prior and Informed Consent** (FPIC)—this right was written into international law under the UN Declaration on the Rights of Indigenous Peoples in 2007.[7] The United Nations Declaration on the Rights of Indigenous Peoples (UNDRIP) lays out in no uncertain terms that Indigenous Peoples have the right to say no to extraction, and yet territorial violence remains a common outcome in the fossil fuel energy sector. Implementation of FPIC remains partial and patchy across extractive sectors, with few countries effectively upholding the accord even as companies working on Native lands must seek and obtain FPIC, and meaningfully involve impacted groups in decision-making.

Historians of liberal thought will point out that criminalization of environmentalism and suspension of habeas corpus rights is not new: liberalism might often make claims to be a political tradition committed to freedom, but evidence shows a preoccupation with order and security.[8] The outcome of structural violence is reproduction of white supremacy in myriad institutions including and beyond governments—perhaps not as obvious is the role of nongovernmental organizations like environmental groups and universities.[9] Institutions that do not directly confront their colonial foundations implicitly replicate bias and normalize discrimination.[10]

In colonizing education systems, people are socialized for bias. **Miseducation** refers to educational systems divorced from the realities of learners, thus reproducing and extending prejudice.[11] Mainstream education often ignores or erases practices of dispossession, reverting to a whitewashed history lacking reflection on how colonizers devalued and severed connections to lands and resources that sustained communities for centuries. For example, the US has broken every treaty signed with Indigenous nations,[12] giving

[7] United Nations General Assembly. (2007). *United Nations Declaration On The Rights Of Indigenous Peoples.*

[8] Grasso, M.T., & Bessant, J. (2018). *Governing youth politics in the age of surveillance.* Routledge. P. 19.

[9] Bosworth, K. (2021). "They're treating us like Indians!": Political Ecologies of Property and Race in North American Pipeline Populism. *Antipode, 53*(3), 665–685.

[10] Stein, S. (2022). *Unsettling the university: Confronting the colonial foundations of US higher education.* Johns Hopkins University Press.

[11] Woodson, C. G. (1933). *The Miseducation of the Negro.* Association Press.

[12] Glick, A. A. (2019). The wild west re-lived: Oil pipelines threaten Native American tribal lands. *Villanova Environmental Law Journal, 30,* 105.

way to outright theft of oil and natural resources. The oil industry has repeatedly violated treaty rights, constitutional rights, and global accords—all the while propagating a destiny narrative that marries oil expansion to nationalism.

Energy violence

Fossil fuel industries are known to perpetuate **gender violence**: physical, sexual, verbal, emotional, and psychological abuse, threats, coercion, and deprivation directed at an individual based on biological sex or gender identity. The intensifying crisis of missing and murdered Indigenous girls, women, and two-spirits (people who are transgender and other genders) necessitates urgent attention it does not receive.[13] Power differentials in the energy sector are readily apparent as inadequate and failed actions do little to stem disappearances near sites of oil and gas in Canada and the US.[14]

Indigenous people are assaulted at epidemic proportions despite many years of awareness of the issue and advocacy from Native communities, exposing lack of political will.[15] Unaddressed violent behavior where pipelines are colocated with Indigenous territories violates human rights.[16] Discrimination is fully evident in dozens of hotspots absent of targeted, effective interventions to stem violence.

To treat Native women's death as inevitable suggests their lives matter less.

In Canada, the Royal Canadian Mounted Police (RCMP) number approximately 1,200 cases of missing and murdered Indigenous girls and women over the past 3 decades; research by the Native Women's Association of Canada suggests those murdered could be as high as 4,000.[17] Most victims were under the age of 30 and the majority were mothers at the time of disappearance or death.[18] This crisis demonstrates uneven

[13] Martin, C. M. & Walia, H. (2019). Red Women Rising: Indigenous Women Survivors in Vancouver's Downtown Eastside.

[14] Anderson, K., Campbell, M., & Belcourt, C. (Eds.). (2018). *Keetsahnak/our missing and murdered indigenous sisters*. University of Alberta Press; Razack, S. H. (2016). Sexualized Violence and Colonialism: Reflections on the Inquiry into Missing and Murdered Indigenous Women. *Canadian Journal of Women and the Law, 28*(2), i–iv.

[15] Morton, K. (2018). Ugliness as Colonial Violence: Mediations of Murdered and Missing Indigenous Women. In *On the Politics of Ugliness* (pp. 259–289). Palgrave Macmillan.

[16] Aubrey, S. B. (2019). Violence against the earth begets violence against women: an analysis of the correlation between large extraction projects and missing and murdered indigenous women, and the laws that permit the phenomenon through an international human rights lens. *Ariz. J. Envtl. L. & Pol'y, 10*, 34.

[17] Levin, J. (2016, May 25). Dozens of Women Vanish on Canada's Highway of Tears, and Most Cases Are Unsolved. *New York Times*.

[18] Native Women's Association of Canada. (2021). NWAC Action Plan to End the Attack Against Indigenous Women, Girls, and Gender-Diverse People.

application of justice:[19] responses are steeped in victim blaming rather than examining unfair threats and risks, which intensify following entry of "man-camps" housing oil industry's transient workers.[20] Using FIFO employment tactics, outside workers frequently disregard local cultures, customs, and laws.[21]

Try this

Research and act*: What organizations demand change to confront gender violence near oil camps? What are their recommendations?*

Oil overwatch

Canada's fossil fuel governance exemplifies a **regime of obstruction**,[22] exhibiting patterns similar to other oil-dependent countries. Where the oil industry secures political and economic power, there is often a tolerance for excess expansion.[23] Firms design public relations campaigns to posit the oil industry as a national public good and obstruct change even when ecological and social costs are exceedingly high. Excluded from decision-making, Indigenous land and water protectors are compelled to use their bodies to block extraction, as seen in Fig. 7.1 during 2016 resistance actions in North Dakota.

Undergirding resistance, science overwhelmingly points to fossil fuel's climate crisis culpability. The necessity of new fossil fuel systems is strongly disputed—so too a blanket interpretation of all oil and gas pipelines as "critical" infrastructure. When infrastructure is deemed critical, states can disregard rights normally guaranteed.[24] **Critical infrastructure** is assets, systems, and networks (i.e., highways, railways, pipelines) considered so vital that their incapacitation or destruction would have a debilitating effect on national economic security, public health, and/or safety.[25]

---

[19] Lavell-Harvard, D. M., & Brant, J. (2016). Forever Loved: Exposing the hidden crisis of missing and murdered indigenous women and girls in Canada. Demeter Press.

[20] Stern, J. (2021, May 28). Pipeline of Violence: The Oil Industry and Missing and Murdered Indigenous Women. *Immigration and Human Rights Law Review*.

[21] Washington, A. E. (2020). Booming Impacts: Analysing Bureau of Land Management Authority in Oil and Gas Leasing Amid the Missing and Murdered Indigenous Women's Crisis. *Administrative Law Review*, 72(4), 719–750.

[22] Carroll, W. K. (Ed.). (2021). *Regime of obstruction: How corporate power blocks energy democracy*. Athabasca University Press.

[23] Burrell, M., Grosse, C., & Mark, B. (2022). Resistance to petro-hegemony: A three terrains of power analysis of the Line three tar sands pipeline in Minnesota. *Energy Research & Social Science, 91*, 102724.

[24] Levine, C. (2021, December 21). New anti-protest laws cast a long shadow on First Amendment rights. The Center for Public Integrity.

[25] McCoy, B. C. (2021). Critical Infrastructure, Environmental Racism, and Protest: A Case Study in Cancer Alley, Louisiana. *Columbia Human Rights Law Review, 53*, 582.

Figure 7.1 DAPL resistance. *(Image credit: "Happi" American Horse direct action against DAPL, August 2016, Desiree Kane, accessed via Wikimedia Commons.)*

Authoritarian control enforces ecocide.[26] Throughout the Americas, oil companies rent state security to protect corporate assets, even as firms break treaties and trespass on Native land.[27] As oil has lost its SLO (Chapter 4), the sector's outright use of **fascism**—centralized autocracy marked by forcible suppression of opposition—is increasing. To defend new expansion of oil, companies rely on physical use of force. Globally, fossil fuel fascism increases the likelihood of planetary collapse within our lifetimes. Locally, community leaders are choosing to risk lives and liberty to stop oil firms from desecrating their territory.

[26] Gobby, J., & Everett, L. (2022). Policing Indigenous Land Defense and Climate Activism: Learnings from the Frontlines of Pipeline Resistance in Canada. In *Enforcing Ecocide* (pp. 89–121). Palgrave Macmillan.

[27] Finley–Brook, M. (2019). Extreme Energy Injustice and the Expansion of Capital in *Organized Violence and the Expansion of Capital*. Paley, D., & Granovsky-Larsen, S. Eds. University of Regina Press. pp. 23–47.

Climate denial and far-right extremism are two heads of the same monster.[28]

Promoted by oil and gas lobbyists, states seek harsher penalties for pipeline protesters. **Criminalization** happens by turning an activity into a criminal offense by making it illegal or by classifying persons as criminals by making their activities illegal.[29] In the US, 39 states have enacted legal actions targeting protest behaviors since 2017, including bans on protesting near oil and gas pipelines.[30]

Criminalization of protest is a global trend.[31] It often rests on **surveillance**, the monitoring of behavior, activities, or data for the purpose of information gathering, influencing, managing, or directing. A problematic asymmetry of power arises between the watcher and the watched, especially when surveillance is racialized.[32] Resource differentials are stark between multinational corporations like Energy Transfer, Enbridge, TC Energy, and the frontline communities caught in harm's way of oil infrastructuring.

Fig. 7.2 demonstrates how technology has come to play a central role in surveillance. **Biometrics** are body measurements and calculations related to human characteristics that can be used for authentication of personal identification and access control.

Promoting dangerous narratives about people of color, surveillance and policing has become a new avenue for corporate risk management.[33] **Geospatial overwatch** refers to using satellites and georeferencing equipment to observe from above or at a distance.[34] The phrase "overwatch" can also confer concern regarding the growing spatial reach of "security" operations. In the case of the Canadian crude and refined oil Transmountain Pipeline, a regime of constant surveillance of daily activities of Indigenous women—the Tiny House Warriors of the Secwepemc territory—resulted in charges of alleged crimes small and large.[35]

Left-leaning governments profess regret for past treatment of Indigenous nations and portend to defend cultural rights. Yet state security operations assist industry partners to

---

[28] Sen, B. (2021, January 20). Fossil-fueled Fascism: how the oil, gas and coal industries fund white supremacy and far-right politics. *Otherwords.*

[29] Ceric, I. (2020). Beyond Contempt: Injunctions, Land Defense, and the Criminalization of Indigenous Resistance. *South Atlantic Quarterly, 119*(2), 353—369.

[30] International Center for Not-for-profit Law. (2022). US Protest Law Tracker.

[31] American Civil Liberties Union. (2013). Take back the streets: Repression and criminalization of protest around the world.

[32] Harb, J., & Henne, K. (2019). Disinformation and resistance in the surveillance of indigenous protesters. In *Information, technology and control in a changing world* (pp. 187—211). Palgrave Macmillan, Cham.

[33] Proulx, C. (2014). Colonizing Surveillance: Canada Constructs an Indigenous Terror Threat. *Anthropologica* 56 (1): 83—100.

[34] Gil, I. C. (2020). The Global Eye or Foucault Rewired: Security, Control, and Scholarship in the Twenty-first Century. *Concepts for the Study of Culture*, 94.

[35] Parrish, W. (2022, May 1). How a major tar sands pipeline project threatens indigenous land rights. Type Investigations.

Technology-aided surveillance
Computer - monitoring of data and traffic on the internet
Cellular phones and telephones - location and recording
Cameras - these are set at important sites, intersections, body cams
Social network analysis - often taken from social media
Biometrics
Aerial surveillance - drones, satellites
Data mining and profiling; identification and credentials (machine readable; tracking)
Wireless tracking (i.e., mobile phones, Radio Frequency Identification (RFID) tags, human microchip)
Geolocation devices - Global Positioning System (GPS)

Human Roles

Stakeouts
Undercover government agents (infiltrators)
Hired security (public, private)

Figure 7.2 Surveillance techniques.

strategically incapacitate Native organizers defending their historic rights.[36] For example, with Idle No More[37]—a network across Canada started by Indigenous women—RCMP orchestrated a surveillance initiative called Project SITKA. Project SITKA tracked 89 Native activists identified as having the potential of criminal threat to the state, all without direct evidence of past harm to critical infrastructure, the alleged motivation for their surveillance.[38] Law enforcement as well as unnamed industry partners received weekly records about the activities of Indigenous targets.

---

[36] Howe, M., & Monaghan, J. (2018). Strategic incapacitation of Indigenous dissent: Crowd theories, risk management, and settler colonial policing. *The Canadian Journal of Sociology/Cahiers canadiens de sociologie*, *43*(4), 325–348.

[37] Preston, J. (2013). Neoliberal settler colonialism, Canada and the tar sands. *Race & Class* 55(2): 42–59.

[38] Crosby, A. & Monaghan, J. (2016). Settler Colonialism and the Policing of Idle No More. *Social Justice*, *43*: 37–57.

There is a trend of profiling people of color in security operations.[39] Private firms share data they gather about organizers with state security forces.[40] Due to infrastructure being classified as critical, covert intelligence exchanges exist without oversight. Bias toward the claims of industry leads the state to violate rights.[41]

In the case of DAPL, Indigenous protesters, now deemed trespassers on their own Native lands, were harassed and arrested. US federal agencies used counter-insurgency tactics against the DAPL protesters, including planting undercover agents[42] and scrambling communication lines to and from Standing Rock.[43] Physical repression involved a range of local, state, and federal troops who employed excessive amounts of tear gas and doused people with water cannons in freezing temperatures, among other cruel and dangerous practices.

In particular instances, policing practices in North America have been found to be rooted in colonialism[44] and the intent to maintain social control following emancipation of enslaved people (Chapter 2). Both digital as well as human technologies and tactics like those enumerated in Fig. 7.2 above can intensify intrusive practices like stalking and harrassing.[45] In turn, protest movements like the Tiny House Warriors become touchstones of resistance involving their own strategic use of technology and social media.[46] These initiatives fit with traditions of community defense, often women-led.[47] Adding to customary local traditions, the legal defense of land and water defenders finds support at the international level: in 2022, the UN General Assembly recognized a safe, secure, and healthy environment as a human right.[48]

[39] Browne, S. (2015). *Dark matters: On the surveillance of blackness.* Duke University Press.
[40] Walby, K., & Monaghan, J. (2011). Private Eyes and Public Order: Policing and Surveillance in the Suppression of Animal Rights Activists in Canada. *Social Movement Studies* 10(1): 21–37; Hansen, H. K. & Uldam, J. (2015). Corporate social responsibility, corporate surveillance and neutralizing corporate resistance. *The Routledge International Handbook of the Crimes of the Powerful,* 186–196.
[41] LeFevre, T. A. (2015). Settler Colonialism. Oxford Bibliographies.
[42] Brown, A. (2018, December 30). The Infiltrator: How an undercover oil industry mercenary tricked pipeline opponents into believing he was one of them. The Intercept.
[43] Brown, A. (2020, November 24). Minnesota tells pipeline company not to run "counterinsurgency" against protestors. The Intercept.
[44] Crosby, A., & Monaghan, J. (2018). *Policing indigenous movements: Dissent and the security state.* Fernwood Publishing.
[45] Mason, C. & Magnet, S. (2012). Surveillance studies and violence against women. *Surveillance & Society* 10(2): 105–118.
[46] Spiegel, S. J. (2021). Fossil fuel violence and visual practices on Indigenous land: Watching, witnessing and resisting settler-colonial injustices. *Energy Research & Social Science,* 79, 102189.
[47] Minuteaglio, R. (2019, November 19). How tiny houses became a symbol of resistance for indigenous women. *Elle.*
[48] UN General Assembly. (2022, July 22). UN General Assembly declares access to clean and healthy environment a universal human right.

Try this

Research and engage: Should surveillance enter the private sphere? Write an opinion piece in a newspaper addressing transparency regarding who can be surveilled, how, and for what reason?

Spatial distribution

Oil companies pick pipeline routes largely based on technical and market criteria divorced from historical social context of transected communities. Once areas are degraded with infrastructure, their downgraded status becomes justification for yet more infrastructure **siting**, meaning to fix or build something in a particular place. Intensification of hotspots and corridors in the oil industry inadvertently bolsters pipeline opposition. For example, a physical **blockade** occupies strategic points that halt or delay momentum or impede access (Fig. 7.3).

Figure 7.3 Strategic insurrection to blockade tar sands oil. *(Image credit: You Shall Not Pass, September 23, 2012 - Tar Sands Blockade, Laura Borealis. Accessed via Flickr.)*

Disrupting operations builds pressure for change. A collective blockage of strategic corridors often targets sites of high importance, like a **chokepoint**: a narrow section of a critical strait or transportation corridor. Other spatially informed opposition strategies (Fig. 7.4) include tree-sits halting construction along pipelines and swarming with groups of participants on bikes or on foot impeding motorists.[49]

Blocking and disrupting

Strike (i.e., worker, hunger, climate, transport, prison, etc.)

Boycott/money deprivation[50]

Civil disobedience; direct action (i.e., sit-in, 'die-in', lock down, meeting disruption, occupation or blockage)[51]

Communication

Door knocking; door-to-door (word of mouth) information sharing

Rename streets and buildings; remove/change commemorative signs, monuments, etc.

Organize rallies, marches, parades, events, street theater, teach-ins, etc.

Worldmaking: create alternative structures

Host a People's Tribunal[51] or citizen assembly

Establish mutual support or assistance[53]

Pass a Bill of Rights for Nature

Divest from fossil fuels and invest in job-producing clean alternatives[54]

Skillshare (barter economy)

Affordable electricity and broadband cooperatives

Figure 7.4 Organizing strategies.[50-54]

[49] Malm, A. (2021). *How to Blow Up a Pipeline*. Verso.
[50] Fossil Free. (n.d.). We can build a fossil free world.
[51] By 2020 We Rise Up. (2019). Ideas for Actions.
[52] Finley–Brook, M., Oba, C., Fjord, L., Walker, R. W., Allen, L. V. R., Harris, F., … & Metts, S. (2021). Racism and Toxic Burden in Rural Dixie. *William & Mary Environmental Law and Policy Review*, 46, 603.
[53] Pflug-Back, K.R. (2019, September 5). Mutual aid for the end of the world. *Briar Patch Magazine*.
[54] DivestInvest. (n.d.). About us.; UNEP (n.d.). Divest-Invest Global Movement.

For opposition movements, spontaneous organization is enhanced by digitization while relying on connective networks and backchannels. These activities tend to be anti-authoritarian.[55] Through the leaderless nature of **horizontalism**—organizing with a focus on equity and collaboration as peers—the need or desire for figureheads is diminished.

"Event activism" includes protest marches and other scheduled protest activities that last a short duration. Tabling, leafleting, press rallies, and symbolic arrests are corollary activities often with limited spatial and temporal reach.[56] These tactics are largely symbolic and require further strategic use of power to create meaningful change. Successful blockades often depend on both effective internal structure as well as geographic strategy. Fig. 7.5 presents physical and digital strategies combined together as hybrid activism.

Unfortunately, many populations around the globe are unable to express their opinions and concerns openly for fear of retaliation. Young people are one of the most intensely regulated groups,[57] and Black and Brown youth receive even more counter-

Physical Tactics	Digital/Remote Strategies
Petitions & boycotts	Online petitions & boycotts
Demonstrations, events	Online commentaries, blogs, posts
Political songs or chants	Social media targeted 'storm'
Strikes/Walk out	Digital strike
Sit-ins/rallies	Virtual display of images from events
Site occupations	Hashtag takeover
Public art or graffiti	Memes, GIFs, TikTok, documentary
Watchdogging	Watchdogging (**Conclusion**)
Countermapping	Countermapping (**Chapter 9**)
Witnessing	Remote sensing (**Chapter 1**)

Figure 7.5 Overt organizing strategies.

[55] Tufekci, Z. (2017). *Twitter and tear gas: The power and fragility of networked protest.* Yale University Press.
[56] McBay, A. (2019). *Full Spectrum Resistance, Volumes I & II.* Seven Stories Press.
[57] Grasso, M.T., & Bessant, J. (2018). Governing youth politics in the age of surveillance. Routledge.

insurgency attention than average.[58] Enhanced vigilance and controls are aimed at BIPOC (Black, Indigenous, people of color) populations of all ages when compared to whites.[59]

Infrapolitics refers to everyday resistance that is quiet, dispersed, disguised, or otherwise seemingly invisible to elites, the state, or mainstream society. Infrapolitics, a prevailing genre of day-to-day politics for most of the world's disenfranchised, is careful and evasive politics that avoids risk.[60] These strategies were developed in authoritarian contexts yet can apply to all areas where activism is discouraged or sanctioned. Marginalized groups may use covert tactics to express their dissent if they don't feel safe (Fig. 7.6).

Organizers develop under-the-radar tactics in response to surveillance, and so surveillance operations continue to evolve as well. Newer forms of surveillance leverage vertical space to watch from above with drones or satellites.[61] The concept of three-dimensional security highlights vertical surveillance geographies.[62]

Effective networks are key components of resistance movements, extending strategies outward. For example, in mass protest events where movement is contained, networked columns of activists might work like "fingers" to move around deterrents. At a national scale, in Canada, protestors blocked train routes in support of Wet'suwet'en resistance to

Physical Tactics	Digital/Remote Strategies
Anonymity, undercover	Encrypted/coded text, pseudonyms, group accounts
Creating autonomous spaces	Closed groups, hidden networks
Sarcasm, humor and jokes	Sharing funny memes or cartoons

Figure 7.6 Infrapolitics.

[58] Springer, D. (2017, May 1). Mask Off: the monopoly on violence and re-invorating an anti-imperialist vision for black liberation. Medium.
[59] Huq, E., & Mochida, H. (2018). *The Rise of Environmental Fascism and the Securitization of Climate Change.* MIT Press.
[60] Scott, J. C. (2012). Infrapolitics and mobilizations: A response by James C. Scott. *Revue Francaise d'etudes Americaines,* (1), 112–117.
[61] Graham, S., & Hewitt, L. (2013). Getting off the ground: On the politics of urban verticality. *Progress in Human Geography,* 37(1), 72–92.
[62] Campbell, E. (2019). Three-dimensional security: Layers, spheres, volumes, milieus. *Political Geography,* 69, 10–21.

the Coastal Gaslink pipeline (Chapter 10) as leaders called for the abusive company to leave unceded territory.[63] With networked amplification to social media, blockades and protests like those at Wet'suwet'en can be rapidly communicated, building pressure in real time for local justice movements.

Temporal analysis

When Native groups build opposition sites, they can become **flashpoints** and influence uprisings and sites of resistance in other locations, discussed further in Chapter 8 as a movement spillover effect.[64] This occurred with the Keystone XL Pipeline and DAPL opposition movements. Flashpoints are sites of demonstrated harm where people are compelled to act with the knowledge that the harm is occurring in other places too, even if felt less acutely. Keystone XL (2011–22) helped inspire DAPL, and both coalitions inspired a surge in Indigenous movement building, further representing a turning point in the anti-pipeline movement.[65]

The DAPL movement grew from a base of Native tribes whose plight motivated thousands of concerned groups and individuals, including youth, elders, veterans, doctors, and environmental organizations. Since Native organizers were successful in slowing construction and garnering wide support in both the Keystone XL and DAPL protests, the response from the private sector was well-funded, leveraging connections with states to outmaneuver grassroots efforts. A case in point was the establishment of **fusion centers**. Organized spatially, there are 79 public–private anti-terrorism fusion centers found across US territories.[66] According to the Department of Justice, fusion centers provide "a mechanism where law enforcement, public safety, and private partners can come together with a common purpose." The US Homeland Security details, "Fusion Centers are state-owned and operated centers that serve as focal points in states and major urban areas for the receipt, analysis, gathering and sharing of threat-related information between State, Local, Tribal and Territorial (SLTT), federal and private sector partners." These protect critical infrastructure, of which private interests own 85%.[67] Yet in the case of DAPL to follow, a multinational energy company leveraged this structure put in place for federal homeland security to advance its own private gain.

[63] Zalik, A. (2020, March 19). TC Energy's name change: Rebooting Canadian pipeline empires. *The Conversation*.

[64] Spice, A. (2018). Fighting Invasive Infrastructures. *Environment and Society*, 9(1), 40.

[65] Frederick, K. (2021, May 24). The Dakota Access Pipeline is a stark violation of Indigenous sovereignty. Prism.

[66] See Electronic Frontier Foundation. (2022, September 28). Atlas of Surveillance.

[67] Brown, A., Parrish, W., & Speri, A. (2017, October 27). The Battle of Treaty Camp. The Intercept.

Globally, concern over the rise of fascism as a political ideology is increasingly common and warranted, particularly when exercised by fossil fuel interests.[68] Extending fossil fascism, **settler colonialism** both destroys and replaces peoples and lands—typically Indigenous peoples and territories[69]—through intrusive infrastructures.[70] Settler colonialism is a phrase that has become commonly used and points to **dispossession** that involves the act of depriving someone of land or property, often from marginalized groups, such as with the severance of social, political, or cultural practices from Native inhabitants. Indigenous territories experience regulatory violence, as protections are weak and state institutions can be manipulated. Settler colonialism is not relegated only to the past but is found to be ongoing, risking the normalization of violence, particularly related to fossil fuel infrastructuring.

Internal colonialism within the same nation-state, like foreign colonialism, often exhibits white supremacy and violates territorial sovereignty.[71] Applied broadly, settler colonialism can center non-Natives and thus reproduce colonial violence.[72] **Wastelanding**—a form of ecocide[73]—is one of settler colonialism's harmful impacts.[74] Native territories become slated as sacrifice zones. Land—long understood as the base of social relations[75]—becomes degraded within property parcels themselves, allowing for extractive resource concessions that are transactional, open to exploitation. In the Canadian tar sands, the state has been accused of permitting industrial genocide[76] as a result of dangerous levels of acute toxicity[77] amidst broad fragmentation of forests and communities.[78]

[68] Satgar, V. (2021). The Rise of Eco-Fascism. *Destroying Democracy: Neoliberal capitalism and the rise of authoritarian politics*, 25.

[69] Kauanui, J. K. (2016). "A structure, not an event": Settler colonialism and enduring indigeneity. *Lateral*, 5(1).

[70] Pasternak, S., Cowen, D., Clifford, R., Joseph, T., Scott, D. N., Spice, A., & Stark, H. K. (2023). Infrastructure, jurisdiction, extractivism: keywords for decolonizing geographies. *Political Geography*, 101, 102763.

[71] McCreary, T., & Turner, J. (2018). The contested scales of indigenous and settler jurisdiction: Unist'ot'en struggles with Canadian pipeline governance. *Studies in Political Economy*, 99(3), 223–245.

[72] Jafri, B. (2017). Ongoing colonial violence in settler states. *Lateral*, 6(1).

[73] Henderson, P. (2019). The Poetics of Settler Fatalism: Responses to Ecocide from within the Anthropocene. *Pivot: A Journal of Interdisciplinary Studies and Thought*, 7(1).

[74] Gross, L. (2019). Wastelanding the Bodies, Wastelanding the Land: Accidents as Evidence in the Albertan Oil Sands. In *Extracting Home in the Oil Sands* (pp. 82–100). Routledge.

[75] Gouldhawke, M. (2020, September 10). Land as a social relationship. *Briarpatch Magazine*.

[76] Huseman, J., & Short, D. (2012). 'A slow industrial genocide': tar sands and the indigenous peoples of northern Alberta. *The International Journal of Human Rights*, 16(1), 216–237.

[77] Finkel, M. L. (2018). The impact of oil sands on the environment and health. *Current Opinion in Environmental Science & Health*, 3, 52–55.

[78] Dando, C. E. (2022). "Tied to the land": Pipelines, Plains, and Place Attachment. *Geographical Review*, 112(1), 66–85.

Dispossession is not a "done deal."[79]

Survivance, an active sense of presence, combines Native resistance with survival. It draws attention to Indigenous worldmaking and generative acts of enlivening cultures. Exploitation of Native ideas and practices in popular culture often misses the struggles experienced, ignoring ongoing territorial violence. The Land Back movement, started in the 1960s, draws attention to ways in which theft from Native groups extends beyond territorial dispossession alone to include stealing knowledge, food, ceremony, language, and much more. Yet when Indigenous Peoples protest to protect their homelands and customary subsistence practices, punishments can be extreme in terms of fines and jail time, as demonstrated in the women-led Giniw Collective case below.

In an era of heightened political tensions, fossil fuel interests often take advantage of increasing polarization and distrust.[80] Through media strategy, fossil fuel interests influence journalists, putting at risk the field's role as watch dog.[81] In a time of declining revenue in many media operations, energy firms strategically influence press messages with corporate advertising dollars.[82] Pressures continue to mount on independent journalists due to consolidation of media outlets under corporate giants. Information in mainstream media is further limited and filtered—often little more than parroting industry bias. Even if overt racist phrases are absent, mainstream journalists rarely attend to colonial history or structural violence in any meaningful way; and they deliver too little coverage addressing Indigenous perspectives.[83]

Try this

Research and reflect: *Examine corporate media coverage of DAPL or a similar oil conflict, how does pro-industry news media amplify misrepresentations of Indigenous Peoples and reinforce environmental racism?*

---

[79] Kauanui, J. K. (2016). "A structure, not an event": Settler colonialism and enduring indigeneity. *Lateral*, 5(1).

[80] Altiparmak, S. O. (2022). Explaining Contentious Energy Policy: Coxian Structural Forces, Environmental Issues, and the Keystone XL Pipeline. *World Review of Political Economy*, 13, 1: 118–143.

[81] Bednar, M. K. (2012). Watchdog or lapdog? A behavioral view of the media as a corporate governance mechanism. *Academy of Management Journal*, 55(1), 131–150.

[82] Zukas, K. J. (2017). Framing wind energy: Strategic communication influences on journalistic coverage. *Mass Communication and Society*, 20(3), 427–449.

[83] Walker, C., Alexander, A., Doucette, M. B., Lewis, D., Neufeld, H. T., Martin, D., … & Castleden, H. (2019). Are the pens working for justice? News media coverage of renewable energy involving Indigenous Peoples in Canada. *Energy Research & Social Science*, 57, 101230.

Illustrative cases

The following illustrated cases show prominent watershed moments in oil industry malfeasance toward Native people and their lands. The gross misallocation of costs and benefits in Indigenous territories and the extensive public and private sector resources elevating oil fascism provide important precautionary lessons for today's energy transition. In earlier cases, legal and physical controls were prominent, while recent examples find Native organizers empowered through the deployment of digital technologies to support their cause.

Since its early commercialization, oil has been tied to predatory tactics. In the early twentieth century, Osage Indian territory in Oklahoma became the nexus of incredible wealth tied to an exploding murder rate.[84] Managed by the Bureau of Indian Affairs (BIA) in conjunction with tribal council (Fig. 7.7), the Osage territory, steeped in newly discovered oil, became the site of a complex legal structure that both advantaged Osage Indians while putting them at great risk of predation from outsiders.

Figure 7.7 BIA Commissioner with the Osage Tribal Council. *(Image credit: Library of Congress, Prints & Photographs Division, photograph by Harris & Ewing, [reproduction number, e.g., LC-USZ62-123456].)*

---

[84] Grann, D. (2017). *Killers of the Flower Moon: The Osage Murders and the Birth of the FBI*. Doubleday.

In 1907, while preparing Oklahoma for statehood, the federal government allotted 657 acres (266 ha) to each Osage person on the tribal rolls. Osage people and their legal heirs (Osage or not) were assigned "headrights" to royalties in newly found oil and its production. The tribe held the mineral rights communally, paying its members a percentage related to their holdings. Congress passed a law in 1921 requiring courts to appoint guardians for each Osage, usually white lawyers or businessmen, who would "manage" their royalties and financial affairs until they demonstrated "competency." Many guardians stole Osage land or royalties, and some were suspected of assassinating their charges to gain headrights. The Osage became statistically the most murdered people in the US. A federal team was sent to the territory to try to solve these crimes. Once there, the Bureau of Investigation, the preceding agency to the Federal Bureau of Investigation (FBI), found a growing market of contract killers murdering Osage people (Fig. 7.8).

In later years, predation became less overt yet more conspiratorial. Following the acquisition of 200,000 acres of Native land in a 1929 oil concession, several oil companies merged.[85] These oil companies—in particular one partner, Koch Industries

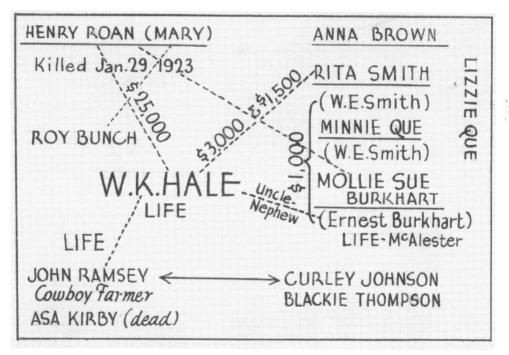

Figure 7.8 Osage FBI murder case. *(Image credit: Hale-Ramsey murder case diagram from the Oklahoman Collection at the Oklahoma Historical Society Photo Archives, photo by Rmosmittens, accessed via Wikimedia Commons.)*

[85] Juhasz, A. (2008). *The Tyranny of Oil.* Harper.

(Chapter 5)—designed a system of overage extraction whereby they routinely extracted more oil than paid for. This systematic oil theft specifically harmed Native Osage. In the late 1980s Koch Industries was finally found guilty following a US Senate investigation.[86] As a result of their experience of oil predation, the Osage were active participants in the 2016–17 DAPL protests.

While the Koch family paid a historic settlement in 1999 for their systematic theft from the Osage,[87] the Koch empire remains deeply committed to both racial and economic injustice. A highly secretive, private company, Koch Industries holds influential sway over the fossil fuel industry at large. Although Charles Koch proselytizes seemingly benign market-based management (Chapter 5), he also interferes behind the scenes to advantage corporate interests. Notorious climate deniers for decades, the Koch Brothers continue to fund networks of deception and obstruction, most visibly through the Heritage Foundation. Charles Koch, nicknamed the "Daddy Warbucks of climate disinformation," was called out with demands that he testify in front of the US House of Representatives Committee on Oversight and Accountability for his role promoting dangerous denialism.[88]

Keystone XL

The proposed Keystone XL Pipeline project, an extension of the Keystone Pipeline mainline, took many twists and turns before its demise in 2021.[89] The mainline Keystone Pipeline carries heavy, crude tar sands oil, to Port Author, Texas, situated on the Gulf Coast (Fig. 7.9), long considered a sacrifice zone due to the extensive illnesses in Black and Brown youth living proximate to oil infrastructure and spin-off chemical and energy plants.[90]

After years on the drawing board, The Obama Administration finally rejected Keystone XL Pipeline construction in 2016.[91] This decision was motivated by concern over greenhouse gas emissions and projected job numbers;[92] research found that pipeline developers likely inflated proposed employment numbers to generate public support.[93] As former President

[86] Leonard, C. (2019, July 22). How an oil theft investigation laid the groundwork of the Koch Playbook. Politico.

[87] Staff. (2000). Blood and Oil: Brothers' Feud Exposes Allegations of Fraud. CBS News.

[88] Negin, E. (2022, March 31) It's time for Charles Koch to testify about his Climate Change disinformation campaign. Union of Concerned Scientists.

[89] Denchak, M., & Lindwall, C. (2022, March 15). What is the Keystone XL Pipeline? How a single pipeline project became the epicenter of an enormous environmental, public health, and civil rights battle. Natural Resources Defense Center.

[90] Genoways, T. (2014). Port Arthur, Texas: American Sacrifice Zone. Natural Resources Defense Council.

[91] Davenport, C. (2015, November 6). Citing Climate Change, Obama Rejects Construction of Keystone XL Oil Pipeline. The New York Times.

[92] Staff. (2013, September 26). Obama Questions Keystone XL Pipeline Job Projections. Huffington Post.

[93] Global Labor Institute. (2011). Pipe dreams? Jobs gained, jobs lost by the construction of Keystone XL. Cornell University.

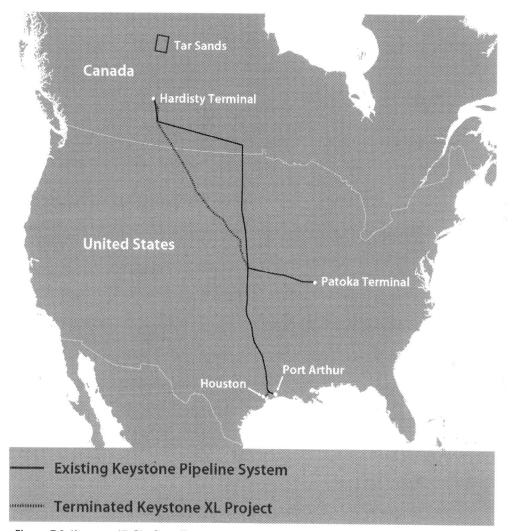

Figure 7.9 Keystone XL Pipeline. *(Sources: TC Energy; U.S. Energy Information Administration (EIA).)*

Trump took office, he resurrected the Keystone XL Pipeline project with boisterous boasts of jobs creation, claiming the pipeline would be built with US steel by US labor:

> **"We will build our own pipes,
> like we used to, in the old days."[94]**

---

[94] Henry, D., & Cama, T. (2017, January 24). Trump takes action to move forward with Keystone, Dakota Access pipelines. *The Hill.*

Trump's US steel pipes proved to be a fantasy. Setbacks continue to delay construction beyond the Trump Administration's departure from office in 2020.[95] Keystone XL was canceled yet again when the Biden Administration entered in 2021, revoking the permit citing research findings that the pipeline was not in the nation's interest.

The long Keystone XL saga ended in major economic losses for TC Energy,[96] which sought to sue the US government for damages but was not successful.[97] In 2021, when TC Energy abandoned the project, only 8% of the pipeline had been constructed.[98] The failure of Keystone XL is considered a major environmental victory—after a decade of sustained activism by thousands of opponents working against the project.[99] Yet a similar struggle continues as TC Energy generates controversy and threatens Indigenous sovereignty with its Canadian subsidiary Coastal Gas Link—a gas pipeline project winding its way across Native territories, destined to export gas offshore to Asia (Chapter 10).

Dakota Access Pipeline

The Dakota Access Pipeline has emerged as a watershed case for Indigenous rights:[100] the Indigenous-founded and Indigenous-led Water is Life Movement sprung forth as thousands of "water protectors" encamped for months, effectively blocking pipeline construction (Fig. 7.10).

In response, TigerSwan, an international security and global stability firm founded in 2008, was hired by Dakota Access, LLC, to administer a public—private intelligence "dragnet" at the Standing Rock protest location. These efforts involved integration with the Bureau of Indian Affairs and the Office of Homeland Security through the North Dakota State and Local Intelligence fusion center. Intelligence operations included aerial observation, radio eavesdropping, and high-tech tools to suppress dissent. On the ground, protestors experienced outright violence, leading to prolonged post-traumatic stress. Protestors were beaten with nightsticks, shot at in close proximity with rubber bullets, sprayed with tear gas to the point of requiring medical attention,

[95] Pulido, L., Bruno, T., Faiver-Serna, C., & Galentine, C. (2019). Environmental deregulation, spectacular racism, and white nationalism in the Trump era. *Annals of the American Association of Geographers, 109*(2), 520–532.

[96] Puko, Timothy (June 9, 2021). Keystone XL Oil Project Abandoned by Developer. *Wall Street Journal.*

[97] Earls. M. (2022, March 14). Keystone XL dispute ends months after project cancellation. *Bloomberg.*

[98] Staff. (2021, March 12). Fact Check: Though Keystone XL Pipeline had secured most of its funding, it was only 8% constructed. Reuters.

[99] Ternes, B., Ordner, J., & Cooper, D. H. (2020). Grassroots resistance to energy project encroachment: Analyzing environmental mobilization against the Keystone XL Pipeline. *Journal of Civil Society, 16*(1), 44–60.

[100] Donnella, L. (2016, November 22). The Standing Rock Resistance is Unprecedented (It's also centuries old). NPR.

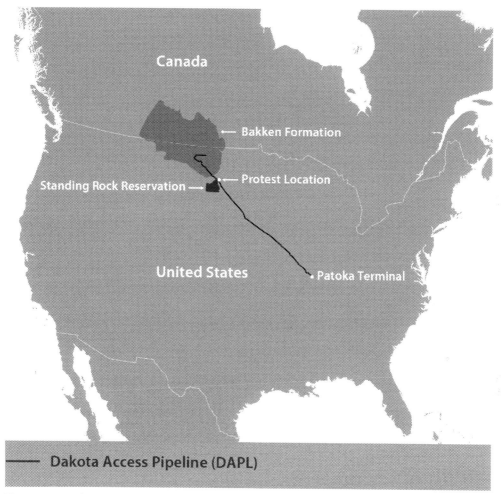

Figure 7.10 Dakota Access Pipeline. *(Sources: USGS, U.S. Census Bureau and U.S. Energy Information Administration (EIA).)*

soaked with water cannons in freezing temperatures,[101] and placed under traumatic stress.[102] Spiritual leaders were treated as "fanatics."[103]

[101] Goldberg, D. (2017). Lessons from Standing Rock—Of Water, Racism, and Solidarity. The New England Journal of Medicine 376(15): 1403–1407; Whyte, K. P. (2017). Standing Rock, Environmental Injustice and U.S. Colonialism. *Red Ink. 17*(1): 154–169.

[102] Isaacs, D. S., Tehee, M., Green, J., Straits, K. J., & Ellington, T. (2020). When psychologists take a stand: Barriers to trauma response services and advocacy for American Indian communities. *Journal of Trauma & Dissociation, 21*(4), 468–483.

[103] Juhasz, A. (2017, June 1). Paramilitary security tracked and targeted DAPL opponents as 'jihadists,' docs show. *Grist.*

TigerSwan asserted that peaceful demonstrators were actual violent extremists deserving suppression by violent measures. Over 900 injuries were recorded, mainly at the hands of armed police forces,[104] including National Guard troops from 10 different states.[105] Assaults on unarmed protesters occurred during prayer circles, without regard for the safety of children and elders.[106] More than 700 arrests were made at a cost surpassing $63 million dollars.[107] Those detained reported inhumane treatment in custody.[108] As charges surfaced that rights were violated, North Dakota sought permission to limit release of information related to these operations.[109]

FBI agents infiltrated camps by going undercover to pose as protestors.[110] Agents also surveilled protestors from the air. For example, National Guard equipment included an Avenger missile with night-vision surveillance.[111] Drones were also used in surveillance (Fig. 7.11).[112] Free from the constraints of a human body (i.e., a pilot), drones enlarge possibilities for overwatch. A **dronescape** is the landscape of an uncrewed aerial vehicle (drone), whose small mobile shape and long reach make it a popular surveillance method.

In spite of the violence, TigerSwan was allowed to influence other energy-related operation locations,[113] including Louisiana[114] and Puerto Rico.[115] For Energy Transfer's part, Standing Rock was just an extreme data point in an already troubled timeline of oil

[104] Levin, S., & Woolf, N. (2016). Dakota Access pipeline: police fire rubber bullets and mace activists during protest. *The Guardian*.

[105] Forum News Services. (2017). ND National Guard costs near $9 million for Dakota Access Pipeline response. *Grand Forks Herald*.

[106] Levin, S. (2017). Dakota Access Pipeline protestors say police used 'excessive force.' *The Guardian*.

[107] Nicholson, B. (2028, October 22). Judge refuses to limit North Dakota's federal lawsuit over DAPL protest policing costs. *The Bismarck Tribune*.

[108] Rogers, J. (2017). Insight into the conflict at Standing Rock: Extractive politics, indigeneity, violence, and local autonomy. Noragric, MA Thesis.

[109] Parrish, W. (2019, February 11). North Dakota seeks to restrict access to public records after Standing Rock reporting exposed law enforcement abuses. The Intercept.

[110] Brown, A. (2018, December 30). The Infiltrator: How an undercover oil industry mercenary tricked pipeline opponents into believing he was one of them. The Intercept.

[111] Lacambra, S. (2016). Investigating Law Enforcement's Possible Use of Surveillance Technology at Standing Rock. Electronic Frontier Foundation.

[112] Parrish, W. (2019, August 25). The US Border Patrol and an Israeli Military contractor are putting a Native American reservation under persistent surveillance. The Intercept.

[113] Brown, A. (2020, November 15). In the mercenaries' own words: Documents reveal TigerSwan infiltration of Standing Rock. The Intercept.

[114] Brown, A., & Parrish, W. (2018, March 30). A TigerSwan employee quietly registered a new business in Louisiana after the state denied the security firm a license to operate. The Intercept.

[115] Brown, A. (2018, March 12). From North Dakota to Puerto Rico, controversial security firm profits from oil protests and climate disasters. The Intercept.

Figure 7.11 Use of drones in surveillance of DAPL protestors. *(Image credit: Standing Rock, National Day of Mourning (Thanksgiving Day), Hrag Vartanian, 2016. Accessed via Flickr.)*

spills and toxic releases from their pipelines.[116] The violence of TigerSwan and Energy Transfer in Bayou Bridge Louisiana, following Standing Rock, is the theme of an EJ Atlas entry in Praxis 3, a demonstration of how community members push back against public–private conflict of interest to advance a controversial pipeline project.[117]

Like Koch Industries, the oil industry at large perpetuates economic inequality and white supremacy,[118] with repercussions extending beyond sites of oil extraction. Kelcy L. Warren, co-founder of Energy Transfer in 1996, retaining his position as largest shareholder and chairman of the board of directors,[119] enraged many with racist tweets during his company's Standing Rock debacle. In 2019, Warren displayed at his resort a prominent statue of confederate leader Robert E. Lee which had been removed from a public

---

[116] Kelly, S. (2018, April 17). For 15 years, Energy Transfer Partner pipelines leaked an average of once every 11 days: Report. Desmog.

[117] EJ Atlas. (2018, October 4). Louisiana's Bayou Bridge Project.

[118] Sen, B. (2021, January 20). Fossil-fueled Fascism: how the oil, gas and coal industries fund white supremacy and far-right politics. Otherwords.

[119] Energy Transfer. (n.d.). Kelsey Warren.

site in Dallas in 2017.[120] This and other evidence reinforces global concerns about oil executives embracing bigotry and supporting repression. In a time when rights of Indigenous People are well established and reparations are part of current legal frameworks, oil companies continue to act with gross negligence yet face minimal penalties.[121]

Enbridge's Line 3 genocide

Enbridge's Line 3 oil pipeline begins in Alberta, Canada, and crosses 200 bodies of water before ending at Enbridge's export terminal in Superior, Wisconsin.[122] As Enbridge violated treaty rights, the pipeline became a site of resistance with tensions carried over from Standing Rock.[123] Water protectors with anti-DAPL experience supported the encampment (Fig. 7.12)[124] as security officers and surveillance experts active at Standing Rock engaged at construction sites of Line 3.[125] Line 3 construction violated Indigenous People's subsistence,[126] bisecting treaty land without tribal consent.[127]

For Native rights advocates, anti-pipeline efforts are only one portion of the struggle. The Giniw Collective operated as a resistance camp aiming to stop Enbridge Line 3 while

| Treaty |
| Concession |
| Encampment for resistance |
| Criminalization |
| Persecution[128] |

Figure 7.12 A timeline of oppression.[128]

[120] Silapatchai, C. (2021, August 12). The fate of one Confederate statue in Texas. An Injustice!.
[121] Regan, S. (2021, February 19). 'It's cultural genocide': Inside the fight to stop a pipeline on tribal lands. *The Guardian*.
[122] Austin, E. (2021, August 30). Line three is cultural genocide at the hands of Enbridge, police and big banks. National Observer.
[123] Veltmeyer, H., & Bowles, P. (2014). Extractivist resistance: The case of the Enbridge oil pipeline project in Northern British Columbia. *The Extractive Industries and Society, 1*(1), 59–68.
[124] Parrish, W., & Brown, A. (2019, January 30). How police, private security, and energy companies are preparing for a new pipeline standoff. The Intercept.
[125] Funes, Y. & Mehrotra, D. (2020, September 18). CBP Drones Conducted Flyovers Near Homes of Indigenous Pipeline Activists, Flight Records Show. Gizmodo.
[126] LaDuke, W. (2019). *Indigenous Food Sovereignty in the United States: Restoring Cultural Knowledge, Protecting Environments, and Regaining Health*. University of Oklahoma Press.
[127] Minnesota Department of Commerce. (2017, August 17). Energy environmental review and analysis.
[128] Nauman, T. (2020, December 11). Native pipeline foes make inroads across Northern Plains. Resilience.

simultaneously teaching and learning about Indigenous sovereignty, traditional knowl-edge, and land defense. Founded by Ojibwe lawyer and activist Tara Houska in 2018 to stop Enbridge Line 3 from crossing Anishinaabe territory,[129] the Giniw Collective is one of several high-profile campaigns that continue to spur resistance to fossil fuel pro-jects through collective action. Houska, like other resistance leaders, was active in the Keystone XL resistance. Targeted and then arrested for protesting Line 3 in 2021, her due process rights were violated along with those of other water protectors.[130] Using funds and resources provided by Enbridge, law enforcement produced 1,000 arrests around Line 3 construction.[131] Enbridge, in partnership with the state, created an escrow account to streamline funding for law enforcement activities designed to control pipeline opposition. This funding reimbursed expenses for security training and patrols, helmets, gas masks, handcuffs, shields, balaclavas, and "protester removal." When a lawyer inves-tigating arrests found evidence of coverups, the sheriff used his office to retaliate.[132]

Water protectors risked arrest to draw attention to legal and regulatory violence. Wa-ter protectors have reason to act to defend resources—see the disaster at Line 6B (Fig. 7.13). Enbridge knew of structural weaknesses in its pipelines yet didn't maintain adequate infrastructure controls—even after prior spills.[133] The Enbridge Line 6B spill saturated 40 miles of the Kalamazoo River with more than a million gallons of tar sands diluted bitumen (dilbit). The rupture went unreported for nearly 17 hours after Enbridge misinterpreted remote alarms.[134] Just days before the rupture, Enbridge representatives under question before the US Congress due to the firm's ongoing poor safety record[135] testified that the company could detect a leak "almost instantaneously."

The spill initially culminated in a first settlement requiring Enbridge to pay nearly $4 million for restoration projects.[136] Further settlements totaled upwards to $177 million as extensive environmental damages caused by the 2010 rupture were complicated by the tenacity of dilbit in waterways, requiring particular remediation methods (Fig. 7.14).

[129] Cardwell, E. (2018). Awake, A Dream From Standing Rock. *Radical Teacher, 112,* 72–74.
[130] Business Human Rights. (2021, July 29). Tara Houska- Giniw Collective.
[131] Beaumont, H. (2021, October 5). Revealed: pipeline company paid Minnesota police for arresting and surveilling protesters. *The Guardian.*
[132] Davis-Cohen, S. (2022, November 16). Sheriff retaliates against lawyers scrutinizing arrests of Water Protesters. Truthout.
[133] Laduke, W. (2014, May 29). LaDuke: A pipeline runs through it. *Duluth Reader;* House, K. (2020, July 24). 10 Years Later, Kalamazoo River spill still colors Enbridge pipeline debate. Bridge Michigan.
[134] Environmental Protection Agency. (2021, March 2). Enbridge Spill Response Timeline.
[135] Wesley, D., & Dau, L. A. (2017). Complacency and automation bias in the Enbridge pipeline disaster. *Ergonomics in Design, 25*(1), 17–22.
[136] Fish and Wildlife Service. (2015, June 8). Enbridge Must restore Environment Injured by 2010 Kalamazoo River Oil Spill.

Figure 7.13 Enbridge Line 6B spill. *(Image credit: Response operations near the source of the spill on Talmadge Creek near the Kalamazoo River. August 1, 2010, U.S. EPA. Accessed via Flickr.)*

Land defenders, water protectors

Land is often a critical resource held by rural low-wealth households, who rely on it for subsistence. Economists who theorize the sale of land as a legal or fiscal relationship miss these deeper relationships and ties to physical and cultural survival. Property rights frameworks are based in and normalize worldviews of epistemic racism. Buoyed by ethnocentric arguments of efficiency and necessity, pipeline companies seize land as a routine act, ignoring deep relationships to land extending back multiple generations and extending forward as important cultural and climate resilience resources.

Oil's continual pattern of spills and loose regulatory enforcement coupled with insignificant, business-as-usual consequences only enrages growing opposition.[137] State agencies violate the territorial rights of Native populations by granting new permits again

---

[137] Kraker, D., & Marohn, K. (2021, March 3). 30 years later, echoes of largest inland spill remain in Line three fight. MPR News.

Figure 7.14 Enbridge Line 6B spill. *(Image credit: Vacuum crews work to remove oil near the actual spill site. August 2, 2010, U.S. EPA. Accessed via Flickr.)*

and again.[138] In this compromised state of affairs, Native organizers understand the need to develop new forms of vigilance combining law and activism with physical sites and digital platforms.[139] For example, the Giniw Collective suggests paths toward decolonization of allyship.[140] Native activists like Elizabeth Archuleta, Winona LaDuke, and Tara Houska articulate an **ethos of responsibility**[141] for dismantling oppression so that people and society can heal.[142] Recognizing differences across cultures, the importance of

[138] Hunsberger, C., & Awâsis, S. (2019). Energy justice and Canada's national energy board: a critical analysis of the line nine pipeline decision. *Sustainability*, *11*(3), 783.

[139] Cragoe, N. G. (2017). Following the green path: honor the Earth and presentations of Anishinaabe Indigeneity. *Wicazo Sa Review*, *32*(2), 46—69.

[140] Sullivan-Clarke, A. (2020). Decolonizing "Allyship" for Indian Country: Lessons from# NODAPL. *Hypatia*, *35*(1), 178-189; Awāsis, S. (2021). Gwaabaw: Applying Anishinaabe harvesting protocols to energy governance. *The Canadian Geographer*, *65*(1), 8—23.

[141] Privott, M. (2019). An ethos of responsibility and indigenous women water protectors in the# NoDAPL movement. *American Indian Quarterly*, *43*(1), 74—100.

[142] Archuleta, E. (2006). "I Give You Back": Indigenous Women Writing to Survive. *Studies in American Indian Literature*, *18*(4), 88—114.

land deserves attention as dispossession is both a physical and cultural process. As environmental justice may not be central to their origins and practices, many scientific and environmental agencies require organizational transformation to deliver effective Free, Prior, and Informed Consent (FIPC) for Indigenous People facing oil infrastructure.[143]

Reparations (Chapter 1) seek to acknowledge and repair injuries caused by wrongs—a fundamental pillar of restorative justice. Indigenous reparation in "Land Back" movements are part of broader discussion about reparations.[144] Comprehensive and effective reparations address both the cumulative harms of racism across generations, as well as contemporary threats such as systemic police violence that cuts BIPOC lives short.[145] Recognizing differences across cultures, struggles past and present share characteristics expressed uniquely across locations and communities. It is fraught to assume similarity of Indigenous genocide and anti-Black racism, yet at the same time maintaining separate lines of analysis ignores relations between the processes.[146] Allyship across differences requires listening to frontline communities.[147] Allyship is most authentic when earned beyond self-designation alone. In organizational settings, rigid ideologies and dogmatic science can reinforce internal racism and in turn counteract allyship.

While examples exist for land reparations, including individual property owners returning land of their own accord, as well as trusts and collectives,[148] the process remains a societal exception.[149] Successful examples include an honor tax to the Wiyot nation and the Sogorea Te Land Trust's urban land "rematriation" (a feminist version of repatriation).[150] In addition to physical transfers of land, language itself can be open to decolonization. Native territorial names, toponyms, and place names are being reintroduced around the world. Similarly, new regional groupings, such as the most affected peoples and areas MAPA (Introduction), prioritize those struggling from inequality in geographies particularly prone to climate disruption.

Large-scale transmission pipelines and related infrastructure can displace people and places along trajectories towards downstream markets. Native land is considered a living

[143] Johnson, S.K. (2018). Leaked Talent: How People of Color are pushed out of environmental organizations. Green 2.0.

[144] Westley, R. (1998). Many billions gone: Is it time to reconsider the case for Black reparations. *BC f World LJ, 19,* 429.

[145] Táíwò, O. O. (2020, June 12). Power Over The Police. *Dissent*; Karteron, A. (2021). Reparations for Police Violence. *NYU Review of Law & Social Change, 45,* 405.

[146] King, T. L., Navarro, J., & Smith, A. (Eds.). (2020). *Otherwise worlds: Against settler colonialism and anti-Blackness.* Duke University Press.

[147] Houska, T. (2019). What listening means in a time of climate crisis. *Literary Hub, 18.*

[148] Resource Generation. (n.d.). Land Reparations and Indigenous Solidarity Toolkit.

[149] Clark, G. (2020). Toward Land Return: Indigenous Environmental Justice and White Resistance in Minnesota. *Tapestries: Interwoven Voices of Local and Global Identities 9*(1), 5.

[150] Sogorea Te Land Trust. (n.d.). https://sogoreate-landtrust.org/

entity constantly shaping and being shaped by other life forms,[151] a relationship very different from the transactional approach of fossil energy infrastructuring. For many long-standing traditional cultures, fossil energy development is an alien encroachment divorced from social and ecological benefit, an involuntary sacrifice on behalf of fossil fuel interests.

Enabled through fossil fuels, our collective relationship with the Earth has changed so fundamentally as to create a socio-ecological metabolic rift (Conclusion),[152] meaning ecological disequilibrium from our vast overuse and destruction of finite resources. A metabolic rift, first identified in application to agricultural production more than a century ago, expands and intensifies with fossil fuel development. Now that we have concrete knowledge of looming climate collapse, will we make necessary changes? Will humans come together in solidarity to address the existential threat of climate disruption? Acts of restorative justice increase our collective potential for survival and do so with moral and practical objectives, establishing healthy, generative, and feasible paths toward climate action and environmental justice simultaneously. Within restorative justice, reparations are a critical pillar,[153] but not without challenges. As current and future damages weigh down our collective financial ledger, opportunities to address the past may seem limited yet are urgently needed to forward environmental justice.

Across frameworks for energy justice,[154] fossil fuels, oil in particular, retain too much power and influence.[155] Without checks and balances to deter conflicts of interest or disinformation, the private sector uses media and government platforms to defend the necessity of its projects and create misleading narratives that justify growth and expansion while dismissing externalities and harmful impacts. It's hypocritical for countries claiming to be democratic—the US and Canada included—to speak of freedom and justice yet violate the rights of Indigenous Peoples and minorities within their borders with blatant disregard for human rights. The connection between state brutality and fossil fuels[156] shown through recent high-profile pipeline battles points to tendencies towards fossil facism primed to intensify with climate disruption. As counteractive, decolonizing processes seek to dissipate fascism and reverse oil violence. Indigenous-led anti-pipeline movements have demonstrated how a more effective environmental movement can

[151] Liboiron, M. (2021). *Pollution Is Colonialism*. Duke University Press.

[152] Schneider, M., & McMichael, P. (2010). Deepening, and repairing, the metabolic rift. *The Journal of Peasant Studies, 37*(3), 461–484.

[153] Táíwò, O. O. (2022). *Reconsidering reparations*. Oxford University Press.

[154] Hess, D. J., McKane, R. G., & Pietzryk, C. (2021). End of the line: environmental justice, energy justice, and opposition to power lines. *Environmental Politics*, 1–21.

[155] Soraghan, M. (2021, January 4). Law: Are pipeline companies buying justice? E&E News.

[156] Herr, A. (2020, July 28). How the fossil fuel industry drives climate change and police brutality. *Grist*; Herr, A. (2020, September 9). An illustrated guide to police brutality and pollution. *Grist*.

leverage power for frontline populations and reverse harmful historical patterns of energy development marked by racism, sexism, and systemic inequalities.

Spatial and temporal synopsis

Space

(1) *Racialized surveillance features technology that extends unfair expressions of power over land and resources.*

(2) *Energy justice movements can engender a collective ethos of responsibility that acknowledges the importance of land, and seeks to end dispossession.*

(3) *Oil opposition takes advantage of spatial tactics, including blockades.*

Time

(1) *Current discriminatory patterns echo age-old racism and ethnocentric behavior.*

(2) *Settler colonialism and its impacts, like wastelanding and dispossession, are ongoing processes tied to infrastructure.*

(3) *In the face of threats from fossil fascism, decolonialization is a strategy to preserve self-determination and collective climate resilience.*

Summary

Unregulated surveillance of people of color and marginalized groups reinforces historical injustices such as breaking treaties with Indigenous populations. Sites of pipeline expansion on Native lands exhibit both outright physical violence and racialized harms. After decades of oil violence normalization throughout education and media sectors, the Standing Rock encampment became a watershed moment with global influence. State and private responses contributed to geospatial overwatch and criminalization, generating a counter-response from women-led resistance efforts focused on systemic change and an ethos of responsibility to dismantle social and ecological harm. Broader society follows the lead of Indigenous movements to demand the end of oil's SLO.

Vocabulary

1. biometrics
2. chokepoint
3. criminalization
4. critical infrastructure
5. environmental racism
6. ethos of responsibility
7. dispossession

8. dronescape
9. fascism
10. flashpoint
11. Free, Prior, and Informed Consent (FPIC)
12. fusion center
13. gender violence
14. geospatial overwatch
15. global white supremacy
16. horizontalism
17. infrapolitics
18. Land Back Movement
19. miseducation
20. regime of obstruction
21. representation
22. security
23. settler colonialism
24. siting
25. surveillance
26. survivance
27. wastelanding

Recommended

Books

Anderson, K., Campbell, M., & Belcourt, C. (Eds.). (2018). *Keetsahnak/our missing and murdered indigenous sisters*. University of Alberta.

Crosby, A., & Monaghan, J. (2018). *Policing indigenous movements: Dissent and the security state*. Fernwood Publishing.

Estes, N. (2019). *Our history is the future: Standing Rock versus the Dakota Access Pipeline, and the long tradition of indigenous resistance*. Verso.

Todrys, K. W. (2021). *Black snake: Standing Rock, the Dakota access pipeline, and environmental justice*. University of Nebraska Press.

Articles

Archuleta, E. (2006). "I Give You Back": Indigenous women writing to survive. *Studies in American Indian Literatures, 18*(4), 88–114.

Houska, T. (2019). What listening means in a time of climate crisis. *Literary Hub, 18*.

LaDuke, W., & Cowen, D. (2020). Beyond *wiindigo* infrastructure. *South Atlantic Quarterly, 119*(2), 243–268.

Film

Killers of the Flower Moon. (2023). *Directed by Martin Scorsese, performances by Leonardo DiCaprio, Robert de Niro and lily Gladstone.*

CHAPTER 8

Methane madness

Contents

Climate Crisis, Energy Violence
ISBN 978-0-12-819501-7,
https://doi.org/10.1016/B978-0-12-819501-7.00005-9

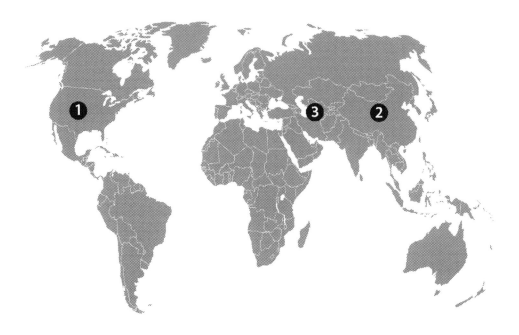

Locations: 1 - North America - Canada, Mexico, United States
              2 - Sichuan Province - China
              3 - Turkmenistan

Themes: Methane; Regulatory violence; Risk

Subthemes: Fixity; Geopolitics; Growth poles; Hub; Infrastructuring

This was not an accident.

Fossil gas' promise … and peril

Activists pushing for climate action have to walk a fine line.[1] Environmentalism itself can be viewed as unwelcome and inconvenient.[2] Energy firms have co-opted the climate agenda by spreading a pervasive myth that fossil gas is clean.[3] A decade ago, industry attempted to sell gas as a bridge fuel to renewable sources—this has been widely discredited.[4] Yet transparency in the gas sector remains elusive. Because methane is regulated narrowly as an air pollutant in permitting,[5] there is insufficient attention to mitigation: without mandates tied to sanctions for noncompliance, reducing methane becomes of secondary importance. Whether their utility provider is government-run or a private entity, many US urban areas suffer leak-prone gas systems that flagrantly leak year after year.[6]

Justified with disinformation about the attributes of gas as clean fuel, government subsidies for gas infrastructure remain staggeringly high. This structural violence is increasingly futile—science clearly shows the combustion of fossil gas and its derivatives intensifies the climate crisis. Coupled with improved monitoring capabilities, dangerous greenhouse gas (GHG) emissions are now visible, putting pressure on industry's practice of obfuscation.

Fossil gas is mainly methane (CH_4)—made up of one carbon atom and four hydrogen atoms. Carbon equivalency is used for reference between GHGs regardless of their composition. **Carbon dioxide equivalent** (CO_2e) is a metric used to compare the emissions from various GHGs on the basis of their **global warming potential** (GWP): a measure of how much infrared thermal radiation the emissions of 1 ton of a gas will absorb over a given period of time relative to the emissions of 1 ton of carbon dioxide (CO_2).[7] GWP is calculated by converting amounts of other gases to the equivalent amount of CO_2. Using a common measure highlights that methane is a potent GHG that deserves escalated importance, in part due to the factor of time. Methane's

[1] Davidson, H. (2021, August 16). 'You follow the government's agenda': China's climate activists walk a tightrope. *The Guardian*.
[2] Finley-Brook, M., Williams, T. L., Caron-Sheppard, J. A., & Jaromin, M. K. (2018). Critical energy justice in US natural gas infrastructuring. *Energy Research & Social Science, 41*, 176–190.
[3] Scanlan, S. J. (2017). Framing fracking: Scale-shifting and greenwashing risk in the oil and gas industry. *Local Environment, 22*(11), 1311–1337.
[4] Howarth, R. W. (2014). A bridge to nowhere: Methane emissions and the greenhouse gas footprint of natural gas. *Energy Science & Engineering, 2*(2), 47–60.
[5] Congressional Research Service. (2022, September 20). The legal framework for federal methane regulation.
[6] Hendrick, M. F., Ackley, R., Sanaie-Movahed, B., Tang, X., & Phillips, N. G. (2016). Fugitive methane emissions from leak-prone natural gas distribution infrastructure in urban environments. *Environmental Pollution, 213*, 710–716.
[7] Environmental Protection Agency (EPA). (2023, April 18). Understanding global warming potentials.

GWP is 86 times greater than that of CO_2 over 20 years.[8] Methane has high heat-trapping potential due to its absorption spectrum, molecular structure, and radiative efficiency.[9]

Fossil gas was promoted as a "diet" or "lite" fossil fuel and labeled "natural" gas.[10] It subsequently became viewed more favorably than other fossil fuels. Misperception of gas as green dissipates with exposure to independent media and scientific findings. Studies show that opinion lowers after shifting from saying "natural" gas to saying "fossil" gas when describing the fuel—meaning using the word "natural" buoys fossil gas' favorability.[11]

On its journey to become a commodity, fossil gas is rife with **methane loopholes**: regulatory gaps that allow companies to avoid transparent reporting, monitoring, and regulation of methane.[12] For example, exploration emissions are commonly excluded. Emissions from processing and transportation are often incomplete and post-metering emissions often ignored.[13] Approximately 70% of methane released from oil and gas operations remains unaccounted at a global level,[14] and existing methane regulations are ineffective.[15] Climate benefits via switching from coal to gas power plants are negated when the methane leakage rate from gas systems exceeds 3.2%.[16]

Leaky gas systems exist due to the lack of enforceable mandates. Firms lobbied against regulations to address fugitive releases for decades. Gas companies tout how they comply with laws without mentioning how they lobby to maintain ineffective regulatory structures. Industry-captured politicians maintain shallow, piecemeal policy, playing into industry's wishes for voluntary targets and self-reporting. In the US, the Obama

[8]   Howarth, R. W. (2020). Methane emissions from fossil fuels: Exploring recent changes in greenhouse-gas reporting requirements for the State of New York. *Journal of Integrative Environmental Sciences*, 1–13.

[9]   Staley, B. (2023, June 19). Unraveling the mystery behind methane Global Warming Potential. *Waste 360*.

[10]   Lacroix, K., Goldberg, M. H., Gustafson, A., Rosenthal, S. A., & Leiserowitz, A. (2021). Different names for "natural gas" influence public perception of it. *Journal of Environmental Psychology*, 77, 101671.

[11]   Kennedy, B., Spencer, A., & Funk, C. (2020, October 19). *Natural gas viewed more favorably than other fossil fuels across 20 global publics*. Pew Research Center.

[12]   Mider, Z. (2022, August 10). Methane loophole shows risk of gaming new US climate bill. *Bloomberg*.

[13]   Gao, J., Guan, C., & Zhang, B. (2022). Why are methane emissions from China's oil & natural gas systems still unclear? A review of current bottom-up inventories. *Science of the Total Environment*, 807, 151076.

[14]   Olczak, M., Piebalgs, A., & Balcombe, P. (2023). A global review of methane policies reveals that only 13% of emissions are covered with unclear effectiveness. *One Earth*, 6(5), 519–535.

[15]   International Energy Agency. (2022, February 23). Methane emissions from the energy sector are 70% higher than official figures.

[16]   Alvarez, R. A., Pacala, S. W., Winebrake, J. J., Chameides, W. L., & Hamburg, S. P. (2012). Greater focus needed on methane leakage from natural gas infrastructure. *Proceedings of the National Academy of Sciences*, 109(17), 6435–6440.

Administration finally tightened methane regulations for oil and gas, starting on public lands in 2016. Industry lobbied against controls, and the Trump Administration rolled them back.[17] In 2022, the Biden Administration proposed new rules arguing that cutting methane would protect communities and spur innovation. When enacted, these rules could cut pollution significantly. Reducing methane is "low-hanging fruit" because the value of the captured methane will pay for abatement.[18] Leaks can often be repaired for a portion of the otherwise lost profit from gas waste.[19] Customers pay for the gas lost,[20] while pipeline operators are reimbursed for fugitive gas through negotiated rates, creating a lack of incentive to fix infrastructure. Since losses are absorbed into gas rates, companies are compensated for leaked gas, even when they delay or avoid repairs.

Regime of measurement

Centralized energy systems imposed from above—a **regime of measurement**—usually have structured accounting to obtain quantifiable outcomes across national territories. To streamline what it means to incur environmental harm, permitting processes tend to focus on a few specific consequences of pipeline construction, such as soil erosion or loss of habitat for endangered species. Thus, lawsuits also focus on these areas. In truth, regulatory frameworks have done little to protect frontline communities from toxic pollution from gas (Chapter 9). While permitting review seems rigorous, regulations focus on standardized measures for select indicators. For example, in the US, hyperfocus on criteria pollutants listed under the National Ambient Air Quality Standards (NAAQS)[21]—six principal pollutants: carbon monoxide (CO), lead (Pb), nitrogen dioxide (NO_2), ozone (O_3), particulate matter (PM), and sulfur dioxide (SO_2)—neglects other toxic elements like **hazardous air pollutants** (HAPs). Hundreds of HAPs are known to cause cancer and have other serious health impacts.[22] US regulations employ assumptions around "**co-pollutants**"[23] found in conjunction with regulated NAAQS pollutants.[24] Laws assume

[17] Davenport, C. (2020, August 10). E.P.A. to lift Obama-era rules on methane, a potent greenhouse gas. *New York Times*.

[18] International Energy Agency. (2021). Driving down methane leaks from the oil and gas industry: A regulatory roadmap and toolkit.

[19] Hendrick, M. F., Cleveland, S., & Phillips, N. G. (2017). Unleakable carbon. *Climate Policy, 17*(8), 1057–1064.

[20] Ma, Y., Wright, J., Gopal, S., & Phillips, N. (2020). Seeing the invisible: From imagined to virtual urban landscapes. *Cities, 98*, 102559.

[21] Environmental Protection Agency (EPA). (2023, March 15). NAAQS table.

[22] Environmental Protection Agency (EPA). (2022, December 19). What are hazardous air pollutants?

[23] Dedoussi, I. C., Allroggen, F., Flanagan, R., Hansen, T., Taylor, B., Barrett, S. R., & Boyce, J. K. (2019). The co-pollutant cost of carbon emissions: an analysis of the US electric power generation sector. *Environmental Research Letters, 14*(9), 094003.

[24] US Environmental Protection Agency (EPA). (2023, March 15). NAAQS table.

that the control of regulated pollutants will mitigate unregulated pollutants at the same time. This conserves resources for regulatory agencies yet provides uncertain public health protection due to the toxicity of some chemicals even in trace amounts as well as variable co-pollutant ratios.

To measure is to know.[25]

Methane vents from upstream (production), midstream (transmission), and downstream (distribution) segments of the fossil gas vector (Fig. 8.1). Fossil gas use has increased dramatically while monitoring and regulations have not kept pace. Regular methane releases are treated as unintentional fugitive "leaks." As a result, gas creates a "methane rift" (Introduction)—an existential rupture in the biosphere induced from methane's temporal and spatial impacts. Gas lock-in counteracts other climate action.

What gets measured, gets managed.[26]

Methane monitoring efforts occur from the sky, from the ground at the facility level, or through a combination of these methods. Instruments have not been adequately

Natural gas production and delivery

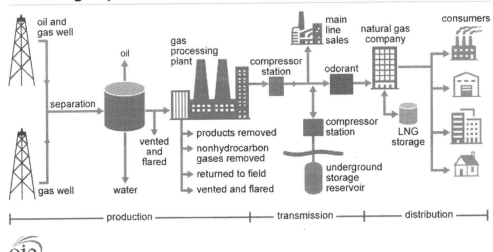

Figure 8.1 Fossil gas infrastructure vector. *(Source: U.S. Energy Information Administration (EIA).)*

---

[25] Lord Kelvin quoted in Saxon, D. (2007, December 17). In praise of Lord Kelvin. *Physics World.*
[26] Peter Drucker quoted in Barnett, P. (2015). If what gets measured gets managed, measuring the wrong thing matters. *Corporate Finance Review, 19*(4), 5.

accurate to this point but critical advances in remote sensing are nascent.[27] For example, the European Space Agency's (ESA) Tropomi instrument captures regional-scale methane levels with improved clarity. Potential breakthrough instruments with more precise resolutions are on the horizon. MethaneSAT—a platform developed by the Environmental Defense Fund—will map and quantify emissions accounting for 80% of global oil and gas production. To understand the scale of the methane problem, MethaneSAT promises **radical transparency**—providing data via a free, accessible, online portal to counter opaque measuring and monitoring practices of the oil and gas sector. This open approach differs from private ventures such as Kayrros, a software-based "environmental intelligence company"[28] that combines and analyzes existing satellite imagery in order to market a "value-added" product designed for investment, carbon trading and industry interests.

Gaseous fictions

Expansion of fossil gas infrastructure after knowledge of harms exposes a captured state. Many suggest we inhabit a "post-truth" era and climate denial is an example of active **obfuscation**—the intentional act of making perspectives unclear or unintelligible.[29] Disinformation impedes timely adoption of safer, cleaner options. Proposed environmental bills are often weakened or gutted during rulemaking. Political agendas are inserted, resulting in codes that are ineffective, contradictory, or incomplete. After implementation, environmental laws face constant attempts at regulatory rollback to streamline procedures or eliminate mandates.

Buildout of gas in both China and the US arose from parallel contradictions: in a race to end coal, pollution from fossil gas was hidden. The push to reduce coal's emissions drove a rapid increase in fossil gas use without full quantification and accounting of the associated costs, including methane leaks. New gas infrastructure proposals intensify our climate crisis, so the gas industry obfuscates by now pretending to be able to remove problematic emissions, in spite of dismal failure in this regard.

A key area of deception exists in net-zero pledges. These are **carbon unicorns**:[30] eco-fantasies that are not in line with reality. Promises of **net-zero**—negating the amount of greenhouse gas emissions produced from human activity—distract from urgent transition to renewable sources. Gas companies with carbon neutrality goals by distant dates often plan to increase GHG emissions in the interim. Nefarious justifications

[27] Brown, D. W. (2023, April 28). A security camera for the planet. *The New Yorker*.

[28] Kayrros (n.d.). The leading environmental intelligence company.

[29] Schmitt, E., & Li, H. (2019). Engaging truthiness and obfuscation in a political ecology analysis of a protest against the Pengzhou Petroleum Refinery. *Journal of Political Ecology*, *26*(1), 579–598.

[30] Friends of the Earth International. (2021). Chasing carbon unicorns.

suggest that to jump-start opportunities firms need to first expand expensive portfolios in fossil gas.

The gas industry seeks infrastructure **lock-in**—when the amount of investment in existing systems is used to justify continued use. Construction of expensive pipelines and import/export facilities creates inertia against transition to alternatives. **Sunk costs**, or money spent that cannot be recovered, are commonly used in arguments for the gas industry. The sunk cost fallacy is a psychological barrier that ties people to unsuccessful endeavors (or dirty energy) simply because they've committed resources to it.

Gas permitting continues in spite of proven risks, inspiring a movement to take away gas' SLO (Chapter 2). Nongovernmental organizations (NGOs), such as Gas Leaks,[31] expose gas industry deceptions, while climate activists organize to block new construction. Anger regarding harms from fossil gas has entered popular culture.[32] Networks of **gastivists**, or anti-gas activists, recognize commonalities in gas-permitting processes where impacted populations are "managed" within formal proceedings and well-evidenced concerns are sidelined. Governments continue to attempt to thwart public contention toward fossil gas, ignoring valid concerns.

Energy violence

Regulatory capture has allowed methane releases to escape effective monitoring and capture, making it impossible to accurately achieve climate goals.[33] GHG intensity in gas operations is a popular measure but tends to be applied relatively, so firms aim to be on par with or slightly better than others, instead of the transformative action required to address an existential threat.

Climate risks include the physical loss and damage from a changing climate—these are split into acute risks, which translate into fast violence, while chronic risks represent slow violence.[34] A prominent climate risk is water-stranded assets (Chapter 4)—these arise from water scarcity (i.e., drought) or overabundance (i.e., flooding). Water is increasingly a limiting factor across the gas life cycle.[35]

Carbon risks must comprehensively reflect present and future loss and damages from GHGs and their attempted mitigation. For many industry definitions, carbon risk is

---

[31] Gas Leaks. (n.d.). Gas leaks but so does the truth.

[32] Ritter, D. (2023). Woodside ecocide. *AQ: Australian Quarterly, 94*(1), 18−27.

[33] Kalen, S., & Hsu, S. L. (2020). Natural gas infrastructure: Locking in emissions?. *Natural Resources & Environment, 34*(4), 3−6.

[34] US Environmental Protection Agency (EPA). (2022, December 1). Climate risks and opportunities defined.

[35] Wang, J., Liu, X., Geng, X., Bentley, Y., Zhang, C., & Yang, Y. (2019). Water footprint assessment for coal-to-gas in China. *Natural Resources Research, 28*, 1447−1459.

narrowly understood in terms of losses from "unexpected" changes in the scope, timing, and speed of mitigation policies. This is an inadequate benchmark because it assumes that regulatory violence can continue in the near term—and suggests climate breakdown is a surprise that wasn't predicted for decades. The gas industry risks building infrastructure that will become stranded when assets lose value before the end of their expected life because of changes in regulation, market forces, societal norms, or innovation.

There will be **transition risks** that emerge from changing fuels like gas that drive climate change. The pace and extent of switching from fossil fuels to renewables determine investment needs. For decades, the up-front costs necessary for renewables have been used as justification to remain locked-in to fossil fuels, disregarding the evidence that the overall costs will be higher the longer we wait. Building climate-ready energy systems will entail unpacking the blatant, long-standing contradictions that hold back necessary mitigation; these are described in the Conclusion as discourses of climate delay.

Fossil gas received an inopportune advantage from policy attention to coal's outsized carbon footprint. However, carbon dioxide reductions are ineffective GHG mitigation if they increase methane emissions.[36] **Carbon-centric** planning involves fetishizing carbon dioxide (CO_2) and ignoring other GHG emissions, such as methane. Overemphasis on the word carbon in regulations and policies formed a regulatory gap and helped the industry greenwash gas as climate-friendly. More holistically, **climate justice** provides a frame for understanding climate change as a political issue and to transform our systems and responses to promote equity—this includes access to transparent data like methane and other emissions as well as transformative regulation and governance. Climate justice requires "**just transition**" to a reformed energy sector that demonstrates ample job creation, climate reparations, and equitable economic flows and power relations (Praxis 7).[37]

Public health crisis

Gas's health harms are significant, particularly near infrastructure sites. Life expectancy is lower near gas wells compared to surrounding areas.[38] Firms promoting gas as clean focus on pollution from smokestacks, where there are significantly lower levels of nitrogen oxides (NOx) compared to coal. Yet, regular exposure to gas increases health

[36] Alvarez, R. A., Zavala-Araiza, D., Lyon, D. R., Allen, D. T., Barkley, Z. R., Brandt, A. R., ... Hamburg, S. P. (2018). Assessment of methane emissions from the US oil and gas supply chain. *Science*, *361*(6398), 186–188.

[37] Feminist Economic Justice for People & Planet Action Nexus. (2021). A feminist and decolonial global green new deal.

[38] Landrigan, P. J., Frumkin, H., & Lundberg, B. E. (2020). The false promise of natural gas. *New England Journal of Medicine*, *382*(2), 104–107.

risks.[39] A review of more than 1,700 studies concludes that fracking poses a threat to air, water, climate, and human health, noting high concern for vulnerable populations including pregnant women, babies, elders, and those with preexisting medical conditions like respiratory disease.[40] Proximity of residences to infrastructure influences the prevalence and severity of negative symptoms.[41] People living near gas infrastructure report respiratory problems, eye and skin irritations, and elevated cancer rates.[42] Preliminary health research on gas in China similarly provides evidence for concern and a call for further study.[43]

There are long-term ecological damages from US gas infrastructure (Fig. 8.2).[44] For example, techniques in gas drilling, such as reinjection of produced water, increase the mobility of radioactive waste.[45–47]

Hydraulic fracturing (fracking) produces unconventional fossil gas by forcing water, chemicals, and sand down a drilled well under high pressure, breaking up geological formations to release gas. Legal loopholes in the US exempted fracking chemicals from

Air	Land	Water
• Methane emissions • Hazardous air pollutants (HAPs)	• Ground-level radium and radon • Technologically enhanced naturally occurring radioactive materials (TENORMs)[46]	• 'Forever' chemicals[47] • Water overuse

Figure 8.2 Key ecological harms from gas infrastructure.

[39] Hendryx, M., & Luo, J. (2020). Natural gas pipeline compressor stations: VOC emissions and mortality rates. *The Extractive Industries and Society, 7*(3), 864–869.

[40] Marusic, K. (2019, June 20). No evidence that fracking can be done without threatening human health. *Environmental Health News*.

[41] Burney, J. A. (2020). The downstream air pollution impacts of the transition from coal to natural gas in the United States. *Nature Sustainability, 3*(2), 152–160.

[42] Concerned Health Professionals of New York/Physicians for Social Responsibility (2022). *Compendium of scientific, medical and media findings demonstrating risks and harms from fracking*, 8th ed.

[43] Ma, Z., Pang, Y., Zhang, D., & Zhang, Y. (2020). Measuring the air pollution cost of shale gas development in China. *Energy & Environment, 31*(6), 1098–1111.

[44] Apergis, N., Mustafa, G., & Dastidar, S. G. (2021). An analysis of the impact of unconventional oil and gas activities on public health: New evidence across Oklahoma counties. *Energy Economics, 97*, 105223.

[45] Zelleke, H. B. (2019). Hydraulic fracturing, radioactive waste, and inconsistent regulation. *Colorado National Resources Energy & Environmental Law Review, 30*, 171.

[46] ALNabhani, K. A., Khan, F., & Yang, M. (2016). Technologically enhanced naturally occurring radioactive materials in oil and gas production: A silent killer. *Process Safety and Environmental Protection, 99*, 237–247.

[47] Horwitt, D. (2021). Fracking with 'forever chemicals.' Physicians for social responsibility.

the Safe Drinking Water Act—known collectively as the "Halliburton Loophole."[48] An example of regulatory violence, energy firms sucessfully argued against transparency and monitoring of fracking fluids, claiming proprietary rights over their unique yet toxic combinations of chemicals forced into, and adjacent, underground aquifers.

Fracking is increasingly unpopular as localities learn firsthand the harm it engenders (Fig. 8.3).[49] The mobility of contamination increases with practices like spreading produced water from fracking on roads or illegal dumping. In many documented cases, residents live near multiple gas wells, compressor stations, or power plants. They can experience **cumulative impacts** from various exposures, magnifying overall body burden synergistically over time.[50] Harms occur downstream as well, where local distribution lines bring gas contaminants directly into residential homes.[51] Gas stoves and appliances create indoor air pollution exposures.[52] Health risks from fossil gas are also endured by workers across gas infrastructures (Fig. 8.4).[53]

- Water over-extraction contributes to scarcity
- Pollution (air, groundwater, surface water, well water)
- Noise, lights and stress

Figure 8.3 Key harms from fracking.

Spatial distribution

The gas industry seeks lock-in with infrastructure, i.e., a type of **fixity** or stationary position. Yet, even with stable infrastructure like a pipeline network, gas mobility can be

[48] Underhill, V., Fiuza, A., Allison, G., Poudrier, G., Lerman-Sinkoff, S., Vera, L., & Wylie, S. (2023). Outcomes of the halliburton loophole: Chemicals regulated by the safe drinking water act in US fracking disclosures, 2014–2021. *Environmental Pollution, 322*, 120552.

[49] Hinojosa, R., Hinojosa, M. S., Fernandez-Reiss, J., Rosenberg, J., & Habib, S. (2020). Unconventional oil and natural gas production, health, and social perspectives on fracking. *Environmental Justice, 13*(4), 127–143.

[50] Environmental Protection Agency. (2022). *Cumulative impacts research: Recommendations for EPA's Office of Research and Development.* EPA/600/R-22/014a.

[51] Michanowicz, D. R., Dayalu, A., Nordgaard, C. L., Buonocore, J. J., Fairchild, M. W., Ackley, R., ... Spengler, J. D. (2022). Home is where the pipeline ends: Characterization of volatile organic compounds present in natural gas at the point of the residential end user. *Environmental Science & Technology.*

[52] Gruenwald, T., Seals, B. A., Knibbs, L. D., & Hosgood III, H. D. (2022). Population attributable fraction of gas stoves and childhood asthma in the United States. *International Journal of Environmental Research and Public Health, 20*(1), 75.

[53] Moridzadeh, M., Dehghani, S., Rafiee, A., Hassanvand, M. S., Dehghani, M., & Hoseini, M. (2020). Assessing BTEX exposure among workers of the second largest natural gas reserve in the world: a biomonitoring approach. *Environmental Science and Pollution Research, 27*(35), 44519–44527.

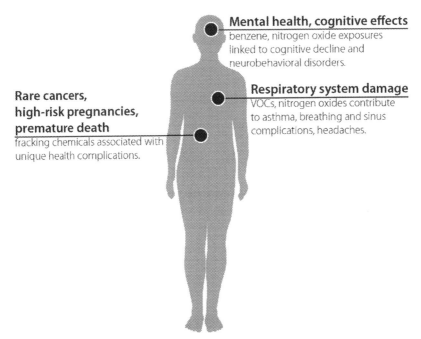

Mental health, cognitive effects
benzene, nitrogen oxide exposures
linked to cognitive decline and
neurobehavioral disorders.

Respiratory system damage
VOCs, nitrogen oxides contribute
to asthma, breathing and sinus
complications, headaches.

**Rare cancers,
high-risk pregnancies,
premature death**
fracking chemicals associated with
unique health complications.

Figure 8.4 Human body and fossil gas chemicals.

blocked during supply disruptions from disasters or conflicts. Spaces of gas infrastructuring are hotly contested. Active processes of **infrastructuring** rely on vertical and horizontal "space grabs."[54] Space grabs can be legal or illegal—or some of both—and entail the use of power to obtain preferential access to resources.

Shale territoriality

Territory is a controlled area of land under the jurisdiction of a ruling power. A gas territory includes subterranean claims. **Territoriality** involves claims to land or water, and includes those driven by incentives for new energy or infrastructure. **Geopolitics** involves power struggles over the control of resources like territory. Shale basins (Fig. 8.5) are strategic geopolitical spaces. Like oil, gas dominance rose from energy colonialism with repressive processes turned both internally and outwards, crossing national borders.

---

[54] Bouzarovski, S., Bradshaw, M., & Wochnik, A. (2015). Making territory through infrastructure: The governance of natural gas transit in Europe. *Geoforum, 64,* 217–228.

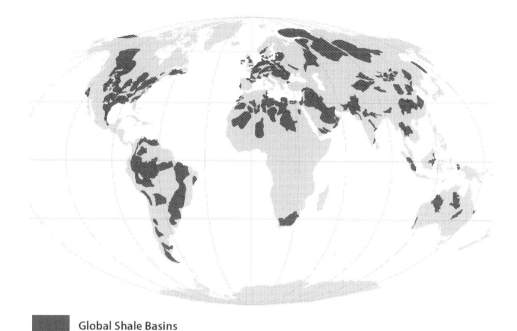

■ Global Shale Basins

Figure 8.5 Global shale basins. *(Sources: Shale Basin and Plays World Dataset, Natural Earth.)*

Traditionally, geopolitical terminology has been used to spotlight international contests. Nevertheless, internal domination can parallel foreign colonization, meaning geopolitics can occur domestically.[55] Geopolitical contests and war distract from responsible governance and induce GHG emissions, whether from releases during combat or as a result of collateral damages, such as the sabotage of infrastructure like the Nord Stream pipelines discussed below.

What about super emitters?

Underestimation within emission inventories occurs in part due to **super emitters**: facilities or sites that emit pollution at exceedingly high rates.[56] Super emitters span all continents, yet most are found in industrialized countries, including Canada, Europe, Japan,

---

[55] Cope, B. (2013). Eco-seeing a tradition of colonization: Revealing shadow realities of marcellus drilling. In *Environmental rhetoric and ecologies of place* (pp. 44–57). Routledge.

[56] Cusworth, D. H., Thorpe, A. K., Ayasse, A. K., Stepp, D., Heckler, J., Asner, G. P., … Duren, R. M. (2022). Strong methane point sources contribute a disproportionate fraction of total emissions across multiple basins in the United States. *Proceedings of the National Academy of Sciences, 119*(38), e2202338119.

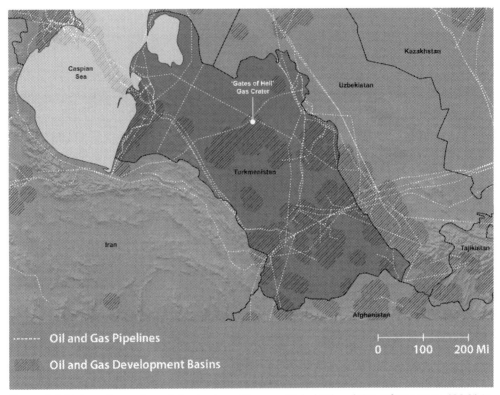

Figure 8.6 Turkmenistan oil gas infrastructure. *(Sources: Global Oil and Gas Infrastructure (GOGI) Inventory, Natural Earth.)*

and the US.[57] Some operations are simply designed to allow high pollution, though emission spikes can also result from equipment malfunctions.

Turkmenistan has the worst global rate of super emitting events.[58] Leaks arise from poorly maintained pipelines and from intentional methane venting after reductions in flaring.[59] There are also naturally high emitting sites—case in point, the "Gates of Hell" gas crater—interspersed amongst burgeoning gas basins and infrastructure (Fig. 8.6 above).[60]

[57] Cozzi, L., Chen, O., & Kim, H. (2023, February 22). The world's top 1% of emitters produce over 1000 times more CO2 than the bottom 1%. International Energy Agency.

[58] Carrington, D. (2023, June 2). US deal could plug Turkmenistan's colossal methane emissions. *The Guardian.*

[59] Irakulis-Loitxate, I., Guanter, L., Maasakkers, J. D., Zavala-Araiza, D., & Aben, I. (2022). Satellites detect abatable super-emissions in one of the world's largest methane hotspot regions. *Environmental Science & Technology, 56*(4), 2143–2152.

[60] Carrington, D. (2023, March 6). Revealed: 1000 super-emitting methane leaks risk triggering climate tipping points. *The Guardian.*

Figure 8.7 Gates of hell. *(Image credit: Darvaza gas crater panorama - "Gate of Hell", 2011 - Tormod Sandtorv. Accessed via Flickr.)*

The "Gates of Hell" crater alone (Fig. 8.7) has been an uninterrupted source of emissions for decades, possibly dating back to the 1960s.

Throughout Turkmenistan's oil and gas fields, methane hotspots (Fig. 8.8) are increasingly exposed through remote sensed data. Even as Turkmenistan develops plans to mitigate its flagrant methane releases,[61] ongoing use and promotion of gas as a preferred energy will translate into future GHG emissions even if excesses from this hotspot are ameliorated.

Turkmenistan has emerged as an outpost for gas pipeline supply to China, which is strategically pivoting away from reliance on Russian gas.[62] The geopolitics of Russian gas highlights another super emitter event, this time on the Baltic Sea, where a fast violence incident created the world's largest recorded methane release.[63] In 2022, unknown perpetrators ruptured the two gas pipelines Nord Stream 1 and Nord Stream 2 with coordinated explosions. Methane releases from Nord Stream 2 are shown in Fig. 8.9.[64]

[61] Carrington, D. (2023, June 13). Turkmenistan moves toward plugging massive methane leaks. *The Guardian.*

[62] Konarzewska, N. (2023, March 10). Energy crisis places Turkmenistan in the geopolitical spotlight.

[63] Reum, F., Marshall, J., Bretschneider, L., Glockzin, M., Huntrieser, H., Gottschaldt, K. D., ... Roiger, A. (2023). Methane emissions from the Baltic Sea 9 days after the Nord Stream explosions (No. EGU23-14,408). *Copernicus Meetings.*

[64] The European Space Agency. (2022, June 10). Satellites detect methane plume in Nord Stream leak.

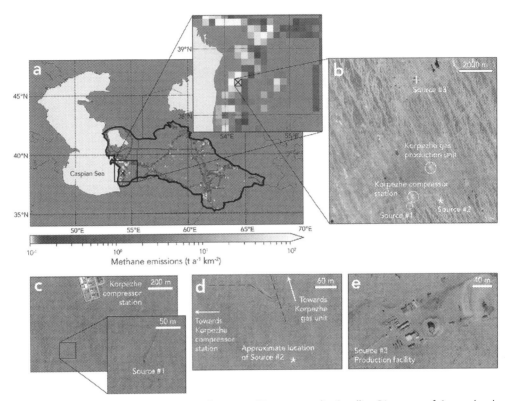

Figure 8.8 Turkmenistan infrastructure hotspot. *(Diagram credit: Satellite Discovery of Anomalously Large Methane Point Sources From Oil/Gas Production, D.J. Varon et al., 2019.)*

More than 220,000 tons of methane were released in 5 days as a result of the pipeline rupture.[65] The resulting gas plume was visible at the ocean surface due to its size and intensity (Fig. 8.10).

This dramatic case sheds light on the vulnerability of underwater infrastructure.[66] Fast violence through terrorism, vandalism, and sabotage requires attention anywhere observation is difficult. The geopolitical dimensions of gas as LNG are found in Chapter 10. The abuse and toxic release of fossil fuels in war and conflict is a core argument for distributed renewable systems as the backbone of a peaceful and equitable world.

---

[65] Jia, M., Li, F., Zhang, Y., Wu, M., Li, Y., Feng, S., … Jiang, F. (2022). The Nord Stream pipeline gas leaks released approximately 220,000 tonnes of methane into the atmosphere. *Environmental Science and Ecotechnology, 12,* 100210.

[66] Soldi, G., Gaglione, D., Raponi, S., Forti, N., d'Afflisio, E., Kowalski, P., … Warner, C. (2023). Monitoring of critical undersea infrastructures: The Nord Stream and other recent case studies. *IEEE Aerospace and Electronic Systems Magazine.*

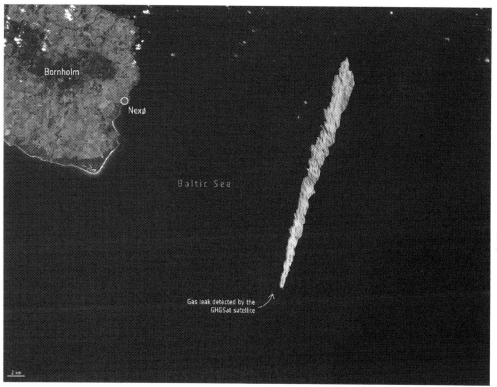

Figure 8.9 Nord stream methane plume. *(Image credit: Emission Plume captured by GHGSat high-resolution Satellite, GHGSat, 2022.)*

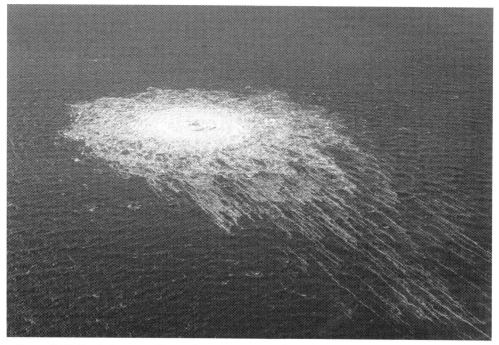

Figure 8.10 Nord stream methane plume at ocean surface. *(Image credit: Danish Defence, 2022.)*

Vertical and horizontal space grabs

Space grabs complicate responsibility for pollution. Mobility of contaminants across property lines adds murkiness surrounding clean-up liability. Unconventional drilling like fracking often involves lateral grabs across vast distances (Fig. 8.11). Adding more complexity, a **split estate** creates different owners of subsurface and surface rights, meaning homeowners are disempowered in matters pertaining to subsurface drilling access.[67]

Fracking and subsequent wastewater disposal contribute to earthquakes. **Induced seismicity** occurs internationally in oil and gas-producing areas.[68] Many induced

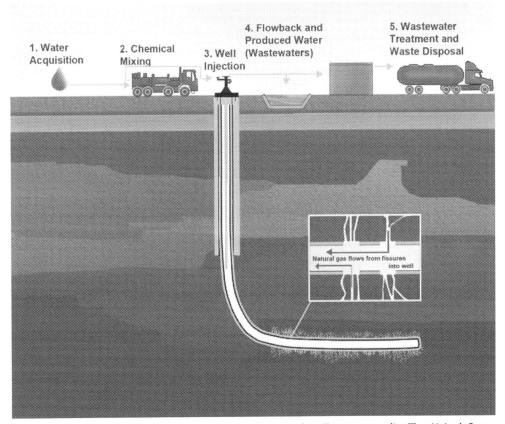

Figure 8.11 Fracking's vertical and horizontal space grabs. *(Diagram credit: The United States Geological Survey, USGS.)*

---

[67] Collins, A. R., & Nkansah, K. (2015). Divided rights, expanded conflict: Split estate impacts on surface owner perceptions of shale gas drilling. *Land Economics*, *91*(4), 688–703.

[68] Myers, S. L. (2019, March 8). China experiences a fracking boom, and all the problems that go with it. *New York Times*.

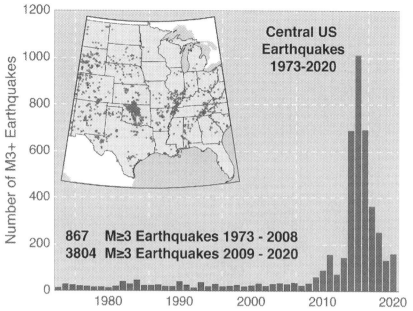

Figure 8.12 US induced and natural seismicity. *(Diagram credit: The United States Geological Survey, USGS.)*

earthquakes are minor (Fig. 8.12 above), but bigger events do occur.[69] For example, in 2011 a 5.7 magnitude earthquake attributed to fracking wastewater injection resulted in damaged homes and infrastructure.[70]

Water worries

Trucking water generates transportation emissions often unconsidered in gas' overall GHG budget. Water impoundments (Fig. 8.13) with storage in open pits creates a risk of spills and leakage of wastewater beyond containment, potentially contaminating soil, surface water, or groundwater. By their design and siting, impoundments result in large land disturbances and generate hazardous air pollution as waste evaporates.

Contamination by gas extraction is not limited to just open pits and the wasterwater they are designed to hold. Communities like Dimock, Pennsylvania, became sacrifice zones after poorly regulated fracking:[71] a documentary film, *Gasland*, raised awareness

[69] Shabarchin, O., & Tesfamariam, S. (2017). Risk assessment of oil and gas pipelines with consideration of induced seismicity and internal corrosion. *Journal of Loss Prevention in the Process Industries, 47*, 85–94.

[70] Campbell, N. M., Leon-Corwin, M., Ritchie, L. A., & Vickery, J. (2020). Human-induced seismicity: Risk perceptions in the state of Oklahoma. *The Extractive Industries and Society, 7*(1), 119–126.

[71] Engelder, T., & Zevenbergen, J. F. (2018). Analysis of a gas explosion in Dimock PA (USA) during fracking operations in the Marcellus gas shale. *Process Safety and Environmental Protection, 117*, 61–66.

Figure 8.13 Fracking water impoundment. *(Image credit: Ted Auch, FracTracker Alliance, 2021.)*

of how everyday water uses including kitchen faucet water was routinely contaminated and often infused with fugitive methane. A 2009 gas explosion in Dimock started a movement to hold companies accountable. Local advocates like Ray Kemble, shown in Fig. 8.14, struggled for years to be heard. The company Cabot Oil and Gas went so far as to sue Kemble.[72]

In 2022, Cabot's corporate criminality was exposed in court as Cabot Oil and Gas pleaded guilty to no less than fifteen criminal charges and nine felonies. Even as this one gas company was proven liable in this particular case—a certain victory—the industry at large has a proven track record of avoidance across its toxic externalities.

Cut gas off at the source

If regulations don't match science, society at large often becomes frustrated and pressures for change.[73] Communities seeking to protect their water often resort to forced immobility through fracking bans. These moratoriums exist across various localities and at national scales.[74]

[72] Phillips, S. (2022, December 13). 'Gasland' driller will pay millions for new water system in Dimock. WHYY.

[73] Maierean, A. (2021). What went wrong? Fracking in Eastern Europe. *Discover Energy*, 1(1), 1—15.

[74] Concerned Health Professionals of New York/Physicians for Social Responsibility (2022). *Compendium of scientific, medical and media findings demonstrating risks and harms from fracking*, 8th ed.

Figure 8.14 Dimock, Pennsylvania. *(Image credit: Susquehanna County residents and supporters urge PA Democratic Gubernatorial candidate Katie McGinty to stop fracking - SustainUS, 2014. Accessed via Flickr.)*

As shown in Fig. 8.15, bans cluster in the Northern Hemisphere. While anti-fracking organizations across Latin America promote bans,[75] Colombia may become the first in the region to ban fracked gas.[76] In 2022, the Colombian government worked with the largest oil union and proposed a ban;[77] petroleum workers supported this fracking ban as they understood extreme energy risks.[78] In 2023, Colombia joined other countries in the Beyond Oil and Gas Alliance.[79] Fracking bans in influential geographies like New York (Fig. 8.16) encourage movement spillover effects to other locations. Yet popular opinion alone has proven insufficient in jurisdictions where industry-captured politicians block passage of bans.

[75] *Infobae.* (6 de abril de 2021). EL sindicato petrolero más grande del país se une a la Alianza Colombia Libre de Fracking.

[76] Griffin, O. (2022, August 10). Colombia green groups try again to ban fracking, hope fourth time's the charm. Reuters.

[77] Ministerio de Ambiente y Desarrollo Sostenible. (10 de agosto de 2022). Radicado el proyecto de ley que cerrará la puerta al fracking en Colombia.

[78] Griffin, O. (2021, April 6). Colombia oil workers join anti-fracking campaign. *Reuters.*

[79] Stackl, V. (2023, August 31). Colombia joins the beyond oil and gas alliance and confirms international climate leadership. Oil Change International.

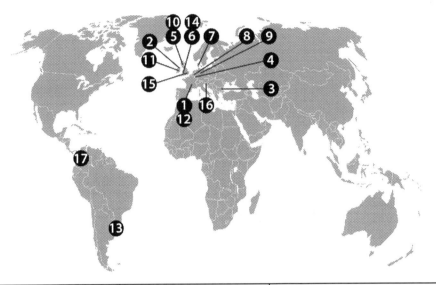

Map Number	Year	Location	Policy | Regulation Type
1	2011	France	Fracking Ban
2	2011	Northern Ireland	Fracking Moratorium
3	2012	Bulgaria	Fracking Ban
4	2013	Luxembourg	Motion Against Exploration
5	2015	Scotland	Fracking Moratorium
6	2015	Wales	Fracking Moratorium
7	2015	Denmark	End to New Applications
8	2015	Netherlands	Fracking Ban to 2020
9	2016-23	Germany	Fracking Moratorium
10	2017	Scotland	Fracking Moratorium

Figure 8.15 National fracking bans.

11	2017	Ireland	Fracking Ban
12	2017	France	Fracking Ban
13	2017	Uruguay	4 Year Prohibition
14	2018	Wales	Intent to Not Approve Permits
15	2019	United Kingdom	Fracking Moratorium
16	2022	Slovenia	Fracking Ban
17	2022	Colombia	Fracking Ban Proposal

Figure 8.15, Cont'd

Figure 8.16 Don't frack New York. *(Image credit: Action & New Yorkers Against Fracking Protest Gov. Cuomo's Plan to Frack New York, August 22, 2012 - CREDO. Accessed via Flickr.)*

Due to the power of energy firms, successful anti-fracking movements often feature collaboration; for example, in the eastern Canadian Province of Nova Scotia, university researchers and Native Mi'kmaq worked together to pressure for a fracking ban[80] in a collaborative process "learning through struggle." To halt new gas infrastructuring, grassroots advocates in dialog with scientists can produce synergistic outcomes.[81]

Try this

Research and spatial analysis: What can we learn from fracking bans at national, regional, and local scales? How can subnational and supranational efforts reinforce efforts at national levels? (See Figure 1.11 for discussion of overlapping governance scales.)

Fossil-fueled fascism

Fossil fascism (Chapter 7) represents authoritarianism in defense of fossil fuel regimes.[82] In response, gastivists employ civil disobedience to make gas infrastructuring more visible, some going beyond **symbolic protest** like rallies or petitions, which are relatively temporary and easy to ignore, to more disruptive and longer duration resistance tactics.[83] These direct actions range from events with banners or protest rallies to multi-year physical occupations blocking strategic construction sites; case in point, tree sits (Chapter 9).

Digital campaigns pressure firms, financiers, and policymakers. In a fast-paced, interconnected world, opposition movements often occupy virtual spaces along with or even instead of physical protests.[84] However, a **digital divide** exists: the gap between people that have access to modern communications technology and those that have restricted or no access, particularly in rural areas. Access is often further influenced by factors such as income and age. This divide shows unevenness in environmental coverage with relative silence to and from marginalized areas.

Virtual spaces create platforms for the exchange of information and increase the ability to share real-time events. Social media creates openings for protest with greater anonymity, and can be used to monitor activist safety for those who choose to be vocal critics. There is demonstrated power in a movement of digitally networked citizens, or

[80] Langdon, J. (2019). An unfractured line: an academic tale of self-reflective social movement learning in the Nova Scotia anti-fracking movement. *Environmental Justice, Popular Struggle and Community Devt*, 83–100.

[81] House, E. J. (2013). Fractured fairytales: The failed social license for unconventional oil and gas development. *Wyoming Law Review, 13*, 5.

[82] Malm, A. (2021). *White skin, black fuel: On the danger of fossil fascism*. Verso Books.

[83] Malm, A. (2021). *How to blow up a pipeline*. Verso Books.

[84] Brunner, E., & DeLuca, K. M. (2020). Affective Winds, decentered knots of world-making, and tracing force: A new conceptual vocabulary for social movements. In *The Rhetoric of social movements* (pp. 156–171). Routledge.

"netizens." However, **internet optimism** assumes overly positive benefits like education will follow from web access. In fact, online environments can be knowledge-poor: for example, the Chinese state fabricates online comments and restrains criticism. Although cyberactivists may create their own modes to circumvent state controls, activities considered threatening certainly draw sophisticated surveillance.

In repressive countries, opposition places activists at greater risk of extrajudicial detention or violence.[85] In these contexts, NGOs skirt sensitive topics and avoid protests. Infrapolitics (Chapter 7) refers to resistance that is dispersed, indirect, or disguised in nature, which emerges in contexts of fascism. With restrictions to protest in China, environmentalists have relied on the use of art to communicate concern. The artist known as Brother Nut installed as a performance piece 9,000 plastic bottles from the ubiquitous Chinese brand Nongfu Spring, but filled with heavily contaminated water culled from a polluted township in China's northwestern Shaanxi province.[86] Previously he produced an installation piece of a fully formed brick composed of vacuumed particulate matter from Beijing's notorious polluted air.[87] The unique impact and educational importance of such keen ecological art is only intensified by the stifling censorship throughout China's media outlets.[88]

Temporal analysis

All fossil fuels remain limited and nonrenewable; all the more so the potential of fracking wells over time: peak production occurs early, followed by continual decline. Like coal, gas has a toxic afterlife following huge waste production.[89] Due to industry capture of politicians, the energy sector exhibits a **regulatory lag**: a delay between when problems are known and when there are laws passed in response.

Performative permitting

Regulatory oversight during permitting gas projects is often performative: important decisions are made in closed spaces lacking transparency before the public weighs in. The private sector's privileged access to politicians and regulators contributes to a biased, unrealistic narrative of growth that ignores externalities and tradeoffs. Job statistics are

[85] Schmitz, R. (2013, April 17). China's toxic harvest. A "cancer village" rises in protest. *Marketplace*.
[86] Mitchell Ryan, O., & Mou, Z. (2018, July 13). With 9000 bottles of dirty 'Spring Water,' a Chinese artist gets results. *New York Times*.
[87] Buckley, C., & Wu, A. (2015, December 1). Amid smog wave, an artist molds a potent symbol of Beijing's pollution. *New York Times*.
[88] Staff. (2022, August 18). The creative ways Chinese activists protest pollution. *The Economist*.
[89] Lees, H., Järvik, O., Konist, A., Siirde, A., & Maaten, B. (2022). Comparison of the ecotoxic properties of oil shale industry by-products to those of coal ash. *Oil Shale, 39*(1), 1–19.

inflated (Chapter 9), while damage to roads, health care costs, and the impacts of spills and disasters aren't adequately deliberated up front.

Regulatory procedures invite public comment yet funnel participation through highly controlled spaces. State processes keep advocates occupied with sanctioned activities that appear meaningful but rarely generate transformative outcomes. Controversial gas project permitting receives hundreds and thousands of comments opposing approval, but this public input is largely disregarded.[90]

In China, the regulatory process has been described as a **"shock absorber,"** a device designed to dampen motion or impulse.[91] This analogy applies to US gas permit hearings as well, where a concerned public expresses outrage over gas-biased approvals.[92] State agencies keep pro-justice advocates busy with formal participation roles (i.e., written comments, hearing testimonies), reducing available energy and resources for disruptive actions and advocacy for systemic, transformative change. Meanwhile, public opposition to permits can engender a governmental management response that downplays issues of contention. One common tactic involves "ringers": experts-for-hire that echo industry talking points and downplay potential ecological or social impacts. Industry-sponsored voices gain media attention, using their platforms to counteract local concerns.

The burden of convincing regulators to change harmful draft permits frequently lies on impacted communities, with legal support from networks of pro-bono environmental law firms whose work can be constrained by limited staff. Nonprofit public interest organizations that litigate environmental issues tend to be small operations when compared to large corporate law firms employed by industry. Oil and gas retains significant influence across the public sector due in part to frequent professional contracting, including lawyers.

In higher education, pronounced corporate influence damages the integrity and reputation of education's role and mission.[93] **Frackademia** points to the gas industry's influence on academic scholarship.[94] A preferred denial tactic mainstreams contrarian views, such as scientists who suggest fracking is safe. These same scientists are often culled

[90] Finley-Brook, M., Williams, T. L., Caron-Sheppard, J. A., & Jaromin, M. K. (2018). Critical energy justice in US natural gas infrastructuring. *Energy Research & Social Science, 41*, 176–190.
[91] Johnson, T. (2020). Public participation in China's EIA process and the regulation of environmental disputes. *Environmental Impact Assessment Review, 81*, 106359.
[92] Scarff, K. (2021). Mapping vicarious proximity: Holistic and dualistic metaphors in a Mountain Valley Pipeline public hearing. *Environmental Communication, 15*(4), 514–529.
[93] Franta, B., & Supran, G. (2017, March 13). The fossil fuel industry's invisible colonization of academia. *The Guardian.*
[94] Ladd, A. E. (2020). Priming the well: "Frackademia" and the corporate pipeline of oil and gas funding into higher education. *Humanity & Society, 44*(2), 151–177.

from industry-friendly and funded departments and programs associated with reputable academic institutions, creating a facade of credibility.[95] For example, the State University of New York (SUNY) was found to have acted improperly in biased research related to fracking.[96] When members of the SUNY Buffalo's Shale Resources and Society Institute were discovered to have serious conflicts of interest with the gas industry, SUNY campus leadership came under scrutiny for their initial decision to refuse investigation.[97] SUNY decision structures have since been altered.

In both democratic and autocratic nations,[98] a well-financed gas industry uses its privileged access to politicians and the media to reproduce messages that underscore positive outcomes while ignoring costly negative tradeoffs.[99] A revolving door between staff from corporate sectors and regulatory agencies is common practice across countries.[100] Due to regulatory capture, the promotion of new fossil gas projects avoids the rigor of sound, comprehensive life cycle analysis (LCA) (Chapter 4).

Regulatory tug of war

Due to regulatory capture, corporate interests are prioritized, leading to a net loss for the public at large. Seated securely through favorable regulation, fossil gas impedes the transition to alternatives that have improved social and ecological outcomes.[101] These unfair political advantages unduly favor gas and allow for the industry to reverse climate action gains. **Mission creep**, or the shift over time of policy or infrastructure away from its original scope, has led to the inclusion of gas in climate policy, which in turn has led to an increase in GHG emissions. Discounting methane from gas is an example of **policy capture**, whereby decisions over policies are repeatedly directed away from the public interest toward a special interest. This is seen in the use of state funds to build more gas plants around the world, coupled with greenwashing that touts mitigation with unproven carbon capture, hydrogen replacement, and other false solutions addressed below.

[95] Cole, N., Ullman, O., Mulvey, K., & Pinko, N. (2021). *Colorado targeted by fossil fuel industry's disinformation playbook*. Union of Concerned Scientists.

[96] Navarro, M. (2012, November 19). SUNY buffalo shuts down its institute on drilling. *The New York Times*.

[97] Navarro, M. (2012, June 29). University will not investigate fracking institute. *The New York Times*.

[98] Hu, Z. (2020). When energy justice encounters authoritarian environmentalism: the case of clean heating energy transitions in rural China. *Energy Research & Social Science*, 70, 101771.

[99] Li, Y., & Shapiro, J. (2020). *China goes green: Coercive environmentalism for a troubled planet*. John Wiley & Sons.

[100] Pons-Hernández, M. (2022). Power (ful) connections: Exploring the revolving doors phenomenon as a form of state-corporate crime. *Critical Criminology*, 30(2), 305−320.

[101] Kemfert, C., Präger, F., Braunger, I., Hoffart, F. M., & Brauers, H. (2022). The expansion of natural gas infrastructure puts energy transitions at risk. *Nature Energy*, 1−6.

Without a fair chance to stop harmful projects as a result of state policy capture by industry, environmental organizations resort to actions targeting gas at its sources. Many jurisdictions experience **regulatory tug of war** with policy proposals such as fracking bans and gas hookup bans. While NGOs and an informed public promote local ordinances to impede new gas hookups, industry seeks bans on said ordinances and provides captured politicians with crafted language to block and overturn hookup restrictions.

Greener alternatives are now cost competitive and provide a greater number of jobs.[102] Another area of employment growth is remediating gas drilling sites. **Capping wells** and retiring lines (Box 8.1) reduces pollution and provides jobs. In leaky systems, states pay to replace whole gas lines even though these are destined to become stranded assets. Local agencies unwittingly pick this more expensive option over a variety of other cleaner renewable options, in large part due to the aggressive lobbying of the gas industry.

BOX 8.1 Spin-off industries: Strategic gas retirement

The retirement of leaking gas lines is a priority for health, safety, and climate reasons. Similarly, abandoned wells expel GHGs and cause adverse health effects and contamination.[103] The value of emitted gas can offset plugging costs,[104] yet policies for idle wells often remain inadequate to cover decommissioning, plugging, and surface reclamation.[105] Extractors can be held responsible for end-of-life costs whereby remediation funds are set aside prior or during extraction.

We can create tens of thousands of good–paying union jobs while also cutting back on emissions.[106]

Site remediation can be an employment growth sector, if mandated. **Orphaned wells**, those abandoned by oil and gas firms,[107] become a problem for a property owner or the government. In 2022, US federal funds were made available for plugging wells in 22 states.[108]

---

[102] Garrett-Peltier, H. (2017). Green versus brown: Comparing the employment impacts of energy efficiency, renewable energy, and fossil fuels using an input-output model. *Economic Modeling*, *61*, 439–447.

[103] Ferrar, K. (2019, April 4). Idle wells are a major risk. FracTracker.

[104] Kang, M., Mauzerall, D. L., Ma, D. Z., & Celia, M. A. (2019). Reducing methane emissions from abandoned oil and gas wells: Strategies and costs. *Energy Policy*, *132*, 594–601.

[105] Raimi, D., Krupnick, A. J., Shah, J. S., & Thompson, A. (2021). Decommissioning orphaned and abandoned oil and gas wells: New estimates and cost drivers. *Environmental Science & Technology*, *55*(15), 10224–10230.

[106] Cumpton, G., & Agbo, C. (2023). *Mitigating methane in texas: reducing emissions, creating jobs, and raising standards.* Texas Climate Jobs Project.

[107] Alboiu, V., & Walker, T. R. (2019). Pollution, management, and mitigation of idle and orphaned oil and gas wells in Alberta, Canada. *Environmental Monitoring and Assessment*, *191*(10), 1–16.

[108] Department of the Interior. (2023, June 5). Through president biden's bipartisan infrastructure law, 24 States set to begin plugging over 10,000 orphaned wells.

Box 8.1 Spin-off industries: Strategic gas retirement—cont'd

Processes of retreat at local and regional scales can occur through strategic limiting of gas expansion. Outright defection by households to cheaper renewable systems poses increased risk to utilities built on fossil fuels. In turn, growing energy burden may be placed particularly on low-income households (Introduction) as utilities continually raise their rates while seeking to maintaining outdated grids built on fossil fuels.

Clustered electrification in the form of block conversions away from fossil fuels to renewable sources and technologies supports circular economies. Too often energy transition is disconnected from community governance or reform movements. Networked or district geothermal enables a transition from gas to geothermal energy.[109] This is a form of **strategic retirement**—accelerated departure from gas using tools of spatial planning alongside commitment to improving justice and equity. Environmental injustice is not addressed effectively if persistent gas leaks remain unfixed and utilities in effect penalize low-wealth households and communities with unbending investments into highly carbonized grid infrastructure.[110]

In fact, incentives often go towards replacement in lieu of outright transition to alternatives. **Split incentives** occur when positives are counteracted with negatives, as is often the case in state subsidy programs. Confusing local governments and homeowners alike, incentives designed to propel renewables are often coupled with gas expansion; one reduces methane while the other undermines methane mitigation.

The urgency of the climate crisis and the harms for low-income communities have led some localities to ban new **gas hook-ups**—the equipment and other infrastructure to connect to distribution lines—redirecting funds for new infrastructure toward electrification. Industry responses include ghostwriting "bans of bans" creating a never-ending regulatory tug-of-war. Nevertheless, the numbers of effective local bans are growing and now include major cities like Boston.

Looking forward, methane leak loss measures have the potential to significantly shift markets. The US Inflation Reduction Act's (IRA) set costs beginning in 2024 for exceedence of methane emissions thresholds.[111] Fees will start at $900 per ton of methane and increase over time. Yet the IRA is riddled with methane loopholes: a charge is assessed on industry-set targets only, and companies can exempt "unavoidably lost" gas such as distribution leaks, creating a failsafe for unregulated methane releases.

With threats from climate, terrorism, and sabotage, private—public partners focus vast resources at hardening of the grid,[112] while adding vulnerable gas and LNG plants. Research on

(Continued)

[109] Gas to geo. (2023, August 15). Main page.

[110] Luna, M., & Nicholas, D. (2022). An environmental justice analysis of distribution-level natural gas leaks in Massachusetts, USA. *Energy Policy, 162,* 112778.

[111] International Energy Agency. (2022, November 8). Inflation reduction Act 2022: Sec. 60,113 and Sec. 50,263 on Methane Emissions Reductions.

[112] Huber, M. (2022). *Climate change as class war: Building socialism on a warming planet.* Verso.

Box 8.1 Spin-off industries: Strategic gas retirement—cont'd

climate and energy resiliency suggests this approach is naive, unreliable, and expensive. Localities are wise to seek a place-based mix of renewable sources. In urban areas, these will be on-grid systems, meaning they are part of the regional electrical **grid**, an interconnected network for electricity provision. Strategic regional approaches employ flexible interconnected renewable supply options backed up with battery storage for electricity and bedrock storage for heat and cooling managed with district heat pumps. Grid-tied semi-autonomous electrical generation in networked **microgrids** advances climate resilience. Rural areas often need to develop an off-grid approach with battery storage in resilience hubs, locations where populations in need of emergency support can receive care. Distributed energy systems can support the collective good.[113]

Try this

Research, analyze, and plan*: What are top priorities for strategic gas retirement and other methane mitigation in your area? How could remote sensing and on-site research assist in transparency regarding leaks?*

Illustrative cases

While gas industry practices in the US and China are distinct, neither is sustainable. Wholly new constructs can emerge that are participatory and transparent (Conclusion), replacing institutional subservience to the gas industry that impedes energy and climate justice.

America's "gaslandia"

The US Natural Gas Act (1938) and revisions (1954, 1992, 2005) advance utility projects defined as national security necessities.[114] US regulatory agencies have been shown to nearly always approve new gas infrastructure—public good is assumed, even when projects do not deliver affordable energy or serve communities through which pipelines pass. Interstate gas transmission lines often lack hookups to the rural localities they pass through and strategically aim to carry gas to export ports for profitable global LNG markets (Chapter 10).[115] Ironically, throughout its expansion across the lower 48 US states, gas' guiding justification was always energy independence. Now that gas exportation

---

[113] Wolsink, M. (2020). Distributed energy systems as common goods: Socio-political acceptance of renewables in intelligent microgrids. *Renewable and Sustainable Energy Reviews, 127,* 109841.

[114] Barikor, F. F. (2021). Studies of the influence of the oil and gas industries on the American energy policy Act of 2005. *South Asian Research Journal of Humanities and Social Sciences, 3*(2): 47—55.

[115] Ciccantell, P. S. (2020). Liquefied natural gas: Redefining nature, restructuring geopolitics, returning to the periphery? *American Journal of Economics and Sociology, 79*(1), 265—300.

Figure 8.17 Gas pipeline. *(Image credit: Ted Auch, FracTracker Alliance, 2017.)*

tops both national and industry agendas, both instability and continual price increases mark domestic gas markets.

Over time, the US has constructed a large network of gas pipelines. A significant portion of the mainline transmission network and local distribution network was installed in the 1950s and 1960s.[116] Construction continues with pipelines superimposed on landscapes (Fig. 8.17 above).

In reality, pipeline construction proposals have guaranteed rates of return, ensuring companies have inherent incentives to construct them. In order to maintain autonomy, companies further create redundant lines. In spite of the nation's incessant pipeline development, the unavoidable fact remains: fossil gas is not renewable, and the US will indeed deplete its profitable and accessible gas reserves (Fig. 8.18).

Firms often submit permit applications using "boilerplate language" mimicking explanations and buzzwords used project to project.[117] Regulatory violence exists in permitting due to failure to account for the unique life cycle and social costs of GHG emissions of each project, and ignoring a project's effect on regional energy mixes.[118] Also left out of the ledger are toxic pollutants interspersed in fracking waste that persist over time. Gas drilling's waste creates a crisis parallel to coal combustion residuals (Chapter 4). By the time impacted communities document the extent of harm, contamination has become concentrated. Remediation costs inevitably fall on taxpayers when companies deny responsibility or escape liability in courts, where burden of proof falls on those experiencing harm and their often pro bono legal support.

Upstream impacts include concentrated gathering fields, forming a uniquely compromised landscape (Fig. 8.19). Well concentration in these zones is particularly impactful as

---

[116] Energy Information Administration. (2019). Natural gas explained.

[117] Scott, T. A., Marantz, N., & Ulibarri, N. (2022). Use of boilerplate language in regulatory documents: Evidence from environmental impact statements. *Journal of Public Administration Research and Theory, 32*(3), 576—590.

[118] Hein, J. F., & Jacewicz, N. (2020). Implementing NEPA in the age of climate change. *Michigan Journal of Environmental & Administrative Law, 10*(1), 1—68.

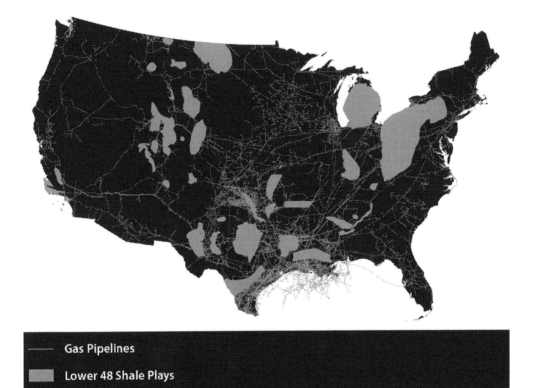

Figure 8.18 Lower 48 states shale plays and major pipelines. *(Source: U.S. Energy Information Administration (EIA).)*

gas firms are exempt from the Clean Water Act, effectively free from disclosure of toxic chemicals used in their fracking fluids.[119] Across basins, ruptured leaky lines and systematic wastewater dispersal spread toxins beyond immediately concentrated hotspots. At individual frack pad sites, intensive services are required, often requiring hundreds of truck trips transporting equipment, chemicals, sand, and water for site development and subsequent fracking (Fig. 8.20).[120] Harms at and near the drill site include emissions, dust and noise of large trucks as well as other construction disruption, not to mention operation risks impacting local residents. While close residential proximity to fracking has been extensively researched and proven harmful across the US, industry norms in China feature virtually no "setbacks".

[119] Abid, Y. (2020). Reflection on shale gas fracking risk assessment and management in the United States. *Washington Journal of Environmental Law & Policy, 11*(1), 1.
[120] Clancy, S. A., Worrall, F., Davies, R. J., & Gluyas, J. G. (2018). The potential for spills and leaks of contaminated liquids from shale gas developments. *Science of the Total Environment, 626*, 1463–1473.

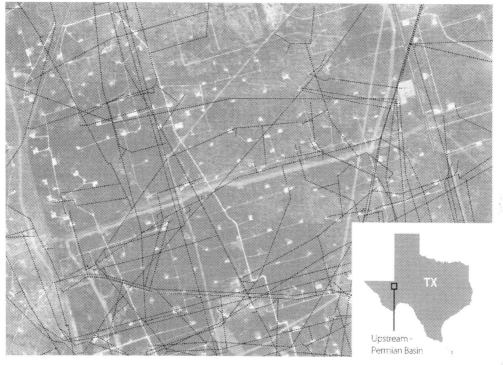

------- **Upstream Gathering Pipelines**

Figure 8.19 Upstream gathering pipelines. *(Sources: Esri, Maxar, Earthstar Geographics, CNES/Airbus DS, USDA FSA, USGS, Aerogrid, IGN, IGP, and the GIS User Community. Gathering line extracted from The Railroad Commission of Texas (RRC).)*

Storage grabs

As extraction sites receive more research attention than other parts of the gas vector, underground storage facilities are relatively poorly understood, in part due to their invisibility. Underground storage involves vertical and horizontal space grabs. Fig. 8.21 shows "stacked" storage in various subsurface geologic formations at different depths. Storage facilities, which cross the US with some regional variation (Fig. 8.22), exhibit risky characteristics, including methane releases[121] and the storage of flammable materials ensuring industrial-scale risk.[122]

[121] Schultz, R. A., Hubbard, D. W., Evans, D. J., & Savage, S. L. (2020). Characterization of historical methane occurrence frequencies from US underground natural gas storage facilities with implications for risk management, operations, and regulatory policy. *Risk Analysis, 40*(3), 588–607.

[122] Gorlenko, N. V., & Murzin, M. A. (2019). Comparative analysis of fire risks in coal and oil and gas industries. *IOP Conference Series: Materials Science and Engineering, 687*(6):, 066009.

Figure 8.20 Frack pad. *(Image credit: Ted Auch, FracTracker Alliance, 2021.)*

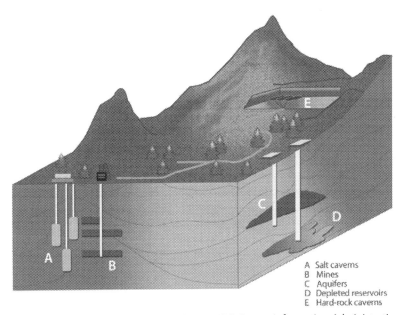

A Salt caverns
B Mines
C Aquifers
D Depleted reservoirs
E Hard-rock caverns

Figure 8.21 Underground storage types. *(Source: U.S. Energy Information Administration (EIA).)*

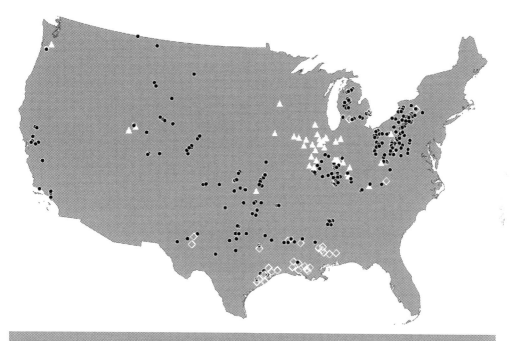

Figure 8.22 Underground gas storage. *(Source: U.S. Energy Information Administration (EIA).)*

Slow violence can also happen at underground storage facilities. In 2015, the worst gas leak in US history culminated as the San Aliso blowout. This calamitous event alone doubled methane emission rates for the entire Los Angeles basin.[123] Even as this facility leaked for years prior, creating a "diesel death zone,"[124] resident complaints of health ailments were continually ignored.

Gas hubbing

US gas markets feature financial future exchanges for trading across numerous delivery points. Trading hubs (Fig. 8.23) are control nodes with digital, financial, and physical

[123] Hendrick, M. F., Cleveland, S., & Phillips, N. G. (2017). Unleakable carbon. *Climate Policy, 17*(8), 1057–1064.
[124] Ramirez, R. (2020, December 17). The fight against fossil fuel infrastructure is the fight for healthy communities. KCET.

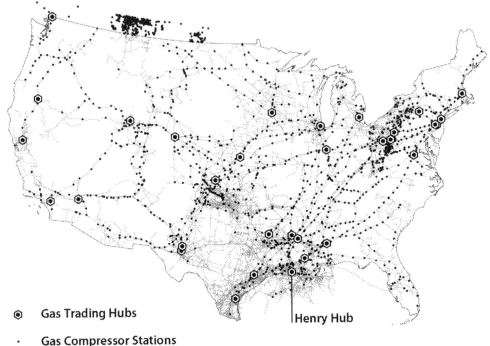

◉   **Gas Trading Hubs**

|Henry Hub

•   **Gas Compressor Stations**

....   **Gas Pipelines**

Figure 8.23 Trading hubs. *(Sources: The Homeland Infrastructure Foundation-Level (HIFLD) Database, U.S. Energy Information Administration (EIA).)*

infrastructures. The Henry Hub, located at the intersection of pipeline systems in Louisiana, is used to benchmark gas markets in North America and globally.[125]

Storage hubs act as underground anchors for long, complex networks of regional gas infrastructure (Fig. 8.24). Accessible to the US Marcellus Shale basin, the planned yet unexecuted Appalachian Storage Hub (ASH) would have featured a massive ethane cracker plant developed by a subsidiary of Shell with significant Chinese investment. After signing a memorandum of understanding in 2017, a US–China trade war in 2018 upended negotiations.[126] Within Appalachia, international investment exploits inequity. Regional poverty remains widespread and foreign interests maintain an inordinate stake

[125] Günther, M., & Nissen, V. (2021). Effects of the henry hub price on US LNG exports and on gas flows in Western Europe. *Gases, 1*(2), 68–79.

[126] Litvak, A. (2018, June 18). China hits pause on Appalachian energy investment citing trade war. *Pittsburgh Post-Gazette.*

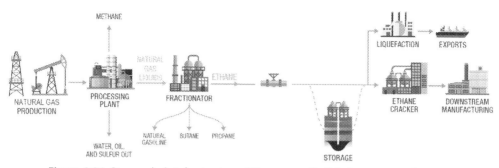

Figure 8.24 Storage hub infrastructure. *(Diagram credit: U.S. Department of Energy.)*

in land used for extraction, exhibiting patterns of exploitation mirroring those of other low-income countries targeted particularly by Chinese investments.

A **spatial "fix"**, like ASH's methane chain, seeks to make property or infrastructure more rigid, often to facilitate extraction. Yet siting gas infrastructure as a spatial fix only underscores climate risks now pervasive throughout fossil fuel infrastructuring.

Try this

Research and spatial analysis*: Examine a portion of gas infrastructure, like a compressor station, power plant, or pipeline. Evaluate the parts of the vector as fixed (i.e., infrastructure) or mobile (i.e., pollutants) and identify any regulatory gaps. Through GIS spatial tools determine proximity and exposure per demographic group and location.*

China's dash to gas

China is both a climate leader and villain:[127] it has surpassed projections for wind and solar energy production while subsidizing fossil gas expansion. Accurate data on China's GHG emissions is difficult to obtain, but satellite observations continually show higher methane emissions from oil and gas operations than reported.[128]

China developed infrastructure to produce and transport gas domestically at an accelerated pace. Transition from coal to gas has lacked holistic life cycle analysis, and even while gas intensifies, new coal permits are still issued across China. Public criticisms of policies and practices designed to transition from coal included gas supply shortages, safety concerns, corruption, and

[127] Meidan, M. (2020). China: Climate leader and villain. In *The geopolitics of the global energy transition* (pp. 75–91).

[128] Chen, Z., Jacob, D. J., Nesser, H., Sulprizio, M. P., Lorente, A., Varon, D. J., … Yu, X. (2022). Methane emissions from China: A high-resolution inversion of TROPOMI satellite observations. *Atmospheric Chemistry and Physics*, 22(16), 10809–10826.

cost.[129] Practices have incurred rapid ecological change with high-intensity extraction, such as drawing from multiple wells at a single drill pad and horizontal drilling across vast distances.[130] Vast regulatory gaps exist even as an elaborate system of environmental assessment has emerged.[131] Chinese energy infrastructure is spatially uneven, and its energy transition can intensify inequality.[132] The extreme proximity of fracking pads to Chinese residential structures shown in Fig. 8.25 denotes a lack of basic spatial regulation within industry practice.

Although press coverage is sporadic, environmental harms in China are alarming. Cancer rates are significant;[133] smog is readily visible and regularly embodied. Normalized, everyday pollution only generates increasing public distrust, coupled with opaque information[134] and low-quality environmental impact assessments.[135] China's Communist Party, in response to coal pollution particularly acute in influential urban centers, adopted severe tactics referred to as "Beijing's green fist."[136] Such tactics are anything but democratic, exacting significant labor and land use costs typically borne unequally by those most vulnerable, often in rural locations. While protests within mainland China may remain subdued, opposition is trending in offshore locations of energy investment,[137] including China's prized, global Belt and Road Initiative (BRI).[138]

[129] Liang, J., He, P., & Qiu, Y. L. (2021). Energy transition, public expressions, and local officials' incentives: Social media evidence from the coal-to-gas transition in China. *Journal of Cleaner Production, 298*, 126771.

[130] Aizhu, C. (2018, June 21). Stepping on the gas: China's homebuilt fracking boom. *Reuters*.

[131] Li, Y., & Shapiro, J. (2020). *China goes green: Coercive environmentalism for a troubled planet.* John Wiley & Sons.

[132] Du, K., Cheng, Y., & Yao, X. (2021). Environmental regulation, green technology innovation, and industrial structure upgrading: The road to the green transformation of Chinese cities. *Energy Economics, 98*, 105247.

[133] Cao, M., Li, H., Sun, D., & Chen, W. (2020). Cancer burden of major cancers in China: A need for sustainable actions. *Cancer Communications, 40*(5), 205—210.

[134] Johnson, T. (2019). Environmental protest in urban China. In *Handbook on urban development in China*. Edward Elgar Publishing.

[135] Liu, L., & De Jong, M. (2017). The institutional causes of environmental protests in China: A perspective from common pool resource management. *Journal of Chinese Governance, 2*(4), 460—477.

[136] Wang, Y. (2022). *Beijing's green fist*. Human Rights Watch.

[137] Staff. (2022, September 23). End to Chinese funding of fossil fuels sought. *Dawn*.

[138] Staff. (2019, September 4). Dozens protest against Chinese influence in Kazakhstan. Reuters.

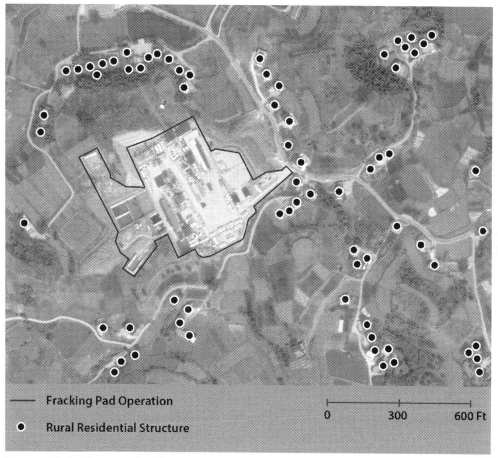

Figure 8.25 Residences embedded with gas infrastructure. *(Source: Google Earth, Maxar Technologies, 2021.)*

Authoritarian spaces

Protesters involved in confrontational riots against land grabs, evictions, or environmental pollution experience state-sanctioned violence or arrests.[139] Protest in China is relatively rare due to inherent risks and deplorable treatment of dissidents, such as in the 1989 Tiananmen Square massacre. Still, concern over plants manufacturing paraxylene (PX) to make synthetic fibers from fossil fuels moved thousands to protest.[140] Chinese media covered the PX protests until 2015, when information starkly declined.[141] The PX protests

[139] Göbel, C. (2021). The political logic of protest repression in China. *Journal of Contemporary China*, *30*(128), 169–185.
[140] Deng, Y., & Yang, G. (2013). Pollution and protest in China: Environmental mobilization in context. *The China Quarterly, 214*, 321–336.
[141] South China Morning Post. (n.d.). Paraxylene (PX).

BOX 8.2 Spin-off industry: Fossil fabrics

China is the world's largest polyester fabric producer. Polyester is spun into fibers and woven to create fabrics. Demand for polyester surpassed cotton in 2002. Global clothing production doubled quickly thereafter.[144] **Fast fashion** is tied to materials like polyester that lack durability, forcing ongoing wardrobe renewal.[145] Subsidies to oil and gas keep prices low, artificially incentivizing consumerism of polyester products while discouraging alternatives. Polyester is the most common fiber in apparel—in spite of extreme environmental and social costs.[146] For example, microplastic pollution is released from polyester as it deteriorates from laundry and from wear, contaminating air, soil, and water. Chemicals in textile production have been shown to cause respiratory damage, cognitive impairment, dermatitis, and cancers.[147] Workers are at risk because many polyester-producing businesses are small, so managers often have less knowledge of risks and fewer occupational protections.[148] Periodic fires in garment factories sometimes make the news, yet daily exposures to toxic substances throughout the production chain are often ignored.[149] PX production creates air pollution,[150] and too often plants have poor ventilation.[151]

PX manufacturing zones were temporarily a hotbed for protest in China.[152] In 2007, large numbers spontaneously gathered in the streets for 2 full days to show opposition to a proposed PX factory in Xiamen. Responding to these "strolling eco-warriors," as they were nicknamed, the government relocated the proposed project.[153] An examination of the spatial diffusion of subsequent PX protests[154] suggests social movement **spillover effects**, where events at one location influence proceedings in one or more new sites (Praxis 2). In 2011, over 12,000 people showed up in Dalian to protest an existing PX plant.[155] The protests appeared successful when the mayor announced the plant would be moved—yet this promise was later reneged.[156] Nonetheless, Dalian protests influenced events in Maoming in 2014.[157] PX protests targeting the local government and Sinopec, a state-owned firm,[158] lasted a full week and became violent.[159] Despite police deleting images from protesters' phones and government controls blocking social media platforms, hacktivists were able to share images and videos beyond the confines of the protest site alone.[160]

were spontaneous in nature, organized digitally,[142] leveraging PX, the raw material in polyester (Box 8.2 above), as persuasive flashpoint to communicate broader woes.[143]

[142] Sun, X., & Huang, R. (2020). Spatial meaning-making and urban activism: Two tales of anti-PX protests in urban China. *Journal of Urban Affairs, 42*(2), 257—277.

[143] Jia, H. (2014, April 16). Para-xylene plants face uphill struggle for acceptance in China. *Chemistry World.*

[144] Wicker, A. (2021, January 19). Why, exactly, is polyester so bad for the environment? Ecocult.

[145] Harding-Rolls, G. (2022, April 12). Changing markets — Deft marketing tricks mask the truth on synthetic fibers. Lampoon.

[146] Palacios-Mateo, C., van der Meer, Y., & Seide, G. (2021). Analysis of the polyester clothing value chain to identify key intervention points for sustainability. *Environmental Sciences Europe, 33*(1), 1—25.

[147] Rovira, J., & Domingo, J. L. (2019). Human health risks due to exposure to inorganic and organic chemicals from textiles: A review. *Environmental Research, 168*, 62—69.

Gaslighting: A great leap backward

The Chinese government gaslit PX protesters in an information-poor environment. **Gaslighting** often includes slandering, disparaging and undercutting scientists and advocates, causing them and others to doubt verifiable claims and scientific findings. In 2014, after PX protests in various localities, more state resources were directed toward co-opting power, diverting attention, and using physical violence to quell opposition. PX protesters were portrayed as an impediment to national and social progress.[161] The framing of protests as harmful agents created a backlash against the activists such that environmentalists' motives and arguments were repeatedly questioned.[162] After a PX plant at the center of earlier protests exploded in 2015, 1,000 angry posts from those who were silenced erupted as "we told you so."[163]

[148] Parvin, F., Islam, S., Akm, S. I., Urmy, Z., & Ahmed, S. (2020). A study on the solutions of environment pollution and worker's health problems caused by textile manufacturing operations. *Biomedical Journal of Scientific & Technical Research, 28*(4), 21831–21844.

[149] Singh, N. (2016). Safety and health issues in workers in clothing and textile industries. *International Journal of Home Science, 2*(3), 38–40.

[150] Lin, B., & Wu, R. (2021). The dilemma of paraxylene plants in China: Real trouble for the environment? *Science of The Total Environment, 779*, 146456.

[151] Qian, W., Guo, Y., Wang, X., Qiu, X., Ji, X., Wang, L., & Li, Y. (2022). Quantification and assessment of chemical footprint of VOCs in polyester fabric production. *Journal of Cleaner Production, 339*, 130628.

[152] Ma, D., Gong, W., Fang, Q., Liu, D., & Liu, Y. (2018). Divergence between general public's risk perception and environmental risk assessment: A case study of p-xylene projects in China. *Human and Ecological Risk Assessment: An International Journal, 24*(4), 859–869.

[153] Staff. (2012, October 23). Viewpoint: The power of China's strolling ecowarriors. BBC.

[154] Zhu, Z. (2017). Backfired government action and the spillover effect of contention: A case study of the anti-PX protests in Maoming, China. *Journal of Contemporary China, 26*(106), 521–535.

[155] Zeng, V. (2020, March 31). *No more toxic than caffeine? The long history of anti-paraxylene protests.* Hong Kong Free Press.

[156] Brunner, E. (2019). *Environmental activism, social media, and protest in China: Becoming activists over wild public networks.* Rowman & Littlefield.

[157] Lee, K., & Ho, M. S. (2014). The Maoming Anti-PX Protest of 2014. An environmental movement in contemporary China. *China Perspectives, 2014*(2014/3), 33–39.

[158] Shanghai, N. D. (2014, April 4). Volatile Atmosphere. *The Economist.*

[159] China Blog Staff. (2014, April 2). Piecing together China's Maoming plant protest. BBC News.

[160] DeLuca, K. M., Brunner, E., & Sun, Y. (2016). Constructing public space: Weibo, wechat, and the transformative events of environmental activism in China. *International Journal of Communication, 10*, 19.

[161] Wang, W., Zhang, H., Han, P., Wang, K., & Cao, M. (2022). Emphasizing actions over words: A Chinese perspective on Thunberg's protest. *American Journal of Economics and Sociology, 81*(2), 287–303.

[162] Yu, X., Wang, J., & Liu, Y. (2021). Civic participation in Chinese cyberpolitics: A grounded theory approach of para-xylene projects. *International Journal of Environmental Research and Public Health, 18*(23), 12458.

[163] Allen-Ebrahimian, B. (2015, April 6). Explosion of once-scuttled chemical plant riles China's web. Foreign Policy.

The people of Xiamen were right.

China's political system lacks vertical and electoral accountability. There's limited political space for NGOs[164]—groups are allowed minor pushback while simultaneously discouraged from more contentious movement-building.[165] While robust alternative news outlets aren't accessible to many, informal communication does play an important role in China.[166]

Methane myopia

The Chinese gas industry has developed relatively quickly,[167] with the majority of infrastructure components developed domestically.[168] Official statements of efficiency may be exaggerated:[169] claims of low methane leak rates (3% or less) are decoupled from upstream extraction and downstream consumption, focusing rather on midstream transmission alone.[170]

Energy investments in China continue to increase carbon emissions,[171] spurred on by cost distortions in the country's gas markets.[172] Existing structures delay energy transformation as upstream and downstream industries remain large, state-owned enterprises. State targets have motivated domestic gas expansion despite the fuel's methane risks,

[164] Zhang, Y. (2020). Broker and buffer: Why environmental organizations participate in popular protests in China. *Mobilization: An International Quarterly, 25*(1), 115–132.

[165] Li, J. (2021). Signaling compliance: An explanation of the intermittent green policy implementation gap in China. In *Local Government Studies* (pp. 1–27).

[166] Chin-Fu, H. (2013). Citizen journalism and cyberactivism in China's anti-PX plant in Xiamen, 2007–2009. *China: An International Journal, 11*(1), 40–54.

[167] Zhen, W., Yinghao, K., & Wei, L. (2022). Review on the development of China's natural gas industry in the background of "carbon neutrality. *Natural Gas Industry B, 9*(2), 132–140.

[168] Tong, X., Zheng, J., & Fang, B. (2014). Strategic analysis on establishing a natural gas trading hub in China. *Natural Gas Industry B, 1*(2), 210–220.

[169] Huang, Z., & Li, J. (2012). Assessment of fire risk of gas pipeline leakage in cities and towns. *Procedia Engineering, 45*, 77–82.

[170] Gao, J., Guan, C., & Zhang, B. (2022). Why are methane emissions from China's oil & natural gas systems still unclear? A review of current bottom-up inventories. *Science of The Total Environment, 807*, 151076.

[171] Li, Z. Z., Li, R. Y. M., Malik, M. Y., Murshed, M., Khan, Z., & Umar, M. (2021). Determinants of carbon emission in China: how good is green investment? *Sustainable Production and Consumption, 27*, 392–401.

[172] Jiang, H. D., Xue, M. M., Dong, K. Y., & Liang, Q. M. (2022). How will natural gas market reforms affect carbon marginal abatement costs? Evidence from China. *Economic Systems Research, 34*(2), 129–150.

[173]and overall GHG reductions from the large-scale transition from coal to gas remain uncertain.[174] Accurate measurements of methane emissions were given low priority, often simply ignored as China's domestic policies promoted fossil gas.[175]

Beyond its energy sector, China has consistently demonstrated the Porter Effect—otherwise known as the Innovation Effect—whereby environmental regulation, as opposed to market-based regulation, positively influences green innovation with efficiency and equity. Researchers found environmental regulation improved the economic performance of firms in 30 Chinese provinces (2003—17); however, Chinese investments in other countries did not share green innovation. Research also showed that strategies like voluntary emissions markets did not drive innovation to the degree that enforceable environmental policy did; in fact, market-based strategies negatively impacted ecological innovation.

Market-based regulation inhibits the improvement of green innovation.[176]

Largely state incentivized, consumption of fossil gas in urban residences and vehicles continues apace in China in spite of the fuel's inherent climate impacts. Ignoring methane's harms in the planning of domestic fuel transition, China's transportation sector experienced an increase in overall emissions after switching to gas derivatives.[177] Policies for urban users provided incentives too: because gas was more expensive than coal, government subsidies supported "last-meter" gas distribution lines bringing the pipe to the end user and retrofits in the residential sector.[178]

[173] Yin, Y., & Lam, J. S. L. (2022). Impacts of energy transition on Liquefied Natural Gas shipping: A case study of China and its strategies. *Transport Policy, 115*, 262—274.

[174] Qin, Y., Tong, F., Yang, G., & Mauzerall, D. L. (2018). Challenges of using natural gas as a carbon mitigation option in China. *Energy Policy, 117*, 457—462.

[175] Zhen, W. A. N. G., Yinghao, K. O. N. G., & Wei, L. (2022). Review on the development of China's natural gas industry in the background of "carbon neutrality. *Natural Gas Industry B, 9*(2), 132—140.

[176] Luo, Y., Salman, M., & Lu, Z. (2021). Heterogeneous impacts of environmental regulations and foreign direct investment on green innovation across different regions in China. *Science of the Total Environment, 759*, 143744.

[177] Pan, D., Tao, L., Sun, K., Golston, L. M., Miller, D. J., Zhu, T., ... Zondlo, M. A. (2020). Methane emissions from natural gas vehicles in China. *Nature Communications, 11*(1), 1—10.

[178] Li, J., She, Y., Gao, Y., Li, M., Yang, G., & Shi, Y. (2020). Natural gas industry in China: Development situation and prospect. *Natural Gas Industry B, 7*(6), 604—613.

Gas and growth

Sichuan province is the center of fracking in China and a growth pole. **Growth poles** are zones planned for regional development. Chinese companies moved west because regional regulations were less stringent than in Beijing and other polluted areas.[179] Rapid internal migration created manufacturing-oriented cities like Chengdu (Fig. 8.26), a springboard for expansion which ties to the concentrated fracking zone to the south. Growth poles, like manufacturing hubs and fracking boom areas, can generate inequity depending on zoning policy, speed of growth, and other factors.

Chengdu is one of the world's top 30 cities for scientific research output[180]—it's a modern megacity with a young educated population that's become a hub for digital industries, especially gaming. Technology manufacturing created a bifurcated economy as

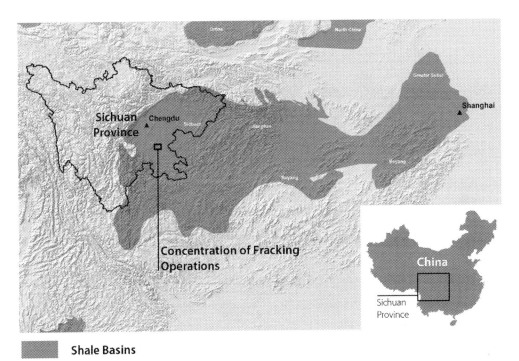

Shale Basins

Figure 8.26 Chengdu, China. (*Sources: Shale Basin and Plays World Dataset, Database of Global Administrative Areas (GADM), Natural Earth.*)

[179] Zhang, D., Fan, F., & Park, S. D. (2019). Network analysis of actors and policy keywords for sustainable environmental governance: Focusing on Chinese environmental policy. *Sustainability*, *11*(15), 4068.

[180] Nature Index. (2022). Top 200 science cities. *Nature*.

Chengdu also hosts low-wage factory jobs with high occupational hazards. In 2020, Chengdu became infamous for a fatal Foxconn factory explosion.[181]

Poor air quality in Chengdu led to public critique in 2016: activists put masks on prominent statues as an environmental protest. The state response was heavy-handed: police in riot gear placed the city center on lockdown and businesses were ordered to contact authorities if anyone sought to copy flyers complaining about smog.[182]

After 2016, politicians directed more attention to air quality in Chengdu.[183] Yet the area still hosts thousands of sites of unregulated pollution, and the transition to gas created yet more air pollution such as volatile organic compounds (VOCs).[184] Furthermore, Chengdu receives air pollution from the west,[185] which gets trapped in the valley around the city by its mountain basin.[186]

Chengdu has a shortage of drinking water—its main source, the Minjiang River, saw flow decrease while water extraction increased with population growth.[187] Water scarcity is becoming dire in China's drier regions with the development of mega-cities. In addition to crop failure, drought reduces hydroelectricity capacity.[188] China's environmental crisis generates global concern given interconnected supply chains.[189] A heat wave in 2022 exposed vulnerabilities as manufacturing plants were forced offline in order to reserve local residents' access to scarce water.[190]

Empire-building

In 2016, 80% of China's oil passed through the Straits of Malacca, a significant structural vulnerability. To address such limiting risk, China's BRI initiative has sought to globally

[181] Lu, E. (2020, September 11). Review: Dying for an iPhone: Apple, foxconn and the lives of China's workers. Labor Notes.

[182] Haas, B. (2016). China riot police seal off city center after smog protesters put masks on statues. *The Guardian*.

[183] Tan, Q., Zhou, L., Liu, H., Feng, M., Qiu, Y., Yang, F., ... Wei, F. (2020). Observation-based summer O3 control effect evaluation: A case study in Chengdu, a megacity in Sichuan Basin, China. *Atmosphere*, *11*(12), 1278.

[184] Xiong, C., Wang, N., Zhou, L., Yang, F., Qiu, Y., Chen, J., ... Li, J. (2021). Component characteristics and source apportionment of volatile organic compounds during summer and winter in downtown Chengdu, southwest China. *Atmospheric Environment*, *258*, 118485.

[185] Scott, M. (2009, February 24). Haze in the Sichuan Basin. *NASA Earth Observatory*.

[186] Riebeek, H. (2008, December 27). Low clouds over central China. *NASA Earth Observatory*.

[187] Gong, Q., Guo, G., Li, S., & Liang, X. (2021). Decoupling of urban economic growth and water consumption in Chongqing and Chengdu from the "production-living-ecological" perspective. *Sustainable Cities and Society*, *75*, 103395.

[188] *Bloomberg*. (2022, October 25). The world's biggest source of clean energy is evaporating fast.

[189] Maizland, L. (2021). *China's fight against climate change and environmental degradation*. Council on Foreign Relations.

[190] Feng, E. (2022, August 20). China battles its worst heat wave on record. NPR.

diversify its transportation routes and enhance supply options.[191] Today, most gas imported into China arrives from its BRI affiliates. BRI now has global reach: member nations grew in number to 140 by 2021.[192]

The BRI initiative has caused significant social and ecological damage internally and abroad,[193] and BRI agreements between countries lack protections for vulnerable populations.[194] BRI has established a system allowing space grabs across the world's most strategic trade routes. China's tactics extend beyond internationally recognized territories.[195] For example, Chinese patrols display military strength at the Spratly Islands by way of oil and gas claims.[196] Another example of encroachment, China's National Offshore Oil Corporation is a key partner for the controversial East African Crude Oil Pipeline (EACOP). China assumed greater control as several other funders divested following environmental opposition.[197]

Key Chinese energy partners also rely on authoritarian practices. Saudi Arabia, China's biggest trading partner,[198] seeks to expand its liquids-to-chemicals capacity with this partnership. Russia sends gas to China through the Power of Siberia Pipeline.[199] Although shrouded in secrecy, China appears advantaged in negotiations for a second pipeline from Russia.[200]

[191] Yin, Y., & Lam, J. S. L. (2022). Impacts of energy transition on liquefied Natural gas shipping: A case study of China and its strategies. *Transport Policy, 115*, 262–274.

[192] Green Finance and Development Center. (n.d.). *Countries of the belt and road initiative.*

[193] Hughes, A. C., Lechner, A. M., Chitov, A., Horstmann, A., Hinsley, A., Tritto, A., … Douglas, W. Y. (2020). Horizon scan of the belt and road initiative. *Trends in Ecology & Evolution, 35*(7), 583–593.

[194] Coenen, J., Bager, S., Meyfroidt, P., Newig, J., & Challies, E. (2021). Environmental governance of China's belt and road initiative. *Environmental Policy and Governance, 31*(1), 3–17.

[195] Staff. (2022, October 3). Japan protests possible gas extraction by China in contested waters. *The Japan Times.*

[196] Tsionki, M. (2021). Imagined mappings of geopolitical power: Liquid borders, military infrastructures and ecological destruction in the South China Sea. In *Visual Culture Wars at the Borders of Contemporary China* (pp. 81–101). Palgrave Macmillan.

[197] Okafor, C.(2023, May 8).China takes over funding of East Africa's oil projects amidst mounting environmental opposition. Business Insider.

[198] Anstey, C. (2022, October 15). China's rich new friend capitalizes on US oil spat. Bloomberg.

[199] Slav, I. (2023, January 3). Russia boosts the export capacity of its natural gas pipeline to China. Oil Price.

[200] Vakulenko, S, (2023, April 18). *What Russia's first gas pipeline to China reveals about a planned second one.* Carnegie Endowment for International Peace.

Missing a mitigation window

For China, the unavoidable fact remains unbending: sea levels at the country's edges have risen faster than the global average.[201] Millions live on land that is projected to be underwater within decades.[202]

China, like the US, has not done enough to reduce methane or respond to the harms from fossil fuel dependencies. New fossil gas creates delays for necessary alternatives. Timing is mismatched with the lethal speed of the climate crisis. New gas and liquefied natural gas (LNG) infrastructure takes years to build before becoming fully operational. As timing of fuel transitions is largely left to prices in distorted markets, greenwashed gas in complex forms proposes to extend the same structural and regulatory violence that occurred with coal and oil. Fossil fascism reinforces intellectual deadening whereby non-fossil fuel innovation is stifled. High rates of science publications implicitly and explicitly advance industry, privileging fossil fuels.[203] In the case of China, direct "return on investment" generated from science spending is expected to materialize with commercialization results.[204] If there had been the same degree of research and development (R&D) attention to renewables, the world we live in today would be very different. To redirect both public and private sector infatuation with gas—its "methane madness"—gastivists attack the gas industry's social license to operate (SLO), undercutting gas' status as a bridge fuel and clarifying its role as a primary climate disruptor.

Spatial and temporal synopsis

Space

(1) *Gas demonstrates fixity through infrastructuring and lock-in and mobility through pollution and transportation.*

(2) *Gas spaces can be highly contested, with territory/territoriality making shale basins strategic geopolitical locations.*

(3) *Regional and global gas buildout involves horizontal and vertical space grabs, with overbuild incentivized by guaranteed returns.*

[201] Staff. (2020, August 27). China is heating up faster than the global average, data shows. *Bloomberg.*

[202] Kulp, S. A., & Strauss, B. H. (2019). New elevation data triple estimates of global vulnerability to sea-level rise and coastal flooding. *Nature Communications, 10*(1), 1–12.

[203] Jung, J. Y., & Zeng, M. (2022). Changing frames: China's media strategy for environmental protests. *Asian Perspective, 46*(3), 423–449.

[204] Gosens, J., Hellsmark, H., Kåberger, T., Liu, L., Sandén, B. A., Wang, S., & Zhao, L. (2018). The limits of academic entrepreneurship: Conflicting expectations about commercialization and innovation in China's nascent sector for advanced bio-energy technologies. *Energy Research & Social Science, 37*, 1–11.

Time

(1) *Methane has far greater climate impacts than carbon dioxide over a shorter period of time yet, this fact is often overlooked in support of gas extraction and consumption.*

(2) *Gas industry employs gaslighting and disinformation to lock-in and prolong the use of fossil gas.*

(3) *Gastivists seek moratoriums or delays on fracking and gas hookups as a step toward ending the use of fossil gas.*

Summary

Fossil gas territoriality involves space grabs, and gas spaces are part of geopolitical strategy. Gas companies use gaslighting and misinformation to deceptively extend the myth of gas as a bridge fuel, even incentivizing expansion at the expense of energy transformation and the climate. Industry power and policy capture have allowed methane loopholes that ignore emissions. In the US and China alike, regulatory agencies downplay toxic and dangerous components of fossil gas since its emissions are less visible and less counted than those of coal. Gas infrastructure expansion and its social license to operate (SLO) are increasingly contested to end lock-in of fossil gas use, particularly as gas exacerbates climate disruption.

Vocabulary

1. capping wells
2. captive ratepayer
3. carbon-centric
4. carbon dioxide equivalent (CO$_2$e)
5. carbon risk
6. carbon unicorns
7. climate risk
8. clustered electrification
9. co-pollutant
10. cumulative impacts
11. deterritorialization
12. digital divide
13. ecocide
14. false solutions
15. fast fashion
16. fixity
17. fossil fascism
18. frackademia
19. fracking
20. fracking ban
21. gas hook-up
22. gaslighting
23. gastivists
24. geopolitics

25. global warming potential (GWP)
26. grid
27. growth poles
28. hazardous air pollutants (HAPs)
29. hub
30. induced seismicity
31. infrastructuring
32. internet optimism
33. just transition
34. lock-in
35. methane loopholes
36. microgrid
37. mission creep
38. net-zero
39. obfuscation
40. off-grid
41. on-grid
42. orphaned wells
43. policy capture
44. radical transparency
45. regulatory lag
46. regulatory tug of war
47. risk
48. shock absorber
49. spatial fix
50. spillover effect
51. split estate
52. split incentive
53. strategic retirement
54. sunk costs
55. super emitters
56. symbolic protest
57. technologically enhanced naturally occurring radioactive materials (TENORM)
58. territoriality
59. territory
60. volatile organic compounds (VOCs)

Recommended

Arboleda, M. (2020). *Planetary mine: Territories of extraction under late capitalism.* Verso Books.
Li, Y., & Shapiro, J. (2020). *China goes green: Coercive environmentalism for a troubled planet.* John Wiley & Sons.
McBay, A. (2019). *Full spectrum resistance* (vols I & II). Seven Stories Press.
Zuboff, S. (2019). *The age of surveillance capitalism: The fight for a human future at the new frontier of power.* Profile books.

CHAPTER 9

Blast zone

Contents

Climate Crisis, Energy Violence
ISBN 978-0-12-819501-7,
https://doi.org/10.1016/B978-0-12-819501-7.00014-X

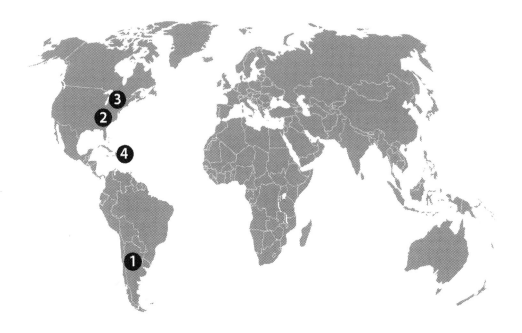

Locations: 1 - Argentina
              2 - Appalachia, United States
              3 - Northeast United States, Canadian Border
              4 - Puerto Rico

Themes: Energy security; Residual risk; Unnatural disaster

Subthemes: High consequence area; Invisibility; Offset; Wokewashing

It's like we are not even here.

Gaseous geographies

Fossil gas is deceptively marketed as clean, climate-friendly, and beneficial to workers and communities, though in actuality it's the opposite. This chapter catalogs on-the-ground harms, starting with the US. Like the intensification of oil sacrifice zones along the Gulf Coast (Introduction), patterns in gas infrastructuring put those with less power at great risk. The supposition of a **blast radius** implicit to fossil gas exemplifies the persistent threat of physical damage to people and property.

A feeling of **invisibility** often resonates throughout impacted communities: lives seem not to matter.[1] Risks accrue to those erased from history, explored herein at Union Hill, Virginia and Vaca Muerta, Argentina. Risk is further compounded in fossil gas geographies through **disaster capitalism**, as is demonstrated in Puerto Rico in the wake of intensifying hurricanes.

Industry obfuscation

Over time, fossil fuel companies have quietly dropped denying harm outright while creating risk management markets.[2] Industry uses strategic disinformation—gaseous lies—summarized in Fig. 9.1 to deflect criticism. Widespread narratives of gas as a clean fossil fuel (a "diet" fuel, so to speak) rely on too-rarely-questioned claims. Taken together, industry narratives equate to **vapor-speak**—misleading claims reifying fossil gas as an essential fuel in spite of risks.

Deceptions range from pedestrian to elaborate: for example, an **astroturf** group is a sophisticated special interest organization operating under the guise of a grassroots environmental group.[3] Fossil fuel firms employ a wide array of consultants to delay policy.[4] Companies launch aggressive public relations campaigns[5] and spend considerable resources lobbying and influencing opinion, for which they in turn ensure ratepayers pay.[6] Some industry trade organizations use concealed tactics like **ghostwriting** legislators' statements of support for projects.[7] Gas companies even propagandize in schools. In Virginia, the Department of Education collaborates with Dominion Energy to produce energy curriculum: in sixth grade, students learn about "cool coal" and "nifty natural gas."[8] An Oregon utility, NW Natural, distributes pro-gas materials in schools.[9] Another energy utility, Eversource, similarly appeals to kids via cartoon friends, Nat and Gus (Fig. 9.2).

---

[1] Gullion, J. S. (2015). Fracking the neighborhood: Reluctant activists and natural gas drilling. MIT Press.

[2] Carroll, W. K. (Ed.). (2021). Regime of obstruction: How corporate power blocks energy democracy. Athabasca University Press.

[3] Lits, B. (2020). Detecting astroturf lobbying movements. *Communication and the Public*, 5(3–4), 164–177.

[4] Franta, B. (2022). Weaponizing economics: Big Oil, economic consultants, and climate policy delay. *Environmental Politics*, 1–21.

[5] Lakhani, N. (2022, October 10). How fossil fuel firms use black leaders to 'deceive' their communities. *The Guardian*.

[6] Joselow, M. (2023, June 21). Three states just barred utilities from charging customers for lobbying. The Washington Post.

[7] D'Angelo, C. (2022, May 2). Gas giants have been ghostwriting letters of support from elected officials. Huffington Post.

[8] Dominion Energy & Virginia Department of Education. (n.d.). Energy in Virginia.

[9] Knoblauch, J. A. (2023, April 20). How we stopped a gas utility's scheme to propagandize children. Earthjustice.

Fossil gas in relation to ...	Industry claim	Evidence
employment	abundant	job predictions inflated
price	inexpensive	costly; dependent on subsidies and externalized costs that burden taxpayers and ratepayers
climate	climate-friendly	potent GHG emissions (methane)
pollution	better than coal	harms continue with different contaminants; methane emissions gravely underreported
safety	low-risk	explosions, fires, spills, flares, leaks and venting
size of reserves	abundant	accessible shale basins are heavily tapped, often with inefficient, leak-prone systems

Figure 9.1 Gaseous lies.

Figure 9.2 Eversource activity book for children. *(Image credit: Gleb Bahmutov, 2021. Accessed via Twitter.)*

The gas industry portrays to its consumers and society at large that gas expansion is both necessary and responsible. Gas firms seek to appear as Good Samaritans in disaster relief scenarios without taking responsibility for their role in radically altering land uses for industrial extraction and emitting significant GHGs that supercharge harm from extreme weather.[10] Gas production and consumption result in the creation of sacrifice zones. As demonstrated in this chapter's illustrative cases, people of color frequently experience disproportionate harm whether in upstream extraction, such as Vaca Muerta in Argentina; in midstream transportation, such as Virginia; or downstream in Puerto Rico.

The gas industry aims to discredit distributed renewable energy, which is more equitable, more resilient, and promotes higher job creation. Clean alternatives to gas can also advance **energy democracy**: participatory representative energy governance, that includes workers, to promote affordable equitable access by and for communities. As an essential commodity, energy managed narrowly to profit investors undercuts its potential as a collective commons benefitting all.

Under increasing pressure to decarbonize its operations, industry invariably seeks technological fixes in place of justice-oriented transition.[11] Democratic management of energy resources—including ownership stake in energy infrastructure itself—is a corrective objective of energy transition.[12] Labor—environment coalitions, termed **blue-green alliances**, are central to preparing the energy sector for the growing extremes of climate change.[13] A decolonial and feminist **Green New Deal** (GND) lays out another path for energy transformation (Praxis 7).[14]

Energy violence

In the rush to replace fossil fuels as climate disruption intensifies, a just transition is not guaranteed in the absence of a popular movement demanding accountability and transparency. Historical precedents in the suffrage, civil rights, and anti-apartheid movements demonstrate that change is indeed possible, but not without brave leaders that spur the public and its institutions to take moral stances. Today, the gas industry holds tight its territorial and market gains, deflecting every call for energy transition; its invasive practices have rapidly intensified across local and global scales. Hyper-extractive

[10] Lehman, J., & Kinchy, A. (2021). Bringing climate politics home: Lived experiences of flooding and housing insecurity in a natural gas boomtown. *Geoforum, 121,* 152–161.

[11] de Onís, C. M. (2021). Energy islands: Metaphors of power, extractivism, and justice in Puerto Rico. University of California Press.

[12] Baker, S. (2021). Revolutionary power: An activist's guide to the energy transition. Island Press.

[13] Huber, M. T. (2022). Climate change as class war: Building socialism on a warming planet. Verso Books.

[14] Feminist Economic Justice for People & Planet Action Nexus. (2021). A feminist and decolonial global green new deal.

BOX 9.1 Spinoff industry: frac sand mining

"Frac sand" is a critical input into the invasive gas extraction method known as fracking.[15] Industry defines sand's potential and value based on size, roundness, and crush resistance.[16] Sand processing (Fig. 9.3) usually occurs close to extraction, significantly expanding the footprint of the fracking operations.

High silica content sand is considered positive for fracking purposes, but interacting with silica is an occupational hazard (Chapter 3). Silicosis from sandblasting[17] occurs when silica becomes airborne and gets lodged in workers' lungs.[18] Laws to protect these workers stalled for years,[19] an example of regulatory violence.

Silicosis is the new asbestosis.

Sand mining operations use flocculants to clean and wash the sand.[20] These additive substances have been shown to infiltrate into groundwater. Long-term exposure to flocculants can lead to nervous system disorders and cancer. Yet government agencies haven't developed regulations for these particular substances in drinking water, further exemplifying regulatory violence.

Sand mining for fracking devastates communities in the central US.[21] Local opposition to sand mining is common.[22] While some landowners are enticed by rents, allowing entry on their land can harm neighbors. Loss of community is commonplace.[23]

industries like sand mining necessary for fracking cause ecological and social displacement to the point that it can be described as wastelanding (Box 9.1 above). Sand is one of several inputs with increasing costs making fossil gas more expensive.[24] Compounding

[15] Bendixen, M., Best, J., Hackney, C., & Iversen, L. L. (2019). Time is running out for sand. *Nature*, *571*(7763), 29–32.

[16] Zdunczyk, M. (2007). The facts of frac. *Pan*, *16*(30), 50.

[17] Staff. (2021, July 1). Silicosis. NHS.

[18] Fedan, J. S. (2020). Biological effects of inhaled hydraulic fracturing sand dust. I. Scope of the investigation. *Toxicology and Applied Pharmacology*, *409*, 115329.

[19] Goad, G. (2013, August 23). Unions applaud as OSHA releases long-stalled worker safety rule. The Hill.

[20] Staff. (n.d.). Frac sand health and environmental impacts. Earthworks.

[21] Holifield, R., & Day, M. (2017). A framework for a critical physical geography of 'sacrifice zones': Physical landscapes and discursive spaces of frac sand mining in western Wisconsin. *Geoforum*, *85*, 269–279.

[22] Hammond, E. A. (2019). Effect of public perceptions on support/opposition of frac sand mining development. *The Extractive Industries and Society*, *6*(2), 471–479.

[23] Pearson, T. W. (2017). When the hills are gone: frac sand mining and the struggle for community. University of Minnesota Press.

[24] Bryan, K. (2023, May 9). Sand shortages push up cost of jam jars and fracking. *The Financial Times*.

Figure 9.3 Frac sand mine and processing facility. *(Image credit: Ted Auch, FracTracker Alliance, 2019.)*

matters, existing high demand from construction and concrete industries only reinforces the sacrifice zones necessary for this material, particularly acute in both the US and China.[25]

Risk "management"

Risk is exposure to danger with possibility of loss or injury. **Residual risk** is the amount of risk or danger associated with an action or event remaining after risks have been reduced by risk controls. **Risk management** is a set of measures aimed at avoiding, minimizing, or decreasing the causes of incidents. Risk management consultants argue what degree of risk is acceptable based on standardized assessments derived from previous events. Using market-oriented equations, they seek to determine whose risk (i.e., families, communities, governments, firms) and what risk (i.e., disease, death, supply disruption) matters more.

Financial systems are set up to tolerate risk, while establishing fines, penalties, and damages to cover unplanned excesses. More importantly, firms seek to avoid lawsuits demonstrating failed risk management. **Risk assessments** involve a systematic evaluation of the potential danger of an action, so that risks can be managed and scaled to a market price. Yet numerically oriented assessments often devalue social and ecological damages, focusing more on market priorities and legal requirements.

[25] Fractracker. (n.d.). Industrial Sands.

In the gas industry, a particular deadly blast in Belgium became its international training case.[26] This 2004 pipeline explosion killed two dozen people and injured more than 100 as debris was projected kilometers outward, and huge fire columns burned people in cars on a highway nearby.[27] As is often the case following gas explosions, those impacted by the Belgian tragedy suffered **post-traumatic stress disorder** (PTSD)[28] as well as collective trauma. PTSD involves individual persistent emotional stress with vivid recall of the experience, nightmares, anxiety, or depression. **Collective trauma**, as the term aptly suggests, is a psychological experience of harm shared by a group of individuals.

A major blast in suburban San Bruno, California, killed eight people in 2000 (Fig. 9.4).[29] Flagrant corruption and misuse of power was found to be persistent by the blast's review process. Assigning the largest fines to date, regulators stated, "We believe this historic penalty sends the right message that gross negligence, corruption and profits-over-safety will no longer be tolerated."[30] Six felony charges arose:[31] utility leadership had diverted funds from safety operations that could have detected deficiencies.[32] "You had the leadership of the California Public Utilities Commission essentially in bed with the utility. Dining, drinking, vacationing—and making deals behind the scenes. [This] behavior has compromised safety."[33]

The large fines in San Bruno, California however did not prove an effective deterrence for yet more harmful industry behavior. For example, in East Harlem (Fig. 9.5)

[26] Mahgerefteh, H., & Atti, O. (2006). An analysis of the gas pipeline explosion at Ghislenghien, Belgium. In 2006 Spring Meeting and 2nd Global Congress on Process Safety.

[27] Perry, S. L. (2012). Development, land use, and collective trauma: The marcellus shale gas boom in rural Pennsylvania. *Culture, Agriculture, Food and Environment, 34*(1), 81—92.

[28] De Soir, E., Zech, E., Alisic, E., Versporten, A., Van Oyen, H., Kleber, R., ... & Mylle, J. (2014). Children following the Ghislenghien gas explosion: PTSD predictors and risk factors. *Journal of Child & Adolescent Trauma, 7*(1), 51—62.

[29] McEntire, D. A., Kelly, J., Kendra, J. M., & Long, L. C. (2013). Spontaneous planning after the san bruno gas pipeline explosion: A case study of anticipation and improvisation during response and recovery operations. *Journal of Homeland Security and Emergency Management, 10*(1), 161—185.

[30] Brooks, J. (2015, April 5). Despite $1.6 billion fine, CPUC President questions ability to police PG&E. KQED.

[31] Larson, K. (2017, January 26). PG&E receives maximum sentence for 2010 San Bruno explosion. ABC 7 News.

[32] Scawthorn, C., Li, S., Davidson, R. A., Kelly, J., Long, L. C., McEntire, D. A., & Kendra, J. (2012). San Bruno California, September 9, 2010 gas pipeline explosion and fire. Disaster Research Center, University of Delaware; Richards, F. (2013). Failure analysis of a natural gas pipeline rupture. *Journal of Failure Analysis and Prevention, 13*(6), 653—657.

[33] Bowe, R., & Pickoff-White, L. (2015, September 8). Five years after the deadly San Bruno explosion: Are we safer? KQED.

Figure 9.4 San Bruno pipeline explosion. *(Image credit: KRON 4 Viewer Pictures of the San Bruno Gas Line Explosion, September 10, 2010 - A Name Like Shields Can Make You Defensive. Accessed via Flickr.)*

Figure 9.5 Fast gas violence in Harlem. *(Image credit: Adnan Islam, March 12, 2014. Accessed via Wikimedia Commons.)*

where a 2014 explosion killed eight people, the utility Con Edison was cited for 11 violations.[34] In another example, in 2018, explosions in Massachusetts killed a person and led to a series of fires[35] following **overpressurization** of distribution lines.[36] The gas utility Columbia Gas was found guilty of violating federal safety laws;[37] a judge approved $143 million dollars in damages[38]—a significant amount that still doesn't fully compensate for the totality of loss and damage.

Following gas explosions, compensatory funds to rebuild are not necessarily guaranteed. Industry routinely goes to great lengths deflecting liability, placing onus on any party not itself—particularly the end consumer. For example, after a 2020 Baltimore, Maryland, explosion, investigations cited a failure of customer equipment after the point of delivery, deeming the company not liable.[39] NGOs were left to gather supplies for victims.[40]

With a gas leak, a spark from a light switch, candles, static electricity, or appliances could ignite an explosion.[41]

Spatial distribution

As administrative and regulatory entities, state agencies tend to focus on formal rights with titles. Jurisdictions are used to control space, including through the use of **boundaries**: the lines demarcating jurisdictions. As is often the case with gas, cross-boundary infrastructure networks (Fig. 9.6) create jurisdictional ambiguity.

Transnational pollution remains a vexing problem. As a case in point, it's hard to assign responsibility for transportation emissions separate from national carbon accounts.

[34] Blain, G. (2015, November 19). Con Ed accused of 11 gas-safety violations in 2014 East Harlem blast after state probe. New York Daily News.

[35] LaFlash, K. (2020, September 20). Gas worker union chief hails DPU's new attitude. Commonwealth Magazine.

[36] Ly, L. (2019, September 24). Merrimack Valley gas explosions were caused by weak management, poor oversight, NTSB says. CNN.

[37] Zaveri, M. & Fortin, J. (2020, February 26). Massachusetts gas company to plead guilty after fatal explosion. New York Times.

[38] Associated Press. (2020, March 12). Judge approves $143M Merrimack Valley gas explosions settlement. NBC 10 Boston.

[39] Hellgren, M. (2020, August 19). 'The houses are gone': 911 calls from Baltimore gas explosion released; residents smelled gas before and after blast. CBS Baltimore.

[40] Burnett, A. (2020, September 14). Five weeks after deadly Baltimore gas explosion, Residents still cleaning up, facing supply chain shortages. CBS Baltimore.

[41] Moran, B. (2018, September 14). What you need to know about gas safety in your home. WBUR.

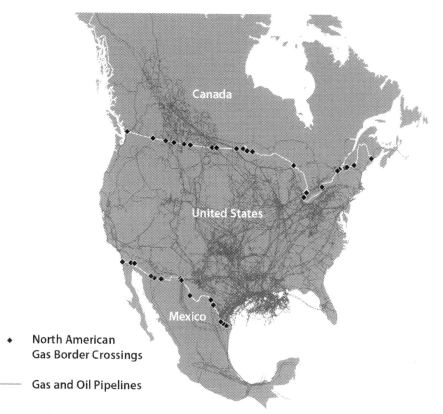

Figure 9.6 North American gas infrastructure networks. *(Sources: Global Oil and Gas Infrastructure (GOGI) Inventory; U.S. Energy Information Administration (EIA).)*

Fuzzy accounting is also evident in foreign finance of US fracking—it isn't clear who is responsible for emissions: the investing financier, the gas producer, or the consumer.[42] With different jurisdictions using unique balance sheets, full and accurate emission inventories are not easily enumerated.

Fractured analysis is institutionalized through a process of **segmentation**—breaking large projects into smaller, more palatable portions. Pipeline permit segmentation hides large negative impacts by breaking up projects so each piece can be argued to cause insignificant harm. If the whole infrastructure system were pooled together for assessment, the extent of damage overall would be undeniable.

---

[42] Bratman, E., Auch, T., & Stinchfield, B. (2022). The fracking frontier in the United States: A case study of foreign investment, civil liberties and land ethics in the shale industry. *Development and Change, 53*(3), 469–494.

Fractured responsibility occurs with splintered authority among governmental jurisdictions. Permits undergo compartmentalized review by different agencies per isolated component (i.e., water, air, solid waste)—so impacts aren't analyzed holistically. There are multiple governance scales (i.e., local, provincial, national) with partial and sometimes conflicting powers. These fractures cause problems for environmental regulation. For example, the US Mountain Valley Pipeline (MVP), discussed below, exemplifies conflicts within courts[43] and between regulatory bodies.[44]

Living in the blast zone

Pipeline logic prioritizes high-impact hubs and strategic connections.[45] In many countries, landowners are entirely dispossessed of land for gas mega-projects. Displacement (Chapter 1) represents the opposite of place and placemaking, discussed in the Introduction.

Pipelines are often sited in lower value real estate markets as easements can cost more when they transect expensive parcels. Regardless land costs, property owners often receive relatively modest indemnification for the seizure of their land.[46] Landowners with easements face new restrictions on their land, yet receive minor property tax relief, if at all. Research has shown negative agricultural ramifications as soil health and crop yields are degraded at right-of-way (ROW) clearances, which leads to devaluation of farming, local land uses, and lower payments for easements.[47] Pipeline easements further exhibit significant disturbances beyond their boundaries with pollution and explosion risk emanating outwards into larger zones.

Many families don't want to accept easements. Firms hire intermediaries, sometimes called landmen or land "sharks," to weaken **holdouts**—those who refuse to sell their property. Companies limit the flow of information to landowners to weaken holdouts and use divide-and-conquer tactics to pit family members, neighbors, and communities against one another.[48]

[43] Alday, K. E. (2021). Givens v. Mountain valley pipeline, LLC and the unresolved circuit split. *Texas A&M Journal of Property Law, 7*, 137.

[44] Tony, M. (2023, April 3). Federal court throws out key DEP water permit for Mountain valley pipeline calling agency's justifications "deficient." Charleston Gazette.

[45] Randolph, N. (2021). Pipeline Logic and culpability: establishing a continuum of harm for sacrifice zones. *Frontiers in Environmental Science, 9*, 216.

[46] Chartier-Hogancamp, V. L. (2019). Fairness and justice: Discrepancies in eminent domain for oil and natural gas pipelines. *Texas Environmental Law Journal, 49*, 67.

[47] Brehm, T., & Culman, S. (2023). Soil degradation and crop yield declines persist 5 years after pipeline installations. *Soil Science Society of America Journal, 87*(2), 350–364.

[48] Ortiz, O. (2018, December 7). How money stokes divide of historic black community in Virginia pipeline battle. NBC News.

Fossil fuel siting ideally involves a planning process to determine **site suitability**: specific criteria or restrictions encompassing existing infrastructure, cultural and historical landscapes, economic impacts, and legacy pollution. Ironically, suitability assessment can cause harm. Once assessment is completed, proposed pipelines are deemed "suitable" and allowed to consolidate needed land, dividing landowners by bartering on a parcel-by-parcel basis (Fig. 9.7).

Some pipeline sections are dirtier and more dangerous, as is the case when sited with above-ground infrastructure like compressor stations (Fig. 9.8), which are necessary to move gas through pipelines.[49] Compressors stabilize the pressure and flow rate of gas within pipeline networks. Fig. 9.9 shows construction of both above and below-ground pipeline infrastructure at a compressor station. Additional above-ground infrastructures include metering and regulation facilities that host "pigging" operations designed to remove contaminants that build up along the interior walls of pipelines.

US compressor station permits authorize the release of criteria pollutants like particulate matter averaged across both 24-hour and 1-year periods. However, larger emission releases occur during **blowdowns**: both scheduled and unplanned "venting," often dur-

Figure 9.7 Midstream transmission. *(Image credit: Ted Auch, FracTracker Alliance, 2018.)*

[49] Zimmerle, D., Vaughn, T., Luck, B., Lauderdale, T., Keen, K., Harrison, M., ... & Allen, D. (2020). Methane emissions from gathering compressor stations in the US. *Environmental Science & Technology*, *54*(12), 7552−7561.

Figure 9.8 Compressor station. *(Image credit: Ted Auch, FracTracker Alliance, 2020.)*

Figure 9.9 Connection to compressor station. *(Image credit: Sullivan County, NY, 2018, Co-Author.)*

ing shut-down and restart episodes.[50] Blowdowns accentuate health risks with concentrated releases of volatile organic compounds (VOCs) and hazardous air pollutants (HAPs). Toxic emissions are often invisible to the naked eye and to traditional imaging (Fig. 9.10). Infrared imaging captures the VOC and methane components of venting and leaks that appear otherwise invisible.[51]

Regulatory "bottom up" monitoring networks rely on point source reporting by industry itself. Yet methane leaks and vented blowdowns are rampant, particularly at

Figure 9.10 Pipeline venting. *(Image Adapted: Pipeline Vent, Epping, ND - Earthworks, 2016. Accessed via YouTube.)*

[50] Fioravanti, A., De Simone, G., Carpignano, A., Ruzzone, A., Mortarino, G., Piccini, M., ... & Bolado-Lavín, R. (2020). Compressor station facility failure modes: causes, taxonomy and effects. Publications Office of the European Union.

[51] Wang, J., Tchapmi, L. P., Ravikumar, A. P., McGuire, M., Bell, C. S., Zimmerle, D., ... & Brandt, A. R. (2020). Machine vision for natural gas methane emissions detection using an infrared camera. *Applied Energy, 257,* 113998.

above-ground infrastructure like gathering and mainline pipeline facilities (i.e., compressor stations). In spite of atmospheric harms from methane in gas (Chapter 8), industry treats fugitive methane as an acceptable and unpreventable cost of doing business.

Emissions detection equipment is improving in accuracy and availability[52] with drones,[53] lidar,[54] and forward-looking infrared (FLIR) cameras. Individual activists and organizations have stepped in with these methane-sensing cameras to visualize leaks, unlit flares, and blowdown events.[55] While visualization strategies that expose infrastructure sites that consistently and repeatedly release methane are critical to awareness of the scope of the problem, regulatory actions toward corrective compliance remain woefully inadequate.

Risk proximity

Pipeline incidents generally emerge from external interference, corrosion, hazards during construction, and negligence. While geohazards make up a minority of pipeline failures, these tend to result in catastrophe and complete loss of service.[56] Steep routes create potentially unstable conditions, increasing the chances of rupture or explosion.[57] Extreme weather events resulting in floods and landslides only contribute additional risk.

During pipeline construction, a **sinkhole** can form as a cavity begins to develop where water erodes the soil surrounding pipes. Alteration of hydrologic systems or vibrations on karst-affected geographies reduces the bearing capacity, so the ground collapses.[58] Land-cover changes and climate change may intensify sinkhole formation.[59]

The use of horizontal directional drilling (HDD) and trench drilling to cross rivers, streams, and wetlands bear potential risks to drinking water, surface water, and/or groundwater. During pipeline construction, a **blowout** can spread drilling "mud," a

[52] Aldhafeeri, T., Tran, M. K., Vrolyk, R., Pope, M., & Fowler, M. (2020). A review of methane gas detection sensors: Recent developments and future perspectives. *Inventions*, *5*(3), 28.

[53] Hollenbeck, D., Zulevic, D., & Chen, Y. (2021). Advanced leak detection and quantification of methane emissions using sUAS. *Drones*, *5*(4), 117.

[54] Rashid, K., Speck, A., Osedach, T. P., Perroni, D. V., & Pomerantz, A. E. (2020). Optimized inspection of upstream oil and gas methane emissions using airborne LiDAR surveillance. *Applied Energy*, *275*, 115,327.

[55] Bussewitz, C. (2022, November 6). Equipment that's designed to cut methane emissions is failing. Associated Press.

[56] Deng, Q., Wang, X., & Wei, C. (2019). Cases and Inspirations of Pipeline Damages Due to Geohazard in China in Recent Years. 대한지질공학회 학술발표논문집, 2019(2), 61–61.

[57] Vasseghi, A., Haghshenas, E., Soroushian, A., & Rakhshandeh, M. (2021). Failure analysis of a natural gas pipeline subjected to landslide. *Engineering Failure Analysis*, *119*, 105,009.

[58] Kim, J. W., Lu, Z., & Kaufmann, J. (2019). Evolution of sinkholes over Wink, Texas, observed by high-resolution optical and SAR imagery. *Remote Sensing of Environment*, *222*, 119–132.

[59] English, S., Heo, J., & Won, J. (2020). Investigation of sinkhole formation with human influence: a case study from wink sink in winkler county, *Texas*. *Sustainability*, *12*(9), 3537.

viscous mixture including bentonite and polymers used to lubricate and cool the drill bit. Blowouts can release fine sediments that bury vegetation and kill fish, among other damages.[60] Fig. 9.11 shows one of many blowouts on the Mariner East 2 Pipeline, constructed by the firm Energy Transfer (Chapter 7) across the state of Pennsylvania.

Energy Transfer's record is riddled with numerous environmental violations and pipeline construction accidents.[61] The firm pleaded no contest in two criminal cases charging systematic waterway harm across hundreds of miles.[62] Further, Mariner East 2 took advantage of restrictions to public accessibility to information following the

Figure 9.11 Mariner East 2 blowout. *(Image credit: Mariner East 2 Pipeline drilling spill into Marsh Creek Lake State Park, Chester County, PA - PK DiGiulio, 2020.)*

[60] Yan, X., Ariaratnam, S. T., Dong, S., & Zeng, C. (2018). Horizontal directional drilling: State-of-the-art review of theory and applications. Tunnelling and underground space technology, 72, 162–173.

[61] Hurdle, J. (2020, September 21). More spills at Lebanon County Mariner East pipeline drill site earn Sunoco more violations. NPR.

[62] Rubinkam, M. (2022, August 5). Pipeline developer pleads no contest in pollution cases. Associated Press.

Figure 9.12 Mariner East 2 construction. *(Image credit: PK DiGiulio and Lora Snyder, 2021.)*

9/11 US terrorist attacks.[63] This led to *Flynn v. Sunoco Pipeline*, known as the "Safety Seven" case, filed by residents along the pipeline. The judge ordered Energy Transfer to include in its written materials the possibility of property damage, injury, and death from a gas leak or explosion.[64] This was a partial win: the judge confirmed safety concerns, yet did not stop the project as plaintiffs had asked. Communities remain at risk with potentially dangerous pipelines sited exceedingly close to residences (Fig. 9.12 above). Families were forced to endure years of prolonged construction with a myriad of induced hazards (Fig. 9.13).

Pipeline corridors create risks that are compounded with multiple industrial or social facilities at one place.[65] An infrastructure **corridor** is a belt linking two features like power plants or export terminals. A **high consequence area** (HCA) is a geographic radius zone along infrastructure where failure—like an explosion—would significantly impact persons and habitat, or result in outright loss of human life and property. In Fig. 9.14, the

[63] Phillips, S. (2020, July 17). More sinkholes develop alongside Mariner East construction in Chester County. StateImpact Pennsylvania.

[64] Phillips, S. (2021, April 14). PUC judge orders Sunoco to improve its public safety guidance, and pipeline safety, on Mariner East project. StateImpact Pennsylvania.

[65] Metts, S. (2019, August 17). The Kentucky Enbridge pipeline explosion: everyday risk buried in American backyards. Medium.

Figure 9.13 Proximity of Mariner East 2 to residences. *(Image credit: PK DiGiulio, 2021.)*

potential impact radius (PIR) is binary; in reality, an actual damage radius may be less or more based on actual conditions at the site of failure.

Infrastructure sited in rural areas is touted as having lower consequence—yet areas historically zoned for agriculture and forestry are not prepared with emergency fire and rescue. Further, rural siting does not eliminate risk—in 2019 a Kentucky mobile home park (Fig. 9.15) was the site of a fatal explosion. Family members of a deceased woman sued the company for failing to maintain the pipeline.[66] Even as prior reports of unusual noises were made by the public, these concerns were dismissed and no investigations took place.

---

[66] Staff. (2020, July 31). 1 year after Lincoln County deadly explosion, more legal action taken against pipeline owner. WHAS11.

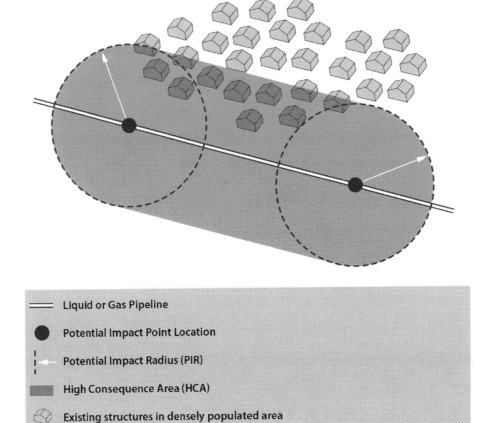

Legend	
═══	Liquid or Gas Pipeline
●	Potential Impact Point Location
←---	Potential Impact Radius (PIR)
▨	High Consequence Area (HCA)
⬡	Existing structures in densely populated area

Figure 9.14 High consequence area. *(Diagram adapted: Pipeline and Hazardous Materials Safety Administration (PHMSA).)*

Figure 9.15 TETCO's fatal blast site. *(Image credit: TETCO pipeline blast site, Lincoln County, Kentucky -*

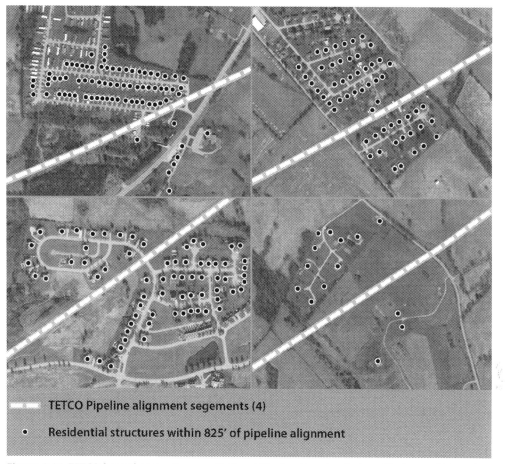

TETCO Pipeline alignment segements (4)

● Residential structures within 825' of pipeline alignment

Figure 9.16 TETCO hazard areas. *(Sources: National Pipeline Mapping System, Pipeline and Hazardous Materials Safety Administration (PHMSA), Microsoft Buildings Footprints dataset, Esri, Maxar, Earthstar Geographics, CNES/Airbus DS, USDA FSA, USGS, Aerogrid, IGN, IGP, and the GIS User Community.)*

This same TETCO pipeline system has a number of higher density areas where fatalities could have been higher given residential proximity, as shown in Fig. 9.16 above. Fig. 9.17 plots the locations of all residences in relation to the TETCO route, a system that has not been adequately maintained.[67] Insufficient maintenance occurs on many pipelines. Companies keep costs low by doing minimal repair and monitoring.[68] After

[67] Metts, S. (2019, August 17). The Kentucky Enbridge pipeline: Everyday risk buried in American backyards. Medium.
[68] Frietas, Jr., G., Blas, J., & Wethe, D. (2021, May 14). Colonial pipeline has been a lucrative cash cow for many years. Bloomberg; Lee, M., & Soraghan, M. (2019, March 4). Deadly pipelines, no rules. E&E News.

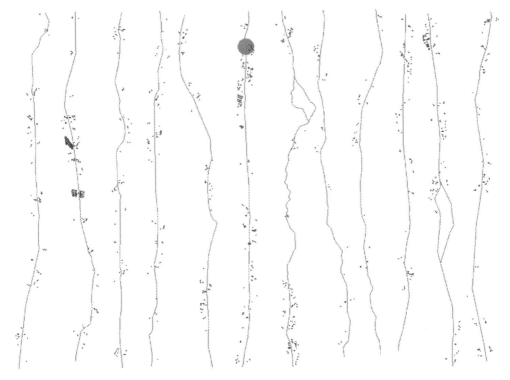

● TETCO Pipeline Explosion Site, August 1, 2019

----- Pipeline alignment sections, separated and organized vertically

· Structures within 825' of pipeline alignment

Figure 9.17 Proximity to TETCO pipeline. *(Sources: National Pipeline Mapping System, Pipeline and Hazardous Materials Safety Administration (PHMSA), Microsoft Buildings Footprints dataset.)*

mergers of companies, pipeline segments of different ages and with varied materials and histories of repair join together.

It's a dangerous industry. These things happen.[69]

Along a pipeline's path, maintanance of its ROW is a space-production process.[70] The risk perception of landowners is distinct from that of the energy or utility company.[71] Inhabited places exist in between gas hubs. Companies focus narrowly on moving gas through and between infrastructure, while impacted communities—perceived by

[69] Loftus, T. (2019, August 1). Gov. Matt Bevin: Lincoln County pipeline explosion a 'tragedy … these things happen.' *Courier Journal*.

[70] Kinyera, P., & Doevenspeck, M. (2023). Governing Petro-(im) mobilities: the making of right-of-way for Uganda's East African crude oil pipeline. *Mobilities*, 1–17.

[71] Jalbert, K., Dickinson, K. L., Baka, J., & Florence, N. (2023). Influence in the right-of-way: Assessing landowners' risk decision-making in negotiating oil and gas pipeline easements. *The Extractive Industries and Society, 14*, 101276.

Figure 9.18 Gas warning signs. *(Image credit: Erica Jackson, FracTracker Alliance, 2021.)*

industry as unavoidable impediments—prioritize the wellbeing of their home. Companies focus more attention to risk assessment than to the mitigation of harms or clean up after disasters.[72] Pipeline markers make underground risks more visible (Fig. 9.18

[72] Chen, C., Li, C., Reniers, G., & Yang, F. (2021). Safety and security of oil and gas pipeline transportation: A systematic analysis of research trends and future needs using WoS. *Journal of Cleaner Production, 279*, 123583.

Figure 9.19 Millennium pipeline right of way. *(Image credit: Orange County, NY, 2018, Co-Author.)*

above) but cannot prevent disasters. Pipeline ROWs are significant ecological disturbances that result in severance and fragmentation of forests and other natural habitats. Fig. 9.19 above shows the abrupt, unnatural logging and clearance pattern at the ROW of the Millennium Pipeline in New York.

Temporal analysis

Politicians seeking to avoid accountability often gravitate to **forward-facing policy** based on promises to do better to avoid addressing past harms. Likewise, in 2022 the US Federal Energy Regulatory Commission (FERC) added new guidelines and an Office of Public Participation.[73] Unfortunately these "reforms" leave structural violence

---

[73] Klass, A., & Welton, S., Wiseman, H., & Goodwin, J. (2022). The Federal Energy Regulatory Commission's New Office of Public Participation: A Promising Experiment in 'Energy Democracy'. Center for Progressive Reform.

largely unresolved.[74] Environmental injustices continue, as demonstrated with the Weymouth Compressor Station in Massachusetts,[75] where state agencies created new pathways for environmental justice claims—and then overrode them.[76]

The Weymouth Compressor Station, located just south of Boston, exemplifies infrastructure mis-placement. Sited near dense low-income housing and a highly trafficked bridge, the facility lies atop a landmass historically inundated with storm surge even before factoring in sea level rise.[77] Yet this placement is strategic as it connects existing pipelines to deliver gas through New England and Canada. Enbridge's Atlantic Bridge Project—of which the compressor station is part—occurred during a crucial time in northeast US gas pipeline construction, drawing protest at multiple locations (Chapter 7).[78]

In Weymouth, many local residents are low-wealth and people of color who experience greater rates of respiratory and heart disease compared to state averages.[79] Heavy industrial facilities at and near the site have polluted for decades,[80] creating a brownfield laden with legacy coal ash, arsenic, and asbestos.[81] Exposures from multiple sites emitting at or near legal limits cumulatively result in impacts above safe levels.

A local advocacy organization FRRACS (Fore River Residents Against the Compressor Station) and allies fought the Weymouth station permit, even risking arrest.[82] In 2016, protestors installed a shipping container at the entrance of the work yard with two activists living inside.[83] Community members set up air testing canisters and watchdog surveillance with video and photos of the site with regular public updates (Conclusion).

[74] Paparo, R. (2021). Not a box to be checked: Environmental justice and friends of Buckingham v. State air pollution control board (4th Cir. 2020). Harvard Environmental Law Review, 45, 219.

[75] Wilson, M. (2021, August 3). Gas projects reveal FERC's environmental justice conundrum. E&E News.

[76] Wortzel, A., & De Las Casas, V. (2021). State laws provide new pathways for environmental justice claims. Natural Resources and Environment, 36(5).

[77] Metts, S. (2019, January 11). The weymouth compressor station: massachusetts' moment of climate truth. Medium.

[78] Thompson, L. (2022, August 10). It's our duty to protect the Great Lakes: groups remain united against Line 5 pipeline. The Manitoulin Expositor.

[79] Wasser, M. (2020, October 13). The controversial natural gas compressor in weymouth, explained. WBUR.

[80] Wasser, M. (2019, December 6). Arsenic and diesel as thick as peanut butter: What's below the future Weymouth compressor? WBUR.

[81] Metts, S. (2020, February 1). The weymouth compressor station: Exhibit a in the persistence of toxic burden 'hotspots.' Medium.

[82] Winters, J. (2021, May 24). The weymouth compressor station. Harvard Political Review.

[83] Staff. (2016, May 26). Protestors blockade planned pipeline site near nuclear plant outside NYC. Ecowatch.

Despite community opposition, the FERC approved the compressor station's construction in 2017,[84] and the City of Weymouth made a capitulatory deal after criticizing the project for years.[85] In 2020, Enbridge completed construction. During a federal review in 2022, a FERC regulator stated that the project never should have been approved.[86] Confusion and conflicting data across regulatory agencies[87] coupled with construction and testing mishaps drew significant local outrage. Nevertheless, in 2023, instead of seeking alternatives to gas, Enbridge sought significant expansion in capacity reliant on gas tied directly to this project.[88]

Sustainable gas?

Employment opportunities in renewable energy outpace job losses in fossil fuels by a large margin.[89] Gas representatives often suggest there exists a necessary tradeoff between the environment and jobs, reinforcing public fear of a transition away from fossil fuels. In fact, gas pipelines do not have a positive track record with either. Industry analyses often conflate direct jobs with indirect jobs. **Direct jobs** are positions within a given industry. Yet construction projects rely heavily on outside workforces that move project to project. By counting ongoing employees as new hires per project, bottom-line full-time equivalents (FTEs) number can be inflated. **Indirect jobs**, most of the employment numbers that gas firms reference, are broad estimates of any connection that supports an industry. Policymakers often assume inflated indirect job numbers are likely to materialize or could not emerge from an alternative source.

An industry argument that casts transitions away from gas as too costly has been an effective delay tactic—a range of GHG mitigation delay discourse is covered in the Conclusion. Additionally, industry promotes gas as a green alternative, always avoiding its devastating methane footprint. For example, **certified sustainable gas** claims to measure below a threshold for emissions intensity and in theory promotes best practices to minimize impacts. Yet common methane measurement gaps in addition to other

[84] Weber, M. (2019, December 2). FERC allows construction of Weymouth gas compressor station, over objections. *S&P Global.*

[85] Trufant, J. (2020, October 30). Weymouth, Enbridge strike deal worth up to $38 million. *The Patriot Ledger.*

[86] Trufant, J. (2022, January 20). Feds: Regulators 'should never have approved' Weymouth compressor, too late to shut it down. *The Patriot Ledger.*

[87] Vardi, I. (2019, April 18). Exclusive: Air Permit ok'd after new evidence of carcinogens at enbridge's planned gas facility in massachusetts left out. De Smog.

[88] Wasser, M. (2023, September 22). Fossil fuel company wants to expand gas pipeline in Northeast. WBUR.

[89] Garrett-Peltier, H. (2017). Green versus brown: Comparing the employment impacts of energy efficiency, renewable energy, and fossil fuels using an input-output model. *Economic Modelling, 61,* 439–447.

logistical constraints create serious weaknesses. Third-party certification processes are typically deployed with no particular universal or industry standard utilized to ensure consistency of findings. For all their fanfare, gas certification schemes serve as a delay tactic.[90]

Carbon capture and storage (CCS) is also used to promote new gas. See Chapter 4 for a discussion of coal CCS. While CCS projects are popular, they are poor at mitigating GHG. CCS project locations demonstrate clusters in North America and Europe (Fig. 9.20).

Active CCS projects only capture a slight fraction of 1% of total global carbon emissions from all fossil fuels on a yearly basis.[91] Heavy industrial sectors may have some potential for capturing and sequestering a portion of GHGs, but retrofitting energy plants

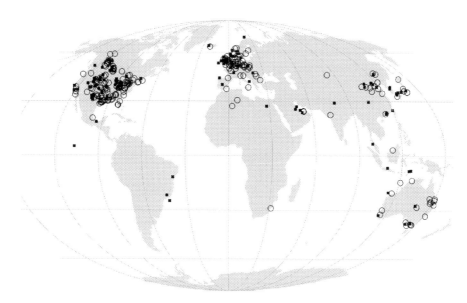

- CCS Site Actively Deployed
- ○ CCS Site in Proposed, Hold or Terminated Status

Figure 9.20 Carbon capture and storage projects. *(Sources: National Energy Technology Laboratory's (NETL) Carbon Capture and Storage (CCS) Database, Natural Earth.)*

---

[90] Brown, H. S. (2023, August 3). The unholy alliance between 'certified' clean natural gas producers and the certifying companies. *The American Prospect.*

[91] As of mid-2023 NETL's CCS Database contains 417 total CCS projects worldwide, including capture, storage, capture and storage, direct air capture and hubs across 37 countries. 10 projects are of unknown status (2.3%); 36 are on hold (8.6%); 54 are potential projects (12.9%); 85 have been completed (20.3%); 63 have been terminated (15.5%) and a total of 169 projects are active (40.5%).

has proven expensive and difficult. Gas projects are merely experimenting with technologies, and not yet consistently storing carbon from pilot attempts.[92] Gas CCS at a commercial scale is yet unproven and even in the best case scenario will not materialize in any near future. After decades of research and development, successful pervasive methodologies are lacking, although there is keen interest by firms seeking profit.[93] In light of CCS' dubious impact, a coalition of countries has implored world leaders to refocus on the urgent need to end oil and gas outright.[94] **False solutions** are proposals to fix something that don't address the scope or timeframe of the problem, or generate a problem of equal or greater proportion. CCS is a false solution as it doesn't address the scope or timeframe of the climate emergency and increases dangerous fossil fuel use in lieu of broader transformation.

Illustrative cases

The gas industry has built its empire on a foundation of violence. Low-wealth populations bear high costs from infrastructuring while experiencing the slow violence of pollution, noise, stress, and climate change. Rural areas, where upstream and midstream infrastructure is often sited, are preferred for low-cost easements for oil and gas transmission corridors with less stringent safety protocols.[95] Peripheral regions are burdened as pollution sinks. Companies route infrastructure without adequate regard to **vulnerabilities** such as those from pre-existing health conditions[96] or **marginality**: the state or condition of being isolated or disadvantaged.

Utility and energy companies are granted powers of **eminent domain**[97] to establish routes.[98] Companies construct energy corridors by exercising eminent domain, termed a "taking."[99] The Atlantic Coast Pipeline (ACP) exemplified contentious land taking: more than 30,000 Indigenous People lived within a mile of the proposed project.[100] The ACP

---

[92] Clark, K. (2023, July 17). Carbon capture pilot launched at natural gas plant in California. *Power Engineering*.

[93] Aronoff, K. (2023, July 24). Are we really going to put polluters in charge of carbon capture? Apocalypse now.

[94] Abnett, K. (2023, July 14). Countries warn against over-reliance on carbon capture tech. Reuters.

[95] Lee, M., & Soraghan, M. (2019, March 4). Deadly pipelines, no rules. E&E News.

[96] Borduna, A. (2021, June 4). Where are the U.S.'s natural gas pipelines? Often in vulnerable communities. National Geographic.

[97] Ely Jr, J. W. (2020). The controversy over energy takings: A tale of pipelines and eminent domain. *Brigham-Kanner Property Rights Journal, 9,* 173.

[98] Ewing, R. (2015). Pipeline companies target small farmers and use eminent domain for private gain. *North Carolina Central University Law Review, 38,* 125.

[99] Winn, A. M., & McCarter, M. W. (2018). Who's holding out? An experimental study of the benefits and burdens of eminent domain. *Journal of Urban Economics, 105,* 176—185.

[100] Emanuel, R. E. (2017). Flawed environmental justice analyses. *Science, 357*(6348), 260.

terminus in North Carolina was sited in an Indigenous Lumbee town, already a sacrifice zone. An ACP lateral line in Virginia terminated in an African American neighborhood with legacy contamination. Three large compressor stations were proposed at 200-mile intervals despite industry standards for smaller compressors within 100-mile intervals.[101]

ACP's Buckingham Compressor Station intensified emissions in the historic Free community of Union Hill, Virginia, a majority Black community. It was sited on a former tobacco plantation benefiting rich absentee landowners while harming descendants of enslaved workers. White plantation descendants received millions for their property, while Black descendants lost property value. Buckingham is one of the south's "burnt counties," where courthouse arson destroyed records of enslavement to avoid restitution.[102] In Union Hill, the ACP hook-up was surveyed on land shared among the family of Taylor Harper, a formerly enslaved man.[103] In granting the permit, state agencies willfully ignored evidence from drone and satellite images[104] showing comprehensive risks for local residents.[105] Community-based counter maps (Fig. 9.21) exposed attempted demographic erasure proximate to the compressor.[106] The ACP siting strategically occurred at a corridor intersection with another midstream pipeline owned by the Williams Companies known as Transco, established in the 1950s. Use of existing easements reinforces damage from infrastructure originally placed without environmental or social impact review.

In a rare reversal,[107] the ACP pipeline project was canceled in 2020 after losing eight required permits.[108] However, the route entailed more than 2,600 easements, and firms held eminent domain takings even after project cancellation.[109] As most easement

[101] Finley-Brook, M., Oba, C., Fjord, L., Walker, R. W., Allen, L. V. R., Harris, F., ... & Metts, S. (2021). Racism and Toxic Burden in Rural Dixie. *William & Mary Environmental Law and Policy Review, 46,* 603.
[102] White, C. W. (2017). The Hidden and the forgotten: contributions of buckingham blacks to american history. *Lamp-Post Publicity*.
[103] Fletcher, K. (2019, July 3). Healthy living and family legacy under threat in Union Hill. Chesapeake Bay Foundation.
[104] Metts, S. (2019, January 7). Dominion energy and environmental racism: A case study in how to lie with maps. Medium.
[105] Ackerman, Z., Musil, R. K., Wakefield, T., McAuliffe, E., Mannion, E., Silva, G., & Bean, D. (2017). Blast zone: Natural gas and the atlantic coast pipeline: causes, consequences and civic action. Rachel Carson Council.
[106] Southern Environmental Law Center. (2021, September 21). Union Hill's historic value recognized in unanimous decision against Atlantic coast pipeline.
[107] Cox, J., & Dietrich, T. (2020, September 4). In Virginia battleground, natural gas pipeline projects face reversals. *Bay Journal*.
[108] Poulos, T. J. (2020). The modern-day case of the lorax within the fourth circuit. *Elon L. Rev., 13,* 291.
[109] Vogelsong, S. (2021, July 27). Federal review recommends leaving canceled Atlantic coast pipeline pipes, felled trees in place. Virginia Mercury.

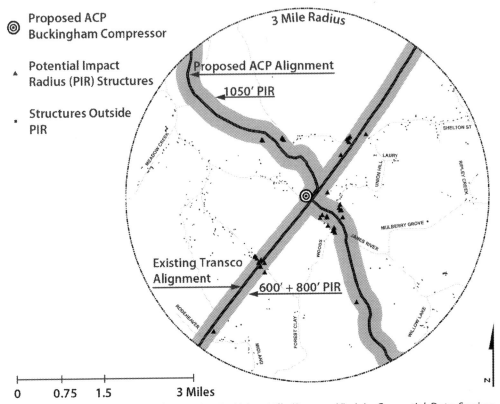

Figure 9.21 Comprehensive risk mapping in Union Hill. *(Sources: Virginia Geospatial Data Services (VGIN), Microsoft U.S. Buildings Dataset, Atlantic Coast Pipeline (ACP).)*

holders are still waiting to get their land back, the name "ghost pipeline" is used to acknowledge harms to families along the route.[110]

Community harm caused by the ACP pipeline project represents an uneven process of **neoliberal multiculturalism**, as ethnic groups can be deemed of differing value to markets.[111] In an attempt to undermine community opposition, ACP used divide-and-conquer tactics.[112] Industry exploits racial conflict[113] and takes advantage of **deference**

[110] Greenfield, N. (2022, June 21). Ghost pipelines: How landowners suffer, long after a project gets canceled. NRDC.

[111] Hale, C. R. (2020). Using and refusing the law: Indigenous struggles and legal strategies after neoliberal multiculturalism. *American Anthropologist, 122*(3), 618–631.

[112] Ortiz, E. (2018, December 7). How money stokes divide of a historic black community in Virginia pipeline battle. NBC News.

[113] Fleischman, L., & Franklin, M. (2017). Fumes Across the Fence-Line: The health impacts of air pollution from oil and gas facilities on African American communities. *NAACP/Clean Air Task Force.*

politics, or the yielding or submitting to existing hierarchies out of respect or self-interest. Political spaces involving industry are seldom neutral—marginalized groups may be made less visible. Deference can be manipulated as racial and gender "uplift": it can feel like "a seat at the table" but in fact can become self-undermining for people of marginalized identities. Deference politics reinforces **elite capture**, meaning a form of corruption whereby public resources are biased for the benefit of a few wealthy well-connected individuals in detriment to the larger population.[114] Elite capture becomes a neoliberal commodity exchange, where identities become capitalism's latest currency. In Union Hill, the company developing the pipeline hired a local elite who used family connections to advance the company agenda.[115]

Mountain valley corrosion

The 303-mile Mountain Valley Pipeline (MVP), an exceedingly expensive, $6.6 billion interstate project, would cross hundreds of water bodies and 225 miles of high slope risk areas in Appalachia—the most among recent long-distance US gas pipelines.[116] Karst geology and an active fault zone along the MVP route further increase the risk of critical failures.[117] In early MVP construction, the project created a slip, putting residential structures in danger of landslide.[118] MVP developers have also been cited for more than 400 water violations, a majority related to erosion. Fig. 9.22 shows MVP construction in sloped terrain, altering landscapes, failing erosion control, and disrupting adjacent farm activities.[119]

Complicating matters, MVP developers resorted to aggressively suing hundreds of landowners to allow survey teams on their land.[120] Local opposition has remained strong since MVP's inception. Construction monitoring is done by watchdog organizations like POWHR (Protect Our Water, Heritage, Rights)[121] and Mountain Valley Watch, involving dozens of street scientists recording hundreds of violations.

[114] Táíwò, O. O. (2022). Reconsidering reparations. Oxford University Press.

[115] Zullo, R. (2022, June 21). In Buckingham's Union Hill, a center of resistance for the Atlantic Coast pipeline, Dominion brings in a ringer. Virginia Mercury.

[116] Hileman, J. D., Angst, M., Scott, T. A., & Sundström, E. (2021). Recycled text and risk communication in natural gas pipeline environmental impact assessments. *Energy Policy, 156,* 112379.

[117] Adams, D. (2016, July 7). Mountainous karst landscape should be a 'no build' zone for pipeline, geologist says. *Roanoke Times.*

[118] Sokolow, J. (2019, August 15). Photos of insanity: Active landslide threatens lives along route of Mountain valley pipeline. Medium.

[119] Bamberger, M., & Oswald, R. E. (2015). Long-term impacts of unconventional drilling operations on human and animal health. *Journal of Environmental Science and Health, Part A, 50*(5), 447–459.

[120] Ashworth, M. (2017, November 15). Mountain valley pipeline sues 300 landowners for their property under eminent domain. RVA Mag.

[121] Hammack, L. (2018, June 2). Environmental watchdogs: A citizens' group monitors the Mountain valley pipeline. *Roanoke Times.*

Figure 9.22 Mountain valley pipeline construction. *(Image source and credit: Mountain Valley Pipeline Weekly Compliance Reports, The Federal Energy Regulatory Commission (FERC), 2018.)*

Since 2017, there have been more than 50 civil actions brought in state and federal courts against the MVP[122]—a high volume of litigation.[123] Environmental lawyers call the barrage of lawsuits they've been forced to bring to stop or delay problematic projects **"legal monkeywrenching"**: an eco-defense tactic serving to cause delay just as physical blockades do by providing time for regulatory gaps to close or markets to shift and potentially stall or outright halt projects.

In situations like these, as people perceive companies as unaccountable, blockades (Chapter 7) become increasingly likely.[124] A strategic encampment, the 932 day Yellow

---

[122] Hammack, L. (2022, January 8). Legal fights continue over the MVP. Roanoke Times.

[123] Moore, D., & Earls, M. (2022, January 26). Mountain valley pipeline's up-and-down legal journey. Bloomberg Law.

[124] Hammond, E. (2019). Toward a role for protest in environmental law. *Case W. Res. L. Rev., 70,* 1039.

Finch tree-sit blocking construction of the MVP[125] drew solidarity across space.[126] In addition to obstruction, activists seek to draw outside attention, informing and educating a larger public of immediate harms.[127] Through this form of activism, impacted communities form resistance solidarity.[128]

After successful resistance from Appalachian communities and allies, MVP received unprecedented intervention,[129] drawing attention to federal corruption. In 2021, led by Senator Joe Manchin of West Virginia,[130] federal lawmakers began discussion of court switching to fast track any pending MVP permits.[131] Manchin, head of the Energy Committee, failed in three attempts to insert an exception for the MVP in federal legislation after receiving major pushback from environmental and racial justice organizations.[132] In 2023, with President Biden's backing, Manchin was finally successful in securing MVP's approvals after positioning the MVP as a political pawn in debt ceiling negotiations. Ultimately, Congress put its finger on the scale for pipeline developers.[133] The Executive Branch overrode judicial and environmental agencies, clearing MVP's path to completion.

It risks opening the floodgates to any pet project to be exempted from environmental protections.[134]

As MVP construction speeds ahead unimpeded, new concerns are emerging over corrosion-related failures as dormant pipes are left exposed in open pipe yards. Pipes are coated with a fusion-bonded epoxy to reduce pipe corrosion and subsequent risk

[125] Dhillon, M. (2021, April 16). Last tree-sitters removed from path of Mountain valley pipeline. The Appalachian Voice.

[126] Finley-Brook, M., Oba, C., Fjord, L., Walker, R. W., Allen, L. V. R., Harris, F., ... & Metts, S. (2021). Racism and toxic burden in rural dixie. *William & Mary Environmental Law and Policy Review, 46*, 603.

[127] Stump, N. (2018, August 9). Legal actions against mountain valley pipeline underscore grassroots activism's importance. OxHRH Blog.

[128] Uveda, R. L. (2022, March 24). Organizing across state lines to stop a pipeline. *Yes! Magazine.*

[129] Cunningham, N. (2023, June 6). U.S. okays Mountain valley pipeline in unprecedented move. *Gas Outlook.*

[130] Goodell, J. (2022, August 4). In exchange for a climate deal, Joe Manchin demanded a terrible price. *Rolling Stone.*

[131] Beyer, A. (2022, August 15). Why does the Mountain Valley Pipeline need Joe Manchin to change the law? Mountain State Spotlight.

[132] Moore, D. (2022, August 2). Mountain valley pipeline shield in Manchin's deal raises hackles. Bloomberg Law.

[133] Harrison, D. (2023, June 25). They fought a pipeline on their land. then congress got involved. *The Wall Street Journal.*

[134] Kail, B, (2023, June 24). Opponents fear Mountain Valley Pipeline fast-tracking could 'open the floodgates.' *Times West Virginian.*

of explosion. This coating degrades in sunlight,[135] and MVP pipes have exceeded industry time limits by years. These outdated pipes are now being placed underground in anticipation of MVP's 2024 in-service target date.

While the MVP project aims for completion, analysts have found MVP would enter a very different gas market from the one in which it was first proposed, likely running only at partial capacity.[136] This expensive, high-risk project is moving forward in spite of both its market and climate implications, demonstrating the oversized political weight of gas companies.[137] The MVP project's lease on life exists as regulators, including the FERC[138] and state agencies,[139] have been willing to amend rules in industry's favor,[140] backended by the federal legislative and executive branches.[141]

Connected to MVP's mainline, an extension project would propel gas into North Carolina from Chatham, Virginia. As in Union Hill, tobacco cultivation has deeply influenced Chatham.[142] Akin to the coloniality of past plantations, the MVP extension would intercept a local geography long dominated by the Transco pipeline corridor.[143] Historic Banister district inhabited by African American and Indigenous families already hosts two existing Transco compressor stations.[144] With high rates of asthma and other health conditions, these same households are targeted with the MVP's proposed new compressor station. The MVP pipeline would transfer output from Appalachia that competes with production from other US sacrifice zones.

[135] Soraghan, M. (2023, June 26). 'Out there rotting': Mountain Valley neighbors fear aging pipe. E&E News.

[136] Staff. (2023, June 23). US Mountain Valley natgas pipe moving forward, capacity may be limited. Reuters.

[137] Vardi, I. (2020, December 3). Virginia lawmaker coordinated support for mountain valley pipeline with project's lobbyist. Energy and Policy Institute.

[138] Satterwhite, E. (2019, April 15). With variance, FERC allows mountain valley pipeline to play it by ear. Virginia Mercury.

[139] Mishkin, K., Ward, Jr., K., & Raghavendran, B. (2018, August 10). What happens when a pipeline company runs afoul of government rules? Authorities change the rules. Propublica.

[140] Agarwal, A. (2021). Regulatory agency capture: How the federal energy regulatory commission approved the Mountain valley pipeline. Yale University.

[141] Greene, S. (2023, June 27). Mountain Valley aspect of debt limit law called unconstitutional. Bloomberg Law.

[142] Walters, M. (2021, November 19). Residents near proposed lambert compressor station push back, cite environmental racism. The Appalachian Voice.

[143] Finley-Brook, M., Oba, C., Fjord, L., Walker, R. W., Allen, L. V. R., Harris, F., … & Metts, S. (2021). Racism and toxic burden in rural dixie. *William & Mary Environmental Law and Policy Review, 46*, 603.

[144] Jones, E. & Shabazz, Q. (2021, July 2). Mountain valley pipeline follows familiar playbook, push pollution into poor and minority communities. Virginia Mercury; Campblin, K., & Adhoot, S. (2021, November 26). On environmental justice, the Mountain valley pipeline is an old story. Washington Post.

Argentine gas hubs

Hewing to international trends, Argentina now promotes its own illusionary promises of fracked gas.[145] The state-run energy company, YPF (formerly named Yacimientos Petrolíferos Fiscales), seeks to create gas hubs for export,[146] in spite of the fact these sites are poorly connected and managed. Dozens of national and multinational companies hold portions of country's massive oil and gas field, the Vaca Muerta Shale Basin, whose carbon intensity is perceived as slightly better than others.[147]

International investors have oscillated between hot and cold in Vaca Muerta.[148] With trending investor sensibility now in a positive cycle,[149] a false sense of exceptionalism is ascendant with claims of ecologically sound extraction.[150] Yet, after a century of oil drilling in Argentina, Vaca Muerta gas operations (Fig. 9.23) are experimental due to the depth of the shale gas and lack of infrastructure across the region.[151]

Internal colonization processes permeate Vaca Muerta, intensifying a now sacrifice zone.[152] Violence toward the Indigenous Mapuche has a long history dating to incorporation into Argentina including a genocidal assault known as the Campaign of the Desert.[153] The Mapuche have continually fought against erasure, seeking the return of ancestral lands. Armed struggle for territorial control against waves of logging, mining, and oil and gas concessions has marked their struggle against extraction practices; bodily barricades are the latest form of resistance to fracking in the basin.[154]

[145] Gilbert, J., & Millan Lombrana, L. (2021, March 10). Argentina is torn between its shale dream and climate goals. Bloomberg.

[146] Lewkowicz, J. (2021, November 3). Argentina gambles on natural gas despite its climate commitment. The Brazilian Report.

[147] McKinsey & Company. (2022, October 21). Vaca Muerta: An opportunity to respond to the global energy crisis. McKinsey.

[148] Cantamutto, F. J. (2020). Vaca muerta and the elusive promises of development in argentina. *Ensayos de Economía, 30*(56), 185–209.

[149] Gouvea de Andrade, M. (2023, March 16). Argentina secures funding boost to kickstart gas exports from 'carbon bomb.' Climate Home News.

[150] Bernáldez, J., & Herrera, R. J. (2020). "94% of the water flows into the sea": Environmental discourse and the access to water for unconventional oil and gas activities in Neuquén, Argentina. In regulating water security in unconventional oil and gas (pp. 155–174). Springer, Cham.

[151] Raszewski, R. (2022, December 27). Analysis: Argentina's Vaca Muerta shale boom is running out of road. Reuters.

[152] Hadad, M. G., Palmisano, T., & Wahren, J. (2021). Socio-territorial disputes and violence on fracking land in Vaca Muerta, Argentina. *Latin American Perspectives, 48*(1), 63–83.

[153] Riffo, L. (2017) Fracking and resistance in the land of fire. *NACLA Report on the Americas, 49*:4, 470–475.

[154] Piñeiro Moreno, N. (2016, March 29). Six indigenous women at the heart of fracking resistance in Argentina. Intercontinental Cry.

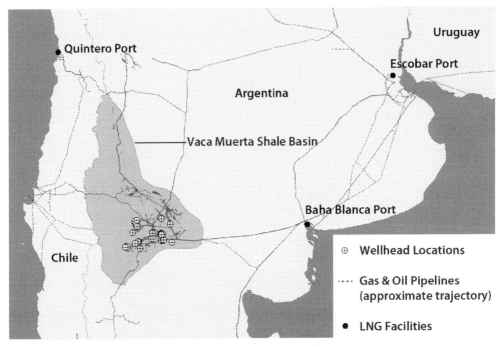

Figure 9.23 Vaca Muerta basin. *(Sources: Global Oil and Gas Infrastructure (GOGI) Inventory; Sources: Shale Basin and Plays World Dataset, Natural Earth.)*

Vaca Muerta communities have experienced all the negatives that come with a gas rush: dangerous frac sand mining,[155] induced seismicity,[156] and vast amounts of produced toxic water.[157] Across the basin, unsafe and illegal dumping has led to watershed-wide damages.[158]

Vaca Muerta gas development draws from a national frontier mentality coupled with insatiable global demand. Cross-border pipeline infrastructure will attempt to feed three LNG ports. Vaca Muerta is vulnerable to predation as politicians allow ill treatment of communities in the path of industrial fracking. Oil workers are brought in from across Argentina[159] to do dangerous, sometimes fatal, work.[160] Laborers have had to strike for better conditions, most recently in 2023 following a series of workplace injuries.

[155] Observatorio Petroleo Sur. (2022, October 28). Río Negro: reclamos por los impactos de la empresa de arenas de fracking NRG.

[156] Gerlo, J., French, G., & Slipak, A. (13 de diciembre de 2022). Tribuna abierta. Vaca Muerta: los financistas detrás de los sismos. La Izquierda Diario.

[157] Staff. (2018, December 17). Indigenous Argentine group sues energy multinationals. France 24.

[158] Forni, L., Mautner, M., Lavado, A., Burke, K. F., & Gomez, R. D. (2021). Watershed implications of shale oil and gas production in Vaca Muerta, Argentina. Stockholm Environment Institute.

[159] Avramow, M. (2022, November 25). Can Argentina balance climate, cash, jobs and justice at Vaca Muerta? Dialogo Chino.

[160] ANRed. (23 de septiembre de 2022). Explosión en la refinería: Vaca Muerta se cobró otras tres vidas.

Discuss and compare: *risk management is often formalized to protect those with resources, leaving margin-alized groups unprotected. Define risk in Vaca Muerta from a Mapuche perspective as compared to gas companies.*

Can't offset injustice

Vaca Muerta represents a **carbon bomb**, a megasite capable of pumping at least a billion tons of CO_2 emissions over its lifetime. Companies operating in Vaca Muerta, such as UK-based Phoenix Global Resources, use deflection tactics such as rainforest conservation tract set-asides to soften the perception of these emissions—yet any carbon offset pales in comparison to the basin's emissions load.[161] Additionally, forest and agricultural carbon offset schemes exhibit known leakage problems.[162] Even after an updated methodology to cover risk and reversal,[163] significant concerns persist about overall net benefits of industrial offset programs.

The intention of an **offset** is something that counterbalances, counteracts, or compensates for something else. Offsetting assumes that it is possible to achieve the result of "no net loss" from development, so long as equivalent conservation value is compensated for or created elsewhere. Offsets are significantly easier and less costly than actual structural change, leading to widespread over-reliance even as offsets have been shown to fall short in actual mitigation.[164] A primary reason for their mixed record is the disparity in emission amounts between a **source**, an origin area for contamination, and a **sink**, an area that holds or stores contamination.

The end goal of offsets is to put a price on pollution, but this is often not an effective means to address damages: the exchange is rarely on par as offsets are functionally lacking.[165] Companies often rely on unproven technologies[166] coupled with narrow LCA

---

[161] Lewkowicz, J. (2022, November 3). Carbon markets in Argentina see growth and concerns as COP27 looms. Dialogo Chino.

[162] Civillini, M. (2023, April 5). Verra's revamped forest offset program comes under fire. Climate Change News.

[163] Verra (n.d.). Area of Focus - Agriculture, forestry, and other land use (AFOLU).

[164] Geels, F. W. (2022). Conflicts between economic and low-carbon reorientation processes: Insights from a contextual analysis of evolving company strategies in the United Kingdom petrochemical industry (1970–2021). *Energy Research & Social Science, 91*, 102729.

[165] zu Ermgassen, S. O., Baker, J., Griffiths, R. A., Strange, N., Struebig, M. J., & Bull, J. W. (2019). The ecological outcomes of biodiversity offsets under "no net loss" policies: A global review. *Conservation Letters, 12*(6), e12664.

[166] Oreskes, N. (2022, August 1). Carbon-Reduction plans rely on technology that doesn't exist. *Scientific American.*

assessments that seldom account for all environmental and social impacts,[167] conducted within offset programs that operate with insufficient transparency.[168]

Mitigation involves acting to reduce the severity, seriousness, or harm of something, such as contamination. Beyond targeting GHG alone, ideally mitigation reduces overall pollution, and enables transformative **co-benefits**: educational and community benefits that occur from environmental initiatives, recognizing that efforts focusing exclusively on GHGs generally have higher tradeoffs and miss important placemaking opportunities.

Offsets allow companies to pay to pollute, while avoiding significant transformation.

An effective offset strategy ideally operates through transparent and fair means, producing real and measurable mitigation results (Fig. 9.24). **Offset leakage** represents shifting emissions across space or time from proposed mitigation, an example of failed offsetting with little to show for actual mitigation.[169] Like leakage, standard accounting systems allow for error as emissions are seldom fully counted. Box 9.2 below addresses offset limitations in specific examples.

Criteria	Description of criteria
real and measurable	reductions are material, quantifiable and traceable
additionality	reductions happened due to the offset project and offset purchaser
transparent	transactions and entities are free from deceit
permanent	offsets are not temporary or easily reversed
registered	formal verification process with reputable independent parties
accounts for leakage	project addresses boundaries, scope, time and technical causes of fugitive emissions or errant accounting
enforceable	process includes sanctions by legal contract for lack of compliance
not double counted	offsets are only claimed by one entity and in one time period
retired	offsets are taken off the market and can't enter back into circulation
contain co-benefits	projects create positive educational, social, economic impacts

Figure 9.24 Principles of high-quality offsets.

[167] Arendt, R., Bach, V., & Finkbeiner, M. (2021). Carbon offsets: an LCA perspective. In Progress in life cycle assessment 2019 (pp. 189–212). Springer, Cham.
[168] Lockhart, A., & Rea, C. (2019). Why there and then, not here and now? Ecological offsetting in California and England, and the sharpening contradictions of neoliberal natures. *Environment and planning e: Nature and space, 2*(3), 665–693.
[169] McLaren, D. P., Tyfield, D. P., Willis, R., Szerszynski, B., & Markusson, N. O. (2019). Beyond "net-zero": a case for separate targets for emissions reduction and negative emissions. *Frontiers in Climate, 1,* 4.

BOX 9.2 Spinoff industry: offsets

Offsets aren't all the same. Equitrans' MVP purchased offsets as the pipeline faced mounting resistance and criticism. The firm exchanged $150 million dollars for coal carbon credits.[170] Because it would offset, not reduce, emissions, and only for a 10 year period,[171] MVP appears to be **virtue-signaling**: the expression of a moral viewpoint with the intent of communicating good character without deserving it. MVP's timing of green investment[172] was suspicious for a project woefully behind and over budget.

In a second offset example, the Philadelphia Eagles football team has agreed to restore sea-grasses and mangroves at National Oceanic and Atmospheric Administration (NOAA)'s Jobos Bay National Estuarine Research Reserve in Puerto Rico, a federally protected tropical ecosystem capturing "blue carbon" stored in the oceans and coastal ecosystems.[173] Notably, Jobos Bay was harmed by coal ash from the US-based AES Corporation (Chapter 3).[174]

Offsets normalize **bandaid solutions** by funding reparative activities after harm instead of preventive solutions. While they can help people feel better about GHG emissions overall, offsets do not address the root causes of climate warming, and detract from transformative structural change.

Firms camouflage GHG pollution as one component of sustainability business scoring, known broadly as **ESG** (Environment, Social, Governance). ESG relies on unclear or partial accounting tools set to different criteria or standards, thus creating a spotty and incomplete **ESG patchwork**.[175] Increasingly, corporate programs are coming under closer scrutiny, and litigation around false offset claims[176] has begun.[177] Net-zero claims relying on offsets are now an emerging legal hazard.[178] Even in the best case scenarios where offsets are real, effective mitigation is not foolproof—wildfires worsened by climate change releasing stored carbon from forest offsets is an ironic example.

Institutions like banks proudly declare "net-zero" intentions (Chapter 8) where the sum of their financed emissions would equal out to zero. Yet commitments frequently

[170] Dodson, W., & Sims, J. (2021, July 16). The greenwashing of the Mountain valley pipeline. Front Porch Blog.
[171] Anchondo, C. (2021, July 13). Pipeline goes CO2 neutral: Innovative or green washing? E&E News.
[172] Cox, N. A. (2021, July 12). Mountain valley pipeline announces plan to offset carbon impacts. BusinessWire.
[173] NOAA Office for Coastal Management. (2023, June 23). NOAA reserve hosts first-ever carbon-offset initiative by u.s. pro football team.
[174] Earthjustice. (2023, May 4). Toxic coal ash in puerto rico:The hazardous legacy of the aes-pr coal plant.
[175] Holland, B. (2022, December 7). Path to net-zero: Lack of emission-reporting standard clouds oil, gas investment. S&P Global.
[176] Quinn, J.B., Alden, A. P., Weiss, I., & Shrader-Frechette, D. (2022, June 23). Offsets: A coming wave of litigation? Quinn emanuel trial lawyers.
[177] Cole, C.A. (2023, June 16). Litigation beginning to challenge carbon offsets. Katten.
[178] Worford, D. (2025, October 22). Evian faces lawsuit over packaging's carbon neutral claims. *Energy and Environment Leader.*

lack detail on how they intend to reach this goal, instead relying on vague carbon storage or offset schemes. These commitments are no substitute for policies restricting fossil fuels or their finance.[179] For example, emissions that are considered Scope 3—released when consumers use a company's products—are routinely ignored by producers and not counted in GHG inventories.[180] Scope 3 emissions are significant for any gas company; leaving off Scope 3 contributes to false claims of carbon neutrality.

Some argue for the use of offsets and carbon storage as integral steps towards a livable climate. However, evidence thus far often shows offset efficiency less than promised, and in many cases foments outright inequality.[181] Companies receive undeserved positive attention as they "pay to pollute,"[182] while places are reduced to spreadsheets, communities and ecosystems fragmented into substitutable quantities.[183] As offset reductions remain voluntary, the least expensive options are frequently chosen. Offsetting has not been shown to be an effective climate action while subsidizing destruction.

Carbon Market Watch recommends: (1) ban net zero and carbon neutrality claims; (2) report absolute emission reductions separately from any emission reductions financed outside of their value chain, rather than a single aggregate number; (3) set targets that cover all of the life cycle emissions and provide details on the reference point used to calculate reductions, i.e. the base-year; and (4) don't balance fossil fuel emissions with carbon stored in non-permanent carbon sinks, like forests.

Redistributing risk

Over time, politicians and regulators have stood by as insecurity and risk have been built into power grids.[184] Fig. 9.25 highlights forms of deception that are deployed to downplay risk in the cases discussed above and below.[185,186] Risk is disproportionately distributed. In the absence of equitable and safe energy governance, impacted communities

[179] Banking on Climate Chaos. (n.d.). Policy score by sector: Oil and gas policy tracker.

[180] New Climate Institute. (2022, February 7). Corporate climate responsibility monitor 2022.

[181] Finley-Brook, M. (2016). Justice and equity in carbon offset governance: debates and dilemmas. In the carbon fix: Forest carbon, social justice and environmental governance. S. Paladino and S. Fiske, Eds. London: Routledge. pp. 74–88.

[182] National association for the advancement of colored people, Environmental and climate justice program. (2021). Nuts, bolts, and pitfalls of fossil fuel pricing: An equity primer on paying to pollute.

[183] Ritter, D. (2018). The coal truth. UWA Publishing.

[184] Chen, S. E., Pando, M. A., Irizarry, A. A., Baez-Rivera, Y., Tang, W., & Ng, Y. (2021). Resiliency of power grid infrastructure under extreme hazards-observations and lessons learned from hurricane maria in Puerto Rico. In civil infrastructures confronting severe weathers and climate changes conference (pp. 1–17). Springer.

[185] Paviour, B., & Cole, A. (2021, September 21). A historically Black town stood in the way of a pipeline so the developers claimed it was mostly white. *The Guardian*.

[186] Ortiz, E. (2018, December 7). How money stokes divide of a historic black community in Virginia pipeline battle. NBC News.

Type	Definition	Examples - Dominion Energy (DE), Phoenix Global Resources (PGR), Equitrans (EQT) and Enbridge (E)
greenwashing	making acts appear more ecologically friendly than they actually are	PGR- purchase carbon offsets for Vaca Muerta EQT- purchase carbon offsets for Mountain Valley Pipeline
whitewashing	painting a narrative of integration while forwarding white privilege	DE - attempt to erase African American households and histories in Union Hill[186]
wokewashing	painting a narrative of equity while advancing unfair systemic violence	DE - structure local investments to divide the community while broadcasting social benefit[186] E - federal regulators like FERC argue for forward-facing policy without addressing damage from Weymouth Compressor Station

Figure 9.25 Three categories of deception.

oppose gas expansion and strive to create alternatives,[187] whether in Argentina, Appalachia, or Puerto Rico (see below).

Try this

Research and compare: *discuss risk redistribution in a "blue-green alliance" that involves collaboration between workers and environmentalists. Use resources in Praxis 7 to explore international cases.*

Puerto Rico's energy battleground

Puerto Rico is a future battleground for LNG dominance.[188] Historically, the territory's grid has been built upon fossil fuels, even though renewable prospects abound and the grassroots struggle for their implementation is incessant (Conclusion). Pressure from the US mainland for gas and LNG is unending, preceded by petroleum and coal. Today's gas only extends and intensifies colonial relationships.[189] Puerto Rican ratepayers are continually held captive with expensive energy. The energy burden (Introduction) is exceedingly high when compared to the mainland: 59% more for residents, 121%

---

[187] Lloréns, H. (2021). *Making livable worlds: Afro-puerto rican women building environmental justice.* University of Washington Press.

[188] Kusnetz, N. (2023, January 26). *Puerto Rico hands control of its power plants to a natural gas company.* Inside Climate News.

[189] de Onís, C.M. (2017, August 21). *For many in Puerto Rico, 'energy dominance' is just a new name for US colonialism.* The Conversation.

more for commercial uses, and 231% more for large industrial uses.[190] Rates are hiked amidst a declining population.[191] Increasing costs with decreasing populations should generate concern about the expense of infrastructure for LNG,[192] particularly as it will be prone to hurricane and flood damage in a heating climate.

A new fortress?

US companies like New Fortress Energy (NFE) influence a race-to-the-bottom (Chapter 2). The company's CEO, the American billionaire Wes Edens, has been deemed the "new king of subprime lending" and a "slumlord" as recently as 2020.[193] Eden has proven willing to embrace risk to capture markets. His fast-paced space grabs around the world feature backroom deals, sometimes connected to non-disclosure or confidentiality agreements.[194]

"Sustainable" has been co-opted and corrupted.

NFE continually makes undeserved claims of sustainability while building new fossil fuel infrastructure. This results in a **green halo effect**[195]—a positive feeling associated with investment in a company as a result of professed environmentalism. Yet NFE points to their support for hydrogen energy, claiming investments in gas are preparing for future hydrogen while building costly infrastructure that detracts from now viable alternatives. The deployment of hydrogen faces significant challenges; it is not immediately suitable to exisitng fossil gas infrastructure,[196] in part due to hydrogen's smaller, more active molecules and high volatility (Chapter 10).

Energy colonialism

NFE, an upstart energy company headquartered in New York, launched a series of operations in Jamaica, 2016, connected to dirty bauxite mines. NFE further gained influence with Jamaica's utility, eventually undercutting it. Profits from NFE's Jamaican activities fund global expansion, exploiting energy-poor countries with limited choices amongst polluting fuels. NFE casts its predation as a modernized energy sector,[197]

---

[190] International Energy Agency. (2023). Puerto Rico territory energy profile.

[191] Schoene Roura, P. (2018, August 3). Editorial: Same old tricks in PREPA circus. Caribbean Business.

[192] Gallucci, M. (2022, July 5). Puerto Rico is pushing LNG when it says it's shifting to renewables. Canary Media.

[193] Feltgen, A. H., & Christensen, D. (2020, September 17). Fortress forgave huge Trump loan. Florida Bulldog.

[194] Bandarage, A. (2021, October 5). New fortress energy, Sri Lanka, and Planet Earth. Island.

[195] I.e., New Fortress Energy. (n.d.). Sustainability.

[196] Pearl, L, (2022, July 22). Hydrogen blends higher than 5% raise leak, embrittlement risks for natural gas pipelines. Utility Dive.

[197] New Fortress Energy. (n.d.). Operations.

peddling gas as a cleaner fuel[198] while burdening low-wealth consumers with expensive new infrastructure.

Like Jamaica, Puerto Rico's debt crisis and ongoing political power vacuum under the Puerto Rico Electric Power Authority (PREPA) (Chapter 3) was of keen interest to NFE. PREPA functioned like a monopoly for decades as the sole utility permitted for electricity generation, distribution, and transmission on the island.

Hurricane Maria threw into relief Puerto Rico's historical colonial vulnerabilities;[199] dramatic infrastructure failure ensued, proving to be a pressing matter of life and death.[200] The island's grid failure now stands as the longest recorded US blackout (Fig. 9.26).[201] Disasters are particularly hard for an older demographic,[202] and during Maria's aftermath, hospital patients couldn't receive treatment without electricity.[203]

Under such duress, disaster capitalism prevailed, culminating in an "**unnatural**" **disaster**[204]—a catastrophe occurring in the Anthropocene worsened by harm from neglect and lack of response afterward.[205] Recovery brings yet more catastrophe as outside investors[206] take advantage of a crisis to privatize public services[207] and the commons in the name of an efficient recovery.[208] The process was particularly hard for youth as suicides increased dramatically. Trauma will remain for a lifetime for the "Maria Generation" growing up in the crisis' aftermath.[209]

[198] New Fortress Energy. (n.d.). Solutions.
[199] de Onís, C.M. (2018). Fueling and delinking from energy coloniality in Puerto Rico. *Journal of Applied Communication Research*, *46*(5), 535–560.
[200] Santiago, R., de Onís, C. M., & Lloréns, H. (2020). Powering life in Puerto Rico: The struggle to transform Puerto Rico's flawed energy grid with locally controlled alternatives is a matter of life and death. *NACLA Report on the Americas*, *52*(2), 178–185.
[201] Lloréns, H., & Stanchich, M. (2019). Water is life, but the colony is a necropolis: Environmental terrains of struggle in Puerto Rico. *Cultural Dynamics*, *31*(1–2), 81–101.
[202] Andrade, E. L., Jula, M., Rodriguez-Diaz, C. E., Lapointe, L., Edberg, M. C., Rivera, M. I., & Santos-Burgoa, C. (2021). The impact of natural hazards on older adult health: lessons learned from Hurricane Maria in Puerto Rico. Disaster medicine and public health preparedness, 1–8.
[203] Lopez-Araujo, J., & Burnett, O. L. (2017). Letter from Puerto Rico: The state of radiation oncology after Maria's landfall. *International Journal of Radiation Oncology*, *99*(5), 1071–1072.
[204] de Onís, C.M. (2018b). Energy colonialism powers the ongoing unnatural disaster in Puerto Rico, frontiers in communication, special issue: Energy democracy, 3, 1–5.
[205] Lloréns, H., Santiago, R., Garcia-Quijano, C.G., & de Onís, C.M. (2018, January 22). Hurricane Maria: Puerto Rico's unnatural disaster. *Social Justice Journal*.
[206] Bonilla, Y. (2018, February 29). For investors, Puerto Rico is a fantasy blank slate. The Nation.
[207] Brusi, R., Bonilla Y., & Godreau, I. (2018, September 20). When disaster capitalism comes for the University of Puerto Rico. The Nation.
[208] Lloréns, H., & Santiago, R. (2018). Women lead Puerto Rico's recovery. *NACLA, Report on the Americas 50*(4):398–403.
[209] Sutter, J. D. (2018, September 17). 'The Maria Generation': Young people are dying and suffering on an island with a highly uncertain future. *The Journal Star*.

Baseline (pre-storm) view of San Juan night lights.

Average San Juan night lights 2 months after Hurricane Maria passed over Puerto Rico.

Figure 9.26 Lights in Puerto Rico after Hurricane Maria. *(Base Image credit: NASA Scientific Visualization Studio.)*

In the midst of a multi-year restructuring process, PREPA's future remains uncertain. Management of the new energy generation company Genera PR was handed directly over to NFE. Genera PR, a subsidiary of NFE, transfers significant risk to Puerto Ricans through its practices. NFE has been found to have sidelined both federal and local governments,[210] building out LNG infrastructure in Puerto Rico without permission prior to construction.[211] The FERC ordered NFE to "show cause"—a rare situation. By the time the FERC found NFE out of compliance, the improper infrastructure was already completed—a clear example of regulatory manipulation.

NFE's ongoing gas coloniality represents a decades-long industrialization of the island territory by the mainland US. By outsourcing control of energy to NFE, a fully privatized entity, the territory's workforce and academic researchers remain sidelined from any meaningful deliberation over its energy future.[212] New gas construction hinders Puerto Rico's transition to renewable energy.[213]

Yet another long-term structural violence, the territory is caught in collective debt bondage.[214] **Shock doctrine** brutally exploits the public's disorientation following a collective shock—such as a debt crisis—to push through pro-corporate measures. These are often called **shock therapy** as privatization and "structural adjustment" programs lessen the economic role of government and promote a neoliberal economy. Increased market control without government regulation intensifies vulnerability to pollution with proliferation of dirty industries in places like Puerto Rico. **Off-shoring** is a spatial pattern of locating polluting industries in places that are separate from areas of consumption. These polluting industries are also often energy "hogs" used to justify building large power plants.

Energy security, energy democracy

In 2019 Puerto Rico embraced a Green New Deal:[215] The government passed a 100% Clean Energy bill,[216] including strategic networks of microgrids[217] for collaborative

[210] Dick, J. (2022, June 15). New Fortress Puerto Rico LNG terminal falls under FERC authority, says court. Natural Gas Intel.

[211] Coto, D. (2020, June 10). Puerto Rico power company worth $1.5B questioned. Seattle Times.

[212] de Onís, C.M. (2017, October 16). Puerto Rican energy researchers excluded from island's energy transition deliberations. Latino Rebels.

[213] Kunkel, K. & Sanzillo, T. (2018, August 16). Plan to turn to imported natural gas will hurt Puerto Rico dearly. The Hill.

[214] Coleman Taylor, A. (2020). Black feminist approaches to the debt crisis. Puerto Rico Syllabus.

[215] Ellsmoor, J. (2019, March 25). Puerto Rico has just passed its own Green New Deal. Forbes.

[216] McKenna, P. (2019, March 26). Puerto Rico passes 100% clean energy bill. Will natural gas imports get in the way? Inside Climate News.

[217] Aros-Vera, F., Gillian, S., Rehmar, A., & Rehmar, L. (2021). Increasing the resilience of critical infrastructure networks through the strategic location of microgrids: A case study of Hurricane Maria in Puerto Rico. *International Journal of Disaster Risk Reduction, 55*, 102055.

hurricane-resilient community-based renewable energy systems (Conclusion).[218] Approved with overwhelming public support, the bill enpowers prosumers—customers that both produce and consume energy—to increasingly produce their own energy. Inspired in part by system collapse and the collective trauma of Hurricane Maria, Puerto Ricans manifested through this bill their anticipation for energy alternatives. Importantly, trade organizations support transformation: the Puerto Rico Electric and Irrigation Industry Workers Union embraces community-based solar electrification on the island. The electrical union aims to contribute to a blue-green alliance for decentralized and distributed energy. These same technologies can work in urban environments—led by community organizations such as UPROSE in New York City,[219] where the Puerto Rican director, Elizabeth Yeampierre, infused an intergenerational, multicultural, and community-led vision for renewable energy and climate justice.[220]

Puerto Ricans and other marginalized groups have to navigate the paradox of resilience:[221] As poor people suffer from a climate crisis caused largely by those with power and resources, they are expected to demonstrate **resiliency**, an ability to recover following disruption. Resilience is a buzzword that is easily romanticized, set out as an uneven and unrealistic expectation. Complicating matters, the concept of **entanglements**[222] recognizes constraints in contexts and moments that do not allow for swift or sweeping change. The politics of entanglement often require subterfuge and fugitivity; at times it resembles guerrilla warfare, in other moments it may appear as a pragmatist art of coping, negotiating, navigating, evading, or subverting what cannot be completely reformed.[223]

Puerto Ricans find themselves resilient yet caught in a state of entanglement. Burgeoning energy democracy operates as a restorative worldmaking practice full of promise. Community organizers push against draconian interference with embodied politics of resistance.[224] Puerto Rico's Green New Deal demonstrates that a foundation for energy transformation can be codified into law. Yet the island's historic debt to fossil fuels invigorates the worst of disaster capitalists and international lenders. A new "legacy" fee has

---

[218] Krantz, D. (2020). Solving problems like Maria: A case study and review of collaborative hurricane-resilient solar energy and autogestión in Puerto Rico. *Journal of Sustainability Research*, 3(1).

[219] UPROSE (n.d.). Welcome to UPROSE!

[220] UPROSE (n.d.). Elizabeth Yeampierre, Executive Director.

[221] Nihad, N. (2022). Navigating the paradox of resilience: Colonial legacies, climate change, and Hurricanes in Puerto Rico. MA thesis, University of Oklahoma.

[222] Bonilla, Y. (2017). Freedom, Sovereignty, and other entanglements. *Small Axe: A Caribbean Journal Of Criticism*, 21(2), 201–208.

[223] Gramsci, A. (1971). Selections from The Prison Notebooks, edited and translated by Q. Hoare and G. N. Smith. p. 276.

[224] Llorens, H. (2018). Imaging disaster: Puerto Rico through the eye of Hurricane Maria. *Transforming Anthropology*, 26(2), 136–156.

been assigned to Puerto Rican ratepayers who, over the course of decades, are on the hook for PREPA-incured debt.[225] At every turn, Puerto Rico's determination for energy security is challenged by fossil fuel interests, past and present. In 2023, NFE's newly formed subsidiary Genera PR, was handed the keys to "modernize" Puerto Rico's grid,[226] replicating with exacting precision coal and oil's original harmful patterns across the island, this time with gas' latest incarnation, LNG, whose international reach is explored further in Chapter 10.

Spatial and temporal synopsis

Space

(1) *A blast radius reverberates and extends risk spatially outward.*
(2) *Infrastructure placement is uneven, creating both rural and urban sacrifice zones.*
(3) *Establishment and maintenance of infrastructure ROWs is a space-production process, causing fragmentation of ecosystems and communities, undercutting resilience.*

Time

(1) *Forward-facing policy and false solutions, like CCS and offsets, delay necessary mitigation and exacerbate climate disruption.*
(2) *Disaster capitalism follows unnatural disasters fueled by years of neglect and lack of response.*
(3) *Energy democracy presents a means of recovery and worldmaking for a livable, equitable future.*

Summary

Agencies tasked with protecting public and planetary health downplay toxic and dangerous components of fossil gas in marginalized areas. The gas infrastructure permitting process enables and enacts structural violence when developers of interstate pipelines treat communities in blast radii and beyond as sacrifice zones situated "in the way" of moving gas from frack pads to markets. Environmental racism through displacement and erasure is common across the gas infrastructure system. The promotion of forward-facing policy and false solutions gives companies erroneous positive images while failing to mitigate harm. Our challenge remains to effectively mitigate climate

[225] Walton, R. (2023, June 12). Puerto Rico utility dept plan could saddle residents with unaffordable power bills for decades, groups warn. Utility Dive. https://www.utilitydive.com/news/puerto-rico-debt-plan-unaffordable-power-bills-FOMB/652654/

[226] NFE Subsidiary Genera Awarded Contract to Manage Puerto Rico's Power Generation System | New Fortress Energy. (n.d.). New Fortress Energy. https://ir.newfortressenergy.com/news-releases/news-release-details/nfe-subsidiary-genera-awarded-contract-manage-puerto-ricos-power

harms while authentically decolonizing vulnerable gas hotspots represented by Appalachia, Vaca Muerta, and Puerto Rico.

Vocabulary

1. astroturf
2. blast radius
3. blowdown
4. blowout
5. blue-green alliance
6. boundary
7. carbon bomb
8. certified sustainable gas
9. collective trauma
10. corridor
11. deference politics
12. direct jobs
13. disaster capitalism
14. elite capture
15. eminent domain
16. energy democracy
17. energy security
18. entanglement
19. ESG (Environmental, Social, Governance)
20. ESG patchwork
21. forward-facing policy
22. frack pads
23. fractured responsibility
24. fragmentation
25. gaslighting
26. ghostwriting
27. green halo effect
28. high consequence area
29. holdouts
30. indirect jobs
31. invisibility
32. legal monkeywrenching
33. marginality
34. neoliberal multiculturalism
35. offset
36. offset leakage

37. off-shoring
38. overpressurization
39. post-traumatic stress disorder
40. regime of measurement
41. residual risk
42. resiliency
43. risk assessment
44. risk management
45. segmentation
46. shock doctrine
47. shock therapy
48. sink
49. sinkhole
50. site suitability
51. source
52. spatial fix
53. unnatural disaster
54. vapor-speak
55. virtue signaling
56. vulnerability
57. wokewashing

Recommended

de Onís, C. M. (2021). *Energy islands: Metaphors of power, extractivism, and justice in Puerto Rico.* Univ of California Press.

Gullion, J. S. (2015). *Fracking the neighborhood: Reluctant activists and natural gas drilling.* MIT Press.

Maldonado, J. K. (2018). *Seeking justice in an energy sacrifice zone: Standing on vanishing land in coastal Louisiana.* Routledge.

Pearson, T. W. (2017). *When the hills are gone: Frac sand mining and the struggle for community.* U of Minnesota Press.

CHAPTER 10

"Freedom" molecules

Contents

Climate Crisis, Energy Violence
ISBN 978-0-12-819501-7,
https://doi.org/10.1016/B978-0-12-819501-7.00015-1

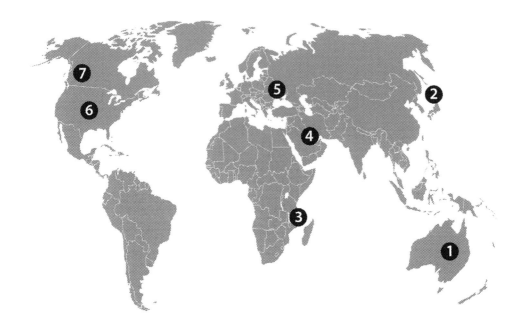

Locations: 1 - Australia
                  2 - Japan
                  3 - Mozambique
                  4 - Qatar
                  5 - Ukraine
                  6 - United States
                  7 - Western Canada

Themes: Lateral violence; Metabolic rift; Spatial emancipation

Subthemes: Accumulation by dispossession; Time-space compression

The US Department of Energy is "doing what it can" for molecules of US freedom to be exported to the world.[1]

[1] O'Neil, L. (2019, May 29). US energy department rebrands fossil fuels as 'molecules of freedom'. *The Guardian*.

LNG's promise ... and peril

Liquefying gas reduces its size, making gas easier to move and store—yet processing liquefied natural gas (LNG) is energy intensive. Liquification of fossil gas, which is mainly methane (CH_4), cools to approximately $-162°C$ ($-260°F$).[2] Regasification then converts LNG back to a gaseous state. This complex processing makes LNG supply chains expensive and intensifies the climate crisis, locking in costly large infrastructure. LNG's promotion as "freedom" molecules,[3] is a romanticized notion that ignores material implications (i.e., contamination, geopolitical conflicts) from LNG expansion. Fig. 10.1 demonstrates gas industry political allyship; a large American flag is prominently displayed at an emerging facility.

In contrast to assuring freedom, LNG creates a methane rift (Introduction) as part of a broader overshoot of ecological thresholds. **Metabolic rift** describes surpassing planetary limits, with metabolism used to describe exchanges between nature and humans and

Figure 10.1 US "freedom" gas. *(Image credit: President Trump Visits the Cameron LNG Export Facility - Official White House Photo - Shealah Craighead, May 2019. Accessed via Flickr.)*

[2] US Energy Information Administration. (2020, July 5). *Natural gas explained: Liquefied Natural Gas.*
[3] Ellsmoor, J. (2019, May 30). Trump administration rebrands fossil fuels as "Molecules of U.S. Freedom." Forbes.

within social spheres.[4] Narrow growth-centric economic policies and structures have driven a rift so wide that technology cannot bind the open gap.[5] LNG only forces the breach wider.

Over time, LNG expansion has gained global traction, resulting in the development of nearly 100 export/import facilities in just the last decade alone (Fig. 10.2). A similar pattern in the mobility of LNG in contrast to domestic, local use is shown in Fig. 10.3. Aggregated import and export activity has only accelerated since 2015. Looking forward, LNG growth is very likely, spurred on by industry's multipronged disinformation campaign designed to portray LNG as a "clean" alternative fuel, much more climate-friendly than its reality (Fig. 10.4).[6]

As with all fossil gas, LNG methane emissions include fugitive releases from seals and equipment connections as well as intentional venting via dedicated outlets.[7] LNG's GHG

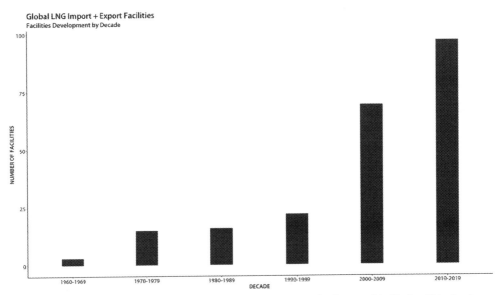

Figure 10.2 Global LNG growth. *(Source: Clear Seas Centre for Responsible Marine Shipping.)*

[4] Lynch, M. J., Long, M. A., & Stretesky, P. B. (2019). Metabolic rift and eco-justice. In *Green criminology and green theories of justice* (pp. 151–192). Palgrave Macmillan, Cham.
[5] Clark, B., & York, R. (2005). Carbon metabolism: Global capitalism, climate change, and the biospheric rift. *Theory and Society*, *34*(4), 391–428.
[6] Dezem, V., Stapczynski, S., & Malik, N. (2020, September 9). Natural gas is losing its luster as "bridge" fuel to renewable energy. World Oil.
[7] Balcombe, P., Heggo, D. A., & Harrison, M. (2022). Total methane and CO2 emissions from liquefied natural gas carrier ships: the first primary measurements. *Environmental Science & Technology*, *56*(13), 9632–9640.

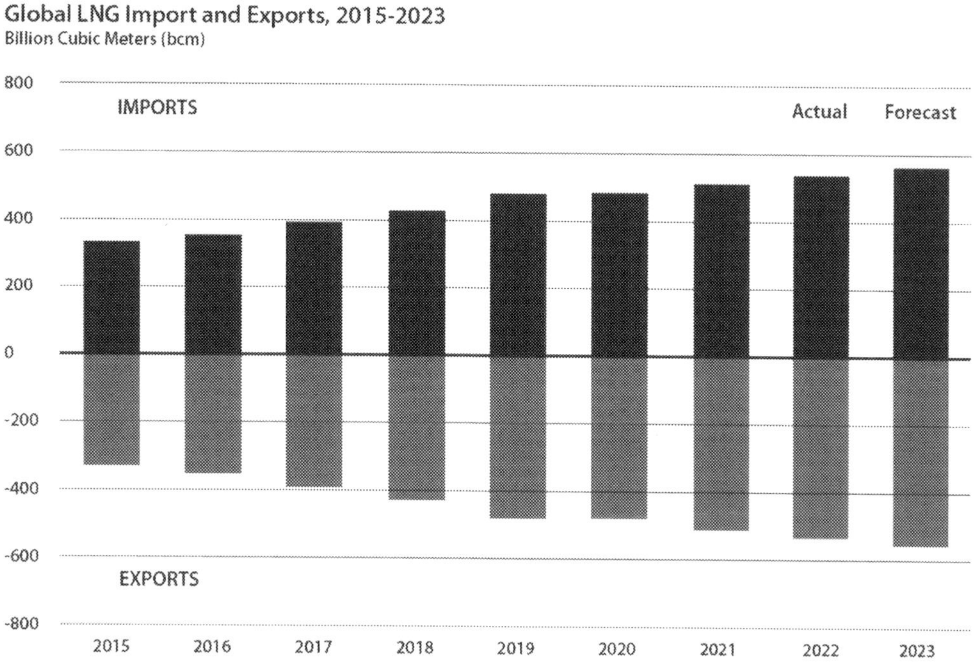

Global LNG Import and Exports, 2015-2023
Billion Cubic Meters (bcm)

Figure 10.3 Global LNG import—export trends. *(Adapted from: IEA, World LNG imports and exports by region, 2015-2023.)*

Methane leakage in fossil gas is undercounted (**Chapter 8**), so LNG emissions are skewed low
Liquefaction and regasification are resource-intensive requiring vast infrastructure
Expensive new LNG infrastructure competes with less expensive renewable alternatives, with risk of becoming stranded by shifting markets and regulations

Figure 10.4 Bridge to nowhere.

assessments are unreliable if they haven't undergone full life cycle analysis across production, processing, shipment, storage, and end use.[8] For LNG, GHG measurements have many stages of collection well-to-wake—emissions produced upstream and midstream (well-to-tank) and downstream (tank-to-wake). LNG emissions from extraction, transport, liquefaction, and regasification have been found to be almost equal to the emissions produced from the burning of the LNG, meaning LNG effectively doubles the climate

---

[8] Agarwala, N. (2022). Is LNG the solution for decarbonised shipping? *Journal of International Maritime Safety, Environmental Affairs, and Shipping, 6*(4), 158—166.

impacts beyond gas alone. In a comparative context, total GHG emissions from exported LNG are likely higher than those from coal.[9] A methane leakage rate of just 1.4—3% is the estimated threshold at which US LNG exports emit as many potent GHGs as coal.[10] Emissions also emerge from the construction of new LNG terminals, regasification facilities, and other equipment.[11]

Energy violence

LNG explosions are relatively rare but can be exceedingly powerful and fatal.[12] Similar to other energy types, LNG regulations have not effectively anticipated, but rather followed, fatal incidents. In 1944, a leakage occurred at an LNG station in Cleveland (US) causing an explosion that spread over 14 blocks, killing 136 people.[13] A series of tank explosions sent fire through sewers and up drains into homes where the intense temperatures vaporized flesh and bone. Following the disaster, firms focused more on safety.

As LNG's accelerated global expansion is afoot, new risks abound. In 2004, an Algerian LNG terminal caught on fire,[14] resulting in 30 deaths and creating concern over residential proximity. Fatalities like these have led to scrutiny over safety protocols.[15] Both expansive and compact, LNG facilities string together tank infrastructure, creating compounding explosive risk.[16] Hazards in LNG facilities include fire, explosion, brittle fracture, asphyxiation, and frostbite. Fig. 10.5 illustrates frame-by-frame vapor expansion just prior to its explosion. A release of LNG facility gas may lead to a domino effect of events with fireball, flash fire, pool fire, or vapor cloud explosions spurring additional damage from shock waves or projectiles.

[9] Morton, A. (2019, July 3). Booming LNG industry could be as bad for climate as coal, experts warn. *The Guardian*.

[10] Natural Resources Defense Council. (2020, December). *Sailing to nowhere: Liquefied Natural Gas is not an effective climate strategy*.

[11] de Jong, M. (2023). LNG: Savior or a new problem in the making? GIES Occasional Paper.

[12] Saloua, B., Mounira, R., & Salah, M. M. (2019). Fire and explosion risks in petrochemical plant: assessment, modeling and consequences analysis. *Journal of Failure Analysis and Prevention*, 19(4), 903—916.

[13] Wuertz, J., Bartzis, J., Venetsanos, A., Andronopoulos, S., Statharas, J., & Nijsing, R. (1996). A dense vapor dispersion code package for applications in the chemical and process industry. *Journal of Hazardous Materials*, 46, 273—284.

[14] Dweck, J., Boutillon, S., & Asbill, S. (2004). Deadly LNG incident holds key lessons for developers, regulators. *Pipeline and Gas Journal*, 39—42.

[15] Congressional Research Service. (2009, December 14). *Liquefied Natural Gas (LNG) import terminals: Sitting, safety and regulation*.

[16] Animah, I., & Shafiee, M. (2020). Application of risk analysis in the liquefied natural gas (LNG) sector: An overview. *Journal of Loss Prevention in the Process Industries*, 63, 103,980.

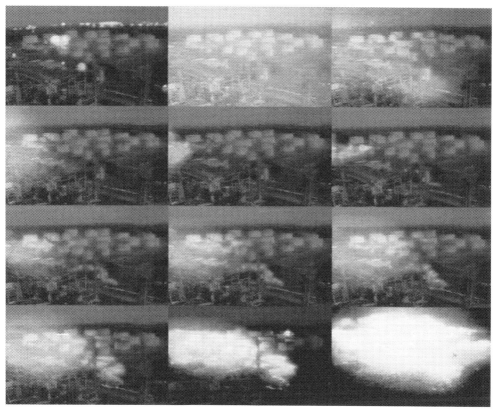

Figure 10.5 Vapor builds up before explosion. *(Image source: Surveillance footage of flame propagation during October 23, 2009 CAPECO explosion, REPORT NO. 2010.02.I.PR, U.S. Chemical Safety and Hazard Investigation Board.)*

With its inherent risks, current LNG safeguards are questionable, particularly exacerbated by rapidly expanding construction, thrusting LNG platforms into new risky geographies—deep offshore and Arctic sites, typical of extreme energy infrastructure. Gasification plants are situated both on land and at sea where **floating rigs** (FLNG) facilitate liquefaction and storage of gas and help to service offshore fields where subsea pipeline structures don't exist. Due to their nascent development, FLNG terminal risks are relatively unknown.[17] To date, LNG tanker collisions during offloading and grounding

[17] Lee, S. (2020). Quantitative risk assessment of fire & explosion for regasification process of an LNG-FSRU. *Ocean Engineering, 197*, 106825.

account for most reported accidents. FLNGs can be unstable in wind and waves and bring unique dangers as they host complex operations in a small area.[18]

LNG injustice

LNG encourages a false sense of energy security for those in power, all the while increasing insecurity for marginalized groups. Geopolitical contests to capture growth markets for LNG force imprudent buildout with significant local impacts. As LNG extends gas territoriality, increases spatial reach, shifts economic balance, and captures political control, **deterritorialization** of local communities is hastened, severing social and cultural practices, and disrupting and disfiguring customary lands.

Dispossession (Chapter 7) deprives people of land or property, typically targeting low-wealth and ethnic groups. **Accumulation by dispossession** allows for the consolidation of wealth among those with privilege gained by the act of dispossessing marginalized individuals or groups of their land and resources: case in point, the customary territory of the Wet'suwet'en, as detailed in the British Columbia LNG illustrative case to follow. Energy violence is expressed through **lateral violence**, where harm is internalized among people of color and other marginalized groups and replicated in acts harming peers; this pattern is also called horizontal harm. Lateral violence is a learned behavior resulting from oppressive and patriarchal methods of governing.[19] In the Wet'suwet'en example, individuals become co-opted by outside developers: tokenized leaders become anchors for infrastructure that harms those of the same culture holding less power. **Market citizenship**—where one's perceived economic value is conflated with their accessible rights—is often demonstrated as leaders overly motivated by self-interest become complicit with the powerful in the devaluing of collective benefit and customary governance practices. The cases presented below exemplify neoliberal multiculturalism as people of color acquiesce to market citizenship, gaining transactional preferential status but undercutting both their own people and other minority populations to the benefit of corporate interests. LNG projects are repeatedly sited in areas marked by trauma, increasing existing vulnerabilities while reinforcing privilege for those with power.

Spatial distribution

Spatial-temporal fixity describes how infrastructure is bound in time and space, including how gas corridors influence territorial claims. Firms seek to beat competitors with vertical and horizontal space grabs. Liquefaction makes it possible to transport gas on tankers,

---

[18] Baalisampang, T., Abbassi, R., Garaniya, V., Khan, F., & Dadashzadeh, M. (2019). Modeling an integrated impact of fire, explosion and combustion products during transitional events caused by an accidental release of LNG. *Process Safety and Environmental Protection, 128,* 259–272.

[19] Native Women's Association of Canada. (2011). *Aboriginal lateral violence.*

trucks, and trains to locations unreachable by pipelines. LNG portends an unique form of **time–space compression**, meaning it is possible to cover more distance in less time, or in the case of LNG, with time as a less constraining factor. Profit-oriented investments have sped up the circulation of capital while simultaneously reducing the significance of being tied to a shale basin. The idea of **spatial emancipation** involving LNG mobility free from prior physical constraints is leveraged by industry, designed to set it apart from competing coal and oil.

While liquefaction extends mobility to new markets, LNG still experiences physical constraints in transportation like pre-existing spatial blockages, such as bottlenecks in strategic shipping straits, intensified with the pandemic. During temporary supply disruptions, such as from extreme weather or geopolitical contests, the practice of price gouging occurs by excessively raising prices above the increase necessary to cover higher costs. Even temporary price increases contribute to a scarcity mindset—a preoccupation with a potential future lack of fossil gas, which motivates support for additional underground storage capacity and more investment in liquefaction "trains" (the name of LNG-producing units), tethering states ever tighter to fossil fuel mega-infrastructure.

LNG has expanded significantly since the first LNG carrier, Methane Pioneer, left the Louisiana Gulf coast in 1959 headed for the UK. This pathway from the US Gulf to Europe is still common and has resurged in importance following Russia's invasion of Ukraine. Today there are more than 400 LNG trade routes and upwards of 50 countries importing or exporting LNG. Fig. 10.6 demonstrates the clustering of facilities in Asia,

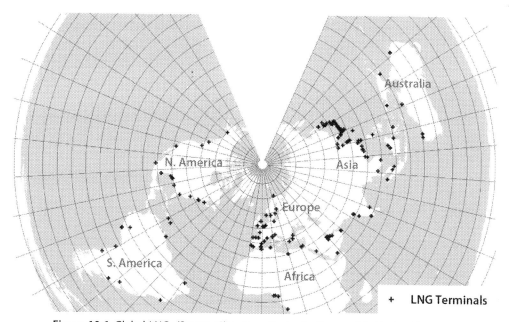

Figure 10.6 Global LNG. *(Source: Clear Seas Centre for Responsible Marine Shipping.)*

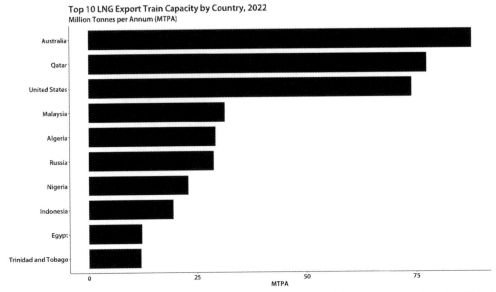

Figure 10.7 Top 10 countries in LNG export capacity. *(Source: Global Gas Infrastructure Tracker, Global Energy Monitor, 2022.)*

Europe, and Africa. Asia accounts for the largest share of LNG imports, and the US and other producing countries compete for long-term contracts to assure access to these dominant markets for decades to come. Fig. 10.7 above shows the top 10 LNG export countries, with Australia, Qatar, and the US being the dominant three. Relationships between these exporters are detailed later in this chapter. From 2016 to 2020, the majority of LNG was exported from the US Gulf, with top destinations including Mexico, South Korea, Japan, and Europe (Fig. 10.8).

Japan has consistently ranked the largest global LNG importer. The development of an oligopolistic, private sector-oriented market following initial Japanese LNG imports in 1967 underscored the nation's determined transition to LNG. Without a robust gas pipeline and underground storage infrastructure, utilities embraced LNG, coupling short-segment, low-pressure delivery with a proliferation of LNG-receiving terminals, more than 30 nationwide.[20]

Globally, LNG markets have been anything but stable, particularly in light of Russia's invasion of Ukraine. With geopolitical tensions rising over decades, superpower nations including Russia, China, and the US now compete with critical LNG producers like

[20] ICLG. (2023). *Oil and Gas Laws and Regulations Japan.*

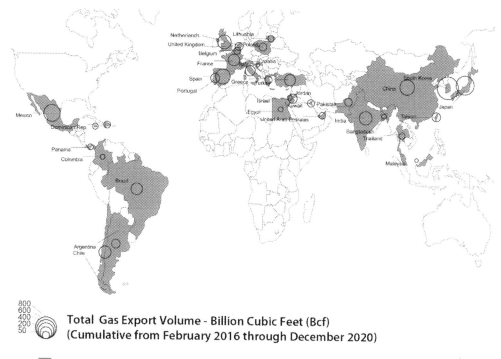

Total Gas Export Volume - Billion Cubic Feet (Bcf)
(Cumulative from February 2016 through December 2020)

Destination Countries

Figure 10.8 US LNG by destination. *(Sources: U.S. Department of Energy Office of Oil & Natural Gas, Natural Earth.)*

Australia and Qatar. In this mix, North America retains its competitive edge due largely to its early infrastructure buildout (Fig. 10.9). LNG facilities have clustered in the US Gulf Coast with its weak regulatory environment, intensifying existing sacrifice zones.

During the Trump Administration, the US heavily solicited LNG exports onto Ukraine, pushing "freedom molecules"[21]—a blatant political manipulation for a self-interested agenda. Former President Trump himself aggressively peddled US LNG exports worldwide, particularly in Europe, especially in Ukraine, as a lynchpin of his American Energy Dominance policy.[22] While an aggrandized notion of US export LNG as spatial freedom was perceived as "silly,"[23] an absurd US-centric imperialist trope, a larger critique was absent to correct willful ignorance about the materiality of LNG geopolitical contests, including its contribution to warfare, not freedom. Acutely, the

[21] Grivach, A. (2020, October 28). 'Molecules of freedom': American LNG Challenges in Europe. *Valdai Discussion Club*.

[22] Cohen, H. (2019, September 30). LNG is the American Aid Ukraine really needs. Forbes.

[23] O'Neil, L. (2019, May 29). US energy department rebrands fossil fuels as 'molecules of freedom'. *The Guardian*.

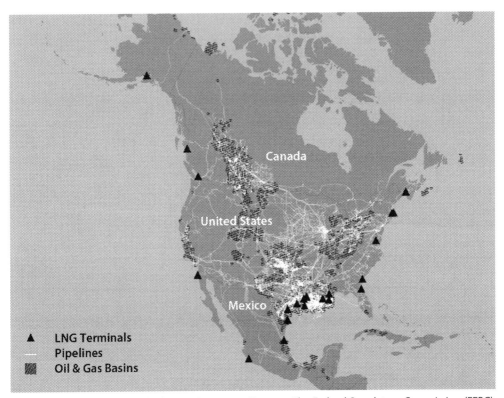

Figure 10.9 US pipelines linked to LNG exports. *(Sources: The Federal Regulatory Commission (FERC), Global Infrastructure Database (GOGI) Inventory, Natural Earth.)*

Ukrainian war serves as a site of energy conflict between exporting Russia and importing western Europe.[24] The suspicious explosion of Nord Stream 2 (Fig. 10.10) in the midst of the Russia—Ukraine geopolitical conflict created a super-emitter event in 2022, with massive physical damage at regional and global scales (Chapter 8).

Gas geopolitics is not only contributing to the conflict between Russia and Europe[25] and within Europe,[26] but ultimately reshaping trans-Atlantic relationships.[27] Political and

[24] Osička, J., & Černoch, F. (2022). European energy politics after Ukraine: The road ahead. *Energy Research & Social Science, 91,* 102757.

[25] Hosseini, S. E. (2022). Transition away from fossil fuels toward renewables: Lessons from Russia—Ukraine crisis. *Future Energy, 1*(1), 2—5.

[26] Gustafson, T. (2020). *The bridge: Natural gas in a redivided Europe.* Harvard University Press.

[27] Bocse, A. M. (2020). From the United States with shale gas: Ukraine, energy securitization, and the reshaping of transatlantic energy relations. *Energy Research & Social Science, 69,* 101553.

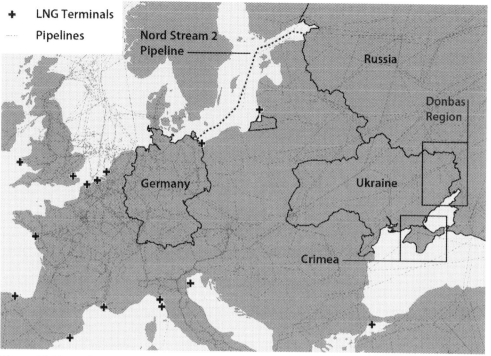

Figure 10.10 Nord Stream 2—strategic im/mobility. *(Source: Global Oil and Gas Infrastructure (GOGI) Inventory, Natural Earth, Gazprom.)*

economic conflicts have emerged over the use of sanctions.[28] Spillover effects from the Russian invasion of Ukraine are occurring across all energy types, particularly in gas and LNG.[29] Yet the manipulation of a scarcity mindset (Chapter 2) as energy firms warned of potential future shortages has allowed for a resurgence in coal and oil promotion as well. In its production and transport, LNG's long supply chain (Fig. 10.11) advantages those with capital and power.[30]

LNG's supply chain creates an environment of productive exclusion (Chapter 1) with low participation of small and medium-sized operators. With expensive megaprojects phased over years, the capital intensity and technical specificity of LNG means control of production is largely inaccessible to smaller producers. Small-scale LNG makes up

[28] Seitz, W., & Zazzaro, A. (2020). Sanctions and public opinion: The case of the Russia–Ukraine gas disputes. *The Review of International Organizations, 15*(4), 817–843.

[29] Savcenko, K. (2023, April 12). How the Russia–Ukraine war is turning natural gas into the 'new oil'. S&P Global.

[30] Bridge, G., & Bradshaw, M. (2017). Making a global gas market: Territoriality and production networks in liquefied natural gas. *Economic Geography, 93*(3), 215–240.

Figure 10.11 LNG supply chain. *(Diagram credit: U.S. Department of Energy (DOE).)*

only about 10% of the LNG market.[31] Meanwhile, imported LNG undercuts domestic regional markets for energy alternatives, including renewables marked by local job creation.

In scaling energy systems to regional needs, longevity is essential. In contrast to expensive LNG mega-facilities, compressed natural gas (CNG) operations in many countries increase local mobility. For populations around the world, CNG provides an affordable, accessible fuel for transportation (Fig. 10.12), assisted by the standardization of equipment hookups. However, CNG operations cannot altogether eliminate the risk of explosion due to tanker or truck rollover, or from leaking or ruptured tanks and lines. These residual risks will remain until vehicles are electrified.

Temporal analysis

Fossil gas systems leak methane, though leakage rates vary across different emission control technologies, places, and timeframes (see details in Chapter 8). LNG is no exception as methane is the core basis of LNG. Emissions of unburned gas are called **methane slip**, and while these releases may seem like a small percentage, climate impact adds up over time, particularly in large-scale operations.[32]

Hidden energy debt has been understood as the uncounted energy that is consumed in goods and services in marginal geographies, like the global South and internally colonized regions.[33] Hidden energy debt is also used here to depict opaqueness in processing stages of complex fuels like LNG, which is erroneously promoted as a clean, ecological fuel, obfuscating a long chain of emissions from fracking, liquefaction, transport, and regasification. These underappreciated releases amount to externalized costs

[31] Tcvetkov, P. (2022). Small-scale LNG projects: Theoretical framework for interaction between stakeholders. *Energy Reports*, 8, 928—933.

[32] Ushakov, S., Stenersen, D., & Einang, P. M. (2019). Methane slip from gas fueled ships: A comprehensive summary based on measurement data. *Journal of Marine Science and Technology*, 1—18.

[33] Akizu, O., Urkidi, L., Bueno, G., Lago, R., Barcena, I., Mantxo, M., … Lopez-Guede, J. M. (2017). Tracing the emerging energy transitions in the Global North and the Global South. *International Journal of Hydrogen Energy*, 42(28), 18045—18063.

Figure 10.12 CNG mobility. *(Image credit: Various CNG Transportation Methods, Bangladesh, India and Vietnam - Michael Davis-Burchat, Shadman Samee, Ajay Tallam, 2007 - 2017. Accessed via Flickr.)*

driving loss, damage, and inequity now and into the future. New infrastructure developments, such as liquefaction plants and export/import terminals, require high financial investment, followed by ecological and social costs, often carried by ratepayers and taxpayers.

Ranging from a year outwards to 2 decades, investors in LNG projects face increasing break-even risk prior to profit:[34] billions of dollars of wasted resources for stalled LNG proposals that never advance to construction. Complex LNG projects are plagued by schedule and budget overruns, extending their **break-even time**—the amount of time required for the discounted cash flows generated by a project to equal its initial cost. If the full costs, including hidden energy debt from each processing stage, were included, break-even time would extend yet further, giving rise to the specter of stranded

---

[34] Rosselot, K. S., Allen, D. T., & Ku, A. Y. (2022). Greenhouse gas emissions from LNG Infrastructure construction: Implications for short-term climate impacts. *ACS Sustainable Chemistry & Engineering, 10*(26), 8539—8548.

assets. While investors exercise pre-set sell orders—a **stop loss point**—to limit their own financial losses, increasing societal costs beg for some form of regulatory cap on expenditures for extreme energy facilities intensifying climate risk.

Energy infrastructure is a major cause of ecological debt owed to the most affected peoples and areas (MAPA), who are often the least responsible for climate change yet experience the greatest loss and damage. Instead of reparation, host countries are bound to debt treadmills (Chapter 6) from expensive energy sector projects funded with massive loans and excessive interest. Low-wealth countries are sprinting to pay off debt related to a constant supply of expensive new technology that drains wealth as expertise and equipment are imported at unfavorable rates. Many of these new technologies—often advancing extraction and processing of yet more coal, oil, and gas—are false solutions (Chapter 8), a demonstration of maladaptation to the climate crisis by making the situation worse.

Being able to differentiate between energy alternatives could encourage leapfrogging to clean technologies[35] versus remaining constrained to climbing a specific energy ladder. In the realm of technology, to **leapfrog** means to bypass traditional stages of development to jump directly to the latest technologies[36] at the scale of consumer, region, and country. The idea of an **energy ladder** (Fig. 10.13) has been challenged as households in various regions and countries have leapfrogged to solar or simultaneously drawn from multiple different energy sources, a situation known as **fuel stacking**.

Actual energy sector transitions are much more complex than a staid energy ladder.[37] The ladder construct suggests rising income will lead to an emancipation of solid fuels; however, it ignores the unfair energy burden (Introduction) for millions of low-wealth populations in industrial countries, who pay more of their income for electricity, sometimes leading to shut-offs when they cannot afford utility bills. This forced scarcity is in part due to the high costs of expensive fossil fuel infrastructure with construction debt pushed onto ratepayers.[38]

While simplistic assumptions of the energy ladder are disproven, LNG and fuels such as hydrogen prompt more nuanced depictions of both their processes and capacities.[39] Like LNG, hydrogen is heavily processed, and much of the promotion of hydrogen today

[35] Yu, Z., & Gibbs, D. (2018). Sustainability transitions and leapfrogging in latecomer cities: the development of solar thermal energy in Dezhou, China. *Regional Studies, 52*(1), 68–79.

[36] Yayboke, E., Crumpler, W., & Carter, W. A. (2020, April 10). *The promise of leapfrogging.* The Center for Strategic and International Studies.

[37] Nguyen, L. T., Ratnasiri, S., & Wagner, L. (2023). Does income affect climbing the energy ladder? A new utility-based approach for measuring energy poverty. *The Energy Journal, 44*(4).

[38] Dawson, A. (2022). *People's power: Reclaiming the energy commons.* OR Books.

[39] Bernard, M. (2020, December 23). Latest "hydrogen economy" round is hype, but there is a place for hydrogen. Clean Technica.

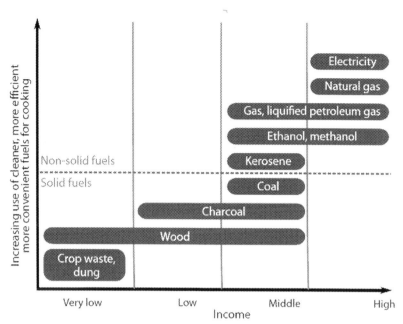

Figure 10.13 Energy "ladder" mis-applied. *(Diagram credit: The Energy Ladder (Paunio, 2018). Accessed via Energypedia.)*

is abetted by confusion. Most hydrogen is produced via "steam-methane reforming," whereby high-temperature steam (700—1000°C) produces hydrogen from a methane source, such as fossil gas.[40] Steam reforming is endothermic, meaning heat must be supplied from other fuels—currently about 95% of commercial hydrogen energy is produced using fossil fuels.

Proponents often tout hydrogen as a fully clean renewable source—attempting to garner a **green halo effect**—as companies gain positive social perceptions based on claimed sustainability efforts, deservedly or not—from the small percentage of truly green hydrogen absent any carbon dioxide footprint. Fig. 10.14 shows two hydrogen generation methods; second-generation green hydrogen is not currently available at large scales.

Hydrogen's uses range from the transport sector to power plants—yet experts encourage caution beyond selective uses.[41] Hydrogen, the lightest of gases, leaks more easily compared to fossil gas and permeates pervasively from high-pressure

---

[40] Department of Energy. (n.d.). *Hydrogen production — Natural gas reforming.*
[41] Le Page, M. (2018, November 22). Hydrogen will never be a full solution to our green energy problems. *New Scientist.*

Phases	Type of Hydrogen
First generation	• Brown (coal and oil) • Blue (gas) • Yellow (nuclear)
Second generation	• Green (renewable)

Figure 10.14 Hydrogen phases and types.

environments.[42] There are climate impacts from hydrogen as an indirect GHG:[43] when hydrogen escapes, most is removed by soils via diffusion and bacterial uptake, with the remaining 20%–30% oxidizing and remaining in the atmosphere for a few years as it increases concentrations of GHGs.

Hydrogen's inherent characteristics of small size, and lack of both color and odor make it difficult to detect and manage. There is the possibility of flaming when mixing with air, even in small percentages. Another consistent environmental concern lies with blue/brown hydrogen lock-in effects,[44] as illustrated in the German and Norwegian illustrative cases to follow. Evidence suggests first-generation hydrogen is not a bridging technology.[45] Derived prominently from fossil gas, the price of blue hydrogen is directly correlated to that of fossil gas.[46] Taking advantage of the green halo effect, fossil gas plants built today increasingly promise a full transition to hydrogen-only technology at a indeterminate period in the future. Such unfounded promotion is misleading given the physical constraints of pipeline corrosion from hydrogen coupled with necessary infrastructure upgrades,[47] as it is not possible to merely swap out fossil gas and LNG with hydrogen fuel.[48]

The ascendancy of new technological "fixes" distracts from energy sector transformation focused on communities and workers. LNG industry conflict with labor unions is

---

[42] Ocko, I. B., & Hamburg, S. P. (2022). Climate consequences of hydrogen leakage. *Atmospheric Chemistry and Physics Discussions*, 1–25.

[43] Ocko, I. B., & Hamburg, S. P. (2022). Climate consequences of hydrogen emissions. *Atmospheric Chemistry and Physics*, 22(14), 9349–9368.

[44] Howarth, R. W., & Jacobson, M. Z. (2021). How green is blue hydrogen? *Energy Science & Engineering*, 9(10), 1676–1687.

[45] George, J. F., Müller, V. P., Winkler, J., & Ragwitz, M. (2022). Is blue hydrogen a bridging technology?-The limits of a CO2 price and the role of state-induced price components for green hydrogen production in Germany. *Energy Policy*, 167, 113072.

[46] Durakovic, G., del Granado, P. C., & Tomasgard, A. (2023). Are green and blue hydrogen competitive or complementary? Insights from a decarbonized European power system analysis. *Energy*, 128282.

[47] Mijim, J. K. A., & Pluvinage, G. (2021). Maximum allowable service pressure for steel pipes used for transport of hydrogen pure or blended with natural gas. *Engineering Science & Technology*, 61–82.

[48] Whitmore, J., & Martin, P. (2022, August 8). Repurposing LNG Infrastructure for hydrogen exports is not realistic. *Globe and Mail*.

growing in Australia, Europe, and the US, among other areas. LNG and hydrogen are more expensive—yet also more profitable—in markets that protect return on investment (ROI) for new infrastructure. This leads to excessive energy infrastructuring that discourages the transition to cheaper and cleaner alternatives. For example, clean energy investments from the US federal government are allocated for hydrogen hubs without clear stipulations to stop both expansion and intensification of fossil gas,[49] leaving taxpayers on the hook for false climate solutions while subsidizing fossil fuels.

Illustrative examples

Taking advantage of geopolitical contests and paths of historical colonialism, LNG's metabolic rift continues to widen, spreading both social and ecological harm. Fossil fascism solidifies as states turn to outright physical control of impacted communities who dare to oppose advancing structural violence. Drawing profit from the construction of expensive facilities to process or move gas, the acceleration of imprudent investments ignores market signals demonstrating a poor performance of fossil fuels when not propped up artificially by politicians.[50]

United States

US gas is dirty by international comparison.[51] US export LNG has been controversial from its inception, initially receiving no more than conditional approval[52] as policymakers wisely questioned LNG exports' advancement of the public interest.[53] Projects emerged largely due to the financial sector's attraction to guaranteed ROI (Chapter 5). Federal stimulus money that was designed to support housing recovery after the 2008 recession was often diverted to major bank investments in extreme energy, with fossil gas and LNG as key beneficiaries.[54]

Cheniere Energy's Sabine Pass was the US' first LNG export terminal, promising a new energy direction with its first export in 2016. Today, Cheniere stands as the preeminent US LNG exporter. Piece by piece, the Sabine Pass facility has grown to consume a

[49] Parkes, R. (2023, August 1). The top ten US hydrogen hubs most likely to win $7bn of government funding. Hydrogen Insight.

[50] Sanzillo, T. (2023, August 1). *Taking stock of the oil and gas sector as the transition to sustainable finance proceeds apace*. Institute for Energy Economics and Financial Analysis.

[51] Gan, Y., El-Houjeiri, H. M., Badahdah, A., Lu, Z., Cai, H., Przesmitzki, S., & Wang, M. (2020). Carbon footprint of global natural gas supplies to China. *Nature Communications, 11*(1), 1–9.

[52] Dotten, M., & Till, D. (2013). Controversy over LNG exports heats up as production expands. *Natural Gas & Electricity, 29*(9), 1–7.

[53] Kim, H. C. (2013). Are LNG Exports in the Public Interest? *Environmental Claims Journal, 25*(2), 154–169.

[54] Sammon, A. (2018, October 22). How the bank bailout hobbled the climate fight. *The New Republic*.

Sabine Pass LNG Terminal

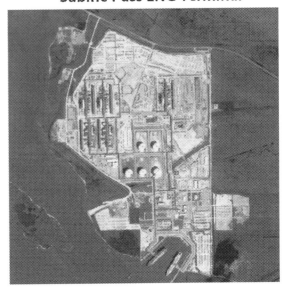

Rose Bowl Stadium ◖🏈◗

|_____|
0.5 Miles

Figure 10.15 Sabine Pass LNG. *(Sources: Esri, Maxar, Earthstar Geographics, CNES/Airbus DS, USDA FSA, USGS, Aerogrid, IGN, IGP and the GIS User Community.)*

massive area, dwarfing the Rose Bowl Stadium footprint—a well-known comparative National Historic Landmark (Fig. 10.15 above). As shown in Fig. 10.16, Sabine Pass' large LNG trains—series of processes and equipment units required for LNG production —were constructed in distinct phases across time.

While LNG investment has gained a reputation as "nervous money" due to known risks,[55] the Ukrainian war spurred on a disaster capitalist mindset, pushing predatory firms and industry groups to leverage geopolitics for the increased uptake of fossil gas. As a result, one LNG proposal after another made its way through the federal regulatory process with ease, so much so that a labor shortage of temporary construction workers—also known as a fly-in and fly-out (FIFO) workforce—now plagues the industry as it tries to complete competing projects.[56]

Sabine Pass was the fastest facility to produce and ship 1,000 LNG cargos, an amount that has since doubled. The site currently features six liquefaction trains, with plans for yet more, some expedited in the wake of the Ukraine invasion.[57] Cheniere recently signed a two-decade contract, benchmarked to the Henry Hub gas price (Chapter 8) plus a liquefaction fee, to send LNG to China with Singapore's EEN. Cheniere's offices outside of the US are

---

[55] Plante, L., & Nace, T. (2021). Nervous money: Global LNG terminals update. Global Energy Monitor.
[56] Kennedy, C. (2023, July 10). Labor shortages could hinder the U.S. LNG boom. Oil Price.
[57] Milman, O. (2022, September 22). How the gas industry capitalized on the Ukraine war to change Biden policy. *The Guardian.*

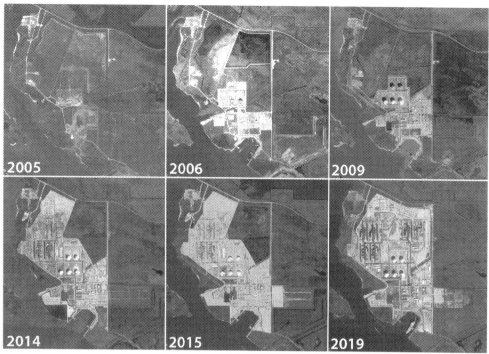

Figure 10.16 Sabine Pass LNG trains by segment. *(Sources: Google Earth, 2021 Maxar Technologies, United States Department of Agriculture (USDA) Farm Service Agency, Texas Orthoimagery Program, United States Geological Survey (USGS).)*

located in Asia and Europe, and even companies like Norway's Equinor that claim to be leading the energy transition seek to take advantage of this Gulf coast sacrifice zone.

In the US Gulf, petrochemical facilities are often co-located with LNG facilities. A common planning practice throughout the industry, multiple site emissions and pollution may be monitored site by site, but rarely as cumulative impacts in dangerous proximity to low-income areas and communities of color common throughout the US Gulf region (Chapter 1).

In its international expansion, the South African petrochemical company Sasol (Chapter 4) took particular advantage of the US Gulf region, also known as Cancer Alley. Sasol produces and sells chemicals in dozens of countries—generating profit off vulnerable geographies at home and abroad. In 2001, Sasol entered Louisiana in one of the poorest and most contaminated areas in the state.[58] Its dirty infrastructure, including numerous ground flares situated in large pit structures (Fig. 10.17) that emit concentrated levels of toxic contaminants,[59] would likely not have been permitted in an affluent area.

---

[58] Sasol. (n.d.). *Sasol North America.*

[59] Wu, X., Li, C. J., Jia, W. L., & Mu, J. C. (2019, March). A risk-effectiveness design method for flare height of natural gas pipeline system. In *International Petroleum Technology Conference*. OnePetro.

Figure 10.17 Ground flaring in a frontline community. *(Image credit: Ground Flare at Sasol's Lake Charles petrochemicals complex in Westlake, Louisiana, on December 10, 2021, Julie Dermansky.)*

Over a decade, Sasol's Lake Charles project developed into a mega-complex, similar to LNG and petrochemical facilities located throughout Cancer Alley.[60] For Sasol, however, this growth was not without significant complications. Due to debt, Sasol had to sell 50% of its chemical plant in a $2-billion partnership with LyondellBasell Industries.[61] An investor-led class action suit was filed after cost overruns, claiming Sasol lied about actual costs.[62]

The Lake Charles project itself has been plagued with operating malfunctions, including fires and shutdowns, and environmental impacts from two large hurricanes made worse by the facility's vulnerable coastal location. Sasol's public stance remains stoic, determined to come "back from the brink" despite ongoing financial and operational challenges. Highlighting its decarbonization efforts, Sasol has recently taken advantage of federal incentives for hydrogen offered in the 2022 Inflation Reduction Act (IRA).[63]

[60] Younes, L., & Bittle, J. (2023, August 2). 'Death stars on sinking land': How liquefied natural gas took over the Gulf Coast. *Grist.*

[61] Theunissen, G. (2020, October 2). A billion here and a billion there: How Sasol came to sell 50% of its Lake Charles Megaproject. News24.

[62] Stoddard, E. (2022, August 31). Sasol reaches $24m class action settlement over US project cost overruns. Daily Maverick.

[63] Helman, C. (2022, October 3). Back from brink, Sasol gets on the path to greener chemicals. Forbes.

Globally expansive petrochemical firms like Sasol promote and benefit from spin-off industries like single-use plastic production (Box 10.1)—a significant contributor to widening metabolic rift, all for a moment of convenience. Plastic produced from fossil fuels creates long and deep harm to the ecosystems on which we rely for life; new research finds accumulating microplastics within many species, including our own bodies.

BOX 10.1 Spin-off industry: single-use plastics

Plastic contamination has spread to all parts of the globe. The US is the world's worst plastic polluter, creating more plastic waste than Europe combined.[64] Successful lobbying from the oil and gas industry to use ever more plastic[65] has created a toxicity crisis.[66] The gas industry argues plastic can secure viability as gas demand from energy transition decreases.[67]

Production of plastics has increased dramatically since the 1950s.[68] Since the 1980s, plastic consumption has risen exponentially.[69] Most US plastic was discarded in landfills until the mid-1980s, with the advent of recycling. Through recycling expansion the majority of plastics and other post-consumer collected materials were exported. China was flooded with contaminated recycled materials and ended recycled materials imports in 2018. When this market disappeared, proposals for incineration and chemical reprocessing surged, creating additional crises. Mechanical recycling results in plastic pelletization and subsequently raw plastic materials, creating microplastic pollution in air, land, and water. Each of us has microplastic pollution in our bodies.

Chemical recycling processes—instead of physical recycling—are toxic and energy intensive. Some problematically argue that making plastic into more plastic could form a circular economy, or even **upcycling**, whereby reused discarded objects or materials form the basis for products of higher quality or value than the original.[70]

The circular economy requires renewable energy as its foundation to avoid the input and recycling of toxic materials, as with coal ash (Chapter 4). There is a crisis of microplastics in air and water around the globe,[71] partly due to recycling operations.[72] Sewage sludge has also

(Continued)

[64] Milman, O. (2021, December 1). 'Deluge of plastic waste': US is world's biggest plastic polluter. *The Guardian*.

[65] Liboiron, M. (2021). *Pollution is colonialism*. Duke University Press.

[66] Tickner, J., Geiser, K., & Baima, S. (2021). Transitioning the chemical industry: The case for addressing the climate, toxics, and plastics crises. *Environment: Science and Policy for Sustainable Development*, *63*(6), 4—15.

[67] Ellinas, C. (2023, March 6). Rising demand for plastics to drive oil and gas use in 2023. Natural Gas World.

[68] Nielsen, T. D., Hasselbalch, J., Holmberg, K., & Stripple, J. (2020). Politics and the plastic crisis: A review throughout the plastic life cycle. *Wiley Interdisciplinary Reviews: Energy and Environment*, *9*(1), e360.

[69] Ball, P. (2020). The plastic legacy. *Nature Materials*, *19*(9), 938—938.

[70] Zhao, X., Korey, M., Li, K., Copenhaver, K., Tekinalp, H., Celik, S., ... Ozcan, S. (2022). Plastic waste upcycling toward a circular economy. *Chemical Engineering Journal*, *428*, 131928.

[71] Tabuchi, H., Corkery, M., & Mureithi, C. (2020, August 30). Big oil is in trouble. Its plan: flood Africa with plastic. *The New York Times*, 30.

[72] Brown, E., MacDonald, A., Allen, S., & Allen, D. (2023). The potential for a plastic recycling facility to release microplastic pollution and possible filtration remediation effectiveness. *Journal of Hazardous Materials Advances*, *10*, 100309.

Box 10.1 Spin-off industry: single-use plastics—cont'd

become a source of microplastics in soil.[73] Microplastic waste has the potential to negatively affect the capacity of oceanic organisms like phytoplankton to capture carbon, harming a crucial global carbon sink. This propels a potential runaway feedback loop where warming itself can negatively impact phytoplankton's mitigation capacities.

Huge amounts of plastic also consolidate into garbage patches in oceans. Pervasiveness of marine litter[74] motivated the UN to deliberate a High Seas treaty.[75] There is a long history of bans, most prolific locally and regionally across Europe. Dozens of countries have banned single-use plastic bags, usually countries without strong plastic industry lobbyists.

Isolated plastic bans are ultimately too fragmented to address the scale of the problem. Other means of reduction are even more piecemeal, often focusing on individuals and their consumption or waste disposal choices.[76] By its nature, plastic pollution has proven best tackled with a global treaty:[77] a UN plastic resolution is currently being negotiated as a critical effort to address the scale of the problem.[78] Plastics pollution extends beyond geography—it is so diffuse that preventative upstream policies are necessary yet require collaborative state and non-governmental action.[79]

The building blocks of plastic are mainly fossil fuels, resulting in a GHG footprint vastly larger than the material itself.[80] Plastic is a hazard to human and planetary health.[81] Bioplastic replacements are indeed viable to nearly, if not all, fossil plastics. However, plastics manufactured from bio-based polymer alternatives need their own rigorous life cycle analysis to avoid negative tradeoffs in energy use, land use, or waste disposal.[82]

Once plastic and other refuse has been created, what are some downstream strategies to reduce its harm? Trash pickers provide value by diverting trash for reuse, yet they are

[73] Stubenrauch, J., & Ekardt, F. (2020). Plastic pollution in soils: Governance approaches to foster soil health and closed nutrient cycles. *Environments*, 7(5), 38.

[74] United Nations Environmental Program. (2021). *Drowning in plastics.*

[75] Kim, J., & Treisman, R. (2023, March 7). *What to know about the new U.N. high seas treaty — and the next steps for the accord.* National Public Radio.

[76] Kolb, N. (2023, June 22). The world needs a global plastics treaty championing urgent action. Will we get one?.

[77] Pekow, C. (2022, February 2). As the world drowns in plastic waste, U.N. to hammer out global treaty. *Mongabay.*

[78] United Nations Environmental Program. (2022, March 2). *What you need to know about the plastic pollution resolution.*

[79] Munhoz, D. R., Harkes, P., Beriot, N., Larreta, J., & Basurko, O. C. (2023). Microplastics: A Review of Policies and Responses. *Microplastics*, 2(1), 1—26.

[80] Zhu, X. (2021). The plastic cycle—an unknown branch of the carbon cycle. *Frontiers in Marine Science*, 7, 1227.

[81] Symeonides, C., Brunner, M., Mulders, Y., Toshniwal, P., Cantrell, M., Mofflin, L., & Dunlop, S. (2021). Buy-now-pay-later: Hazards to human and planetary health from plastics production, use and waste. *Journal of Pediatrics and Child Health*, 57(11), 1795—1804.

[82] Rosenboom, J. G., Langer, R., & Traverso, G. (2022). Bioplastics for a circular economy. *Nature Reviews Materials*, 7(2), 117—137.

frequently scorned.[83] The important value of their role expands due to the climate crisis,[84] yet they are forced to fight for recognition and dignity.[85] Energy projects can displace trash pickers,[86] who are vulnerable to limited access to garbage due to chemical recycling and incineration. How can communities support workers who divert waste to establish healthier and safer jobs in the reuse economy while recognizing their existing contributions and protecting their rights?

Try this

Plan and act: *what is the most effective means to end single-use plastics? Would you focus on limiting supply or demand? What solutions to reduce plastics do you propose for where you reside?*

Situated below extraction and transmission but before plastic's consumption and refuse, a type of manufacturing plant called an ethane cracker transforms ethane—a component of fossil gas—into ethylene, a critical building block of plastics. Often clustered in hotspots, crackers take advantage of "wet" gas that contains natural gas liquids (NGLs) and water vapor along with methane. NGLs (i.e., ethane, propane, and butane) are utilized in downstream industries creating feedstock for plastic.

Cracker plants exhibit high energy use and bring toxic burdens to an area. Along the US Gulf coast, LNG facilities and cracker plants cluster in discernible spatial patterns adjacent to vulnerable populations, natural resources, and waterways (Fig. 10.18). Port Arthur, situated below Beaumont, TX in Fig. 10.18, was discussed in the Introduction as a hub and sacrifice zone facing a public health crisis. Today, the surrounding area hosts a number of LNG sites and massive cracker plants—nicknamed the Titans of Plastic.[87] Petrochemical firms were attracted to Texas by tax incentives—leading to the withholding of billions of dollars from budgets of local schools over decades. Texas' Chapter 313, a two-decade-old tax abatement, operated as a form of corporate welfare.[88] In real dollar terms, petrochemical industries have been the largest Chapter 313 beneficiaries. Promised jobs and other benefits designed to offset tax abatements never fully materialized. Expensive handouts were paid for by taxpayers until the unpopular program finally ended in 2022.

---

[83] International Alliance of Waste Pickers. (n.d.). *Home.*

[84] Chang, J. (2008, March 25). Climate change improves image of trash pickers. *Spokesman.*

[85] Boudreau, C. (2022, April 13). The waste picker fighting for global recognition. *Politico.*

[86] O'Hare, P., & Fernandez, L. (2022, November 30). Waste pickers risk their lives to stop plastic pollution—Now they could help shape global recycling policies. Phys.org.

[87] Marusic, K. (2022, September 15). These are the new Titans of Plastic. *Sierra Magazine.*

[88] Woods, E. (2023, February 2). Critics say state tax break helps petrochemical companies and hurts public schools. *The Texas Tribune.*

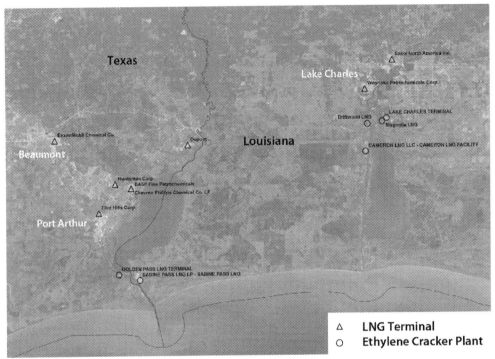

Figure 10.18 LNG and cracker plant co-location. *(Sources: Esri, Maxar, Earthstar Geographics, CNES/ Airbus DS, USDA FSA, USGS, Aerogrid, IGN, IGP and the GIS User Community. U.S. Energy Information Administration (EIA), The Federal Energy Regulatory Commission (FERC), U.S. Census Bureau.)*

Cracker plants are exceedingly toxic operations, most in the US exhibit abysmal environmental track records, with more than 50% in the country being tracked by the EPA as "high priority violators"—many of these are located in Cancer Alley.[89] Cheap plastics carrying high health and environmental costs represent a broad regulatory failure.[90] Foreign firms are often end recipients of incentive structures designed to bolster particular regions seeking to become hubs. Although consistently concentrated in the Gulf, the Marcellus Shale glut (Chapter 8) has recently shifted plastic production northward.[91] Shell's Ethane Cracker Plant in Beaver County, PA, took more than a decade to develop

[89] Bernhardt, C. (2022, September 20). Plastics industry boom brings flood of new ethylene "cracker" plants, despite frequent environmental violations. Oil and Gas Watch.

[90] Sicotte, D. M. (2020). From cheap ethane to a plastic planet: Regulating an industrial global production network. *Energy Research & Social Science, 66*, 101479.

[91] Fuhr, L., & Franklin, M. (Eds.). (2020). *Plastic Atlas 2109* (2nd ed.). Heinrich Böll Foundation/Break Free from Plastic. Heinrich Böll Foundation.

Figure 10.19 Shell polymers Monaca Complex. *(Image credit: Ted Auch, FracTracker Alliance, 2019.)*

(Fig. 10.19 above). Building this plastics production hub was controversial from its planning inception in 2009;[92] regardless, Pennsylvania committed $1.65 billion in tax breaks to attract the $3 billion dollar plant.

As of 2023, the plant stands completed (Fig. 10.20), but during its first 6 months of operation, a troublesome number of unscheduled contamination events occurred repeatedly. The plant was taken off-line for a few weeks after being cited for flaring and wastewater violations.[93] Even prior to its completion, violations were issued to Shell, and within the first year of operation dozens of malfunctions were reported.[94]

---

[92] Murtazashvili, I., Rayamajhee, V., & Taylor, K. (2023). The Tragedy of the Nurdles: Governing Global Externalities. *Sustainability*, *15*(9), 7031.

[93] Pierce, R. (2023, May 18). Shell Cracker Plant in Beaver County will be in 'shutdown mode' for weeks to come. Channel 11 News.

[94] Glabicki, Q. (2023, August 8). Inside Pennsylvania's monitoring of the shell petrochemical complex. Inside Climate News.

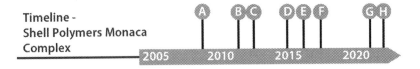

A - Company website launch
B - Shell Chemical Appalachia LLC announces Shell Polymers
    Monaca (SPM) petrochemical complex
C - Delays due to supply concerns
D - Air and wastewater permits granted
E - Final approval; Pennsylvania gives largest state subsidy
    ($1.65 billion tax credit) to the complex project
F - Construction begins
G - Complex operational
H - 43 Startup and full operation malfunctions plague complex
    within 1+ year operation; complex fined $10 million

Figure 10.20 Infrastructuring the cracker timeline.

A local community watchdog organization and monitoring program called "Eyes on Shell" (Conclusion) grew out of concern for local family safety in proximity to the plant.[95] Residents created a platform to live-feed air quality information and receive crowdsourced data documenting incidents. Grassroots efforts like these are increasingly common in jurisdictions where regulators and politicians fail to protect public health and wellbeing. Industry's capture of those formally responsible for monitoring means that impacted communities often incur additional burdens to document pollution and sound the alarm.

Extreme Qatari LNG

Qatar defies physical limits and historical constraints to compete for global LNG markets.[96] A relative newcomer to LNG markets, Qatar LNG is exceptionally expensive, requiring major investments. Fig. 10.21 shows Qatar's primary megaplex where its LNG activities have been concentrated and intensified.

[95] Beaver County Marcellus Awareness Community. (n.d.). *Eyes on shell.*
[96] Wright, S. (2017). Qatar's LNG: Impact of the changing East-Asian market. *Middle East Policy, 24*(1), 154—165.

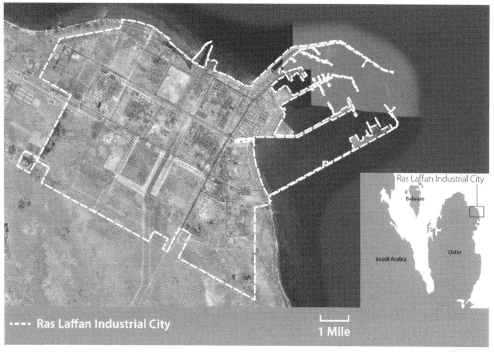

Figure 10.21 Qatar's LNG megaplex, Ras Laffan. *(Source: Natural Earth, Esri, Maxar, Earthstar Geographics, CNES/Airbus DS, USDA FSA, USGS, Aerogrid, IGN, IGP, and the GIS User Community, Ras Laffan Industrial City extracted via OpenStreetMap (Overpass Turbo API).)*

Qatar has leveraged its LNG as part of a strategic plan to move past human rights and labor boycotts that emerged around the hosting of the 2022 FIFA World Cup.[97] Boycotts of the international soccer tournament emerged from concerns about Qatar's systematic abuse of labor, as more than 6,500 migrant workers were reported to have died during construction for the World Cup. LNG helped Qatar avoid debilitating costs incurred as a result of the boycotts. In this process, Qatar struck a deal with China,[98] which helped Chinese energy enterprises acquire more control over the global market at a time of high LNG competition. Chinese buyers then resold many of the cargoes to the highest bidders in Europe and Asia.[99]

[97] Oxford Analytica. (2021, March 26). Qatar will lean on core LNG strength post-boycott. Expert Briefings.

[98] Mills, A., & El Dahan, M. (2023, June 20). Qatar strikes second big LNG supply deal with China. Reuters.

[99] Stapczynski, S. (2023, February 20). China taking control of LNG as global demand booms. Rigzone.

During this same period, "Qatargate" became a leading scandal, upsetting prior notions of the integrity of the European Union (EU). Corruption was revealed for all to see as unethical relationships including bribery between member countries only complicated Europe's ongoing crisis in energy geopolitics.[100] With Russian gas deemed unethical due to Russia's war with Ukraine, European countries, particularly Germany, have looked to Qartar as a quick fix. As political scandal clashes with Europe's reliance on gas, the very democratic system on which the whole of EU democracy rests appears weakened.[101]

Beyond Europe alone, Qatar is set to massively increase LNG exports, requiring lock-in of LNG buyers tethered to long-term contracts,[102] even as global markets become oversaturated with LNG production.[103] QatarEnergy—a state-owned petroleum operation, and one of the largest global oil and gas companies —is consolidating production through a partnership with Chevron Phillips via an ethane cracker at the Ras Laffan site; similarly, a joint venture between the two is slated to open in 2026 in Orange, Texas.[104]

QatarEnergy's push for new global partners has relied on greenwashing (Conclusion), claiming to have a greater positive impact than evidence demonstrates. Qatar is doubling down on LNG, advertising gas as climate-friendly in spite of its high GHG emissions. Satellite data demonstrates that Qatar has been emitting heavily and continues routine flaring,[105] as depicted in Fig. 10.22.

After building quickly for a decade to secure LNG market share, Qatar began its greenwashing campaign as a national climate strategy that would, among other goals, improve the carbon intensity of gas operations.[106] This fanfare in advance of the UNFCCC COP26 climate change summit mimicked other Gulf Arab states, including Saudi Arabia, which announced a net-zero emission target by 2060. These efforts operate as strategic moves to keep gas in play in spite of calls from most countries to fully phase out fossil fuels.

Qatar's north field expansion reportedly will include a carbon capture and storage (CCS) facility (Chapters 2 and 10 discuss CCS further). This vision is by no means reality any time soon, taking years to develop and demonstrate viability. Academics receiving funding from Qatar Petroleum and Shell are using models to simulate geological

[100] Cooper, C., & Zimmermann, A. (2022, December 19). Qatar scandal gives Europe a big gas headache. *Politico*.

[101] Lynch, S. (2022, December 19). Anatomy of a scandal: How 'Qatargate' crisis shook EU to its core. *Politico*.

[102] Calabrese, J. (2023, April 10). Qatar doubles down on LNG amid energy market volatility. MEI.

[103] Stapczynski, S. (2023, May 23). Clock is ticking for Qatar to sell its LNG. Bloomberg.

[104] Conoco Philips. (n.d.). *Ras Laffan Petrochemical Project*.

[105] World Bank. (n.d.). *Global Gas Flaring Data*.

[106] Mills, A. (2021, October 28). Qatar targets 25% cut in greenhouse gas emissions by 2030 under climate plan. Reuters.

Figure 10.22 Qatar LNG. *(Image credit: Ras Laffan LNG Terminal, Qatar - Matthew Smith, 2012. Accessed via Flickr.)*

conditions in the hopes of bringing experiments to material reality.[107] Media coverage unquestioningly repeats future storage promises and Qatari claims of sequestration success. Whether or not the actual tons of carbon are successfully stored, the timing of this positive spin creates a "green halo effect" to promote additional LNG expansion. Yet it's naive to assume benefits unless demonstrated through rigorous independent evaluation.

Try this

Research, monitor, and act*: International accords aim to end routine flaring by 2030 but compliance is uncertain.[108] What remote or in-person strategies do you recommend to monitor efforts toward this goal? How will you communicate your findings?*

[107] Qatar Carbonate and Carbon Storage Research Centre. (n.d.). *Welcome to the QCCSRC.*
[108] Wen, Y., Xiao, J., & Peng, J. (2023). The effects of the "Zero Routine Flaring by 2030" initiative: International comparisons based on generalized synthetic control method. *Environmental Impact Assessment Review, 100,* 107095.

Mozambique's blood LNG

When local populations outright rejected an LNG facility sited in a notorious conflict zone, the country's president asked for,[109] then received, international support to instead expedite the project.[110] This power grab represents a too-common pattern in Africa, as exemplified with Nigerian oil (Chapter 6). LNG is no exception to energy violence.[111]

Prior to putting a shovel in the ground, Mozambique's World Bank-financed mega-LNG plant[112] harmed villages,[113] contributing to instability and violence in a region long plagued by conflict as demonstrated in Fig. 10.23.[114] Conflict near the border with Tanzania was well known before the multi-billion-dollar investment was proposed. Since 2017, an Islamist-linked violent insurgency has attacked facilities in the area, including two armed attacks on LNG operations. First, in 2019, armed men opened fire on an Anadarko Petroleum convoy driving on a road near its LNG project. Soon after, Anadarko sold its holdings in Mozambique to Total Energy. Insurgents ambushed and killed eight Total subcontractors traveling by van that same year.[115] The violence against energy companies had many contributing factors.[116] This was a zone of food insecurity experiencing a cholera outbreak. Local populations were physically and economically displaced. Windfalls that were promised hadn't arrived, contributing to hostility toward an external workforce.

As shown in Fig. 10.24, the Mozambique LNG project will occupy a huge swatch of acreage; it also stands as the largest debt financing project in Africa's history.[117] To

[109] Oxford Analytica. (2020, February 20), Mozambique may look to partners to bolster security. Expert Briefings.

[110] Whitehouse, D. (2020, November 27). Mozambique LNG prospects lifted by international response to Islamist insurgency. The Africa Report.

[111] Staff. (2021, May 30). Myanmar: Chevron, Total suspend some payments to Junta. Eurasia Review.

[112] Touissant, E. (2020, December 23). Climate and the environmental crisis: Sorcerer's Apprentices at the World Bank and IMF. CADTM.

[113] DeAngelis, K., Rawoot, I., Lemos, A., Rabeiro, D., & Bhatnagar, D. (2019, March 19). Mozambique LNG destroys villages and the environment. Rainforest Action Network.

[114] Macuane, J. J., Buur, L., & Monjane, C. M. (2018). Power, conflict and natural resources: The Mozambican crisis revisited. African Affairs, 117(468), 415–438; Hanlon, J. (2021). Ignoring the Roots of Mozambique's War in a Push for Military Victory. Conflict Trends, 2021(2).

[115] Nhamire, B., & Hill, M. (2020, December 25). Mozambique repels insurgent attack near Total's 20b LNG facility. World Oil.

[116] Makonye, F. (2020). The Cabo Delgado Insurgency in Mozambique: Origin, ideology, recruitment strategies and, social, political and economic implications for natural gas and oil exploration. African Journal of Terrorism and Insurgency Research, 1(3), 59–72.

[117] Staff. (2020, July 17). Total signs 14.9 billion debt financing for huge Mozambique LNG project. Reuters.

Figure 10.23 Mozambique conflict. *(Sources: The Armed Conflict & Event Data Project (ACLED) Dataset, Natural Earth, Esri Maxar, Earthstar Geographics, CNES/Airbus DS, USDA FSA, USGS, Aerogrid, IGN, IGP, and the GIS User Community.)*

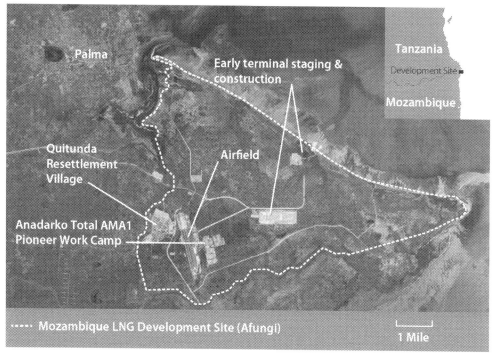

Figure 10.24 Mozambique LNG development site. *(Source: Natural Earth, Esri, Maxar, Earthstar Geographics, CNES/Airbus DS, USDA FSA, USGS, Aerogrid, IGN, IGP, and the GIS User Community, Development Site extracted via OpenStreetMap (Overpass Turbo API).)*

develop the project amidst ongoing conflict, the project employed French ex-military personnel,[118] an integral component of fossil fascism. Not unexpectedly, continued acts of terrorism stifled the project's timeline.[119] Total also damaged its reputation with local communities by formalizing an association with a government whose security forces are known to violate human rights. Outcry over extrajudicial executions—including the shooting of a woman by forces allegedly tied to the LNG site security—were met by yet more violence from the state and allied South African troops.

Mozambique's mega-LNG project is just one instance where international law for human rights falls short on effective corporate accountability. Corporations and their actors assume the protections of **extra-territoriality**, meaning diplomatic immunity achieved through international law, overriding local jurisdiction. Investors and partners consider themselves "*adjacent to*" criminality or immorality. This position allows investors to reap benefits while ducking responsibility for corruption and repression. Like so-called "blood" diamonds harvested through violence across global conflict regions, fossil gas embodies a similar form of violence in its sites of global operation.

Research and compare: What justification was given to build Mozambique's mega-LNG facility in an existing conflict zone, displacing local populations? What are the best and worst case scenarios for energy facilities in areas of violent conflict?

Canada LNG blockades

Historically, Canada has been one of the world's most expansive energy extraction locations, yet its global LNG sector has lagged in development.[120] A decade ago, as many as 20 export facilities were proposed in Canada; only a few have moved forward, particularly along its west coast.[121] Indigenous communities maintain their active presence, dating back millennia, surviving trauma induced by residential schools dating back to the 19th century coupled with campaigns of cultural genocide.

The aggressiveness of large-scale construction and repeated backsliding on environment and climate policy has only broadened social critique and resistance within Canada,

[118] Bowker, T. (2020, September 4). Mozambique projects adjust to life in a war zone. *Petroleum Economist*.

[119] Reed, E. (2021, April 20). East African LNG outlook seems positive in 2021 but fluctuating prices, political instability and terrorism risk stifling growth. Energy Voice.

[120] Nie, Y., Zhang, S., Liu, R. E., Roda-Stuart, D. J., Ravikumar, A. P., Bradley, A., … Bi, X. T. (2020). Greenhouse-gas emissions of Canadian liquefied natural gas for use in China: Comparison and synthesis of three independent life cycle assessments. *Journal of Cleaner Production, 258*, 120701.

[121] Incorrys. (2022, March 7). *West Coast Canada LNG Export Projects*.

1997 - Delgamuukw Supreme Court hands down landmark decision recognizing that aboriginal title is not extinguished in the areas claimed by Wet'suwet'en and Gixtsan
2012 - LNG Canada selects TC Energy (TransCanada) to design, build, own and operate Coastal GasLink[128]
2018 - Environmental permits granted and Coastal GasLink announces it will proceed with project construction
2019 - BC Supreme Court grants Coastal GasLink an injunction calling for the removal of any obstructions to work sites the company has been authorized to use.
2020 - The Wet'suwet'en First Nation serves Coastal GasLink with an eviction notice as workers are trespassing on their unceded territory, backed by national protests[129]
2022 - TC Energy puts in place an 'equity' agreement with a few Bank Councils

Figure 10.25 Timeline of the campaign.

fortified by global solidarity. Locally, the Wet'suwet'en matriarchal leadership[122] has utilized infrastructure chokepoints for both Native resistance with connections to international solidarity in the face of the Coastal GasLink Pipeline Project.[123] Part of a regional LNG infrastructuring plan, the 670-km Coastal GasLink pipeline is designed to fuel an LNG Canada liquefaction plant at Kitimat driven by a Shell-led investment and construction consortium with state-controlled Malaysian, Chinese, and Korean companies.

Caught in harm's way, Indigenous Peoples' subsistence territories including trapping lines and smokehouses are unvalued and erased by colonial energy firm practices. In the Coastal GasLink project, lead company TC Energy has repeatedly tried to manipulate and circumvent territorial rights—a clear extension of colonial violence (Fig. 10.25 above), leading to several standoffs and a reinvigoration of Native self-governance at a bioregional scale.[124] Resistance camps have been erected along the proposed pipeline route. One of the oldest, Unist'ot'en Camp, erected in 2009 as a checkpoint, has expanded into a full camp, with a healing center and bunks for visitors, who are invited according to terms of Free, Prior and Informed Consent (FPIC) (Chapter 7).[125]

---

[122] van Meijeren Karp, A. (2020). *More than a pipeline.* Doctoral dissertation, Lund University.
[123] Diabo, R. (2020). Canada's Colonial Web Entraps Wet'suwet'en Nation. *Indigenous Policy Journal,* *31*(1).
[124] Unist'ot'en Camp. (2019). *Yinka Dini — People of this Earth/Unist'ot'en — People of the Headwaters.*
[125] Unist'ot'en. (n.d.). *Come to the Land.*

For more than a decade, the Canadian state has claimed superiority in defense of private interests, even as violations of Indigenous and human rights have occurred. To form, the state issued an injunction order in 2019 against a Wet'suwet'en blockade. In response, Wet'suwet'en hereditary chiefs issued their own eviction notice to the Coastal GasLink project.[126] Despite competing orders, Canadian police enforced just the state injunction, arresting dozens during a raid.[127] The Crown declined to prosecute because of lack of protocol in the conduct of arresting officers. Ongoing protests from Wet'suwet'en and allies persist as the standoff continues (Fig. 10.25 above).[128,129]

The struggle in Wet'suwet'en territory connects to broader Land Back movements.[130] TC Energy (formerly TransCanada) has a long track record of unethical operations with historic and recent environmental racism and water violations.[131] In 2020, a **domino effect** occurred—a flashpoint that sets off a chained, cumulative reaction—in this case shutting down ports and blocking rail lines connecting the major cities (Fig. 10.26). Solidarity with Wet'suwet'en showed an emergence of network-level tactics not before seen at this scale.[132]

Police arrested 43 protesters in this solidarity action, who were preventing access to Vancouver ports. Other activists occupied government buildings and held rallies in Newfoundland, Labrador, and Alberta. Native pipeline organizers jumped scale to catalyze a **boomerang effect** through transnational advocacy networks.[133] A boomerang effect happens when, in spite of national attempts to clamp down on a social movement, its message appeals to international advocacy and human rights networks that captures media and policy attention—thus swinging back to magnify local efforts.

Nevertheless, like other examples in this chapter, the Canadian state doubled down on fossil fuels in spite of public outrage and protests, and the Royal Bank of Canada

[126] Unist'ot'en Camp. (2019). *Timeline of the campaign.*
[127] Democracy Now!. (2021, November 19). *Canadian police raid Wet'suwet'en pipeline blockage, arrest 15 land defenders.*
[128] Coastal GasLink. (n.d.). *About Coastal GasLink.*
[129] Gobby, J., Temper, L., Burke, M., & von Ellenrieder, N. (2022). Resistance as governance: transformative strategies forged on the frontlines of extractivism in Canada. *The Extractive Industries and Society, 9,* 100919.
[130] Barrera, J. (2020, November 25). Beyond the Barricades. CBC News.
[131] Trumpener, B. (2021, December 6). Coastal GasLink failed to fix nearly 2 dozen environmental violations along pipeline route, B.C. officials say. CBC News.
[132] Staff. (2020, February 11). Indigenous pipeline blockages spark Canada-wide protests. BBC News.
[133] Keck, M. E., & Sikkink, K. (1999). Transnational advocacy networks in international and regional politics. *International Social Science Journal, 51*(159), 89–101.

Figure 10.26 Solidarity protests. *(Image source: Wet'suwet'en solidarity banner near Vaughan, Ontario, February 15, 2020. Accessed via Wikimedia Commons.)*

appears an unbending funding stalwart.[134] Activists are not backing down either—masked assailants attacked LNG site equipment causing millions of dollars in damages.[135]

A continuation of past acts associated with oil and gas alone, Canada's pipeline problems now extend to LNG development.[136] In response, frontline organizers have been effective in their use of social media to protect protesters (Conclusion). Funders increasingly face reputational damage;[137] and solidarity movements continue to exert pressure on financial institutions to divest funds.[138] A particular target, China, has been named by Native resistors and networked allies as an active instigator—the greatest volume of LNG is destined for the Asian country.

In 2022, after years of conflict throughout Indigenous territory, TC Energy established agreements with formal First Nations band councils along the route, signing option

---

[134] Staff. (2023, April 13). Canadian bank named world's largest fossil fuel financier. Al Jazeera.

[135] Singh, A., & Nickel, R. (2023, February 1). The cost of a controversial B.C. pipeline keeps rising, leading to plunging share prices. CBC News.

[136] Lauerman, V. (2021, December 21). LNG Canada hits pipeline problem. Petroleum Economist.

[137] Janzwood, A., Neville, K. J., & Martin, S. J. (2022). Financing energy futures: the contested assetization of pipelines in Canada. *Review of International Political Economy*, 1—24.

[138] Stand.Earth. (2022, November 1). *150+ organizations demand banks befund Coastal GasLink, respect Wet'suwet'en rights.*

deals for potential sale of a 10% stake to just two Indigenous groups representing a total of sixteen.[139] Not representative of all, unfitting to customary Indigenous cultures or geographies, TC Energy deemed the deal "historic," spinning its accomplishment as a true net positive.[140] In fact, TC Energy resorted to negotiation only after years of aggressive confrontation. Referencing conflict with industry and state, Wet'suwet'en hereditary Chief Woos spoke to other chiefs and representatives at the Onondaga Longhouse at Six Nations of the Grand River—he described harm created by Coast GasLink as,

> **"… trauma, it has come in the form of having
> our own people fight amongst each other."[141]**

Coastal GasLink's benefit agreements with selected band councils along the project's route in 2018 is a textbook example of **lateral violence**. Wet'suwet'en hereditary leaders have declared that band councils do not have authority over traditional territories beyond reserve boundaries.[142] The band council system is a product of the Indian Act.[143] As such, governments of BC and Canada claim agreements with elected band councils constitute consent. However, Supreme Court cases recognized traditional governance forms, including the hereditary chief and clan system, on traditional territories. Elected band councils have limited jurisdiction only over reserve lands. Hereditary chiefs have jurisdiction over traditional territories.[144] Chiefs, clans, and house groups are responsible to the land and the people and can be removed if they fail to fulfill their duties.[145]

Like Tar Sands oil precedence discussed in Chapter 5, Canada continues to shows its hand as a petro-state and perpetrator of genocide through LNG development, disregarding the essential role of Native guidance. The revolving-door (Chapter 5) tendencies of firms like TC Energy and government regulators bring into question the independence of state assessments and commitments.[146] Beyond its social conflicts and colonial

[139] Staff. (2022, July 28). Cost of Coastal GasLink pipeline jumps 70% to 11.2B as TC Energy settles dispute. Money Market Advisor.

[140] TC Energy. (2022, March 9). *Press Release: TC Energy signs equity option agreements with Indigenous Communities across the Coastal Gaslink project corridor.*

[141] McCullough, K. (2022, August 2). 'Landmark discussion' at Six Nations begins Wet'suwet'en hereditary chiefs' nation-to-nation tour. *The Hamilton Spectator.*

[142] Trumpener, B. (2022, July 7). Wet'suwet'en leader charged with criminal contempt over Coastal Gaslink Pipeline blockade. CBC News.

[143] Suzuki, D. (2019, January 17). Pipeline blockage is a sign of deeper troubles. The David Suzuki Foundation.

[144] McPhail, S. (2018, September 26). The wrong chiefs are signing pipeline benefit agreements. Terrace Standard.

[145] Lee, J. (2016, August 17). Haida strip two hereditary chiefs of titles for supporting Enbridge. *Vancouver Sun.*

[146] Meyer, C. (2022, October 26). How oil and gas lobbyists build 'very close relationships' with politicians and governments. The Narwhal.

infractions, Canada's LNG sector is facing financial headwinds[147] and dissension within; a dispute between TC Energy and LNG Canada has resulted in yet another costly over-run.[148] Like all fossil fuel projects in the era of climate crisis, Canada's LNG infrastructure faces the possibility of stranded asset status as its breakeven time is extended ever outward.

Try this

Research and compare: Identify hidden energy debt and productive exclusion from LNG Canada. How does infrastructuring this project and related sites of extreme energy create a metabolic rift involving social, cultural, and ecological harms?

Australia's coastal LNG hubs

Australia is a top-tier LNG exporter and the global leader in LNG capacity. Even so, spatial imbalance exists across the country, where most LNG export projects are concentrated in the north, while cities southward often rely on high-priced import gas.[149] Construction offshore and along distant shores has energy firms reliant on an outside workforce, while these fragile coastal environments are poorly suited for mega LNG facilities (Fig. 10.27).

Like many transnational firms dead set on continued production by any means, Australia's Gargon LNG project stands as a test case for carbon capture and storage (CCS).[150] The $3 billion dollar CCS development, which received $60 million in federal funding, was delayed for years due to technical setbacks, and it ultimately missed its 5-year target.[151] This particular LNG plant now faces calls to shut down.[152]

The Gargon LNG facility is similar to others due to the spatial and landscape incongruence of extreme energy in fragile environments creating ecological harm as well as social conflict. The project was initially proposed at $37 billion, but the budget jumped to $54 billion, making it the costliest private sector project in Australia's history.

[147] Singh, A., & Nichol R. (2023, February 1). The cost of a controversial B.C. pipeline keeps rising, leading to plunging share prices. CBC News.
[148] Tuttle, R. (2022, July 28). TC Energy raises gas pipeline cost nearly 70% to $9 billion. Bloomberg.
[149] Hepburn, S. (2019, October 19). Australia has plenty of gas but our bills are ridiculous. The market is broken. The Conversation.
[150] Fernyhough, J. (2022, February 8). Australian LNG group unveils giant outback carbon capture reservoir. *Financial Times*.
[151] Morton, A. (2021, July 19). 'A shocking failure:' Chevron criticized for missing carbon capture target at WA gas project. *The Guardian*.
[152] Cox, L. (2021, January 14). Western Australia LNG plant faces calls to shut down until faulty carbon capture system is fixed. *The Guardian*.

Figure 10.27 Mega LNG facility on the Australian coast. *(Image credit: Lock the Gate Alliance, 2013. Accessed via Flickr.)*

From its inception, the megaproject was risky due to the selection of a marine site. Costs grew for navigation aids and pens to contain boat traffic following an incident with a submarine collision in the project corridor. Difficult logistics and supply to the area slowed down operations; ultimately workers were scapegoated for a 2-year delay.[153] Building multiple large LNG facilities simultaneously invariably pressures project workforces, and wage pressure increases on construction firms themselves.

Gargon, like international peers, witnessed the effects of the coronavirus pandemic, with worsening conditions as states encouraged "flexibilities," reducing standards for protection. Labor unions were demonized for communicating concerns over safety; in the case of Gargon, the loss of effective site management led to workers experiencing mercury poisoning.[154] During this time, lack of coordination caused frequent delays across

[153]   Ellem, B. (2021). Labor and megaprojects: Rethinking productivity and industrial relations policy. *The Economic and Labor Relations Review, 32*(3), 399–416.

[154]   Lewis, J. (2021, May 3). Second mercury poisoning incident reported at Chevron's Gorgon LNG project. Upstream.

projects with companies responding by reducing workforce numbers.[155] With unequal access to media, union blaming was all too easy amongst delayed contract agreements to restart projects.

As short-term disruptions like the coronavirus pandemic hamper project construction schedules, global sites, including Australian LNG hubs, are being developed without any meaningful climate guidance in a rapidly heating world.[156] Managing climatic heat stress on construction sites is increasingly difficult—companies are unprepared to keep workers safe in extreme weather, most simply by increasing break times.[157] As the planet's temperature rises, workers' break times are instead cut as part of a neoliberal squeeze that ignores worker wellbeing to the exclusive benefit of profit margins.

As in other locations where workers are insecure, labor unions try to defend domestic jobs in Australia. Firms developing offshore LNG in Australia received significant pushback from organized labor in the Offshore Alliance, a partnership between the Australian Workers' Union and the Maritime Union of Australia.[158] Labor disputes and strikes are ongoing at Shell's Prelude LNG facility located offshore of Australia.[159] In response, Shell planned a lockout to disrupt worker power—another form of lateral violence—designed to win its labor battle.[160] Prelude is co-owned by Shell along with Asian partners including Korea Gas Corp and a subsidiary of Chinese Petroleum Corp. Complex investment partnerships like that of the Prelude disperse liability and insulate energy firms from local or national pressure to improve social impacts.

LNG strikes have taken place in Europe as well, particularly France. Starting in 2016 as an extension of labor disputes at nuclear and gas plants, unions began to block LNG terminals.[161] Later, in 2022, workers went on strike at an LNG facility due to complaints over conditions, temporarily shuttering the facility for several days.[162] In 2023, strikes expanded, and this time were directed at the government's pension overhaul, as workers halted production at oil refineries and blocked imports from entering LNG facilities for weeks on end.[163]

[155] Ogge, M., & Campbell, R. (2021). When the going gets tough … the gas industry sacks workers. The Australian Institute.

[156] Jotzo, F., & Mazouz, S. (2019, June 18). Australia's energy exports increase global greenhouse emissions, not decrease them. The Conversation.

[157] Jia, A. Y., Rowlinson, S., Loosemore, M., Gilbert, D., & Ciccarelli, M. (2019). Institutional logics of processing safety in production: The case of heat stress management in a megaproject in Australia. *Safety Science, 120*, 388–401.

[158] Offshore Alliance. (n.d.). *We're here for you.*

[159] Staff. (2022, July 26). Australian unions extend strike at Shell's LNG facility. Reuters.

[160] Paul, S. (2022, July 20). Shell pay battle escalates at Prelude LNG of Australia. Reuters.

[161] Staff. (2016, May 22). French union to block LNG terminals. *Energy Intelligence.*

[162] Staff. (2022, June 30). Unions say strike halts flows at France's second largest gas depot. Reuters.

[163] Crellin, F. (2023, March 29). Factbox: French strikes hit refining, LNG imports. Reuters.

In spite of industry claims to promote good, stable jobs, LNG is a culmination of coal, oil and gas before it as workers are harmed, deemed low priority relative to international profit. Firms continue practices which position workers as replaceable units, using strike-breaking tactics as opportunities arise. This pernicious form of energy colonialism extended through LNG is as common now as it was in epochs prior. Fossil capital seeks to outmaneuver calls for just transition by discursively re-aligning justice with the "common good" of fossil fuels[164]—even while eroding possibilities for actual human survival. As countries like Australia intensify infrastructuring in a global race to secure LNG market share, behind the scenes the LNG industry often undercuts host nations. The fact is unavoidable: Australian LNG is not necessarily a domestic good; it is majority foreign-owned, extracting and producing a product designed specifically as export beyond host borders.[165]

Try this

Discuss and compare: *Research LNG union action in France, Canada, and Australia. What solutions would you propose for these labor concerns in the context of energy transition and climate risk?*

Freedom molecules?

Extreme energy sites across the globe are marked by either too little or too much water, either a constraint or emancipation of their extractive and market potential. Apart from the sloganeering of "Freedom" molecules, the molecular scale matters too. In the case of fossil gas and its liquefied state as LNG, escaping CH_4 rapidly intensifies a heating atmosphere. For hydrogen production, its method of separation literally makes a world of difference. Globally, the melting Arctic creates new spaces to extend violence and metabolic rift. As one of many amplifying feedback loops (Chapter 1) in this text, the liquid Arctic makes way for yet more gas extraction. Such feedback loops contain cascading effects which pose emerging global risks, including existential risks of ecological and societal collapse.

Until this point, the one large-capacity LNG plant operating in the Arctic region— Yamal LNG (Chapter 1)—is being joined by others as a result of an increasingly liquid Arctic. The liquid Arctic is both a knowledge and infrastructural frontier of accumulation, a so-called trillion-dollar ocean with logistical spaces of extraction, circulation,

---

[164] Goods, C. (2022). How business challenges climate transformation: an exploration of just transition and industry associations in Australia. *Review of International Political Economy, 29*(6), 2112—2134.

[165] Australian Workers Union. (2017). *Shipping away our competitive advantage.*

and securitization.[166] LNG changes traditional gas markets by extending beyond the 20th-century's pipeline model.[167] Arctic LNG sites seek to take advantage of melting ice to open new routes.[168]

Extreme energy based in hyper-extractivism has major territorial implications.[169] The Yamal project required colonization of Indigenous territories.[170] In 2014, Yamal became strategically important for Russia because of sanctions.[171] Further sanctions due to the Ukraine war slowed down supply to other planned Russian LNG sites. China stepped in to supply the Yamal LNG site after disruptions from sanctions.[172]

Crossing permafrost makes operations more challenging and requires extra equipment, often expensive and unproven, including gigantic FLNGs and LNG transshipment terminals.[173] With extreme energy of unprecedented size and scope, the liquid Arctic is not only an expression of metabolic rift, but a material fact allowing fossil gas speculation.

Other so-called solutions relying on gas also mislead. In contrast to blue hydrogen from fossil gas, dark green hydrogen—made from water molecules—represents high-quality renewables, a preliminary goal under Germany's push toward hydrogen. Germany's subsequent backsliding serves as a cautionary example; promotion of hydrogen energy prior to sound and tested technology leads to confusion.[174] The country plans to launch a hydrogen network with more than 1,800 km of pipelines by 2030, partially through funding expected from an European Union member state industrial funding venture—The Important Projects of Common European Interest (IPCEI). Designed primarily to invigorate microelectronics and communication technologies, this sprawling infrastructuring plan points to vague notions of "greener" technologies and transition,

[166] Watts, M. J. (2022). Arctic Abstractions. *Arctic Abstractive Industry: Assembling the Valuable and Vulnerable North, 5*, 187.

[167] Yermakov, V., & Sharples, J. (2021). *A phantom menace: is Russian LNG a threat to Russia's pipeline gas in Europe?* Oxford Institute for Energy Studies.

[168] Wang, F., Li, G., Ma, W., Wu, Q., Serban, M., Vera, S., … Wang, B. (2019). Pipeline—permafrost interaction monitoring system along the China—Russia crude oil pipeline. *Engineering Geology, 254*, 113—125.

[169] Watts, M. J. (2021). *Hyper-extractivism and the global oil assemblage: Visible and invisible networks in frontier spaces.* https://doi.org/10.4324/9781003127611-16.

[170] Gavrilova, K. (2023). Arctic LNG production and the state (the case of Yamal Peninsula). In *The Siberian World* (pp. 273—284). Routledge.

[171] Lavrenteva, N. (2020). Energy Policy in the Arctic: Yamal LNG in Russian International and domestic political agenda. *The Journal of Cross-Regional Dialogues/La Revue de dialogues inter-régionaux.*

[172] Humpert, M. (2023, May 22). China to supply key turbines to Novatek's Arctic LNG 2. *High North News.*

[173] Humpert, M. (2022, September 9). Western sanctions delay ppening of arctic LNG 2 project by one year. *High North News.*

[174] Alkousaa, R., & Kraemer, C. (2023, July 26). Germany's updated hydrogen strategy sees heavy reliance on imported fuel in future. Reuters.

not particular and sound energy policy. In place of domestically sourced wind and solar the plan opens the door to hydrogen imports by ship. By signing hydrogen cooperation agreements with exporting countries including Canada, United Arab Emirates and Australia, Germany becomes party to hydrogen made from fossil fuels.

Germany is also developing a direct blue hydrogen pipeline from Norway, which Norway promotes as less leaky and more climate-friendly than US gas.[175] Ironically, in 2023, Equinor, a company majority owned by the government of Norway, recently signed a long-term contract for LNG from the US Gulf.[176]

Norway's Equinor has made a name in carbon capture. Norway aims to capture carbon from blue hydrogen and store it in the seabed. This could change the ocean chemistry when CO_2 mixes with seawater, and may harm marine life. Norway's Longship carbon capture and storage (CCS) project at the Northern Lights experiences unpredictable subsurface conditions.[177] Demonstration areas have seepage. Yet technology is not only the challenge—a key impediment is exceedingly high costs. Hype around the Northern Lights CCS is used to prop up new fossil fuels expansion, but the site appears unlikely to achieve its proposed mitigation.

German leaders plan to roll out subsidies for renewable hydrogen, offering to make a percentage from electrolyzers, but these percentages are relatively low and unlikely to offset the harm from blue hydrogen containing more GHG emissions than the direct use of gas.[178] While countries like Norway and Germany accelerate hydrogen from offshore wind, regardless the initial source, due to hydrogen's inherent properties, leakage and volitility are currently not fully accounted.

Through its partnership with Norway, Germany has cited preferences for carbon to be severed from hydrogen imports and stored. This strategy erroneously suggests that CCS is a truly viable technology, a fact not yet demonstrated at a significant scale necessary for climate mitigation (see Chapters 4 and 9). Germany's plan expands fossil gas, thus abnegating GHG mitigations proposed from transition to hydrogen. This case points to the need for clear restrictions against the continual intensification of fossil fuels in the transition to renewables.

Extreme energy invariably extends metabolic rift, and false solutions, including CCS, do not repair the split. Based on scientific evidence, LNG and blue hydrogen—both claimed as ameliorations of climate risk and pathways to energy freedom—only intensify

[175] Damman, S., Sandberg, E., Rosenberg, E., Pisciella, P., & Johansen, U. (2020, February 14). *Largescale hydrogen production in Norway — Possible transition pathways toward 2050*. SINTEF.

[176] Reuters. (2023, June 21). *Cheniere signs LNG supply agreement with Equinor for Sabine Pass expansion*.

[177] Hauber, G. (2023, June 14). *Norway's Sleipner and Snøhvit CCS: Industry models or cautionary tales?*

[178] Newborough, M., & Cooley, G. (2020). Developments in the global hydrogen market: The spectrum of hydrogen colors. *Fuel Cells Bulletin, 11*, 16—22.

climate disruption and stymie energy transformation. In the conclusion section to follow, five strategic interventions are explored that both confront the mechanisms of energy violence and lay the foundation for energy justice.

Spatial and temporal synopsis

Space

(1) LNG is an extreme energy type with firms competing for its infrastructure lock-in.
(2) Liquefaction lengthens supply chains while contributing to metabolic rift, and in particular methane rift.
(3) LNG makes fossil gas mobile yet still faces spatial and social constraints like planetary limits, labor shortages, and regional conflict.

Time

(1) LNG markets and technologies move much more quickly than coal did in an example of time-space compression.
(2) Hidden energy debt includes external costs over time and space.
(3) False solutions accelerate a methane rift while delaying energy democracy and sector-wide transformation.

Summary

LNG has become a preferred global fuel due to its mobility across space. Climate disinformation about LNG aids its lock-in as a particularly costly infrastructuring while expanding adjacent false solutions including CCS and blue hydrogen, accelerating the methane rift even further. LNG perpetuates hidden energy debts over space and time that are ignored in rapidly changing markets and technologies. Over a longer term, LNG is subject to spatial and time constraints, i.e., climate disruption, despite its enhanced storage properties. In the end, LNG is not a viable solution to address the climate crisis, rather serving to further delay energy transformation.

Vocabulary

1. accumulation by dispossession
2. boomerang effect
3. breakeven times
4. domino effect
5. energy ladder
6. extraterritoriality
7. floating rigs

8. fuel stacking
9. green halo effect
10. hidden energy debt
11. lateral violence
12. leapfrog
13. metabolic rift
14. methane slip
15. resettlement
16. spatial emancipation
17. stop loss point
18. time-space compression
19. upcycling

Recommended

Liboiron, M. (2021). *Pollution is colonialism*. Duke University Press.
Maddow, R. (2019). *Blowout*. Crown.
Shaw, C. (2023). *Liberalism and the challenge of climate change*. Taylor & Francis.

Conclusion: Mapping fossil energy's deadly grasp

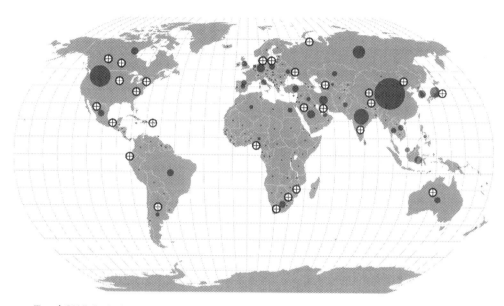

Total GHG Emissions per Country, 2021

MtCO₂e

— 12000
— 5000
— 1000
— 100

⊕   **Research Sites**

Research Sites + Emissions. *(Sources: Climate Watch data, Climate Watch, 2022, GHG Emissions - Washington, DC: World Resources Institute. Natural Earth.)*

Mapping fossil energy's deadly grasp

Decisions over the fate of the world are too often left in the hands of fossil energy interests—those most responsible for damaging emissions—all the while shifting blame and the worst consequences onto those least responsible: a vast majority of humans and other sentient beings. If it seems unfair, it is.

Climate change is no longer seen as just an "issue"—our climate is the context within which all else now unfolds. To date, climate change has often been treated as a niche interest to be considered and contained within fields like geography, sustainability, or environmental studies, when in fact climate breakdown is relevant to everything. **Climate justice** involves fairness in mitigation and adaptation, providing a framework for understanding climate change as a political problem rather than merely technical or material challenges. A justice-centered approach requires transforming our current systems to address inequity. An essential component of transformation is the reduction of unfair harm to frontline communities living and working in unsafe pollution hotspots.[1] Transforming our energy supply from fossil fuels to renewable sources is not only essential for a livable future, where we have an opportunity to live safely and healthier; when done well, renewable systems benefit workers and all species relying on air and water. Placing human and species rights at the top of climate mitigation is necessary as climate change doubles as ecocide.

The Most Affected Peoples and Areas (MAPA)—the millions who will be most exposed to unlivable heat and destructive rising seas—have been exploited for centuries. Previously colonized people and areas are unprepared to survive the ravages of climate change,[2] even with GHG mitigation commitments laid out in the 2015 Paris climate accords (which virtually no countries are presently meeting).[3] Yet world leaders enter climate negotiations touting energy transition, only to leave with watered-down goals and a new round of handouts for industry's expensive false solutions. Thousands of global climate activists pressure tirelessly for urgent reforms with necessary ethical safeguards; yet negotiators fail to deliver accords capable of addressing the crisis at hand.

We cannot have a safe planet without climate justice.[4]

For MAPA areas to succeed in a time of climate breakdown, reparative justice is essential.[5] Regardless of immense ecological debt owed to MAPA, financial agencies still draw vast quantities of debt interest monies from MAPA countries as the hold of colonialism and imperialism remains strong. Low-wealth countries have little choice outside loans to keep their economies afloat—an economic practice that has increased dramatically since 2010 with more than 50 countries in debt crises, forced to spend multiple times more on repayments than on addressing the climate crisis. Meanwhile, international

[1] Lenferna, G. A. (2018). Can we equitably manage the end of the fossil fuel era? *Energy Research & Social Science, 35*, 217–223.
[2] Táíwò, O. O. (2022). *Reconsidering reparations.* Oxford University Press.
[3] Rockström, J., Gupta, J., Qin, D., Lade, S. J., Abrams, J. F., Andersen, L. S., ... & Zhang, X. (2023). Safe and just Earth system boundaries *Nature*, 1–10.
[4] Hood, M. (2023, May 31). 1.5C of warming is too hot for a just world: study. Phys.org.
[5] Zodgekar, K., Raines, A., Jacobs, F., & Bigger, P. (2023). A Dangerous Debt-Climate Nexus: In a warming world, technocratic fixes are inadequate responses to vulnerability. Achieving climate and economic justice in the Caribbean calls for reparative solutions. *NACLA Report on the Americas, 55*(3), 319–326.

finance institutions (IFIs) such as global banks require indebted countries to commit repayment through future fossil fuel revenues, perpetuating the insidious cycle of dirty energy.[6]

Climate injustice is rooted in time as well as location. As top income earners produce more GHGs than the rest of the world, climate injustice ensues: the wealthiest reap benefits from their outsized emissions, while the rest of the world faces greater instability and growing risk. People who are older than 50 have seen a majority of the GHGs in the atmosphere released during their lifetimes and have reaped the benefits. While youth experience gains along with others from prior and existing infrastructure, they bear gigantic costs, including the prospect that their lives will be shortened dramatically. Scientific evidence of this irrecoverable loss has been publicized for more than a decade,[7] but has yet to spur a moral or practical response from leaders. While industrialized countries have worked to extend human life in wealthy nations, this translates directly into fewer resources for the rest of the world and for future generations.[8]

Lack of investment in youth
symbolizes abandonment of our future.

Inadequate response from public leaders damages public trust.[9] Young people experience higher rates of eco-anxiety than older generations.[10] A global survey found that 59% of adolescents and young adults report feeling "very" or "extremely" worried about climate change, and over 45% report that this interferes with their daily functioning.[11] Distress about climate change correlates with perceptions that governments are failing to respond adequately, culminating in a sense of betrayal. Youth also report that their

[6] Ahmed, K. (2023, August 21). Rich countries 'trap' poor nations into relying on fossil fuels. *The Guardian.*

[7] Hansen, J., Kharecha, P., Sato, M., Masson-Delmotte, V., Ackerman, F., Beerling, D. J., & Zachos, J. C. (2013). Assessing "dangerous climate change": Required reduction of carbon emissions to protect young people, future generations and nature. *PloS one, 8*(12), e81648; Ehrlich, P. R., & Ehrlich, A. H. (2013). Can a collapse of global civilization be avoided? *Proceedings of the Royal Society B: Biological Sciences, 280*(1754), 20,122,845.

[8] brown, A.M. (2022, June 23). *On Being* podcast: We are in a time of new suns.

[9] Hickman, C., Marks, E., Pihkala, P., Clayton, S., Lewandowski, R. E., Mayall, E. E., ... & van Susteren, L. (2021). Climate anxiety in children and young people and their beliefs about government responses to climate change: a global survey. *The Lancet Planetary Health, 5*(12), e863−e873.

[10] Dooley, L., Sheats, J., Hamilton, O., Chapman, D., & Karlin, B. (2021). Climate change and youth mental health: Psychological impacts, resilience resources and future directions. See Change Institute.

[11] Hickman, C., Marks, E., Pihkala, P., Clayton, S., Lewandowski, R. E., Mayall, E. E., van Susteren, L. (2021). Climate anxiety in children and young people and their beliefs about government responses to climate change: A global survey. *The Lancet Planetary Health, 5*(12), e863−e873.

concerns are dismissed[12]—even as worries about climate change shape life decisions regarding where to live, what careers to pursue, and whether to have children.[13]

Effective energy justice can be laid out broadly in five key interconnected methods (Fig. C.1). A compressed timescale for its fulfillment is now necessary as decades of insufficient mitigation have passed, resulting in accelerating ecological breakdown. Facing our predicament requires envisioning steps to reduce harm and lessen anxiety and helplessness. Energy democracy (Chapter 9) requires an informed, active electorate. Education is at the heart of energy justice; societal embrace of disinformation contradicts accurate historical and scientific fact, both integral to climate justice. Action prompts integrated throughout each chapter of this text as well as the Praxis provide suggestions to continue learning in line with a living laboratory approach.[14] Just as energy injustice is part of everyday life, its solutions are too.

Method	Procedures to advance energy justice
1	Acknowledge systemic denial and delay
2	Expose structural violence
3	Identify viable alternatives, educate and mobilize
4	Divest-reinvest
5	Confront fossil fuel malfeasance

Figure C.1 Energy justice—Five interconnected methods to rewire society

Method 1—Acknowledge denial and delay

How did we get here? Prior chapters show repetitive structural violence in coal, oil, and gas operations over the past century, duplicating similarly oppressive and extractive patterns with each fuel. With widespread systemic violence, the current predicament should surprise no one. **Climate breakdown** is based upon willful ignorance. The trigger and multiplier is rising temperatures driven by burning fossil fuels. Breakdown translates as all-encompassing ecological disruption with the prospect of societal collapse, all driven by industrial GHG emissions. Reckless political leaders are feigning ignorance of **overshoot**—we are surpassing planetary limits through air pollution, water extraction,

[12] Hickman, C. (2020). We need to (find a way to) talk about... eco-anxiety. *Journal of Social Work Practice*, 34(4), 411–424.
[13] Wray, B. (2022). *Generation Dread: Finding Purpose in an Age of Climate Crisis*. Knopf Canada.
[14] Bouzarovski, S., Damigos, D., Kmetty, Z., Simcock, N., Robinson, C., Jayyousi, M., & Crowther, A. (2023). Energy justice intermediaries: Living Labs in the low-carbon transformation. *Local Environment*, 1–18.

chemical use, and more.[15] Ecological footprint calculations show we are using resources at a rate that would necessitate what we do not have—access to multiple planet Earths.[16]

Overshoot implies more than overusing resources alone—we have poisoned and made less resilient the very ecosystems we rely on for life. In an ecosystem, resiliency refers to the ability to withstand and bounce back from typically singular disturbances, such as fire. Management practices that prioritize profits can undercut entire ecosystems; for example, Canadian forest fires that displace communities and burn megatons of stored carbon are intensified by decades of industrial forestry with grooming for timber species and use of herbicides like glyphosate.[17]

Across the globe, species are undergoing mass extinction.[18] Biodiversity is being lost rapidly with an **ecological cascade** effect: a series of secondary extinctions that are triggered by the primary extinction of a key species in an ecosystem. A clear example of the damages of overshoot is found in our oceans, which make up the majority of the Earth's surface. Oceans in our era have operated as unmanaged dump sites, leading to the collapse of marine biodiversity, ocean acidification, and loss of necessary processes of carbon sequestration.

Planetary overshoot considers what we remove or diminish as well as what we add—introduction of novel entities (chemical pollution) is a risky and poorly understood form of planetary overshoot.[19] Chemicals are difficult to study given our incomplete knowledge of components and their synergistic effects, compounded by the high pace of their production, outstripping the capacity for assessment and monitoring.[20]

Global boiling is the current state of our planet where many terrestrial and marine locations experience brutally hot temperatures. It is no longer adequate to say global warming or even global heating when the planet is literally burning with thousands of wild fires blazing out of control and ocean heat spiking at an unprecedented rate. Fast violence is evident in sudden fires, like occurred in Maui, Hawaii (US) in 2023.[21] Victims

[15] Wackernagel, M., Schulz, N. B., Deumling, D., Linares, A. C., Jenkins, M., Kapos, V., … & Randers, J. (2002). Tracking the ecological overshoot of the human economy. *Proceedings of the National Academy of Sciences*, *99*(14), 9266–9271.

[16] Global Footprint Network. (n.d.). What is your ecological footprint?

[17] English, J. (2019, November 24). Grooming forests could be making fires worse, researchers warn. *CBC*.

[18] Kolbert, E. (2014). *The sixth extinction: An unnatural history*. A&C Black.

[19] Rockström, J., Steffen, W., Noone, K., Persson, Å., Chapin III, F. S., Lambin, E., … & Foley, J. (2009). Planetary boundaries: Exploring the safe operating space for humanity. *Ecology and Society*, *14*(2).

[20] Persson, L., Carney Almroth, B. M., Collins, C. D., Cornell, S., De Wit, C. A., Diamond, M. L., … & Hauschild, M. Z. (2022). Outside the safe operating space of the planetary boundary for novel entities. *Environmental Science & Technology*, *56*(3), 1510–1521.

[21] Mai-Duc, C. & Caldwell. A. A. (2023, April 16). In maui wildfire, many fear children are large share of the dead. *The Wall Street Journal*.

experience post-traumatic stress disorder (PTSD) after such deadly episodes.[22] Extreme heat—the new normal—also represents slow violence with increasing droughts leading to crop loss.[23] Losses across multiple grain-producing areas are predicted with large-scale sea-surface temperature oscillations now present in ocean waters.[24]

Extreme heat is a period of high heat and humidity with temperatures above 90° for at least 2—3 days. When a persistent ridge of high pressure traps heat over an area, it can create deadly **heat domes**.[25] While this extreme heat is temporary, the larger trend towards both increased heat generally coupled with frequent extreme heat events is now well established.

Increasing temperatures make jobs potentially fatal for workers during periods of high heat.[26] Even healthy and young agricultural, construction, and delivery workers are dying from heat stroke.[27] Heat exhaustion involves overheating to the point of heavy sweating, dizziness, nausea, and fainting. Heat stress occurs when the body cannot rid itself of excess heat. **Wet-bulb temperature** is a measure of heat stress; high wet-bulb temperatures, when the air is so saturated with water that sweat can no longer effectively cool the body, can be fatal after several hours.[28] Hyperthermia occurs when the body is absorbing or generating more heat than it can release. These conditions have great implications for people who labor outside and live where air conditioning is unobtainable. Children are at risk of heat-related illness because of the increased surface area of their body relative to their weight,[29] and the inability to gauge indicators.[30] Like children,

[22] Dooley, L., Sheats, J., Hamilton, O., Chapman, D., & Karlin, B. (2021). Climate change and youth mental health: Psychological impacts, resilience resources and future directions. See Change Institute.

[23] Kornhuber, K., Lesk, C., Schleussner, C. F., Jägermeyr, J., Pfleiderer, P., & Horton, R. M. (2023). Risks of synchronized low yields are underestimated in climate and crop model projections. *Nature Communications, 14*(1), 3528.

[24] Hasegawa, T., Wakatsuki, H., & Nelson, G. C. (2022). Evidence for and projection of multi-breadbasket failure caused by climate change. *Current Opinion in Environmental Sustainability, 58*, 101217.

[25] Henderson, S. B., McLean, K. E., Lee, M. J., & Kosatsky, T. (2022). Analysis of community deaths during the catastrophic 2021 heat dome: Early evidence to inform the public health response during subsequent events in greater Vancouver, Canada. *Environmental Epidemiology, 6*(1).

[26] Spector, J. T., Sampson, L., Flunker, J. C., Adams, D., & Bonauto, D. K. (2023). Occupational heat-related illness in Washington State: A descriptive study of day of illness and prior day ambient temperatures among cases and clusters, 2006—2021. *American Journal of Industrial Medicine.*

[27] Goodall, J. (2023). *The heat will kill you first: Life and death on a scorched planet.* Little, Brown and Company.

[28] Timperley, J. (2022, July 31). Why you need to worry about the 'wet-bulb temperature'. *The Guardian.*

[29] Piccone, J. (2021, June 3). Take heat-related illness seriously—especially in children. OSF Healthcare.

[30] Dotan, R. (2021). Determinants of exertional heat stroke: Are children and youth indeed more vulnerable? *Journal of Athletic Training, 56*(8), 801—802.

the elderly too are susceptible to heat-related deaths with increased risk due to cardiovascular, respiratory, or other preexisting health conditions.

Mitigation to reduce overall GHGs has not happened at the pace necessary, and the repercussions of our shortcomings are playing out on land and water, including in episodic crop failures and lower yields.[31] However, land-based heat has received more attention in comparison with oceans, which store heat and release it slowly.[32] The biosphere and atmosphere are undergoing rapid heating. Interconnected ramifications range from coral bleaching, to ocean acidification, to increased forest fires, to melting glaciers and sea level rise. Each degree of heating means more moisture is deposited into the atmosphere during increasingly severe weather events like storms, leading to flash floods and other forms of fast violence. Meanwhile, soil is depleted of moisture, a cruel slow violence for subsistence agriculturalists in particular. Another relatively fast violence event is a bomb cyclone, a powerful, rapidly intensifying storm associated with a sudden and significant drop in atmospheric pressure.[33] Changes in global atmospheric circulation portend increasingly dramatic implications, such as severe atmospheric rivers delivering fast violence to coastal regions.

Discourses of climate delay

Started in 1995, the United Nations Framework Convention on Climate Change (UNFCCC) Conference of Parties (COPs) (Chapter 2) has set global climate policy. For more than a decade COP has been critically dubbed the "Conference of Polluters."[34] Skirting transformational commitments, wealthy countries stall progress.[35] The UN's failure to agree on equitable rules delayed necessary action for the first 2 decades of climate negotiations. This trend continues, but now the US, China, and Saudi Arabia—a select group of fossil fuel superpowers—repeatedly stall negotiations to phase out dirty energy. To solidify inaction, UNFCCC meetings have been held in countries without freedom of expression where protest is illegal—special spaces hidden from the general public are set off for sanctioned symbolic rallies that take the place of independent, disruptive demonstrations and actions. Concerns were widespread regarding human rights violations and climate delay by the 2023 host government, the United

[31] Rees. W. E. (2023). The human ecology of overshoot: Why a major 'population correction' is inevitable. *World*, 4(3), 509−527.

[32] Talukder, B., Ganguli, N., Matthew, R., Hipel, K. W., & Orbinski, J. (2022). Climate change-accelerated ocean biodiversity loss & associated planetary health impacts. *The Journal of Climate Change and Health*, 6, 100114.

[33] Mandel, K. (2022, December 23). 'Bomb cyclone' and other weird weather words for our climate change era. *Time*.

[34] Langelle, O., & Petermann, A. (2011, December 3). Looking back: Global day of action against UN conference of polluters. Global Justice Ecology Project.

[35] Clarke, R. (2022). Climate COP-up, COP-out & hyCOPrisy. *Theory & Struggle*, 123(1), 34−45.

Arab Emirates (UAE). Making matters worse, the CEO of an oil company—Sultan Al-Jaber of Abu Dhabi National Oil Company (ADNOC)—served as the UNFCCC COP 28 president. More than 100 members of the European Parliament and the US Congress signed a letter asking for his withdrawal as president of the 2023 COP 28.[36] Following the lead of Al-Jaber, the UAE announced a goal that is strategic greenwashing as it is not materially or economically feasible—Al-Jaber twisted around the former UNFCCC goals to phase out fossil fuels and instead invented a new delay tactic by announcing an unrealistic alternative goal to eliminate emissions while maintaining fossil fuels—even as costly technologies like CCS continue to fail. Seen as a strategic reversal, the timing of fossil fuel phaseout has always been a central issue at prior COPs. This wordsmithing by Al Jaber and the UAE garnered criticism by many, including the UN Secretary General,[37] who has constantly argued for greater ambition to avoid climate catastrophe.

Self-defeating myths that detain us from climate action have been labeled **discourses of climate delay** (Fig. C.2).[38] For example, a common delay strategy is to emphasize the downside to energy transition without considering the loss and damage from inaction.

The hope that some soon-to-be-developed innovation will miraculously save us without the hard work of structural change—termed **technological optimism**—is part of a culture of "soft climate denial." This version of denial relies on mythology about how we might decarbonize fossil fuels, contributing to false optimism, whereby people who understand full well the catastrophic implications of climate breakdown carry on blissfully, postponing urgently needed action. The **technological hubris** of the energy industry—a dangerous arrogance that firms will be able to control technology and use it to their advantage—lies in the assumption that technological "fixes" can make up for poor resource management and weak governance. Industrial society's naive keenness for new technology and its belief in easy solutions have been exploited as a dangerous form of delay. The overconsumption by wealthy countries and households is often hidden in emissions that have been off-shored to less privileged locations. The technological promise of the electric vehicle (EV)—ostensibly a mitigation strategy that nonetheless requires extensive mining for materials only to be plugged into a grid burning fossil fuels—is a prime example of technological optimism fallen short of full structural change.

Dirty energy hovers in every sector,
even the "green" economy.

[36] van Halm, I. (2023, May 24). EU and US politicians call for the removal of Al Jaber as COP28 president. *Energy monitor*.
[37] Nichols, M. (2023, June 15). UN chief to fossil fuel firms: Stop trying to 'knee-cap' climate progress. *Reuters*.
[38] Lamb, W., Mattioli, G., Levi, S., Roberts, J., Capstick, S., Creutzig, F., … Steinberger, J. (2020). Discourses of climate delay. *Global Sustainability, 3*, E17.

Figure C.2 Discourses of climate delay. *(Adapted from: Discourses of climate delay. Global Sustainability, 3, E17. Lamb, W., Mattioli, et al., 2020.)*

GHG mitigation—when distributed and holistic—works. Conversely, ineffective, technocentric fossil fuel-centered mitigation is a false solution. As discussed previously with coal (Chapter 2) and gas (Chapter 9), carbon capture and storage (CCS) is a global example of a false solution that underscores the need to reduce overall emissions rather than rely on net-zero fantasies. A popular proposal is to rely on CCS technologies to attempt to remove emissions from the atmosphere and return GHGs to a livable proportion. This suicidal plan relies on ignorance of the precarity of an even 1.5° temperature increase over pre-industrial periods, and on disinformation suggesting CCS is viable (Chapters 2 and 9). In reality, CCS is estimated to have about a 7% effectiveness, but even this low number doesn't fully account for the technology's high energy usage, canceling out potential benefit.[39] CCS has been demonstrated as ineffective at a large scale after ongoing expensive failures.[40] The majority of CCS used to date has been in enhanced oil recovery (EOR), with the goal to increase production and to extract yet more oil.[41] CCS costs more, increases overall fossil fuel use,[42] and grants the fossil fuel industry an undeserved social license to continue to operate.

Like CCS, global shipping exemplifies an industry riddled with contradiction and rebound effects. The sector's 2020 sulfur dioxide emission regulations brought immediate health benefits for those living near busy shipping lanes. Yet, ironically, the very sulfur particles polluting shipping routes produced actually helped offset global scale warming by blocking solar radiation. New shipping rules reducing sulfur in marine fuels created a "termination shock," in this case a temperature increase from global brightening that immediately replaced previous global dimming from atmospheric particulate matter (i.e., shipping's sulfur pollution). In 2018, prior to setting shipping rules reducing sulfur, scientists warned of the need to increase GHG mitigation to make up for this predicted termination shock.[43] Unsurprisingly, policy makers failed to accelerate the reduction of GHGs as recommended—an alarming, risky pattern of global climate policy.

Adaptation is an act or process of changing to better suit a situation. A growing tension around the term adaptation now exists as it's increasingly unclear if humans and the world's species have the capacity to avoid extinction in the face of rapid climate warming. Meanwhile, many climate mitigation and adaptation policies conveniently ignore structural violence,[44] meaning these limited responses to climate breakdown laden with trade-offs often intensify harms as much as, or even more, than they ameliorate impacts. Quantitative assessments to standardize vulnerability miss bias within their equations, dangerously undermining adaptation efforts for the most at risk.[45]

Maladaptation occurs when actions intended to encourage climate adaptation create more overall harm than the benefits they portend—also termed **impact paradox**.

[39] Science and Environmental Health Network. (n.d.). Carbon Capture and Storage Facts.
[40] Drugmand, D., & Muffett, C. (2021). Confronting the myth of carbon-free fossil fuels: Why carbon capture is not a climate solution. Center for International Environmental Law (CIEL).
[41] Joshi, K. (2021, April 14). Carbon capture's litany of failures laid bare in new report. *Renew economy.*
[42] Jacobson, M. Z. (2023). *No miracles needed: How today's technology can save our climate and clean our air.* Cambridge University Press.
[43] Roberts, K. B. (2018, February 6). Cleaner ship fuels will benefit health, but affect climate too. *Udaily.*
[44] Finley-Brook, M. & Holloman, E. L. (2016). Empowering energy justice. *International Journal of Environmental Research and Public Health, 13*(9), 926.
[45] Robinson, S. A., Roberts, J. T., Weikmans, R., & Falzon, D. (2023). Vulnerability-based allocations in loss and damage finance. *Nature Climate Change,* 1—8.

Maladaptation is all too common.[46] Here technological hubris and disaster capitalism pervade investment strategies that prioritize high profit margins over lessening climate risk, particularly in infrastructure-centered projects. More effective long-term strategies that challenge the status quo are needed. Those that center ecologically focused mitigation and behavioral adaptation, lowering GHG emissions, and improving equity and justice outcomes, avoid intensifying maladaptation.[47]

Long-term observers of climate negotiations have been forthright about the failure of governments to work together for the common good. With climate risks blinking red, scientist and educator messengers have been silenced and ignored for decades. The Introduction of this book focuses on the importance of education to break through disinformation. Reflection is central to education[48]—otherwise we miss important lessons and remain unable to move to the next steps of designing solutions and taking effective actions.[49] A **teachable moment** occurs when a person is open to learning and reflection, such as after a difficulty or challenge. Today, the degree of risks and who bears the greatest loss and damage is obvious. How we collectively learn from the unfolding climate crisis will in large part determine the extent to which we select effective strategies for mitigation and adaptation or continue prioritizing false solutions and maladaptation.

Try this

Analyze and act: *Record the words and actions of your government representatives during negotiations at the most recent UNFCCC COP. Would you describe these national and international commitments as sufficient? Contact the chief negotiator of your country and UN officials to express concerns. Write a letter to the editor to make this position public.*

Method 2—Expose structural violence

As GHG emissions and capitalism are inextricably linked, so too is the widening gap in inequality. Those who are paying daily for climate loss and damage on the frontlines are not those who profit most. For perspective, the wealthiest 10% of the US population are responsible for 40% of its national emissions—this is in large part due to the high GHG

---

[46] Reckien, D., Magnan, A. K., Singh, C., et al. (2023). Navigating the continuum between adaptation and maladaptation. *Nature Climate Change*. https://doi.org/10.1038/s41558-023-01774-6

[47] Singh, C., Reckien, D., Magnan, A. K., Orlove, B., Schipper, L., & Coughlan de Perez, E. (2023, August 21). Guest post: Gauging the success of climate change adaptation. *Carbon Brief*.

[48] Chatterjee, P., & Maira, S. (Eds.). (2014). *The imperial university: Academic repression and scholarly dissent*. University of Minnesota Press.

[49] Macy, J., & Johnstone, C. (2022). *Active hope (revised): How to face the mess we're in with unexpected resilience and creative power*. New World Library.

emissions rooted in stock market investments.[50] The carbon footprints of just a handful of billionaires alone, through "investment emissions," equal the carbon footprints of entire countries.[51] Another prime example of climate injustice occurs as home insurance companies withdraw from markets due to climate impacts.[52] A family's home, often a foundation of security, is now exposed to predatory markets in insurance alternatives, or simply no insurance at all.

Market-driven environmentalism reinforces inequality.

Climate justice is fundamentally different from greener capitalism.[53] **Market environmentalism** refers to the use of economic incentives and signals to determine waste and pollution management, monitoring our relationship with the Earth through a price lens. This transactional, self-interested approach sets **artificial boundaries** around households, institutions, and countries, ignoring the ecological and social connections that make us interdependent. Artificial boundaries are simplifications of the world largely for accounting and measuring purposes. This is one means of greenwashing, the use of **elastic metrics**, selecting parameters with bias to warp measurements and target a particular outcome or agenda.

When one's place is determined by market value, marginalized individuals and groups are forced to compete for space and agency, while those with power and resources use their privilege to continue to advance often at expense of the greater good. **Green neoliberalism** treats nature as capital and allows the private sector to voluntarily design responses to the climate crisis, which is the driving force behind the proliferation of ineffective offset mechanisms and narrow, technology-driven incrementalism. One of its oft-repeated terms—"sustainable development"—is frequently reduced to no more than oxymoronic sloganeering.

Myth of the level playing field

As politicians protect fossil fuel markets, geopolitical tactics, including sanctions and boycotts, are undermined. Following the start of Russia's 2022 invasion of Ukraine,

---

[50] Paddison, L. (2023, August 17). America's richest 10% are responsible for 40% of its planet-heating pollution, new report finds. CNN Business.

[51] Taino, S. (2022, November 7). Billionaires emit a million times more greenhouse gases than the average person, study finds. CNN.

[52] Copley, M., Hersher, R, & Rott, N. (2023, July 22). How climate change could cause a home insurance meltdown; Bogage, J. (2023, September 3). Home insurers cut natural disasters from policies as climate risks grow. Washington Post.

[53] Dawson, A. (2010). Climate justice: The emerging movement against green capitalism. *South Atlantic Quarterly, 109*(2), 313–338.

countries have continued to trade energy with Russia—either directly or through third parties—in spite of espoused commitments to international sanctions on Russia.

Policy makers repeatedly allow themselves great latitude for exceptions, conveniently avoiding hard decision-making and negotiation. One global example involves countries discarding GHG mitigation goals during discussions of war.[54] When world leaders are at war, which occurs continually, environmental accords are consistenly undermined. Furthermore, military readiness doesn't adequately consider extreme heat or climate planning, worsening exposure to fatal conditions from lack of preparation.[55] Essential conversations on climate change are not happening with the urgency necessary in military and defense sectors, with troubling practical and ethical implications. Direct emissions from war make up an estimated 6% of global GHG totals.[56] The indirect emissions of relocations and collateral damages are much higher. Buildout from military operations sends pricing and political signals for prolific investment, advantaging military contractors and governments which they service. Top on the list is the US, the largest arms seller, whose buyers are spread out amongst a hundred countries.[57] Unsurprisingly, the US military also tops the emissions list as the biggest single polluter.[58]

At the UN, vulnerable groups with the least political power tend to remain unfairly exposed to climate disruption. This imbalance is exemplified in the lack of political will to create a legal category for **climate refugees**—those forced to leave their homes due to sea level rise and other climatic shifts—a type of displacement now impacting every continent. Tribal relocation of Indigenous Peoples from homelands has happened all over the world, including the US, acutely in Louisiana's Isle de Jean Charles.[59] Temporary adaptations (i.e., sea walls, lifting homes up on stilts, etc.) are just that—temporary. A lack of effective adaptation intensifies **climate migration**—the primarily voluntary movement of peoples driven by fast or slow climate-exacerbated disasters, including abnormally heavy rainfalls, prolonged droughts, desertification, environmental degradation, sea-level rise, and cyclones.

---

[54] Kashwan, P. (2023). Ecocide. In *Dictionary of ecological economics* (pp. 143–143). Edward Elgar Publishing.

[55] Best, K., Stephenson, S., Resetar, S., Mayberry, P., Yonekura, E., Ali, R., … & Wolf, V. (2023). *Climate and readiness understanding climate: Vulnerability of US joint force readiness*. Rand Corporation.

[56] Claußen, A. (2022, January 8). War is a climate killer. *International Politics and Society*.

[57] Buccholtz, K. (2022, March 14). The world's biggest arms exporters. Statistica.

[58] Crawford, N. C. (2022). *The pentagon, climate change, and war: Charting the rise and fall of US military emissions*. MIT Press.

[59] Munster, J. (2023). Oil, indifference, and displacement: An indigenous community submerged and tribal relocation in the 21st century. *American Indian Law Journal, 11*(2), 5.

Faced with these serial disasters, climate migrants are made further vulnerable by narrow legal frameworks and criteria for refugee status. Under the UN 1951 Convention framework for refugees, status can be granted when climate adversity is coupled with armed conflict and violence, but not necessarily without this compounding condition of risk.[60] Within increasing environmentally induced migration, vulnerability is shaped and reproduced through histories of militarism, racial capitalism, and the geopolitics of fossil fuels. Colonial foundations are hidden through policy bias towards those able or willing to relocate, leaving many communities and traditional villages at an even more precarious disadvantage.[61] Women and children are often forgotten casualties of disasters as well.[62]

Today's climate policies insidiously benefit unscrupulous players. Without transformative governance, energy democracies often struggle with democratization while greening—often not doing either well.[63] Private sector market environmentalism also lacks democratic process.[64] Government avoidance of direct spending on environmental and social goods supports opaque incentive systems, such as tax credits. For example, in the US, federal credits "pay" private investors to invest in public infrastructure on the government's behalf, for which firms can then claim legal tax shelters. This closed-loop privatization model has often benefited the fossil fuel industry while imposing problematic costs that stymie renewable energy.[65]

A **fog of enactment** which involves a gap between policy goals and actual results,[66] is "a long, quiet process in which the language of bills is converted into the specificity of laws, and where interest groups and other actors can organize to gut even the strongest legislation." Wins become losses as historic legislative achievements can be turned into desultory, embarrassing failures.[67] Recently, the passage of the US Inflation Reduction

[60] United Nations High Commissioner for Refugees. (2020, October 1). Legal considerations regarding claims for international protection made in the context of the adverse effects of climate change and disasters.

[61] McVeigh, K. (2023, January 16). Lost for words: fears of 'catastrophic' language loss due to rising seas. *The Guardian.*

[62] Cutter, S. L. (2017). The forgotten casualties redux: Women, children, and disaster risk. *Global Environmental Change, 42,* 117–121.

[63] Bruggers, J. (2021, February 28). A legacy of the new deal, electric cooperatives struggle to democratize and make a green transition. Inside Climate News.

[64] Sovacool, B. K., Bell, S. E., Daggett, C., Labuski, C., Lennon, M., Naylor, L., ... & Firestone, J. (2023). Pluralizing energy justice: Incorporating feminist, anti-racist, Indigenous, and postcolonial perspectives. *Energy Research & Social Science, 97,* 102996.

[65] Knuth, S. (2021). Rentiers of the low-carbon economy? Renewable energy's extractive fiscal geographies. *Environment and Planning A: Economy and Space,* 0308518X211062601.

[66] Stokes, L. C. (2020). *Short circuiting policy: Interest groups and the battle over clean energy and climate policy in the American States.* Oxford University Press.

[67] Klein, E. (2020, October 6). How a climate bill becomes a reality. Vox.

Act (IRA) represents historical investments in renewables, but tradeoffs abound, propagating fossil fuels for at least another decade, eroding the gains of cleaner energy.[68]

Expanding fossil fuels during a climate crisis is increasingly viewed as a literal crime against humanity and **ecocide** — the destruction of the natural environment by deliberate or negligent human action.[69] In the current state of necropolitics, where market buoyancy of fossil fuel firms receives more attention than famines or droughts, the scope of our **future deathprint**—the predicted number of fatalities from contamination and global boiling—begins to emerge, and numbers are in the billions.[70] In Chapter 1, we presented the concept of an energy deathprint as a way to measure the lives put at risk by construction of a coal or gas plant. This measure can be extended to climate change. Fossil fuel companies like Exxon Mobil and broader industries like Koch Industries (Chapter 5) are responsible for extreme weather and other climate change-induced deaths. The excessive release of GHG with the knowledge of global warming represents no less than a crime against humanity.

The privilege of private fossil fuel interests over the collective fate of humanity is anything but equal. Continued expansion of fossil fuels will launch us into an unlivable future; not one major coal, oil, or gas company can present an adequate climate plan for even the next decade, many already reversing stated renewable targets. The status quo of fossil fuel expansion will continue apace if corporate climate goals rather than proactive deeds continue to find safe haven in regulatory gray areas.[71] Coupled with as much as $640 billion in government subsidies annually, energy companies are inordinately incentivized to continue "**business-as-usual**" (BAU), the continuation of current status quo financial practices and market trends, demonstrating a profound economic and power imbalance against alternative, sustainable energy development. BAU has led to the current potentially unsurmountable economic reality: extreme weather spurred by the climate crisis is costing millions of dollars per hour.[72]

[68] Brown, M., & Phillis, M. (2022, August 18). Climate change? The Inflation Reduction Act's surprise winner, the US oil and gas industry. *USA Today*.
[69] Jones, O., Rigby, K., & Williams, L. (2020). Everyday ecocide, toxic dwelling, and the inability to mourn: A response to geographies of extinction. *Environmental Humanities*, *12*(1), 388—405.
[70] Pearce, J. M., & Parncutt, R. (2023). Quantifying global greenhouse gas emissions in human deaths to guide energy policy. *Energies*, *16*(16), 6074; Lenton, T. M., Xu, C., Abrams, J. F., Ghadiali, A., Loriani, S., Sakschewski, B., … & Scheffer, M. (2023). Quantifying the human cost of global warming. *Nature Sustainability*, 1—11.
[71] New Climate Institute. (2022, February 7). Corporate climate responsibility monitor 2022.
[72] Carrington, D. (2023, October 9). Climate crisis costing $16m an hour in extreme weather damage, study estimates. *The Guardian*.

Method 3—Identify viable alternatives, educate, and mobilize

The Secretary General of the UN, Antonio Guterres, points out:

Solidarity is self-interest.

If we fail to grasp that, everyone loses.

While this text has laid out evidence of fossil energy's harmful impacts across global hotspots, it also operates as a toolbox for climate action—including legal action to break through the lack of enactment. At the same time, reparative actions can attend to discrimination, serving as benchmarks to assure past harms are not replicated in proposed "solutions." Black and Brown activists and professionals working or volunteering with green organizations often face discrimination.[73] At the same time, frontline populations can be exploited for fundraising purposes. Mainstream nonprofit organizations receive more of the funds for environmental justice work, while not-funded are the frontline community leaders who bear the burden of evidence gathering to counter pollution permits. These trends are part of an expansive fog of enactment that settles around stale structures of inequality.

Antiquated legal systems continue to inadequately address the climate crisis, inherently linked to unequal structures and privileges. Historically, climate litigation has been impeded in many jurisdictions by judges unwilling to grant people—even minors of age considered protectorates of the state—the legal standing to sue their government. While the court of public opinion is more proactive and justice-oriented than state judiciaries, climate malfeasants are finally beginning to be tried in the courts (Chapter 5). Most significantly, Portuguese youth have secured a case before the European Human Rights Court, where the court's rulings are legally binding on the 32 member countries; compliance failure results in hefty fines decided by the court.[74] Today, there are thousands of climate cases around the world, the majority of which have been filed since 2020.[75] Hundreds of Swedish youth, including Greta Thunberg, are plaintiffs in a lawsuit against their own government. In the US, former President Trump's Department of Justice (DOJ) filed petitions six times to impede the *Juliana v. The United States* case—a climate lawsuit filed by youth. Subsequently, the Biden Administration's DOJ has sought to dismiss the case[76] in light of the *Held v. Montana* (youth v. government) decision which found that minors do indeed possess a constitutional right to protection as trusts of the

[73] Johnson, S. K. (2019). Leaking talent: How people of color are pushed out of environmental organizations. Diverse Green.

[74] Girit, S. (2023, September 27). Climate change: Six young people take 32 countries to court. *BBC*.

[75] Dickie, G. (2023, July 26). Climate change lawsuits more than double in 5 years. *Reuters*.

[76] Corbett, J. (2023, July 7). 'Nothing short of outrageous': Attorneys for youth climate plaintiffs blast Biden DOJ. *Common dreams*.

state. Although *Held v. Montana* was a state-level case, the precedent stands nonetheless for other and future US climate lawsuits based on guaranteed children's protection.[77]

An international legal and judicial framework for action was strengthened with the UN's 2022 recognition of a right to a clean and healthy environment. In 2023, the UN pushed for even greater protections for young people. After consulting with more than 16,000 children in more than 120 countries, the UN Committee on the Rights of the Child updated their 1989 Convention on children's rights to confirm that children have a right to a clean, healthy, and sustainable environment.[78] Legal cases targeting governments and fossil fuel firms have been slow to gain speed, but with dozens of lawsuits in various courts,[79] are now shifting the conversation regarding climate reparations. Opportunities now abound for law schools themselves to gain experience and offer curricula in climate litigation.[80]

The title of Naomi Klein's award-winning book surveying climate change, "This Changes Everything," is often conveniently decoupled from its subtitle "Capitalism vs. The Climate."[81] Free markets are anything but free as the actual accumulating costs of GHG emissions are hidden behind greenwashing. The evidence of growing emissions cannot be ignored due to corporate propaganda and measurement regimes that fail to fully convey the scale of the problem. Even with the promise of advanced monitoring, the stakes for survival are being continuously raised. To describe our predicament, even the term Anthropocene that seemed so apt just a few short years ago is now challenged; the use of Thermocene, the era of fossil fuel-based environmental change, or the Necrocene, the age of mass extinction, are increasingly apt descriptors in this epic of global boiling.[82]

Today's carbon-intensive liberal societies privilege market controls for runaway emissions in spite of risks.[83] Political norms serve to maintain social and psychological guardrails that prevent necessary reforms in light of the climate crisis.[84] Politicians remain

---

[77] Stiffman. (2023, August 18). Behind landmark climate ruling in mont., a trailblazing nonprofit law firm and an army of youth activists.

[78] Conley, J. (2023, August 28). UN affirms climate crisis threatens children's rights, potentially bolstering youth-led lawsuits.

[79] Sabin Center for Climate Change Law. (n.d.). Global climate change litigation.

[80] Holder, J. (2015). Fossil free: Linking divestment campaigns in universities with legal education. *Environmental Law Review, 17*(4), 233−236.

[81] Klein, N. (2015). *This changes everything: Capitalism vs. the climate.* Simon and Schuster.

[82] Dawson, A., & Paik, A. N. (2023). Germinations: An introduction. *Radical History Review, 2023*(145), 1−11.

[83] Demaria, F., Kallis, G., & Bakker, K. (2019). Geographies of degrowth: Nowtopias, resurgences and the decolonization of imaginaries and places. *Environment and Planning E: Nature and Space, 2*(3), 431−450.

[84] Shaw, C. (2023). *Liberalism and the challenge of climate change.* Taylor & Francis.

trepidatious about challenging capitalism's hegemony and thus cling to technological optimism and compromised regulatory targets, all the while directing essential resources away from community-based and collaborative solutions. Recognizing that fossil fuels stand in the way of a livable future, and that political interests are too captured to lead, decentered approaches demand action, like the fossil fuel nonproliferation treaty—started among frontline Pacific nations—that continues to expand its global reach.[85] With energy transformation, opportunities for education, democracy, equity, and justice can fill gaps in status quo practices where interactions remain largely transactional.

Past is present and future

In many US cities, racially discriminatory lending policies called **redlining** created clusters of minority households living in areas limited to predatory investment options and deteriorated conditions. Redlined areas were also targeted with industrial uses resulting in dangerous air pollution.[86] Insurance companies and federal flood relief programs repeat discriminatory actions to protect white and wealthy communities and help them recover from disaster while abandoning communities of color. These trends were pointed out more than a decade ago[87] yet they persist, undermining local efforts to build resilience.

Persistent patterns of pollution and spatial racism exist in different geographies (i.e., MAPA, global South, urban, etc.). Racial and gender inequity further hamper the climate movement.[88] Whiteness dominates mainstream environmental organizations. Black expressions are undervalued, severing critical connections to history,[89] and Black ecologies and relevant lived experience are often ignored or silenced.[90] Acknowledgment of this discrimination is fundamental to reparation and restitution[91] and is integral to an equitable future.[92]

[85] Earth Island. (n.d.). Fossil fuel non-proliferation treaty.

[86] Cushing, L. J., Li, S., Steiger, B. B., & Casey, J. A. (2023). Historical red-lining is associated with fossil fuel power plant siting and present-day inequalities in air pollutant emissions. *Nature Energy*, 8(1), 52–61.

[87] Bullard, R. D., & Wright, B. (2012). *The wrong complexion for protection: How the government response to disaster endangers African American communities.* NYU Press.

[88] von Woerden, W. (2022, July 29). Why the climate justice movement should put decoloniality at its core. Resilience.

[89] McKittrick, K. (2006). *Demonic grounds: Black women and the cartographies of struggle.* University of Minnesota Press.

[90] Roane, J. T., & Hosbey, J. (2019). Mapping black ecologies. *Current Research in Digital History*, 2.

[91] McCutcheon, P., & Kohl, E. (2019). You're not welcome at my table: Racial discourse, conflict and healing at the kitchen table. *Gender, Place & Culture, 26*(2), 173–180.

[92] WECAN International. (2023). Gendered and racial aspects of the fossil fuel industry in North America and complicit financial institutions.

Whether a small municipality or even country, lack of capital is a common feature for communities of color facing the climate crisis. When Black townships in the US have access to loans, inherent higher interest rates mean capital costs more.[93] This same form of disparity in lending is a major global driver of structural violence. In 2022, an effort to reform IFI lending in the wake of the climate crisis—the Bridgetown Initiative 1.0—was introduced by Prime Minister Mia Mottley of Barbados.[94] It identified unfairness in "a great finance divide" whereby developing countries pay up to eight times more than developed countries for borrowing.[95]

Like most climate negotiations, financial aspects hold power over social, ecological, and political safeguards. A subsequent Bridgetown Initiative 2.0 was made palatable to big donors with ample status quo financialization, additionally undercutting the requirement of climate finance to be "new and additional" to existing aid.[96] As a debt-based model, the current policy increases the vulnerability of those most climate-impacted while diluting the rationale for climate finance rooted in the historical and ongoing obligations of GHG intensive, industrialized countries.

Information access, data emissions

Access gaps between rich and poor are entrenched and widening throughout the world. Inequality permeates society, but this is especially acute in the technology sector, inextricably dependent on electricity. Modern technology and the internet are how many of us receive information about climate—all the while the dirty manufacturing for computers and telecommunications equipment is offshored and disregarded. Similar to the material unevenness of the telecommunications sector—where those with the latest gadgets do not bear the worst costs and harms for these privileges—economic inequality defines access to information. Even when people have access to the internet, verifiable information is frequently locked behind paywalls.

A limiting factor of access to climate research and information about energy alternatives has been the control of information by big publishing houses that own the majority of academic journals[97] with an oligarchic structure organized around profit. Corporate

[93] Smull, E., Kodra, E., Stern, A., Teras, A., Bonanno, M., & Doyle, M. (2023). Climate, race, and the cost of capital in the municipal bond market. *Plos One, 18*(8), e0288979.
[94] Rasool-Ayub, H. (2023, April 4). Barbados's urgent call for a global climate finance plan. *New America*.
[95] United Nations Secretary-General. (2023). SDG Stimulus to deliver agenda 2030.
[96] Feminist Action Network for Economic and Climate Justice. (2023). Unpacking the bridgetown initiative: A systemic feminist analysis & critique.
[97] Jha, A. (2012, April 9). Academic spring: How an angry math blog sparked a scientific revolution. *The Guardian*.

capture of science education impedes the open access dissemination of knowledge (Box C.1).[98]

BOX C.1 Spinoff industry: Climate science publishers

Fee structures in scientific publishing have been proven "bad for science."[99] Researchers pay to publish in journals, and then other researchers pay to view the publications, so publishing companies profit twice. Paywalls limit access to those without the means to pay, including low-wealth readers and scientists from MAPA who have data to publish and who themselves experience heightened climate risk as a result of their constrained resources. Climate-oriented publications to and from low-wealth populations would have benefitted the world, yet the pressing need for climate action and education as part of broader structural transformation—climate justice—is late entering academic cultures.[100] This is in part due to the privileging of certain information as a result of profit and markets being granted credence to determine value and impact. For example, for-profit motives influence product offering and data access. The Dutch company Elsevier publishes leading climate journals, including the *Lancet* and *Global Environmental Change*, while simultaneously publishing materials that facilitate fossil fuel exploration and drilling.[101] Elsevier also markets data services and research portals that are used to expand oil and gas infrastructure and production. Such controversies with science and data publishers point to a broader concern about how research and publishing is too-often captured—there are often ties between oil companies, public universities, and regulatory agencies.[102] Repetition of industry talking points in some scientific literature, sometimes funded by industry firms or associations themselves, provides fodder for climate denial.

Pay walls cause misuse of government funds when publicly funded research is unavailable. The University of California and other schools have pushed back against Elsevier based on justification of the need for open access to data and findings.[103] In fact, Elsevier's fees structure and its negative repercussions for access to science information sparked an international boycott,[104] as scientists stated that they would not review nor publish in Elsevier journals.[105] Many

---

[98] Resnick, B & Belluz, J. (2019, July 10). The war to free science. Vox.

[99] Burayni, S. (2017, June 27). Is the staggeringly profitable business of scientific publishing bad for science? *The Guardian*.

[100] Kinol, A., Miller, E., Axtell, H., Hirschfeld, I., Leggett, S., Si, Y., & Stephens, J. C. (2023). Climate justice in higher education: A proposed paradigm shift towards a transformative role for colleges and universities. *Climatic Change, 176*(2), 15.

[101] Westervelt, A. (2022, February 24). Revealed: Leading climate research publisher helps fuel oil and gas drilling. *The Guardian*.

[102] McGee, K. (2022, October 18). Two-thirds of board members overseeing public universities are Abbott donors. They are not shy about wielding influence. *KUT*.

[103] McWilliams, J. (2019, March 12). Why should taxpayer funded research be put behind a paywall? Pacific Standard.

[104] Cook, G. (2012, February 12). Why scientists are boycotting a publisher. *Boston Globe*.

[105] Khoo, S. (2019, October 22). Opinion: Boycotting Elsevier is not enough. *The Scientist*.

BOX C.1 Spinoff industry: Climate science publishers—cont'd

scientists quickly broke their promise because they felt they had no other option.[106] The parent publishing firm Reed Elsevier, one of the largest in the world, has a complex and highly profitable structure. For Elsevier, the ability to exploit data access has extended to new sectors and markets in recent decades. Originally a paper manufacturer, Reed International PLC is now a holding company with a half-interest in publishing giant Reed Elsevier PLC, which was formed in 1993 to combine operations of Reed International with Elsevier NV. Reed International co-owns companies in the areas of scientific, professional, business, and consumer publishing. Reed subsidiary RELX creates the ability to mine data. Private firms sell raw data and also structured information from data. For example, RELX creates profitable research metrics and products by tracking the activities and associations of its authors and surveil who accesses articles. These data metric products predict the researchers and research projects who'll have the most "impact," a prized metric in academia. Impact rankings influence grants and employment, potentially reinforcing access and prestige.

Universities and grant funders have become profitable data sources for analytics companies. These data metric markets increasingly take decisions out of the hands of trained experts. The results have not benefited the advancement of knowledge necessary for addressing the climate emergency. At the expense of broader, more equitable, and more sustainable fields of knowledge and education, the market-driven publishing houses sideline climate science and focus on profit from sales tied to petroleum engineering and other fossil fuel industries.

These data products raise ethical and equity concerns. RELX and other data firms exploit a lack of data laws to build profitable data products to sell to surveillance firms, employers, landlords, insurance companies, and others.[107] Companies funnel unfiltered, unvetted data through algorithms,[108] often leading to biased, including racist, outcomes as institutions use "risk management" products to inform decision-making.[109]

While "big data" remains popular and promising for its potential to address the world's problems, including environmental crises,[110] the range of ways big data is filtered and managed by corporations does not receive sufficient scrutiny.[111] In fact, researchers often ignore how data are limited by access, partly as a result of cultural assumptions, including

[106] Heyman, T., Moors, P., & Storms, G. (2016). On the cost of knowledge: Evaluating the boycott against Elsevier. *Frontiers in Research Metrics and Analytics, 1*, 7.

[107] Landam, S. (2022, November 9). The quiet invasion of 'big information.' *Wired*.

[108] Verma, P. (2022, July 15). The never-ending quest to predict crime using AI. The Washington Post.

[109] Heaven, W. D. (2020, July 17). Predictive policing algorithms are racist. They need to be dismantled. *MIT Technology Review*.

[110] Lahsen, M. (2022). Evaluating the computational ("Big Data") turn in studies of media coverage of climate change. *Wiley Interdisciplinary Reviews: Climate Change, 13*(2), e752.

[111] Lamdan, S. (2022). *Data cartels: The companies that control and monopolize our information.* Stanford University Press.

xenophobic biases. Often, expensive big data efforts reinforce inequality and injustice without openly acknowledging or understanding built-in bias. A particularly relevant aspect of big data research is its outsized emissions footprint, creating "institutional super emitters." Left unaddressed, universities tied to big data operate as energy "hogs," perpetuating harm, undercutting climate action and contributing to a consumption deathprint on par with the extraction and production deathprint discussed in Chapter 1.

Like institutional super emitters, the emergence of cryptocurrencies carries a vast yet unaccounted dirty energy throughput. Even as crypto's initial allure may have worn off, its hugely outsized emissions continue unabated, in the background, unscrutinized, and unmitigated.[112] Digitization itself is tied to material consequences—GHG emissions, toxic metals, and excessive water use top its long list of impacts. The cloud's digital storage infrastructure contains outsized ecological costs with heavy reliance on fossil fuels (Praxis 5).

As emerging technologies themselves create new mitigation challenges, persistent, historical emitters continue apace in the background. Case in point: the steel industry. Used in cars, trains, wind turbines, and power stations, among other widespread uses (Box C.2), steel is central to modern economies yet its emissions are largely unmitigated. Mining of iron ore used for making steel is extractive and often toxic to nearby communities.[113] While steel can be made without coal, the vast majority is indeed made with coal.[114] Steel plants release toxic air and water pollution, though the industry is now facing pressure to produce cleaner steel.[115]

BOX C.2 Spinoff industry: Steel bicycles

China is the world's largest steel producer. The Taiwanese company Giant Bicycles created a conundrum when it developed a plant in China to take advantage of lower labor costs by using coal energy, in turn utilizing dirty energy to create expensive bikes sold in Asia, the US, and Europe — all marketed as a cleaner alternative to cars.

While bikes can use recycled metals,[116] and even have vegan seats,[117] these are niche options and are not mass produced around the world. Most Giant bikes, made of steel, are

(Continued)

---

[112] Marris, E. (2023, March 22). Crypto is mostly over. Its carbon emissions are not. *The Atlantic.*

[113] Nicholas, S., & Basirat, S. (2022, June 28). Iron ore quality a potential head wind to green steelmaking—technology and mining options are available to hit net-zero steel targets. *IEEFA.*

[114] Owens, J. (2020, September 24). Is it possible to make steel without fossil fuel? GreenBiz.

[115] Kuykendall, T., & Holzman, J. (2021, March 16). Steel industry under rising pressure to produce greener products. SP Global.

[116] Stott, S. (2020 October 30). How green is cycling? Riding, walking, ebikes and driving ranked. Bike Radar.

[117] Mowery, L, (2022, May 31). Priority bicycles: making bikes simpler to buy, ride and own. Forbes.

BOX C.2 Spinoff industry: Steel bicycles—cont'd

destined for Europe and Asia.[118] The Tuoketuo Power Station located in Inner Mongolia, China, the largest coal-fired power station in the world—an incredibly large 6720 MW unit—supports manufacturing of Giant bicycles.[119] Giant receives subsidies for this production.[120] The company's success represents an example of disaster capitalism as economic hardship opened this market. After workers at the Schwinn bicycle plant in Chicago went on strike in 1980, Giant was able to gain business and become a key Schwinn supplier. Giant was making more than two-thirds of Schwinn bikes by the mid-1980s, representing the majority of Giant's sales.[121] For labor rights, this is an example of the global race to the bottom discussed in Chapter 2.

Consumers do have choices. For example, when purchasing a bike, consider a used model, if available. For new purchases, examine the full lifecycle of the product, including the treatment of the workforce of the company and any ecological tradeoffs from production, transportation, and sales.[122] Choices do exist.

Equitable transition to renewables will require discarding historical discrimination against blue collar workers who are currently experiencing artificial intelligence (AI) displacement. For sustainable energy development, AI operates as a double-edge sword: risking the recreation of fossil energy's inequities while gaining breakthrough information and network securities, as well as new data tools for climate mitigation.[123] In the existing manufacturing sector, gender discrimination often continues. United Steel Workers (USW) has ameliorated some forms of harm; case in point, "Women of Steel," a program designed to instill gender equity.[124] With inequitable practice across its workforce, industrial organizations pit workers against environmental regulations designed to protect workers themselves and the communities in which they live. Exceptions do exist: again, the USW has made environmental commitments dating to the early 1990s, then leading the labor movement with the 2006 Policy Statement "Securing Our Children's World." Also in 2006, the USW co-founded the "Blue-Green Alliance," partnerships

[118] Sutton, M. (2021, May 14). Giant revenues up 55% in first quarter, E-bikes 30% of sales. Cycling Industry News.

[119] Central News Agency. (2019, October 17). Taiwan Taichung power plant could partially suspend operations over pollution. *Taiwan News*.

[120] Li, Y., & Li, C. (2019). Fossil energy subsidies in China's modern coal chemical industry. *Energy Policy*, *135*, 111015.

[121] Spinney, J., & Lin, W. I. (2019). (Mobility) Fixing the Taiwanese bicycle industry: The production and economisation of cycling culture in pursuit of accumulation. *Mobilities*, *14*(4), 524–544.

[122] The Good Trade. (2020, August 18). 10 brands making sustainable bicycles & conscious biking gear.

[123] Thomas, E. (2023, September 8). Predictive AI offers energy security amid changing industry. Energy Monitor.

[124] Fonow, M. M. (1998). Women of Steel: A case of feminist organizing in the United Steelworkers of America. *Canadian Woman Studies*, *18*(1), 117–22.

between labor organizers and environmentalists like Sierra Club (Chapter 9). Organized labor has huge potential to advance sustainable energy. Yet today it is still rare to see labor and environment combined in policy, a glaring regulatory oversight. One exception came after a United Auto Workers (UAW) strike in the US: General Motors will now bring EVs under the union's master agreement.[125] Alliances between environmentalists and labor organizers have never been more important as they hold real promise to maintain living standards on offer by fossil fuels industries.[126]

Prioritize the end of energy violence

As renewables urgently replace out-of-date and polluting fossil fuel systems, it is necessary to prioritize energy transformation with systemic change. Renewables with violent tendencies aren't a solution. Regardless of the fuel type, energy mega-projects have been shown to prey on insecurity as a common method to access land. Similarly, "fast" energy deployed in a quick growth spurt often generates conflict, whether fossil fuels or renewables. During the 2022 Winter Olympics in China, a green energy rush occurred as Beijing sought to spotlight its technological prowess. As citizens complained about fast deployment, they faced imprisonment.[127] Oversized, ill-suited energy deployments such as mega-solar projects involve displacements, a form of structural violence, including outright territorial violence. Renewable sources are not immune to **energy sprawl**— land uses converted to low wattage output in similar fashion to low-density suburban development that results in wasteful sprawl. As forested land and fertile farmland are rapidly converted to energy production—in particular large solar farms—locally determined "best uses" are uprooted, causing harm, creating social upheaval and backlash. This can create green sacrifice zones when connected to violence.[128] With rapid deployment of any form of energy infrastructure, quick land grabs can intensify poverty.

---

[125] Feliz Leon, L. (2023, October 6). The UAW just secured a landmark win in the fight for a pro-worker green transition. *Jacobin*.

[126] Pollin, R. (2023). Fossil fuel industry phase-out and just transition: Designing policies to protect workers' living standards. *Journal of Human Development and Capabilities*, 1–30.

[127] Staff. (2021, December 21). Human cost of China's green energy rush ahead of Winter Olympics. *France 24*.

[128] Zografos, C., & Robbins, P. (2020). Green sacrifice zones, or why a green new deal cannot ignore the cost shifts of just transitions. *One Earth*, 3(5), 543–546.

Another example of problematic utility-scale renewables exists with various mega-solar projects in India.[129] As with fossil fuels, harms are felt most by those without power, particularly poor communities and ethnic minorities.[130] Energy sprawl can mean land grabbing from traditional uses, such as wind farms usurping common property *ejido* lands in southern Mexico. In addition to land loss, minorities become targets for repression.[131] When dispossession is done for supposedly environmentally friendly fuels, it's known as **green grabbing**. Under current models, renewable energy infrastructure requires extracting mineral and fossil fuel resources—often from MAPA—for manufacturing, transportation, construction, and operation. Renewable energy atop existing practices amounts to yet more burden. Additive energy models that dominate current energy transitions typically do not address structural violence. By adding in renewable energy, or adding in women and members of minority groups, existing exploitative, top-down, profit-driven energy operations are not transformed, nor even diluted; rather, they extend violence into and onto new systems and communities.

Similar to narrow additive energy models, green economies and markets can extend harm if power relations and economic inequality are left unaddressed.[132] One such platform that seeks full restoration within a greened economy is A Feminist and Decolonial Global Green New Deal (GGND). In this platform, ten advocacy methods directed at the US Green New Deal are articulated.[133] Innovative approaches such as the GGND are required that increase jobs, improve working conditions, lower occupational risks, and build communities — all the while delivering equity.[134] Solar projects built into existing urban or rural landscapes (i.e., interspersed with crops, on school roofs) create more jobs than commercial solar farms alone. In addition to rooftop solar on government offices, hospitals, schools, parking lots, stadiums, and other buildings, "floatovoltaics" can be installed on catchment ponds or reservoirs, including those behind dams. Polluted

[129] Stock, R. (2022). Triggering resistance: Contesting the injustices of solar park development in India. *Energy Research & Social Science, 86*, 102464.

[130] Temper, L., Avila, S., Del Bene, D., Gobby, J., Kosoy, N., Le Billon, P., … & Walter, M. (2020). Movements shaping climate futures: A systematic mapping of protests against fossil fuel and low-carbon energy projects. *Environmental Research Letters, 15*(12), 123004.

[131] Dunlap, A. (2021). Does renewable energy exist? Fossil fuel+ technologies and the search for renewable energy. In *A critical approach to the social acceptance of renewable energy infrastructures* (pp. 83–102). Palgrave Macmillan.

[132] Almeida, D. V., Kolinjivadi, V., Ferrando, T., Roy, B., Herrera, H., Gonçalves, M. V., & Van Hecken, G. (2023). The "greening" of empire: The European green deal as the EU first agenda. *Political Geography, 105*, 102925.

[133] Muchhala, B. (2020, August 24). Towards a decolonial and feminist global green new deal. Rosa Luxemburg Stiftung; Feminist economic justice for people & planet action nexus. (2021). A feminist and decolonial global green new deal; Reyes, E. (2021, August 16). A decolonial, feminist Global Green New Deal. *The Ecologist.*

[134] Baker, S. (2021). *Revolutionary power: An activist's guide to the energy transition.* Island Press.

areas like designated **brownfields**, or lands that are underused because of concerns about contamination, are particularly good candidates for solar deployments.

As solar is deployed, its scale and placement matters. As the oldest colony in the world, Puerto Rico exemplifies a solar battleground: grassroots community efforts struggle to gain necessary state support,[135] while private sector models overseen by the US-based New Fortress Energy (Chapter 9) increase burden on the island population living with high energy costs and poor services.[136] Puerto Rico also demonstrates tension around large-scale industrial models that are creating energy sprawl and contributing to devastating floods, particularly with "trickle down" solar among energy conglomerates like the AES Corp[137] whose practice is to add solar to existing coal sites (Chapter 4). Efforts to solarize from the mainland undercut innovative local grassroots efforts to meet community needs while promoting resilience to disasters.[138] However, in spite of ongoing structural violence, Puerto Ricans continue to advance local alternatives and cultivate hope through resistance. At its universities, programs now exist to train students to mobilize for change.[139]

The never-ending disaster recovery in Puerto Rico is an ideal opportunity to envision change through participatory placemaking practices.[140] The sustainable, fair energy that has been developed in Puerto Rico[141] shows that improving gender equality is achievable while building new solar systems.[142] Training women provides a sustainable model for renewable energy employment[143] while reducing discrimination.[144] Communities in Puerto Rico also present active examples of microgrids for energy commoning. Grassroots

[135] Bentley, C. (2023, September 17). Puerto Rico hopes solar project will secure electric grid for future hurricanes. NPR.

[136] Acevedo, N. (2023, January 25). Puerto Rico officially privatizes power generation amid protests, doubts. NBC News.

[137] Santiago, R. (2021, July 26). Puerto Rico's future is solar. Recovery funds should go there, not to its outdated grid. Grist.

[138] Gallucci, M. (2022, June 8). Puerto Ricans are powering their own rooftop solar boom. *Canary Media*.

[139] Deil-Amen, R., Cammarota, J., Zayas Cruz, Y., & Pérez, G. (2022). Cultivating hope through creative resistance: Puerto Rican undergraduates surviving the disasters of climate and colonization. *International Journal of Qualitative Studies in Education, 35*(8), 830–842.

[140] Llorens, H. (2021). *Making livable worlds: Afro-Puerto Rican women building environmental justice*. University of Washington Press.

[141] Umpierre, A. (2022, February 23). On Puerto Rico's roofs, renewable energy brings employment for women. Direct Relief.

[142] Cruz Mejias, C. (2022, January 9). Women help fuel region's solar energy revolution. *Global Press Journal*.

[143] Umpierre, A. (2022, February 23). On Puerto Rico's roofs, renewable energy brings employment for women. *Direct Relief*; Jaramillo-Nieves, L. (2021, December 1). Diversifying Puerto Rico's solar workforce: advancing opportunities for women and the lgbtqtti+ community. Interstate Renewable Energy Council.

[144] Cruz Mejias, C. (2022, January 31). Fighting for Puerto Rico's solar revolution—and against sexism. WorldCrunch.

organizations teach families to produce affordable power for themselves and increase self-reliance, particularly in the periods following repeating disasters, particularly hurricanes. The electrical needs of most Puerto Rican homes fit within community-based renewable energy models and can be located on rooftops[145] or brownfields. In contrast, poorly designed solar "farms,"[146] overtake valuable fertile terrain, compounded by rising food insecurity across Puerto Rico due to the coupling of climate change, centuries of colonization, and misuse and contamination of the island's agricultural regions. Highly selective siting of solar farms targeting brownfields or within community-based agrivoltaic systems is required given the high stakes of survival where hunger and malnutrition are ever more common across Puerto Rico, its island geography particularly exposed to climatic shifts.

Visualizing and intervening

Activism, which has been central to the climate movement across decades, posits collective goals through action. In turn, sound mental health of both individuals and the groups to which they belong, work with, and work for, is often reinforced through action—a necessary resource for the existential challenge of our climate emergency.[147] **Mobilizing** describes the process of organizing for action in terms of preparing, assembling, and readying to build capacity for movement and response. Mobilizations are occurring at all scales from the local to the global—these sometimes disparate efforts can be harnessed even further into a collective force to "upend" the status quo. Civil **resistance** can be understood in the energy sector as governance from the ground up.[148] Resistance is the refusal to accept or comply in an attempt to prevent something, in this case climate breakdown. Resistance can take six basic forms: (1) physical disruption of resource flows: occupations and blockades (i.e., the Wet'suwet'en illustrative case in Chapter 10), (2) boycotts and other financial pressure, (3) enacting Indigenous sovereignty, law, and governance, (4) winning the battle of ideas: media, communications, and new imaginaries, (5) transformative alliances: building support across cultures, sectors, movements, and regions, and (6) multi-pronged approaches: building power by combining different

[145] Santiago, R., de Onis, C., & Llorens, H. (2022, September 21). Another hurricane makes clear the urgent need for rooftop solar in Puerto Rico. NACLA.

[146] Santiago, R., Llorens, H., & de Onis, C. (2022, February 17). The devastating costs of Puerto Rico's Solar "farms." NACLA.

[147] Thrive. (2022). The times are uncertain. Our commitment to mental health is not.

[148] Gulliver, R., Fielding, K., & Louis, W. R. (2021). *Civil resistance against climate change*. Washington, DC: International Center on Nonviolent Conflict.

strategies. In reality, movements are often strongest with a multi-pronged approach rather than picking and choosing between essential tactics.[149]

Climate organizers often create action from a myriad of emotional responses, including some seemingly opposed—love and rage, anxiety and hope, joy and despair. Singularly, **climate rage**—the expression of intense anger over runaway emissions and the threat of ecocide—is often a core motivation.[150] Mega projects that create sacrifice zones are often targets of public opposition driven by collective climate rage. In the process, activists are undeservedly scapegoated for the very situations they protest—a form of victim blaming. For example, the firms behind the Mountain Valley Pipeline (MVP), running from the Marcellus and Utica Shale Basins southeasterly across the Appalachian mountains towards urban centers, have repeatedly gone to court seeking injunctions against grassroots protest organizations for impeding what is asserted as "essential" energy progress. These firms fail to highlight the illegality of the project itself as it crossed federally protected forest lands (Chapter 9). Most recently, the legislative branch of the federal government has inserted itself through the actions of coal baron Senator Manchin to override opposing courts,[151] setting a precedent that severely weakens the balance of powers between US branches of the federal government.[152]

Not limited to fiascos in the US alone, governments around the world are failing to effectively lead the energy transition, unwilling to curtail the corrupting influence of fossil fuel interests. As regulators and courts have their hands tied by captured politicians, as happened with the MVP, citizens are compelled to directly challenge fossil fuel's social license. As one slice of media work done in environmental alliances, independent "watchdogs" publicize illegal and harmful behavior. A **watchdog** uses its eyes and ears to document harm and loudly pronounces danger to warn a community. In the environmental realm, warnings often expose illegal or unethical conduct, waste, theft, or undesirable practices. A compounding use of social media shows state repression of Indigenous land defenders,[153] where people around the world have instantaneous access to video footage or photographs of police raids or of rogue state security agents. The

[149] Gobby, J., Temper, L., Burke, M., & von Ellenrieder, N. (2022). Resistance as governance: Transformative strategies forged on the frontlines of extractivism in Canada. *The Extractive Industries and Society*, *9*, 100919.

[150] Stanley, S. K., Hogg, T. L., Leviston, Z., & Walker, I. (2021). From anger to action: Differential impacts of eco-anxiety, eco-depression, and eco-anger on climate action and wellbeing. *The Journal of Climate Change and Health*, *1*, 100003.

[151] Southern Environmental Law Center. (2023, June 27). Press release: Environmental groups argue attempt to throw out MVP lawsuits unconstitutional.

[152] East Daley Analytics. (2023, August 24). In MVP Permit Battle, the Court Finally Yields.

[153] Richards, K. S. (2022). Tiny houses, treesits, and housing on the front lines of the TMX pipeline resistance. *Canadian Theatre Review*, *191*, 38—45.

Figure C.3 *"Ash the elf" watches Weymouth compressor station. (Image credit: Fore River Residents Against the Compressor Station.)*

importance of watchdogs to spotlight Indigenous resistance and keep protestors safe is documented in Chapter 7, with additional digital resources in Praxis 12.

Remarkable, determined examples of watchdogging exist worldwide; in the northeast US, "Ash the Elf" has pervaded social media since the inception of a highly controversial gas pipeline compressor station situated south of Boston in Weymouth, Massachusetts. With a documented portfolio of mishaps, violations, and affronts, "Ash the Elf" has put the transnational conglomerate Enbridge Inc. under a microscope, with its social license repeatedly undercut by its own behavior, broadcast online (Fig. C.3 above).

Collective direct action movements like the UK-founded Extinction Rebellion, which started in 2018, find their footing in online broadcasts designed to reinforce on-the-ground actions. Tactics are becoming increasingly bold, such as interrupting major sporting events. Also from the UK, Just Stop Oil, launched in 2022,[154] has undergone extensive litigation for its direct actions, resulting in increased sanctions against protesters themselves. Disruptive climate action increased in 2023 in the US after the start of the

---

[154] Just Stop Oil. (n.d.). What if the Government doesn't have it under control?

youth-led organization Climate Defiance, known for inconveniently drawing attention to complicit actors by bird dogging.[155] **Bird dogging** involves calling out politicians at public gatherings and in public spaces and can be done consistently by following a particularly important or offensive decision-maker relentlessly; or alternatively, the action can be deployed sporadically and unexpectedly, such as popping up at a fundraising dinner or public relations event.

An unfortunate consequence of its own making, climate distress only encourages climate activism against fossil fuel interests.[156] The **Climate Necessity Defense** is a political-legal tool increasingly used by climate activists to justify protest actions taken in defense of the climate.[157] This legal defense is used in certain situations where a person's actions that would normally be considered unlawful are justified because they were necessary to prevent greater harm or evil. It has rarely been applied effectively—yet—to acquit. However, the necessity plea has influenced some judges to give more lenient penalties for actions, including blocking construction that industry and governments deem "essential." This is a counter trend to the more dominant tendency toward criminalizing environmental protest discussed in Chapter 7. Across decades, not only have the stakes of the climate crisis intensified, but acute retribution for direct action has as well.[158]

Against a backdrop of lawfare and aggressive prosecution of activists, decision-makers including politicians are often poorly equipped to effectively respond to concurring climate, water, and energy crises. When the use of science coupled with community organizing is deployed, the balance of power can shift away from corporate interests controlling the narrative. In the city of Boston, Massachusetts, grassroots efforts that involve both academia and communities have gained the support of local government at the mayor's office.[159] Utilizing science-based equipment and methods, the fossil gas leaks that pervade urban centers like Boston are documented by community-based teams. The results are undeniable for all to see (Fig. C.4)[160]; utilities are forced to reckon

[155] Influence Watch. (2023). Climate defiance.

[156] Ballew, M., Myers, T., Uppalapati, S., Rosenthal, S., Kotcher, J., Campbell, E., Goddard, E., Maibach, E., & Leiserowitz, A. (2023). Is distress about climate change associated with climate action? Yale Program on Climate Change Communication.

[157] Climate Disobedience Center. (n.d.). The climate necessity defense: A legal tool for climate activists.

[158] Farrell, B. (2009, December 8). Prostitutes await UN climate delegates, while steel cages await protesters. Waging Nonviolence.

[159] McKain, K., Down, A., Raciti, S. M., Budney, J., Hutyra, L. R., Floerchinger, C., ... & Wofsy, S. C. (2015). Methane emissions from natural gas infrastructure and use in the urban region of Boston, Massachusetts. *Proceedings of the National Academy of Sciences, 112*(7), 1941—1946.

[160] Phillips, N. G., Ackley, R., Crosson, E. R., Down, A., Hutyra, L. R., Brondfield, M., ... & Jackson, R. B. (2013). Mapping urban pipeline leaks: Methane leaks across Boston. *Environmental Pollution, 173*, 1-4.

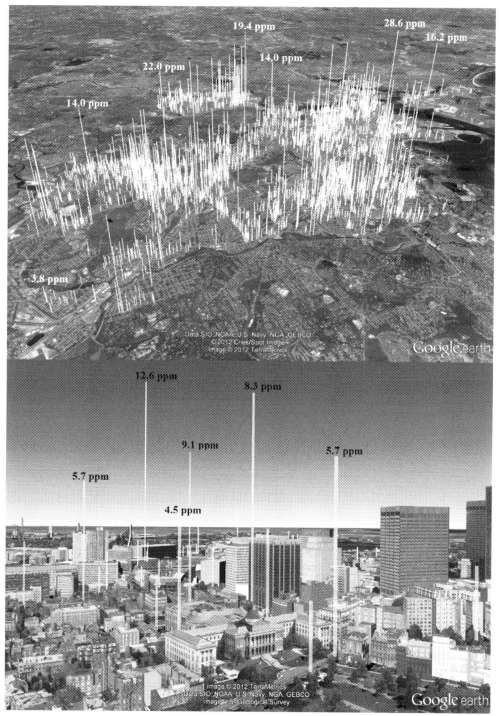

Figure C.4 Street-level methane mapping, Boston, Massachusetts. *(Image credit: The Boston Climate Action Network (BostonCAN), Boston Gas Leaks.)*

Figure C.5 *(S)hell Bus* exhibition. *(Image credit: Darren Cullen.)*

with previously hidden, pervasive leakage hotspots harming local urban environments and contributing to climate warming.

Historically, fossil fuels, particularly gas, have taken advantage of infrastructure placement—upstream and underground—coupled with the invisibility of carbon dioxide, methane, and hazardous pollutants. Visualization of the climate crisis is a critical tactic to expose industry subterfuge while making new analytical and emotive connections.[161] Visual artists increasingly recognize the existential threat of climate crisis and generate insightful, often searing political critiques that creatively focus the public's attention.

In the UK, artist Darren Cullen's traveling satirical art exhibition deconstructs Royal Dutch Shell's veneer of greenwashing to hilarious and insightful ends (Fig. C.5 above). Titled *(S)hell Bus,* Cullen's work became the target of cease and desist demands from Shell, with Cullen responding online: "Does SHELL consider HELL to be their trademark? If they do not currently trade under this name, are they considering rebranding as HELL in the near future?" Royal Dutch Shell finally gave up their demands, likely realizing Cullen's online virality only intensified brand vulnerability.[162]

[161] Lescaze, Z. (2022, March 25). How should art reckon with climate change? *New York Times.*

[162] Walfisz, J. (2022, May 7). Spelling Mistakes Cost Lives: the political artist who uses fun to fight the establishment. Euro News.

Figure C.6 Artist Joanie Lemercier at RWE coal mine site. *(Image credit: 10 Lemercier, J. (n.d.). Studio Joanie Lemercier, Picture by Benjamin Jung from Blast.)*

Recently, Cullen visited Ogoniland (Chapter 6) in the Niger Delta, producing a video documentary *Hell in the Niger Delta* that both shocks and elucidates not through satire but the horrific spectacle of Shell's flagrant ecocide.[163]

French Artist Joanie Lemercier extends his digital, multimedia practice to include visualizations that emphasize the scale of extractive harm, particularly of the coal industry across Europe (Chapter 4).[164] In a longstanding campaign that highlights the complicity of US software company Autodesk—seemingly unrelated directly to fossil fuels—films, online bird-dogging, and multimedia installations upturn the company's cool, clean, greenwashed image to reveal an intimate relationship with coal, particularly with German coal giant RWE.[165] Lemercier harkens back to traditional portraiture, creating literal *Faces of Coal*—CEOs, local politicians, and bankers tied to RWE—through ink made of lignite coal.[166] Fig. C.6 shows Lemercier with a coal ink portrait of Autodesk CEO Andrew Anagnost.

Vocal climate scientists warning the public of impending crisis have realized—albeit late—that climate data visualization on its own is not the most effective tool given the scale of threat. In this gap, artists see opportunity for evidence-based work that conjoins fact with artistic practice. US artist Jess Irish takes on the ubiquity of plastics through an intimate, profound experience in her own life—the dying and death of her mother. Through her film *This Mortal Plastik* (Fig. C.7), Irish unravels the alluring promise and

---

[163] Spelling Mistakes Cost Lives. (2023). Hell in the Niger Delta.

[164] Lemercier, J. (n.d.). Studio Joanie Lemercier.

[165] Ford, B. (2023, April 12). One of tech's cleanest companies is making tools for coal mines and oil drills. *Bloomberg.*

[166] Lemercier, J. (2023, March 2). Faces of coal. Studio Joanie Lermercier.

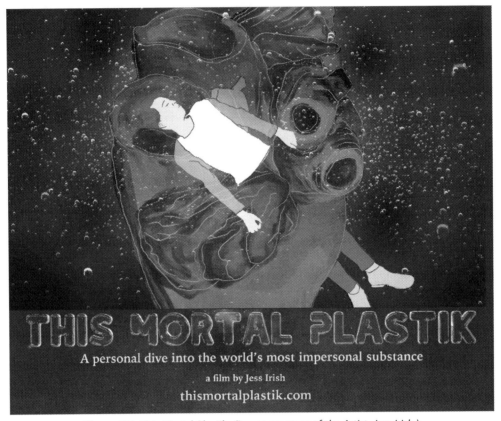

Figure C.7 This Mortal Plastik. *(Image courtesy of the Artist, Jess Irish.)*

devastating consequence of plastics in all its dimensions intimately related to our own time and those we love on Earth. By drawing historical fact, scientific evidence together with human emotions, unexpected connections can inspire and propel authentic activism.

Method 4—Divest—reinvest

With overwhelming evidence of deep environmental impact linked to the climate crisis, the argument for proactive, transformative change is strengthened. In BAU scenarios, local, regional, and global economies show a deep preference for harmful subsidies to fossil fuel interests. These subsidies, including outright tax breaks, rarely consider harmful externalities to ecosystem services. As governmental subsidies advantage fossil fuel interests,[167] repercussions are global in scope with planetary overshoot and widespread destruction increasing odds of looming financial loss across all scales of the economy.[168]

Subsidies in hand, economies and marketplaces remain steadfastly ignorant to adequate climate risk assessment practices. In spite of ample available data, industries are continually unprepared for repeated crop failure and supply chain disruptions. Lacking effective research protocols, even simple curiosity, BAU is unwilling to take climate risk seriously of its own accord. In the end, ignoring science is fundamentally an unsound long-term investment position.

Through continued BAU practices, GHG exposure intensifies investor risk as companies rife with super emitters and high emission-driven revenues refuse to decarbonize. Investment in unfavorable sectors or polluting companies will increasingly become a liability as markets shift. An exceptional, if underappreciated, accomplishment of the environmental movement is the elimination of fossil fuel's social license to operate (Chapter 2), as discussed in earlier chapters addressing coal, oil, and gas. Even as boosterist rhetoric from elite legal and financial experts still garners attention, analysis reliant on traditional measures that do not encompass rapid climate warming are simply inaccurate to the crisis at hand. With losses and damages tallied in billions, continued investment in the very driver of the climate crisis amounts to significant financial risk.[169]

As high income earners blithely disregard their extreme carbon footprint, all life and humanity bears their indifference. Through both consumption and investment patterns, the wealthiest segments of society have disproportionate climate influence and responsibility—for better and worse.[170] As society's top-tier seeks an escape hatch from climate crisis through wealth and privilege—however futile—the opportunity to bind all segments of society together in common cause is tragically forfeited.

Closely associated with privilege, many prestigious academic institutions face acute demands for change.[171] A significant win of the climate movement has occurred on college campuses as students, faculty, and alumni have forced administrations to divest from fossil fuels. While campus leaders signal climate-related research in their institution as a sufficient contribution to solving a global crisis, members of these academic communities

[167] The B Team. (2022, February 16). The world is spending a least 1.8 trillion a year, equivalent to 2% of GDP, on subsidies that are driving the destruction of ecosystems and species extinction.

[168] Future Earth. (2020). 2020 world risk perceptions.

[169] Vondrich, (2019, November 26). NY, divest from fossil fuels already. *NY Daily News*.

[170] Nielsen, K. S., Nicholas, K. A., Creutzig, F., Dietz, T., & Stern, P. C. (2021). The role of high-socioeconomic-status people in locking in or rapidly reducing energy-driven greenhouse gas emissions. *Nature Energy*, 6(11), 1011–1016.

[171] Finley-Brook, M., & A. Krass (2016). Higher Ed's carbon addition. *Human Geography*. 9(1): 83–87.

are left unsatisfied, demanding authentic decoupling from fossil fuels.[172] Targeting both the literal and figurative foundations of education, strides are being made even at the inception of education in preschool to rectify academia's poor, haphazard approach to preparing students for a future of climate challenges.[173]

Educators Bill McKibben and Naomi Oreskes point out how the Teachers Insurance and Annuity Association's (TIAA) portfolio of $1.2 trillion is heavily invested in fossil fuels. TIAA is the largest manager of retirement funds and invests on behalf of five million professionals in academia, medicine, culture, and other nonprofits.[174] Through its in-house asset management firm Nuveen, millions of TIAA dollars are invested directly in Adani coal bonds (Chapter 4).[175] Do these teachers and other professionals know their retirement is tied to the world's fourth-largest holder of coal bonds? As the retirement market has almost entirely shifted away from pensions to employee-owned 401(k)s, both white and blue collar workers are invariably invested in fossil fuel extraction and production.[176] As our climate crisis intensifies, these funds face significant, yet undisclosed, risk from stranded fossil fuel assets, exposing the financial insecurity in which so many workers have literally banked their own future.[177]

Divest—reinvest is a paired concept, linking withdrawal from structural energy violence to proactively support communities and workers that otherwise could be stranded in the transition away from fossil fuels. As finance is removed from fossil fuels, new investment opportunities emerge that prioritize remediation, energy efficiency, and other sustainability initiatives.

Growth of empathy and solidarity

Excess consumption costs lives: the true cost of energy has been conceptualized throughout this book as both a current and future deathprint. As a response to overshoot and global inequality, the degrowth movement (Introduction) seeks balance within planetary boundaries as a basis for transformational change, arguing that we can live more complete and purposeful lives while using fewer natural resources, sharing what we

[172] Le Billon, P., & Kristoffersen, B. (2019). Just cuts for fossil fuels? Supply-side carbon constraints and energy transition. *Environment and Planning A: Economy and Space,* 0308518X18816702.

[173] Fahlén, E., Högberg, A., Ingelhag, G., Ljungstedt, H., Lundström, H., & Perzon, M. (2020). Hoppet-the first fossil-free preschool. *IOP Conference Series: Earth and Environmental Science, 588*(4), 042065.

[174] Mckibben, W.E., & Oreskes, N. (2023, March 7). How Harvard Can Help Solve TIAA's Climate Problem. The Harvard Crimson.

[175] Solensky, S. (2023, September 1). TIAA must stop funding fossil fuels and climate destruction. CT Mirror.

[176] Spear, S. (2023, May 16). New tool shows retirement plans financing fossil fuel expansion through bond holdings, putting savers at risk. As you sow.

[177] Invest Your Values. (n.d.). Fossil fuel bonds in 401ks.

Number	Transformation
1	democratizing the economy
2	redistribution and social security
3	democratizing technology
4	revaluation of labor
5	democratization of the social metabolism
6	international solidarity

Figure C.8 Transformation agnostic to financial growth.

use more equitably (Fig. C.8 above). An "agrowth" position is agnostic to growth—meaning not for or against growth per se—while decentering growth and profit as core objectives. In this platform, priority shifts towards reparative justice with specific focus on democratic process, health, and equity.

Illustrative cases in this book demonstrate real-life examples as a launching pad to look deeply at our own energy systems. Strategic investments in good energy are much more effective than divestment alone. How can utility ownership be more equitable, and how can electricity delivery be climate-informed and resilient? Identify climate justice and reparations in your local context and identify ways to participate.

Try this

Research, map, and discuss*: Where is your energy from? How do you connect to local, national, and global energy markets, corridors, and networks? What are some social and ecological implications?*

Community-based research

As presented in Chapters 1, 6, and 9, minority and low-wealth populations are often unfairly burdened with polluting facilities.[178] There lies what has proven to be an important watchdog role for scholar-activists.[179] Participatory science involving impacted

---

[178] Cranmer, Z., Steinfield, L., Miranda, J., & Stohler, T. (2023). Energy distributive injustices: Assessing the demographics of communities surrounding renewable and fossil fuel power plants in the United States. *Energy Research & Social Science, 100*, 103050.

[179] Chatterjee, P., & Maira, S. (Eds.). (2014). *The imperial university: Academic repression and scholarly dissent.* University of Minnesota Press.

populations often captures evidence that otherwise slips through regulatory gaps. Often academic institutions are wedded to status quo outcomes, hampering transformative practice while maintaining structural inequality.[180] Individual scholars and research groups respond with community-based practices that provide high-impact educational opportunties for their students while constructing connections with partner commuinities.

Particularly, but not limited to academia, social norms encourage civility in professional spaces even as employees face increasing climate anxiety, often in silence. Widespread inaction from academia has negative consequences within and beyond campuses.[181] Scholars who directly confront peers and supervisors with inconvenient information are often ostracized, and their career advancement can suffer; some leave academia altogether.[182]

Universities and colleges along with primary and secondary education are ideal springboards for green equitable solutions. Even after three industrial revolutions, we are far from a post-extractive world[183] where fossil energy's harms are significantly diminished.[184] At present, grassroots movements around the world are establishing connections between the oppression of women and oppression of the Earth.[185] Globally gendered perspectives have long been silenced; through daily struggle for subsistence,[186] or during political and geopolitical strife,[187] women continue to disproportionately bear the brunt of the climate crisis.

Examining current extractive and exploitative patterns through an ecosocialist and ecofeminist lens is well-suited to imagining alternatives.[188] Critical analyses of energy

---

[180] Von Bulow, C., & Simpson, C. (2021). What matters most? Deep education conversations in a climate of change and complexity. In *Deep adaptation: Navigating the realities of climate chaos* (pp. 224–249). Polity Press.

[181] Thierry, A., Horn, L., Von Hellermann, P., & Gardner, C. "No research on a dead planet": Preserving the socio-ecological conditions for academia. In *Frontiers in Education* (Vol. 8, p. 1237076).

[182] Read, R. (2023, August 10). UEA humanities cuts reflect inability of academia to confront climate crisis. *Times Higher Education*.

[183] Arboleda, M. (2020). Planetary mine: Territories of extraction under late capitalism. Verso Books.

[184] Gilmore, R. W. (2022). *Abolition geography: Essays towards liberation*. Verso Books.

[185] Nelson, A. (2022). Ecosocialism from a post-development perspective. In *Post-capitalist futures* (pp. 31–40). Palgrave Macmillan.

[186] McCutcheon, P. (2019). Fannie Lou Hamer's freedom farms and Black agrarian Geographies. *Antipode, 51*(1), 207–224.

[187] Binoy, P. (2018). Darly and her battle with the sand-mining mafia: Tracing a feminist geopolitics of fear in the production of nature. *Human Geography, 10*(2): 37–53.

[188] Chattopadhyay, S., Gahman, L., & Watson, J. (2019). Ecosocialist pedagogies: Introduction. *Capitalism Nature Socialism, 30*(1), 26–30.

politics receive too little attention.[189] Ecofeminist and environmental historian Stefania Barca provides four targets to undo climate breakdown: (1) colonial relations that privilege the West, (2) gender relations forged out of historical bias toward technology and industry, (3) class relations built on inequality and oppression, and (4) species relations delegating non-humans as of lesser value.[190]

Try this

Apply Stefania Barca's four targets to a specific industry like the steel industry and its byproducts. What could be done to advance "just" transition for jobs and climate justice within this industry?

Method 5—Confront fossil fuel malfeasance

In aggregate, today's and tomorrow's damages to the atmosphere and biosphere are really beyond count; at the least, companies owe billions of dollars in **climate reparations**.[191] Climate reparations are payments for the loss and damage of climate change, and payments must be additional to, or an extension of, rectification of harms from colonialism and imperialism alone. Meeting the need for historic reparations while addressing the modern climate emergency is a contemporary, collective challenge.

Those most responsible for climate loss and damages are not paying the costs or helping adapt when and where possible.[192] Billions of dollars of damages result annually from fossil fuel's structural violence—one scholarly study assessed the amount owed at $209 billion dollars per year and the top 21 fossil fuel companies owe at least $5.4 trillion for the devastating global harm they have caused.[193] Oil heirs increasingly acknowledge their personal harms. For example, using the Climate Emergency Fund and the Equation Campaign, oil heirs finance disruptive direct action in organizations like Just Stop Oil and Climate Defiance—using direct economic transfers for urgent climate action as a type of "solidarity dividend."[194]

---

[189] Finley-Brook, M., Williams, T. L., Caron-Sheppard, J. A., & Jaromin, M. K. (2018). Critical energy justice in US natural gas infrastructuring. *Energy Research & Social Science, 41,* 176—190.

[190] Barca, S. (2020). *Forces of reproduction: Notes for a counter-hegemonic anthropocene.* Cambridge University Press.

[191] Lakhani, N. (2023, May 19). Fossil fuel firms owe climate reparations of 209 bn a year, says study. *The Guardian.*

[192] Sultana, F. (2023, August 3). Fossil fuel exec is slated to lead COP28. We must decolonize climate governance. Truthout.

[193] Grasso, M., & Heede, R. (2023). Time to pay the piper: Fossil fuel companies' reparations for climate damages. *One Earth, 6*(5), 459—463.

[194] Buckley, C. (2022, August 10). These groups want disruptive climate protests. Oil heirs are funding them. *New York Times.*

Our climate emergency and the urgent necessity to stop burning fossil fuels provides a window of opportunity for restorative socio-ecological transformation. Fossil fuel firms have emitted the majority of the carbon and methane in our atmosphere—this malfeasance is grounds for more than just payments for loss and damages (Introduction). The "Corporate Death Penalty" means **judicial dissolution**, a legal procedure in which a corporation is forced to dissolve or cease to exist following the revocation of its charter due to its significant harm to society—in this case ecocide.[195] The fossil fuel industry is liable for harmful deception and warrants severe punishment.[196] Fossil fuel firms whose leadership knew about climate change decades ago and still knowingly committed ecocide warrant repercussions for their intentional recklessness. Furthermore, as detailed in Chapter 6, specific companies like Shell (Nigeria) and Texaco-Chevron (Ecuador) have been found legally culpable for acutely inflicting damage upon local Indigenous and ethnic populations.

In hand with reparations, solutions to the loss and damage of fossil fuels are an active process of energy commoning. **Energy commoning** means valuing distributed energy systems as common goods. In contrast, the fossil fuel industry, led by investor-owned utilities (IOUs) and captured public entities, has retained control of commodities, holding public ratepayers hostage to monopolistic power that favors short-term profit through dirty and unsustainable energy. Lack of access to transmission lines still limits renewable energy. Lobbyists have defeated bills by claiming high costs. Stalled transmission lines for renewable deployment represent a strategic physical blockage holding back more deployment of solar and wind that would outcompete fossil fuels based on price. Even after the cost of renewables has fallen lower than that of fossil fuels, particularly when considering climate resilience,[197] a **spatial imbalance** in renewable transmission lines as compared to fossil-oriented systems remains. This imbalance creates a particular form of uneven development that demonstrates the strong influence of politics on infrastructuring.

In order for effective energy transition, local geographies—communities, towns, cities, provinces, and regions—must take on leading roles in energy distribution. Energy and its development is a fundamental function of human geography, encompassing our relationship with the Earth and fellow citizens. Beyond transition alone, energy transformation further recenters communities for justice and restorative placemaking.

---

[195] Bandiera, R. (2021). Ecocide: Kill the corporation before it kills us. *International Journal for Crime, Justice and Social Democracy, 10*(1), 159–161.

[196] Wentz, J., & Franta, B. (2022). Liability for public deception: Linking fossil fuel disinformation to climate damages. *Environmental Law Report, 52*, 10995.

[197] Osman, A. I., Chen, L., Yang, M., Msigwa, G., Farghali, M., Fawzy, S., ... & Yap, P. S. (2023). Cost, environmental impact, and resilience of renewable energy under a changing climate: A review. *Environmental Chemistry Letters, 21*(2), 741–764.

Energy commons value local ecologies, resulting in less harmful impacts compared to traditional, top-down distribution systems that treat electricity as a commodity to be extracted, developed, transported, and traded for profit.

To break fossil energy's cycle of violence, energy commoning holds strategic promise for effective energy transition. By positioning local communities as rightful decision-makers, a myriad of transactional, profit-driven imbalances can be rectified. Best known in France, **people's assemblies** are gaining popularity as long-standing customs of traditional societies are used to shore up energy, climate, and environmental literacy.[198]

Try this

Discuss and act: *Is there an opportunity for your campus, town, or city to switch to fossil fuel-free investments? See Praxis 8 for additional divestment planning. How would you allocate reinvestment funds for your community?*

Energy commoning

Active verbs describing infrastructuring across this book demonstrate how coal, oil, and gas companies expand self-interested extraction. Energy commoning is a radical departure, embracing traditions of the commons found in producers like subsistence agriculturalists and waste pickers (Chapter 10) who work with the Earth and support collective GHG mitigation. Successful grassroots responses from Indigenous and ethnic communities become recognized and supported rather than ignored. In local energy commons, there is no one-size transformation approach (Fig. C.9), yet there are mutually reinforcing objectives.

In place of hierarchical operations reliant on highly militarized and authoritarian practices, horizontal and cooperative systems operate beyond the easy capture of industry and states.[199] Publicly owned and operated power companies provide viable alternatives to top-heavy IOUs. However, public guiding boards alone don't necessarily go far enough to be deemed transformative; internal power structures must often be further transformed. In the case of many public utility commissions (PUCs), oversight authority tends to replicate trends in private oversight: a makeup of predominantly white males

---

[198] Yeung, P. (2020, November 20). 'It gave me hope in democracy': How French citizens are embracing people power. *The Guardian.*
[199] After Oil Collective. (2022). *Solarities: Seeking energy justice.* University of Minnesota Press.

Epistemic Justice	Transformative Processes	Restorative Exchanges
transparency	decolonization	reparations
access to information	ecosocialism	mutual support
participatory science	ecofeminism	public power
accountability	direct action	degrowth / agrowth

Figure C.9 Core elements in energy transformation.

possessing higher education and similar professional connections.[200] Conflicts of interest can exist on oversight commissions where there is a revolving door between energy firms and regulators. Public sector energy can also become oppressive and ecologically harmful if there is a lack of transparency and accountability working explicitly across both dimensions of "people power."[201] Without reform, pro-industry practices can easily become normalized, hurting ratepayers by poor decision-making, incomplete policy implementation, and unnecessary rate hikes.[202]

Effective energy reparations cover loss and damage caused by climate change in co-ordination with payments for historical wrongdoing.[203] The assumption of high GHG emission regions or endless future growth and extraction from former colonies, where reparations are overdue, is out-of-touch and intrinsically unjust.[204] Climate justice targets and tactics can and should be tailored to specific geographies (Fig. C.10).[205–210]

Calls for reparation from the Caribbean Community (CARICOM) deserve attention. As a multinational bloc representing 15 Carribean member states, CARICOM has proposed reparations that range from a formal apology for slavery to governing autonomy,

[200] Pervost, L. (2023, July 5). Who decides where we get electricity and how much we pay? Mostly White, politically connected men. Energy News Network.

[201] Dawson, A. (2022). *People's power: Reclaiming the energy commons.* OR Books.

[202] Schladen, M. (2023, February 27). Ohio utility regulator front and center in massive bailout scandal. *Ohio Capital Journal.*

[203] Caribbean Community. (n.d.). CARICOM Reparations Committee.

[204] D'Alisa, G., Demaria, F., & Kallis, G., Eds. (2014). *Degrowth: A vocabulary for a new era.* Routledge.

[205] Táíwò, O. O. (2022). *Reconsidering reparations.* Oxford University Press.

[206] Darity Jr, W. A., & Mullen, A. K. (2020). *From here to equality: Reparations for Black Americans in the twenty-first century.* UNC Press Books.

[207] Buckley, C. (2022, August 10). These groups want disruptive climate protests. Oil heirs are funding them. *New York Times.*

[208] Baerga Aguirre, J.L., Director. (2021). El Poder del Pueblo.

[209] Freedom to Thrive. (n.d.). Freedom to Thrive.

[210] Forge Organizing. (n.d.). The Forge Organizing Strategy and Practice.

Type (target and/or tactic)	Description	Justification	Examples
target	unconditional cash transfers	disparate wealth and income; cumulative impact of slavery, poverty and exploitation	reparations to African American descendants of enslaved; universal basic income
target	global climate funding	loss and damage in the billions, especially in MAPA	solidarity dividend
target	community control	participatory democracy	Puerto Rican community solar in Guayama
tactic	divest-reinvest	transfer funds from fossil fuel corporations to communities	Freedom to Thrive
tactic	knowledge is power	'the Flint Strategy'	Community based participatory science; academia-community research alliances
tactic/target	deciding together	distribution of power	Forge Organizing; collective bargaining for the common good

Figure C.10 Climate reparations.

debt cancellation, and monetary compensation over decades. Sixty percent of the population of CARICOM's region is under the age of 30 and most communities are highly vulnerable to hurricanes, flooding, and sea level rise.

The reparations movement is a call for partnership.[211]

In the face of climate breakdown, ethical partnerships are needed more than ever. **Mutual support** is horizontal assistance among equals that benefits widely, in contrast to unequal lending or debt diplomacy.

Try this

Research, envision, and act: *Envision a livable and just society built through transformational degrowth. What elements are most needed now and in the future? What do you think will be the most challenging? How can you collaborate to actualize this worldmaking? What do we all gain through mutual support?*

[211] CARICOM. (2023, March 9). Historic reparative apology guided by the UWI and CARICOM reparations commission.

Deep space and deep time

Deep ecologists point out inherent species bias in humanity-based approaches to climate justice. One of the greatest forms of energy violence is the violence done to other species—many of whom existed long before *Homo sapiens*. Humans have been alive for just a small part of geologic time. As we face our climate crisis, we are invariably caught in a very short moment of geologic time that we can claim as human. Planning for a future of transformation is our collective challenge as we get ever closer to the hard, cold facts of climate overshoot. But thinking about humanity's and planet Earth's possibility beyond the narrow confines of modern human history; that is, placing ourselves in **deep time**, meaning geological time or cosmic time by using the billions of years of the Earth to factually frame our existence, can be imaginative and hopeful.[212]

Beyond ourselves and planet Earth lie galaxies. The Earth is a tiny portion of an extensive galaxy. The concept of deep space allows current populations facing existential threat to realize our position in schematics of time and space broader than compressed 24-hours-in-a-day or 365-days-in-a-year production cycles of modern society. Extended time and space before and after us can be terrifying, but also comforting as it allows respite from our always-on, 24-7 media-frenzied clock demanding our constant attention. The framework of this book—organized to explore the influence of time as pertinent to energy sector spaces and to the climate crisis itself—broadens energy context. Too often permits or research depicting a small slice of time and space are given priority over the more complex reality that looks at change over extended time and space.

Anishinaabe theorist Gerald Vizenor gives us the framework of survivance, highlighting that today's Indigenous struggles represent so much more than tragedy alone. Survivance (Chapter 7) represents an active and self-determined presence, enabling continuation and creativity. Indigenous Potawatomi philosopher Kyle White defines collective continuance for Native peoples as the community's capacity to be adaptive. Native voices are particularly essential in repairing their contaminated territories as regenerative land; they are elemental for Indigenous Peoples taking back sovereignty.[213] In Praxis 9 to follow, a Survivance Guide details the ways in which we can learn from Indigenous peoples for adaptive capacity during climate crisis. The climate justice organization in the US Gulf region Taproot Earth issues a call to climate action drawing on "replicable gravity"—meaning anchoring and working collectively to shift what is possible.[214]

---

[212] Macy, J., & Johnstone, C. (2012). *Active hope: How to face the mess we're in without going crazy*. New World Library.

[213] Reeves, E. (2021). Taking back sovereignty: The importance of native voices in addressing environmental harms to native land. *California Western International Law Journal, 52*, 615.

[214] Taproot Earth. (n.d.). Liberation Horizon.

We fail outright by censoring the consequences of climate inaction. If we are told our home rests on an eroding foundation; that it contains corroding pipes ready to burst; that the roof is days away from collapsing; and yet do nothing in response even though we have resources to fix these problems, most reasonable people would say we are negligent, hurting ourselves in the process. In a similar way, we handicap ourselves by giving credence to normalized, controlling narratives filled with misogynistic, anti-poor, and harmful invective towards the most vulnerable, including those of color. Authoritarian, fascist discourse has no place in effective energy transformation.[215]

Forging the way into our uncertain future requires "people power"—large social movements willing to demand and uphold transparency—and freeing ourselves from the enduring grasp of energy violence. Our innate desires to feel hopeful, to look on the bright side, to abide by social norms of positive thinking and avoid the negative, play a part in our collective climate inaction.[216] Living fully includes accepting our fatality. In the spirit of bell hooks (Praxis 11), living fully includes learning how to "die well,"[217] whether we have decades remaining or much less. The Dedication of this book is to frontline defenders for energy justice from Nigeria, Honduras, and South Africa, who died fighting for their people's right to an energy commons. The transformative change they have sought is increasingly urgent.

When conducting research and working towards climate solutions, we would do well to remind ourselves that climate is not a narrow technical problem that we alone are destined to fix. Change remains shallow if self-interested agendas override comprehensive inclusion of historically exploited communties. **Research and development** (R&D)—the search for innovation, and improvement in products and processes—is part of transition and transformation, and attending to the risk of structural violence can be integrated as a design outcome. Never-ending "development" focused solely on market measurements of economic growth has in large part brought us to an ecological edge; the adoption of energy transformation principles into the R&D sector can interrupt the status quo that keeps profits in the hands of the few corporations most responsible for the climate crisis.

Through a restorative shift in R&D, energy projects can become more equitable and suitable to their siting geographies. A collective energy commons—one that is free from structural violence—is a necessary mechanism to loosen the enduring grasp of fossil fuels, liberating us once and for all for restorative action as we collectively face the climate crisis.

[215] Deep Adaptation Forum. (2022, June 16). Deep adaptation forum's charter.
[216] Wilson, P. J. (2021). Climate change inaction and optimism. *Philosophies*, 6(3), 61.
[217] hooks, b. (2014). *Teaching to transgress*. Routledge.

Try this

Research and action: *What energy sector research gaps require more attention? What resources (time and money) exist to advance scholarly and/or community-based innovation for energy transformation and climate justice?*

Conclusion: Spatial and temporal synopsis

Space

(1) *Persistent patterns of pollution and spatial racism are tied to fossil fuels and can exist even with renewable energy.*

(2) *State financial structures extend climate violence and incentivize out-of-date fossil fuels; renewable alternatives require political transformation for fair access.*

(3) *Renewable energies are most effective with bioregional and collaborative approaches; consideration of energy sprawl; and attention to the discontinuation of structural violence.*

(4) *Postextractive models like networked district geothermal energy promise high efficiency and low risk.*

Time

(1) *Payments for climate reparation fall to the polluters, including fossil fuel firms and their leadership.*

(2) *Urgency for survival generates conditions that may lead to disruptive direct action.*

(3) *A blue-green alliance of labor and environmentalists is at a critical juncture in energy transformation.*

(4) *A compressed timescale is necessitated by insufficient mitigation over decades, and the resulting acceleration of ecological breakdown.*

Summary

Energy continues to be an inadequately understood sector with great influence and consequence for the climate crisis. This book provides a foundation to understand and analyze global energy sites, as well as a toolbox for transformative energy change. Authentic public power includes participatory ownership and control. Blue-green alliances consisting of workers and environmental activists together are instrumental for energy transformation. Energy commoning mitigates the worst impacts of global boiling while ensuring energy is shared equitably. Through five suggested methods, readers are presented with extended examples to confront fossil energy's grasp on power.

Vocabulary

1. adaptation
2. artificial boundaries
3. bird dogging
4. brownfields
5. business-as-usual (BAU)
6. climate breakdown
7. carbon exposure
8. climate justice
9. climate migration
10. Climate Necessity Defense
11. climate rage
12. climate refugee
13. climate reparations
14. deep time
15. discourses of climate delay
16. divest—reinvest
17. double standard
18. ecological cascade
19. elastic metrics
20. energy commoning
21. energy sprawl
22. extreme heat
23. fiduciary responsibility
24. fog of enactment
25. future deathprint
26. global boiling
27. global brightening
28. global dimming
29. green grabbing
30. green neoliberalism
31. heat dome
32. impact paradox
33. maladaptation
34. market environmentalism
35. mobilizing
36. mutual support
37. overshoot
38. people's assemblies

39. rebound effect
40. redlining
41. research and development (R&D)
42. spatial imbalance
43. technological hubris
44. technological optimism
45. voluntary standards
46. watchdog
47. wet-bulb temperature

Recommended

Books

Bendell, J., & Read, R. (Eds.). (2021). *Deep adaptation: navigating the realities of climate chaos.* John Wiley & Sons.

Dawson, A. (2022). *People's power: Reclaiming the energy commons.* OR Books.

Johnson, A. E., & Wilkinson, K. K. (Eds.). (2021). *All we can save: Truth, courage, and solutions for the climate crisis.* One World.

Llorens, H. (2021). *Making livable worlds: Afro-Puerto rican women building environmental justice.* University of Washington Press.

Macy, J., & Johnstone, C. (2022). *Active hope: How to face the mess we're in with unexpected resilience and creative power.* New World Library.

Acronyms

AAR	Association of American Railroads
ACNR	American Consolidated Natural Resources
ACP	Atlantic Coast Pipeline
AIA	Access to Information Act (Canada)
AIM	American Indian Movement
AMD	acid mine drainage
API	American Petroleum Institute
ASH	Appalachian Storage Hub
ASU	Air Separation Unit
b/d	barrels per day
BAU	business-as-usual
bcm	billion cubic meters
BIA	Bureau of Indian Affairs
BIPOC	Black, Indigenous, People of Color
BLEVE	Boiling Liquid Expanding Vapor Explosion
BNDES	Brazilian National Bank for Economic and Social Development
BOP	blowout preventer
BRI	Belt and Road Initiative
BRICS	Brazil, Russia, India, China, South Africa
BTEX	benzene, toluene, ethylbenzene, and xylenes
C&D	construction and demolition
CAIA	Chemical and Allied Industries Association
CalEnviro Screen	California screening tool
CBPAR	community-based participatory action research
CCP	Chinese Communist Party
CCR	coal combustion residuals
CCS	carbon capture and storage
CDM	Clean Development Mechanism
CEO	chief executive officer
CH$_4$	methane
CHEXIM	Export-Import Bank of China
CNG	compressed natural gas
CO	carbon monoxide
CO$_2$	carbon dioxide
CO$_2$e	carbon dioxide equivalent
COICA	Coordinator of Indigenous Organizations of the Amazon River Basin
CONAIE	Confederation of Indigenous Nationalities of Ecuador
CONFENIAE	Confederation of Nationalities of Ecuadorian Amazon

COP	conference of parties
COPD	chronic obstructive pulmonary disease
COPINH	Council of Honduran Popular and Indigenous Organizations
CPP	Clean Power Plan
CRT	Critical Race Theory
CSR	Corporate Social Responsibility
CWP	Coal-Workers Pneumoconiosis
DAPL	Dakota Access Pipeline
DEIJ	diversity, equity, inclusion, justice
DEQ	Department of Environmental Quality
DIY	do-it-yourself
DOE	Department of Energy
DOJ	Department of Justice
EACOP	East African Crude Oil Pipeline
EFF	Electronic Frontier Foundation
EIA	Energy Information Administration
EIS	Environmental Impact Statement
EJ	Environmental Justice
EJAtlas	Environmental Justice Atlas
EJSCREEN	EPA environmental justice screening tool
EP	Equator Principles
EPA	Environmental Protection Agency
EQT	Equitrans
ESG	Environmental, Social, Governance
ETP	Energy Transfer Partners
EU	European Union
FBI	Federal Bureau of Investigation
FDI	foreign direct investment
FERC	Federal Energy Regulatory Commission
FIFO	fly-in, fly-out
FLIR	forward-looking infrared
FLNG	floating liquefied natural gas terminal
FOIA	Freedom of Information Act (US)
FPIC	Free, prior and informed consent
FRRACS	Fore River Residents Against the Compressor Station
FSO/FPSO	floating production storage and offloading unit
FTE	full-time equivalent
GBR	Great Barrier Reef
GCC	Global Climate Coalition
GGFR	Global Gas Flaring Reduction Partnership
GHGs	greenhouse gases

GIF	Graphics Interchange Format
GIS	Geographic Information Science
GND	Green New Deal
GPS	Global Positioning System
GW	gigawatt
GWP	global warming potential
H₂S	hydrogen sulfide
HAPs	hazardous air pollutants
HCA	high consequence area
IACHR	Inter-American Court of Human Rights
IFI	international finance institutions
IOU	investor-owned utility
IPCC	Intergovernmental Panel on Climate Change
IRA	Inflation Reduction Act
km	kilometer
kv	kilovolt
kWh	kilowatt hour
KZN	KwaZulu Natal
LCA	lifecycle analysis
LG&E	Louisville Gas and Electric Company
LLC	limited liability corporation
LNG	liquified natural gas
LPG	liquid petroleum gas
M&R	metering and regulation station
MAPA	Most Affected Peoples and Areas
MCEJO	Mfolozi Community Environmental Justice Organisation
MEND	Movement for the Emancipation of the Niger Delta
MMA	Montreal, Maine and Atlantic
MOP	Maximum Operating Pressure
MOSOP	Movement for the Survival of the Ogoni People
MSHA	Mine Safety and Health Administration
MVP	Mountain Valley Pipeline
MW	megawatt
NAACP	National Association for the Advancement of Colored People
NAAQS	National Ambient Air Quality Standards
NDA	Niger Delta Avengers
NDAs	non-disclosure agreements
NEPA	National Environmental Protection Act
NFE	New Fortress Energy
NGO	non-governmental organization
NOAA	National Oceanic and Atmospheric Administration

NO$_x$	nitrogen oxides
NSW	New South Wales
NUMSA	National Union of Metalworkers
NYSE	New York Stock Exchange
O$_3$	ozone
OCT	Our Children's Trust
OPEC	Organization of the Petroleum Exporting Countries
P2G	power to gas
PAHs	polycyclic aromatic hydrocarbons
PAR	participatory action research
Pb	lead
PEMEX	Petróleos Mexicanos
PFAS	per- and polyfluoroalkyl substances
PGR	Phoenix Global Resources
PHMSA	Pipeline and Hazardous Materials Safety Administration
PIR	potential impact radius
PM	particulate matter
PMF	probable maximum flood
POWHR	Protect Our Water, Heritage, Rights
PPA	power purchasing agreement
PPE	personal protective equipment
PREPA	Puerto Rico Electric Power Authority
Promesa	Puerto Rico Oversight, Management & Economic Stability Act
PTSD	post-traumatic stress disorder
PX	paraxylene
R&D	research and development
RCMP	Royal Canadian Mounted Police
RCRA	Resource Conservation and Recovery Act
RFID	radio frequency identification
RNG	renewable natural gas
ROI	return on investment
RoN	rights of nature
ROW	right-of-way
RPT	rapid phase transition
RTU	Remote Terminal Unit
SAR	synthetic aperture radar
Saudi Aramco	Saudi Arabian Oil Company
SCADA	Supervisory Control and Data Acquisition
SDWA	Safe Drinking Water Act
SLTT	State, Local, Tribal and Territorial
SMCRA	Surface Mining Control and Reclamation Act

SNG	synthetic natural gas
SO$_2$	sulfur dioxide
SUNY	State University of New York
SWDA	Solid Waste Disposal Act
TENORMs	technologically enhanced naturally occurring radioactive materials
TETCO	Texas Eastern Transmission
TVA	Tennessee Valley Authority
UK	United Kingdom
UMWA	United Mine Workers of America
UN	United Nations
UNDRIP	United Nations Declaration on the Rights of Indigenous Peoples
UNFCCC	United Nations Framework Convention on Climate Change
UNHCR	United Nations High Commission on Refugees
US	United States
USGS	US Geological Survey
VCE	vapor cloud explosion
VEPCO	Virginia Electric and Power Company
VOCs	volatile organic compounds
WHO	World Health Organization
WV	West Virginia
WVMA	West Virginia Manufacturing Association

Praxis

Contents

Praxis enacts theory in an engaged application of ideas often referred to as a "living laboratory." While learners are particularly encouraged to examine their own energy grid, each exercise can extend as a global application. Where applicable, mixed-method and cross-disciplinary approaches are encouraged in particular praxis prompts. Effective climate defenders (Introduction) combine cultural and technical knowledge with understanding of culture and power. A geographic information systems (GIS) exercise can be found in **Praxis 5**. Geospatial and remote sensing tools can be deployed across many other prompts, with a word of caution to be sensitive to eco-social risks during quantitative spatial analysis and mapping.[1] Throughout all 12 prompts, environmental justice research approaches encourage community partnership.[2] **Praxis 12** provides additional web resources to extend the reach of this praxis guide.

---

[1] Monmonier, M. (2018). *How to lie with maps*. University of Chicago Press.
[2] Boda, P. A., Fusi, F., Miranda, F., Palmer, G. J., Flax-Hatch, J., Siciliano, M., ... & Cailas, M. (2023). Environmental justice through community-policy participatory Partnerships. *Journal of Environmental Protection*, *14*(8), 616–636.

Praxis 0—Acknowledgment and positionality

Land acknowledgments are an important first step[3] in showing respect, gratitude, and support for Indigenous-led grassroots change and commitment to returning land,[4] though they often do not go far enough.[5]

Reflect: Decolonial praxis can help regain our humanity and reconnect to nature.[6] Before starting, reflect on your intentions for land acknowledgment. Consider your **positionality**, referring to how social networks and power shape options and viewpoints. How does your identity and position influence your perceptions of energy? Solidarity and work alongside native communities requires a continuous consent process to ask and receive permission prior to each interaction.[7]

Take action

- Support Indigenous organizations through transparent and agreed purpose, including work and financial support.
- Support Indigenous-led movements and campaigns and encourage others to do so.[8]
- Commit to returning Indigenous land.

Recommended resources

- Giniw Collective.
- Native Governance Center.
- Resource Generation (toolkit).
- Red Nation podcast.

Try this

Research and act: Write a land acknowledgment. Identify opportunities to support Landback Movements and other social justice initiatives for healing and reparation.

Praxis 1—Mapping power and visualizing money flows

Power mapping is particularly effective for creating strategic plans by solidifying alliances, targeting actions, and assuring equity and relevance. This technique can help change agents to document and communicate power relationships. There is no standard methodology but general steps include:

---

[3] Native Governance Center (n.d). The land We're On.
[4] Resource Generation. (n.d.). Land Reparations and Indigenous Solidarity Toolkit.
[5] Native Governance Center. (2019, October 22). A guide to Indigenous land acknowledgement.
[6] Desai, K., & Sanya, B. N. (2016). Towards decolonial praxis: Reconfiguring the human and the curriculum. *Gender and Education, 28*(6), 710–724.
[7] Yellowhead Institute. (2019). Land Back.
[8] Landback. (n.d.). Manifesto.

Identify a focus: Depending on the objective, power maps can focus on a variety of policies, projects, institutions, sites, or problems.

Map major groups: Identify key decision-making groups, institutions, or blocs. Make a diagram placing them around the identified focus.

Map individuals: Identify key individuals associated with each of those groups and position them on the diagram.

Determine relational power lines: Make the components in your diagram proportional and illustrative. Draw lines connecting people and institutions that have relations to each other. Provide a key (e.g., solid lines show direct relationships; if there are financial flows, put a dollar amount and make the size of the line proportional).

Determine and visualize money flows: Research money flows. Show connections so they become known and visible to highlight responsibility.

Target priority relationships: Analyze relationships and connections to identify nodes of power and influencers with money. Move to the center the people/groups that have the most relational lines drawn to them. Discuss results. Plan action steps. Identify ways to access the individuals or institutions on the map.

Visualize and communicate results: This could be done in various formats, including relationship circles. Most approaches are aspatial, meaning they do not map location. However, spatial approaches georeference powerscapes, which is helpful to show the specific locations where power is held, and from whom and to whom money flows.[9]

Mobilize, reflect, and revisit: Mobilizing describes effective grassroots organizing for an ever-shifting purpose (Conclusion). Reflection allows change agents to improve tactics. Pay attention to diffuse **capillary power**—interactions influencing thoughts and ideation as well as actions. The ways people think, feel, and act are fundamental to building grassroots responses.

Recommended resources

- As You Sow
  - ⊛ (Racial Justice Score Card)
- (Corporate Mapping Project)
  - ⊛ Fossil Power Top 50 (i.e., Enbridge, TC Energy)
- (Public Accountability Database)
- (Toolkit for Youth Eco-Activism)

Try this

Map and discuss: *Make a power map of a local energy policy, such as your city or university campus. Show financial connections between groups or organizations. If you have access to spatial data, map the spatial locations of power and finance.*

---

[9] LittleSis. (n.d.). LittleSis* is a free database of who-knows-who at the heights of business and government. *opposite of Big Brother.

Praxis 2—EJAtlas and zine makers

Due to censorship and a widespread lack of independent venues free from corporate control and bias, environmental and social movements have long published information in creative outlets. Today a popular method for do-it-yourself (DIY) publishing is a **zine**, a small-circulation booklet historically created by grassroots movements without access to mainstream media platforms. EJAtlas represents another independent publishing platform; it highlights global cases collected through **crowdsourcing**, the act of collecting services, ideas or content through the contributions of a large group of people.[10] The online atlas embraces transparency; each entry is documented and sourced extensively. Many entries have been developed by teams at universities, with NGOs, or in grassroots movements. Fig. P.1 includes a sample of EJAtlas entries tied to cases in this textbook.

Location	EJAtlas entry title
Dedication	Proyecto Hidroeléctrico Agua Zarca, Honduras
Introduction	Yamal Mega natural gas project, Arctic Russia
Chapter 1	The 1986 catastrophic nuclear accident in Chernobyl, Ukraine
Chapter 2	Somkhele coal mine owned by Tendele, KwaZulu-Natal, South Africa
Chapter 3	Urban construction sites, pneumoconiosis crisis in Shuangxi, Hunan, China
Chapter 4	Paiton Baru Coal-fired Power Plant, East Java, Indonesia
Chapter 5	Deep Horizon oil spill USA
Chapter 6	NNPC pipeline explosions, fires and spills in Ijegun, Nigeria
Chapter 7	Louisiana's Bayou Bridge Pipeline Project, United States
Chapter 8	Enbridge natural gas compressor in North Weymouth, MA, USA
Chapter 9	Loma de La Lata y Vaca Muerta en Neuquen, Argentina
Chapter 10	Chevron's Gorgon gas extraction meet with labor unions' opposition, Australia
Conclusion	Sea Level Rise and Tribal Relocation in Isle de Jean Charles, Louisiana, USA

Figure P.1 Sample entries from EJAtlas.

---

[10] Temper, L., Del Bene, D., & Martinez-Alier, J. (2015). Mapping the frontiers and front lines of global environmental justice: The EJAtlas. *Journal of Political Ecology, 22*(1): 255—278; Martinez-Alier, J. (2021). Mapping ecological distribution conflicts: The EJAtlas. *The Extractive Industries and Society*.

Crowdsourcing supports inclusion of grassroots perspectives that are often not heard beyond a local area; this is a ongoing deficiency of global coverage that contains significant information gaps.[11] Nearly 4000 cases have been detailed at EJAtlas, yet there are sites of energy violence that remain undocumented or underdocumented. In authoritarian contexts, the EJAtlas represents a critical dissemination platform. For example, as discussed in Chapter 8, PX protests spread across China yet faced repression. These listings in Fig. P.2 have additional value as effective censorship within China makes it hard to uncover evidence of PX protests and other dissent. EJAtlas listings from China continue to provide one of the only records of social concerns about PX in China.

Protest	Location in China
1	Jiujiang, Jiangxi
2	Dalian, Liaoning
3	Pengzhou, Sichuan
4	Ningbo, Zhejiang
5	Maoming, Guangdong
6	Zhangzhou and Fujian
7	Kunming, Yunnan
8	Jinshan, Shanghai

Figure P.2 EJAtlas-listed PX protests across China.

Try this

Research and share: If you have experienced a site of energy violence and have detailed evidence, follow the EJAtlas format to catalog concerns (Atlas, n.d.). Other online platforms such as storymaps, or creative outlets like zines, can also be utilized to elucidate energy violence in your own region.

Praxis 3—Climate justice in commemoration

Reclamation and reparations require changing commemoration and naming practices for streets, buildings, and features across the landscape that no longer express what society

---

[11] Environmental Justice Atlas. (n.d.). Become a collaborator.

values (Chapter 3). **Toponyms** are the names of places—they can be built from natural surroundings or by honoring people who are considered important. Much of our landscape is named after colonizers, slaveholders, corporations, and business leaders past and present. A **toponymic workspace** describes how names can be updated to reflect current knowledge and values.[12] As the social license to operate (Chapter 2) of the fossil fuel industry is removed, judicial dissolution (Conclusion) represents a process to dismantle firms that have been convicted of crimes against humanity.

Transformation of our landscape is always attached to societal values. In place of history's colonizing patterns, the contribution of environmental defenders can be recognized by renaming plazas, transportation stations and public buildings to highlight climate leaders like Tom Goldtooth, Vanessa Nakate, Elizabeth Wathuti, James Hansen, or Bill McKibben.

Try this

Research and leverage change: *Identify an existing location commemorating a climate malefactor* (Wright) *— a powerful leader using their access to finance and influence to subvert climate action — and discuss how to rectify misinformation they circulated or to publicize harmful deeds they commited. Alternatively, discuss how you could rework a current site of commemoration, or establish a new site, to recognize a climate hero.*

Praxis 4—Exposing virtue signaling

An exacting means to expose virtue signaling is the comparison of what is being said with what is actually being done. This method effectively cuts through empty corporate promises to create actions in line with promises. The uses of artificial boundaries and elastic metrics (Conclusion) for climate action are a form of greenwashing by warping parameters and measurement strategies to target a particular agenda. Elastic methods allow entities to accentuate small or symbolic commitments like recycling or offsets yet leaving much bigger problems untouched. This allows for the practice of paying "lip service" — words without action — to social responsibility claims rather than taking substantial measures to improve performance. Promotion of the use of voluntary standards without regulatory oversight and compliance has been manipulated to advance deception such as those listed in Fig. P.3 (i.e., greenwash, whitewash, wokewash). These constructs can help elucidate specific claims missing key information, that are embellished or outright untrue. Many other specific forms of greenwashing are defined in the Introduction.

---

[12] Alderman, D. H., & Reuben, R. R. (2020). The classroom as "toponymic workspace": Towards a critical pedagogy of campus place renaming. *Journal of Geography in Higher Education, 44*(1), 124—141.

Deception	Definition	Examples
greenwash	intentional deception to embellish environmental positives or obscure negatives	Fossil gas companies use a bait and switch strategy to make blue hydrogen seem ecologically positive in spite of higher emissions (**Chapter 10**).
whitewash	painting a narrative of racial integration while forwarding white privilege	'Just' transition strategies focus on sustainable consumption options for Europe and North America and extend harm to MAPA from mining and resource overuse (**Conclusion**).
wokewash	painting a narrative of equity while advancing unfair systemic violence	AES Corporation promotes renewables for the global North yet maintains coal combustion in Latin America and the Caribbean (**Chapter 4**).

Figure P.3 Typology of deception.

Empty ESG pledges (Chapter 9) are too common. With elastic ESG methods, many strategies seek to gloss over concerns and diffuse criticism and maintain the status quo. A lack of transformative change often can be tied to conflicts of interest. Special interest ties can be exposed with methods like "follow the money" and power mapping as discussed in **Praxis 1**.

Try this

Identify an example of deception. *What would you say in a letter to the company or other responsible agent for a particular instance of corporate virtue signaling? Explain potential harm resulting from misleading or untrue claims.*

Praxis 5—Hidden energy in cloud storage

Contradictions of **cloud storage**—a term utilized to define typically off-site and online digital storage—are particularly hard to decipher due their seeming invisibility. Just because energy is used in and for digital spaces doesn't deem it clean or low carbon. While their interiors are hyperclean,[13] most data centers draw the majority of their power from dirty electricity grids powered by fossil fuels and use diesel generators for backup power supply. Digital or virtual space has physical implications, including vast energy requirements (Box P.1). Power-hungry air handlers are staples of most advanced data centers, where cooling accounts for a large portion of electricity usage.[14]

---

[13] Taylor, A. R. E., & Velkova, J. (2021). Sensing data centres. In Klimburg-Witjes, N., Poechhacker, N., & Geoffrey C. Bowker, G.C. (Eds.). Sensing in/security: Sensors as transnational security infrastructures (p. 287). Mattering Press.

[14] Moro, J. (2021, April 26). Air-conditioning the internet: Data center securitization as atmospheric media. *Media Field Journal*.

BOX P.1 Spinoff industry: cloud storage

Hyperscale data centers that feed 5000 or more servers and are 10000 square feet or greater in size are part of a global trend.[15] Microsoft, Amazon, and Google account for over half of today's hyperscale data centers.[16] Firms pledge to customers that data services will be available anytime, so operations are designed to be redundant—if one system fails, another is ready to take its place to prevent disruption in user experiences.

The US hosts hundreds of hyperdata centers.[17] Drought-stricken communities are pushing back against these facilities,[18] so developers have shifted geographies, moving operations east to areas with more water availability. In particular, a "data center alley" in Virginia has formed next to the federal government, the world's largest data customer.[19] State tax breaks made the site attractive, creating an intense hotspot or cluster (Fig. P.4).[20]

Firms received millions of dollars in incentives to build data centers in Virginia. Newer data center models advertise increases in energy efficiency and other upgrades[21] but most expansion has used traditional technology, which is predicted to become obsolete quickly due to the vast energy and water consumption, as well as the physical size of equipment. Virginia's hotspots now experience pressure to limit further expansion.[22] Town officials are seeking more diverse revenue streams as data centers are increasingly prone to instabilities with risks of global chip shortages or other supply chain disruptions.

Try this

Define and analyze: *What is the hidden energy and water use in hyperscale data centers like that of Ashburn, Virginia? Use GIS and remote sensing to quantify distances to key data markets, electricity transmission lines, and water supplies. Assess possibilities for a transition to renewable energy sources for Ashburn's data centers.*

Praxis 6—Ending plastic requires global solidarity

Particular challenges — climate change and reducing plastic waste — are so expansive they exist beyond any one geography. Even as scientists work to define the technical language for plastic bans and regulations, the effects are years away. A non-proliferation treaty for fossil fuels is also

[15] Kaur, D. (2021, August 5). Here's why India is witnessing a data center boom. Tech Wire Asia.

[16] Synergy Research Group. (2021, January 26). Microsoft, Amazon and Google account for over half of today's hyperscale data centers.

[17] Ashtine, M., & Mytton, D. (2021, October 19). We are ignoring the true cost of water guzzling data centres. *The Conversation*.

[18] Solon, O. (2021, June 19). Drought-stricken communities push back against data centers. NBC News.

[19] Weingarten, D. (2021, June 22). How your cloud data ended up in one Virginia County. CS Monitor.

[20] Cole, A. (2022, May 18). Why is Ashburn known as Data Center 'Alley'? Upstack.

[21] Gillin, P. (2020, December 9). New concepts of the sustainable data center arise amid an evolving market. Data Center Frontier.

[22] Augenstein, N. (2022, March 3). Loudoun Co. considers preventing data centers along Va. State Route 7. WTOP.

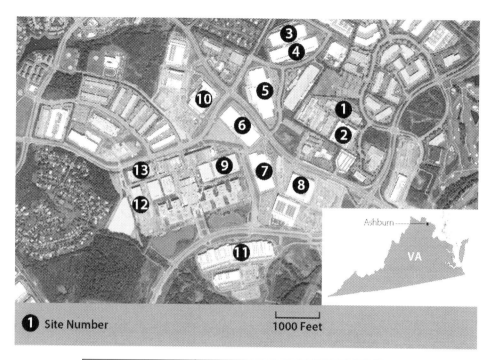

Site Number	Data Site
1	21715 Filigree Ct Data Hub
2	Equinix DC11 Data Center
3	DFT Data Centers
4	Digital Realty Northern Virginia ACC5
5	Digital Realty Northern Virginia ACC7
6	Digital Realty Northern Virginia ACC9
7	ACC10 Digital Realty
8	LC2 - CloudHQ Data Center
9	Aligned Data Center - Ashburn IAD-01
10	NTT Global Data Centers Americas - VA3
11	Digital Realty Northern Virginia IAD39
12	Aligned Data Center - Ashburn IAD-02
13	Equinix Data Center

Figure P.4 Data Center Alley, Ashburn, Virginia.

in the works, but would not offer an immediate effect. If you were planning a global strategy, would you look to cut off production at the source, perhaps as plastic or fracking ban? What holds promise as an effective tactic to end fracking? The history of fracking bans was discussed in Chapter 8; just recently one such ban in Colombia has received international attention due to its location in the global South and support of national oil workers.[23]

Recommended resources

- (The Chemical Footprint of a Plastic Bottle)
- (Fossil Fuel Non-Proliferation Treaty)
- (UNEP Plastic Pollution Resolution)

Try this

Research and engage*: Can a global treaty be successful in ending fossil fuel expansion? In what other ways can people show power in numbers to cut off coal, oil and gas supply? How can we end plastic use in the most efficient, quick and equitable way?*

Praxis 7—Building blue-green-brown alliances

Energy is a critical battleground for redistributing wealth and power to mitigate the worst harms of climate change.[24] Blue-green-brown alliances—referring to the inclusion of environmental justice in labor and worker safety movements—hold significant promise for climate action;[25] yet energy transition focused narrowly on profit rarely brings equitably shared benefits for workers and people of color. In particular, there are disproportionate harms from greener energies for the most affected peoples and places (MAPA). Collaborations coordinated across efforts for change are fundamental to "just" energy transformation.[26] A Green New Deal (GND) has the potential to be a game changer—with more jobs and greater equity.[27] Yet GND approaches cannot rely on business as

[23] Griffin. (2021, April 6). Colombia oil workers join anti-fracking campaign. Reuters.

[24] Kolinjivadi, V. (2019, December 7). Why a Green New Deal must be decolonial. Al Jazeera.

[25] Bakari, M. E. K. (2021). Setting up the three sides of the sustainability triangle: The American blue-green-brown alliances revisited. *Social Sciences & Humanities Open*, 3(1), 100134.

[26] Galvin, R., & Healy, N. (2020). The Green New Deal in the United States: What it is and how to pay for it. *Energy Research & Social Science*, 67, 101529.

[27] Economic Commission for Latin America and the Caribbean (ECLAC). (2022, November 7). Latin America and the Caribbean: The green transition can be an economic and social game changer, says new report.

usual.[28] As GND initatives become increasingly widespread, decolonial configurations are emerging (Fig. P.5).[29-45]

Increasingly, students, faculty and staff at universities are pushing for GND platforms beyond administrative commitments.[46] The youth-led Sunrise Movement calls for accountability and transparency across campus climate action, particularly targeting university endowments as a means to drive change regionally.[47] Bottom-up movements seek to build community power for a "just" transition.[48] For example, a national network of electrical cooperatives has emerged in Spain referred to as Som Energia.[49] The organization teaches people how to look for and build energy options—a model that can be replicated

[28] Féliz, M., & Melón, D. E. (2023). Beyond the Green New Deal? Dependency, racial capitalism and struggles for a radical ecological transition in Argentina and Latin America. *Geoforum, 145*, 103653.

[29] The Red Nation. (n.d.). The Red Deal.

[30] Feminist Economic Justice for People and Planet. (2021). A Feminist and Decolonial Global Green New Deal.

[31] European Commission. (n.d.). Next Generation EU.

[32] Lee, J. H., & Woo, J. (2020). Green New Deal policy of South Korea: Policy innovation for a sustainability transition. *Sustainability, 12*(23), 10191.

[33] Tienhaara, K., Yun, S. J., & Gunderson, R. (2022). South Korea's Green New Deal 2.0: Old wine in new bottles? In *Routledge Handbook on the Green New Deal.* Taylor & Francis.

[34] US Congress. (2019, February 7). House Resolution 109.

[35] Ocasio-Cortez, A. (2023). The Green New Deal Implementation Guide.

[36] MacArthur, J. L., Hoicka, C. E., Castleden, H., Das, R., & Lieu, J. (2020). Canada's Green New Deal: Forging the socio-political foundations of climate resilient infrastructure? *Energy Research & Social Science, 65*, 101442.

[37] Chang-Diaz, S. (2022, February 1). A Green New Deal for Massachusetts.

[38] Clavier, F. (2021, May 19). Singapore and South Korea's Green response to 'build back better'. Nextrends Asia.

[39] Gulf South for a Green New Deal Alliance. (n.d.). GulfSouth4GND Policy Platform.

[40] Renew New England Alliance. (n.d.). Fighting together for a just New England.

[41] Boston Green New Deal Coalition (n.d.). Boston Green New Deal Coalition; City of Boston (2023, September 5). Join the Work to Become a Green New Deal City.

[42] Furnaro, A., & Kay, K. (2022). Labor resistance and municipal power: Scalar mismatch in the Los Angeles Green New Deal. *Political Geography, 98*, 102684.

[43] City of Boston. (2023, October 4). A Green New Deal for Boston Public Schools.

[44] California Green New Deal Coalition. (n.d.). About. https://greennewdealca.org/ (23 July 2024)

[45] UC Green New Deal Coalition. (n.d.). About us.

[46] UC Green New Deal Coalition. (2021). University of California Policy Green New Deal Platform.

[47] Schwartz, J. (2023). Sunrise AU proposes Green New Deal to Increase transparency with sustainability efforts. The Eagle.

[48] Furnaro, A., & Kay, K. (2022). Labor resistance and municipal power: Scalar mismatch in the Los Angeles Green New Deal. *Political Geography, 98*, 102684.

[49] Pellicer-Sifres, V., Belda-Miquel, S., Cuesta-Fernández, I., & Boni, A. (2018). Learning, transformative action, and grassroots innovation: Insights from the Spanish energy cooperative Som Energia. *Energy Research & Social Science, 42*, 100—111.

Decolonial	Federal	State/ Territory	Regional	City/ School
The Red Deal[29] A Feminist and Decolonial Global Green New Deal[30]	US GND Implementation Guide[34,35] Canada's Pact for a Green New Deal[36] South Korea Green New Deal[32,33]	A Green New Deal for Massachusetts[37] California Green New Deal Coalition[44] Singapore's Green Plan[38]	Next Generation EU[31] Renew New England Alliance[40] Gulf South for a Green New Deal[39]	Los Angeles Green New Deal[42] Boston Green New Deal Coalition[41] A Green New Deal for Boston Schools[43] UC Green New Deal Coalition[45]

Figure P.5 Green New Deal proposals.

broadly.[50] An alternative urban transportation cooperative in Quito combines feminism[51] with new visions for transport.[52] This movement reminds us that infrastructuring is not just about expensive equipment, and that a social experiment can also be a form of infrastructure building.[53]

Energy transformation requires political, economic, and social change that goes well beyond narrow fuel switching. To envision a process where you live, what would you first target as an optimum local opportunity?

Try this

Research and engage: *Analyze an energy transformation project or a GND policy. What are the best strategies to provide jobs and promote equity during decarbonization? How would you reduce conflicting negative trade-offs, such as between urban and rural areas or between marginalized groups?*

---

[50] Belda-Miquel, S., Pellicer-Sifres, V., & Boni, A. (2020). Exploring the contribution of grassroots innovations to justice: Using the capability approach to normatively address bottom-up sustainable transitions practices. *Sustainability*, *12*(9), 3617.

[51] Gamble, J. (2019). Playing with infrastructure like a Carishina: Feminist cycling in an era of democratic politics. *Antipode*, *51*(4), 1166–1184.

[52] Gamble, J. (2020). A Transit Manifesto for Quito: Citizen-led, low-carbon alternatives, from cycling to informal transportation, can infuse sustainable transit plans with solutions that already meet popular sector needs. *NACLA Report on the Americas*, *52*(2), 199–205.

[53] Gamble, J. (2017). Experimental infrastructure: Experiences in bicycling in Quito, Ecuador. *International Journal of Urban and Regional Research*, *41*(1), 162–180.

Praxis 8—Climate reparations with divest-reinvest

Divestment from fossil fuels that is paired with investment can replace lost jobs and create new jobs, a process known as 'divest-reinvest.' The Movement for Black Lives has taken a broader approach, seeking fossil fuel divestment coupled with decriminalization through the reduction of funds for incarceration, prison systems, and surveillance. Reinvestment can target health, education, job training, and other beneficial initiatives for populations experiencing environmental injustice.[54] How would you address equity gaps summarized in Fig. P.6?

Equity Gaps
Business as usual investment reinforces historical patterns of structural violence
Fossil fuel dependent communities experience job loss without reinvestment
Fossil fuel divestment campaigns can inadequately incorporate social justice
'Tied aid' is popular with donors as it allows for control and extends privilege
Underrepresented groups that seek access to capital can become beholden to funders

Figure P.6 Equity gaps in investment.

Various tactics to remove investments from fossil fuels have been deployed, whether targeting tar sands oil, coal, or all fossil reserves—some institutions aim broadly to end all new finance of fossil fuels altogether. Institutions have a range of options for how and when to initiate divestment. **Coal exposure** highlights investments in companies that are significantly active in the coal sector, formally determined as more than 25% of revenue coming from thermal coal production.

Billions of dollars pledged towards new fossil fuel projects are harmful to cheaper and cleaner renewable energy alternatives, which don't have adequate access to investment streams as banks have historically prioritized fossil fuels. Divestment campaigns call on banks to reconsider their exposures. **Fossil banks** are banks that are disproportionately financing climate chaos.[55] Despite increased pressure for divestment, fossil banks continue to double down on fossil fuel expansion as a significant portion of their portfolios.[56] Across the banking sector, massive investments in fossil fuels show little sign of abatement even after the Paris Climate Accords Fig. P.7 highlights resources and cases of fossil banking activity.[57]

---

[54] Bratman, E., Brunette, K., Shelly, D. C., & Nicholson, S. (2016). Justice is the goal: Divestment as climate change resistance. *Journal of Environmental Studies and Sciences, 6*(4), 677–690.

[55] Fossil Banks: No Thanks. (n.d). Banks are financing climate Chaos.

[56] Endicott, M. (2019, March 28). US banks pledged to fund renewable energy, but they still spend way more on fossil fuels. *Mother Jones.*

[57] Kelly, S. (2019, March 20). Global Banks, led by JPMorgan Chase, invested $1.9 trillion in fossil fuels since Paris Climate Pact. DeSmog.

Research Databases	Information Available
Bank Track's Focus Banks	Coverage of prominent banks specifies influential global ties
Fossil Bank's Fuel Finance Report	The 60 largest banks have invested trillions in fossil fuels since the 2015 Paris Climate Agreement
Illustrative Studies	**Key Findings**
"Capitalizing on Collapse"	From 2009 to 2023 eight major banks funded the majority of Amazonian oil and gas extraction (**Chapter 6**)
Mountain Valley Pipeline	Bank of America, JPMorgan Chase and Wells Fargo finance the MVP (**Chapter 9**) as well as other controversial global pipelines

Figure P.7 Breaking up with fossil banks: databases and illustrative cases.

Try this

Research and engage*: Conduct a review of your university endowment or retirement system. Does a divest-reinvest plan exist? Do ethical investment guidelines exist? Does the institution you assessed meet its carbon action or social justice guidelines? Why or why not? If there are shortcomings, what steps do you recommend?*

Carefully crafted divestment strategies that advance social justice are imperative.[58] With optimism in hand, be keenly aware that institutional promises do not invariably meet reality. With universities, there can be a complex arrangement of direct and indirect investments over different timeframes and under various funds. Institutions increasingly face reputation and financial risks through fossil fuel investments. To inspire change, promote investment opportunites that outright beat the returns of fossil-intensive funds. Fossil-free research requirements ensure the independence of educational institutions.[59]

Recommended resources

- Faculty for a Future
- Fossil Free Funds
- TIAA Divest

Try this

Define and analyze*: Climate reparations are incomplete, a work in progress. Pick a location and assess how much has been lost and damaged. What information gaps exist? How might you use GIS, or other data analyses to visualize and communicate loss and damage?*

[58] Hestres, L. E., & Hopke, J. E. (2019). Fossil fuel divestment: Theories of change, goals, and strategies of a growing climate movement. *Environmental Politics*, 1—19.
[59] Fossil Free Research. (n.d.). No more fossil fuel research money.

Praxis 9—Survivance guide

Narratives based on lived experience can be powerful tools for change, particularly for marginalized groups who are not fully or accurately recognized in dominant narratives.[60] Survival and culturally based resilience (Survivance, Chapter 7) from grassroots origins often carries households and communities forward following disasters, particularly in cases where government response is bungled or compounds original harms. A now classic example, 3 months after the Chernobyl accident, the Ukrainian Ministry of Health issued a pamphlet to residents exposed to radioactive fallout:

> Dear Comrades! Since the accident at the Chernobyl power plant there has been a detailed analysis of the radioactivity ... The results show that living and working in your village will cause no harm to adults or children...
> Please follow these guidelines: Limit fresh greens. Do not consume local meat and milk. Wash down homes regularly. Remove topsoil from the garden and bury it in specially prepared graves far from the village...[61]

If you were writing a modern-day survivance guidebook for a particular disaster event—or even the climate crisis itself—what should it include and what should it exclude? This mental exercise based on knowledge of place and infrastructure can encourage necessary change. As an example, picture this: in a virtual blackout after a major disaster in your town, residents find themselves wading through waist-deep flooding in search of food, clean water, and high ground. Some die in this search. Some die in their homes, waiting for provisions, electricity, and food to arrive. With limited and expensive gas supplies, the community centers and schools shutter their doors to local residents.[62]

As the past teaches us about the future, an invariable lesson from disasters is that human behavior is malleable and variable, generous or horrific. As poly-crises increasingly surround us, how can we encourage people to work together and maintain a commitment to justice and equity rather than devolving into fear and chaos with only self-interest as a guide?

Try this

Research and act*: Research probable threats of climate change in your community, such as extreme weather. Armed with this information, how might you prepare and encourage local institutions to do so as well? How can you inform or help others in your community? What local survivance should be prioritized?*

---

[60] Climate Alliance Mapping Project. (2023). Climate Justice Story Maps.
[61] Quoted from Brown, K. (2019). *Manual for survival: A Chernobyl guide to the future*. Penguin UK.
[62] Revised from Baker, S. (2021). *Revolutionary power: An activist's guide to the energy transition*. Island Press.

Praxis 10—Eco-defense and counter-tactics

Eco-defense tactics are actions to disrupt damage to the environment. Some examples are civil disobedience, direct actions including property damage, and sabotage. These activities garner unprecedented surveillance, or geospatial overwatch (Chapter 6). Surveillance occurs in everyday power relations people encounter at home, at work, and in schools and communities.[63]

Green surveillance targeting environmentalists has been aggressively deployed against anti-logging groups and animal rights activists as well as anti-pipeline activists, discussed throughout the DAPL case in Chapter 7. Forest and water defenders are targets of deadly violence (Dedication). For example, in the US in 2023, police actions directed at the Defend the Atlanta Forest protesters included intimidation, repeated shows of force, and murder of a protester—Manuel "Tortuguita" Terán.[64] Body-camera footage and two different autopsies show police shot Terán more than a dozen times, while they were sitting cross-legged with both hands up.[65]

Surveillance of environmental protestors has been known to feature private surveillance (**gray surveillance**), including corporate-financed espionage with professional sleuths (i.e., former or current state intelligence agents). These agents are paid to spy on activists and NGOs. **Permanent surveillance**, a threat to democracy, involves constant observation, continuously creating and archiving data streams.[66] The Electronic Frontier Foundation has created a primer on self-defense relevant to researchers, activists, and impacted communities.[67]

Countersurveillance techniques are undertaken by the public to prevent surveillance, including covert surveillance, and include technical methods of detecting tracking devices. Encampments or networks of eye witnesses stand as more traditional methods. For example, a devoted community science group monitors harm from Enbridge pipelines.[68] Waadookawaad Amikwag (Those Who Help Beaver) use drones and thermal imaging to monitor breaches. Environmental organizations are training the public to

[63] Lee, A. (2022). Hybrid activism under the radar: Surveillance and resistance among marginalized youth activists in the United States and Canada. *New Media & Society*, 14614448221105847.

[64] Pellow, D. N. (2023). Confronting Institutional Violence in the Context of Climate Justice Politics. *Social Media+ Society, 9*(2), 20563051231177913.

[65] Akbar, A. A. (2023). The Fight Against Cop City. *Dissent, 70*(2), 62–70.

[66] Zuboff, S. (2019). *The age of surveillance capitalism: The fight for a human future at the new frontier of power.* Profile Books.

[67] Electronic Frontier Foundation. (n.d.). Surveillance Self Defense.

[68] Marohn, K. (2022, August 6). Line 3 Aquifer breach is leaking more groundwater. MPR News.

spot construction violations[69] and appealing inadequate pipeline permits by reporting violations and initiating requests for investigation of violation documentation.[70] Watchdogs include local observers taking photos and videos,[71] testifying in court, running for office, even spending time in prison to protect their communities.[72] The work of watchdogs can assist the state with criminal investigations.[73] If necessary, citizen risk assessment can be started and maintained with a small budget paid for by crowdsourcing.[74]

Try this

Research and communicate: What patterns do you see related to surveillance in DAPL, Bayou Bridge, and other Energy Transfer or Tigerswan examples? How might in-person or remote pipeline monitoring contribute to energy transformation?

Praxis 11—Ecological grief and grounding

Psychoterratic feelings about the Earth range from great appreciation and awe to growing eco-distress. People often avoid talking about what makes them uncomfortable, yet research proves discussion of suppressed anxieties is proactive and valuable. Engaging in ecological and social grounding is part of assessing the threats we face. Defense mechanisms we use to convince ourselves things can't be as bad (i.e., the climate breakdown is somehow manageable) are pervasive. This contributes to **disavowal**: a subtle but powerful form of soft denialism whereby impacts aren't fully recognized due to subconscious psychological defenses. **Foreclosure** involves shutting oneself off preemptively. A commonly observed example is when people don't talk or learn about climate change because they find it upsetting, fear conflict, or believe they lack education or training to understand complex energy. This form of self-censorship can be subconscious. As an example, foreclosure occurs in the very consideration of fossil-free alternatives;

---

[69] Dominion Pipeline Monitoring Coalition. (2019, April 25). Visit the new pipeline CSI website.

[70] Williams, J. E., Rummel, S., Lemon, J., Barney, M., Smith, K., Fesenmyer, K., & Schoen, J. (2016). Engaging a community of interest in water quality protection: Anglers monitoring wadeable streams. *Journal of Soil and Water Conservation*, 71(5), 114—119A.

[71] Fractracker. (2019, May). The Falcon Public Monitoring.

[72] Griswold, E. (2018, October 26). A pipeline, a protest and the battle for Pennsylvania's political soul. *New Yorker*.

[73] Templeton D., & Hopey, D. (2019, January 28). State conducting criminal investigation of shale gas operator. *Post Gazette*.

[74] Staff. (2018, July 6). 'Citizen's risk assessment' of Mariner East Pipeline project moves forward. *Daily Local*.

decades of energy disinformation have paid off for the fossil fule industry, claiming transition as too costly, too complicated, too ambitious—none of which are actually true.

The climate crisis is no longer a distant prospect—it is here and it is now. Yet our collective preference for avoiding topics that appear negative or uncomfortable ostracizes those either willing or simply needing to speak. This can result in a tragic subjugation of very normal, important emotional responses to both threat and loss.[75] Across the range of emotional responses to climate breakdown, eco-distress is at present poorly known to clinicians and often pathologized.[76] A new field of expertise on the psychology of climate anxiety is however now developing.

How do we face the urgent need for change, without becoming paralyzed with fear or apathy?

As the intensity of climate breakdown grows, paying attention to these emotional challenges can help to build internal capacity and encourage us to support others in a solidarity model that recognizes we are all in this together. Ample opportunites now exist and will only increase to help re/build the world we wish to live in. Tools to process the climate crisis, such as the practice of **ritual** in repeated ceremonial practices, are grounding and healing.

I'm no longer so rattled. I am more resolved about the need for complex change and how I can bring it forth.[77]

Resources do exist to help prepare for a future of great risk where our action now will significantly influence our future options. The 4Rs of deep adaptation are pertinent here—resilience, relinquishment, restoration, and reconciliation: (1) What do we value most that we want to keep and how? (2) What could we let go of so as not to make matters worse? (3) What could we bring back to help us with these difficult times? (4) With what and whom shall we make peace as we awaken to our mutual mortality?[78]

Try this

Design, reflect, and engage: *How could you develop community-building practices that are grounding and healing? Write an essay exploring what it would mean to die well and what you would wish to accomplish prior.*

---

[75] Thierry, A., Horn, L., Von Hellermann, P., & Gardner, C. "No research on a dead planet": Preserving the socio-ecological conditions for academia. In *Frontiers in Education* (Vol. 8, p. 1237076).

[76] Van Nieuwenhuizen, A., Hudson, K., Chen, X., & Hwong, A. R. (2021). The effects of climate change on child and adolescent mental health: Clinical considerations. *Current Psychiatry Reports*, 23.

[77] Wray, B. (n.d.). *About*. Gen Dread. Substack.

[78] Bendell, J., & Read, R. (Eds.). (2021). *Deep adaptation: Navigating the realities of climate Chaos*. John Wiley & Sons.

Praxis 12—On-site and online learning resources

Given the pervasiveness of energy all around us, research and group learning opportunities are abundant. The noted resources below focus primarily on digital, online data sources that can be used as ideal launch points for research projects, both local and global in scope. These are just a small selection—what else would you add? Find other evidence-based websites to inform yourself and those with which you engage.

Recommended resources

- Carbon Brief World's Coal Power Plants
- Carbon Mapper
- Climate Alliance Mapping Project
- Climate TRACE
- Geoengineering Monitoring
- Global Coal Plant Tracker
- Global Coal Mine Tracker
- Global Energy Monitor
- Global Methane Tracker
- Global Registry of Fossil Fuels
- The Gas Index

Organizations

- Faculty for A Future
- Fractracker Alliance
- Fridays for Future
- Insure Our Future
- Rise Up Movement

Climate action necessitates life-long learning both in workplaces, campuses and communities, as well as online. Spaces for engagement aren't confined to just formal classrooms. For those without digital access, newspaper articles, books, articles, videos, and other formats are at hand in public archives and libraries. Research increasingly finds that people are anxious about climate change, finding it hard to locate and participate in frank and honest discussion following years of broad social denial coupled with disinformation.[79] The more anxiety a topic causes, the harder it can become to broach the topic. This is a collective challenge: initiate climate

---

[79] Koslov, L. (2019). Avoiding climate change: "Agnostic adaptation" and the politics of public silence. *Annals of the American Association of Geographers, 109*(2), 568–580.

discussion and action not filled with fear but with camaraderie, sound facts and intent to help prepare for what our future holds.

Try this

Research and engage*: How can data transparency support decarbonization and energy justice? Which transparency reports would you share as recommended resources?*

Vocabulary

1. capillary power
2. cloud storage
3. coal exposure
4. crowdsourcing
5. countersurveillance
6. disavowal
7. eco-defense tactics
8. ecological grief
9. foreclosure
10. fossil banks
11. gray surveillance
12. green surveillance
13. permanent surveillance
14. positionality
15. psychoterratic
16. ritual
17. spatial imbalance
18. toponyms
19. toponymic workspace
20. zine

References

Atlas, E.J. (n.d.). How to enter a case? Step by step. https://ejatlas.org/backoffice/cms/en/how-to-enter-a-case-step-by-step/ (Accessed 22 July 2024).

As You Sow. (2024). Racial Justice Report. https://www.asyousow.org/reports/racial-justice-dec2023 (Accessed 16 April 2024).

Canadian Centre for Policy Alternatives. (n.d.). Corporate Mapping Project. https://www.corporatemapping.ca/ (Accessed 16 April 2024).

Defend Our Health. (2023). Hidden Hazards: The Chemical Footprint of a Plastic Bottle. https://defendourhealth.org/campaigns/plastic-pollution/hiddenhazards/ (Accessed 16 April 2024).

Earth Island. (n.d.). Why we need a Fossil Fuel Non-Proliferation Treaty. https://fossilfueltreaty.org/ (Accessed 16 April 2024).

ECO-UNESCO. (n.d.). Toolkit for Youth Eco-Activism. https://learn.ecounesco.ie/courses/toolkit-for-youth-eco-activism/ (Accessed 16 April 2024).

Griffin, Oliver. (2021). Colombia oil workers join anti-fracking campaign. https://www.reuters.com/article/idUSKBN2BT2XG/ (Accessed 16 April 2024).

Resource Generation. (n.d.). Land Reparations and Indigenous Solidarity Toolkit. https://resourcegeneration.org/land-reparations-indigenous-solidarity-action-guide/ (Accessed 16 April 2024).

The Center for Public Integrity. (n.d.). The Accountability Project. https://publicaccountability.org/ (Accessed 16 April 2024).

UNEP. (2022). What do you need to know about the plastic pollution resolution. https://www.unep.org/news-and-stories/story/what-you-need-know-about-plastic-pollution-resolution (Accessed 16 April 2024).

Wright, G., Olenick, L., & Westervelt, A. (2021). The dirty dozen: Meet America's top climate villains. https://www.theguardian.com/commentisfree/2021/oct/27/climate-crisis-villains-americas-dirty-dozen (Accessed 16 April 2024).

Recommended

Books

Kaza, S. (Ed.). (2020). *Joanna Macy - A Wild Love for the World and the Work of Our Time*. Shambhala.

Servigne, P., Stevens, R., & Chapelle, G. (2020). *Another End of the World is Possible: Living the collapse (and not merely surviving it)*. John Wiley & Sons.

Wallace-Wells, D. (2020). *The Uninhabitable Earth: Life after Warming*. Tim Duggan Books.

Wray, B. (2022). *Generation Dread: Finding Purpose in an Age of Climate Crisis*. Canada: Knopf.

Curricula

Archiving the Anthropocene.
Burning Ashes: Puerto Rican Voices (documentary).
Climate Clock World.
Degrowth - Foreign Editions (books).
Divided We Dance (documentary).
Fox in the Hen House - Getting Oily Hands Off Climate Research (podcast).
Juliana Curriculum.
El Poder del Pueblo/The Power from the People (documentary).
The Drawdown Roadmap.

Glossary

Abandonment (noun) feeling at a loss; cut off from a crucial source of sustenance that has been withdrawn, either suddenly or through a process of erosion; (verb) the intentional act of empowered institutions in which their capital and resources are withdrawn from sacrifice zones

Accumulation by dispossession consolidation of wealth among those with privilege gained by the act of dispossessing marginalized individuals or groups of their land and resources

Acid mine drainage the formation and movement of highly acidic water rich in heavy metals from a mining site

Acknowledgement (land) statements of indebtedness that often involve recognition and respect for the Indigenous Peoples who first inhabited an area

Adaptation an act or process of changing to better suit a situation, such as climate change

Afterlife material and ethical burden remaining after use of a product or completion of an activity (i.e., heavy metals from battery recycling, radioactive waste, "blood diamonds")

Amplifying feedback loops when the product of a positive reaction leads to an increase in that reaction; it is a positive feedback loop that exacerbates the effects

Anthropocene the current geological epoch dating from the commencement of significant human impact on Earth's geology and ecosystems, including anthropogenic climate change

Anti-trust measures legislation breaking up, preventing, or controlling trusts or other monopolies, with the intention of promoting competition in business

Artificial boundaries simplifications of the world for accounting and measuring purposes

Astroturf a special interest group deceptively attempting to appear to outside observers to be a legitimate political movement; contrast with grassroots

Bankruptcy a legal process through which people, corporations, or other entities who cannot repay debts to creditors may seek relief from some or all of their debts

Beneficial reuse state-sanctioned repurposing and sale of industrial byproducts

Biometrics body measurements and characteristics used as identification

Bioregion areas with their own ecologies and cultures, in which humans and other species are rooted, actively participating at various scales beyond the immediate locale; also known as biocultural regions by Indigenous federations

Black lung coal workers' pneumoconiosis (CWP)

Blackout loss of electricity due to various factors including grid failure

Blast radius Euclidean distance from an explosive source location where impacts will occur

Blindspots areas of significant harm overlooked or misunderstood by the public and policy makers; often propagated by industry deception tactics

Blockade strategic protest locations of physical occupation designed to halt, delay, and/or impede access

Blowdown venting of concentrated emissions during scheduled and unplanned shutdowns of compressor stations

Blowout an inadvertent return of drilling mud during pipeline construction

Blue-green alliance cooperation between labor organizations and environmental groups

Body burden amount of concentrated chemical and toxic pollutants in the human body

Bomb trains term used by rail industry to identify rolling stocks of highly volatile oil or gas

Boom a rapid growth or increase (i.e., oil boom)

Boomerang effect (in multi-scale organizing) refers to when local or regional opponents of a project or issue who are block out of formal spaces are able to "jump scale" to draw attention at national or international levels, thereby drawing support for their cause

Boomtown a town undergoing rapid growth, usually due to the location of a popular resource or market

Boosterism excessive promotion to privilege a place or sector toward a particular agenda; the overly enthusiastic promotion of a person, organization or cause (i.e., "clean" coal boosterism)

Bottleneck a narrow junction that impedes traffic flow; it can represent a constraint or a control point (i.e., chokepoint) that can cause a strategic supply point to become backlogged; key to strategies for socio-spatial and economic control

Boundary a border; a line between jurisdictions or zones

Break-even times the amount of time required for the discounted cash flows generated by a project to equal its initial cost

Brownfields land where redevelopment or reuse is complicated by the presence of hazardous materials, pollution, or contaminants

Brownout sporadic loss or provision at reduced voltage or amperage of electricity; usually short-term, whether planned or spontaneous

Buen vivir (Spanish) see sumak kawsay (Quechua); this worldview and constitutional law in Ecuador evokes a thriving collective based on the fullness of life of the community

Bureaucratic violence intentional use of state rules and agencies to make decisions or enact policies that result in or have a high likelihood of resulting in injury, death, psychological harm, or deprivation for a person, group, or community

Business-as-usual (BAU) continuation of status quo financial practices and market trends

Bust an economic downturn following a phase of rapid growth (i.e., boom and bust)

Capillary power circulates throughout society as part of personal and political interactions and serves to influence thoughts, beliefs, and actions

Capitalism the economic and political system currently dominant in most of the world in which the private owners of productive property (capital) focus power and decision-making toward the purpose of maximizing profits

Capping wells plugging and maintaining closed oil and gas drill sites

Captive ratepayers customers without a choice in utility provider because of a monopoly in service provision and contracts

Captured state a type of systemic political corruption in which private interests significantly influence a state's decision-making processes to their own advantage

Carbon bomb a project producing at least a billion tonnes of CO_2e emissions over its lifetime

Carbon capture and storage (CCS) an unproven suite of technologies to reduce the amount of carbon dioxide (CO_2) by containing it in such a way that it is unable to affect the atmosphere

Carbon dioxide equivalent (CO_2e) a metric measure used to compare emissions of various greenhouse gasses on the basis of their global warming potential (GWP) by converting amounts of other gases to an amount of emitted CO_2 capable of trapping the same amount

Carbon risk financial risk associated with the transition to a low-carbon economy (carbon asset risk, carbon exposure)

Carbon unicorn future net zero pledges that justify current fossil fuel use

Carbon-centric fetishizing carbon dioxide (CO_2) and ignoring other GHG emissions, such as methane; dominant carbon rhetoric overly focused on some forms of carbon

Carcinogenic potentially cancer producing

Censorship restrictions on collection, display, dissemination, and exchange of information, opinions, ideas, and imaginative expression

Certified sustainable gas variety of business standards alleging improved oversight, methane intensity, and control of fugitive emissions

Chokepoint a narrow section of a critical strait or transportation corridor

Circular economy a system of production and consumption underpinned by renewable energies that involves sharing, reusing, refurbishing, and recycling existing materials and products as long as possible, thus decoupling the economy from consumption of finite resources

Civil disobedience the active, professed refusal of a citizen to obey certain laws, demands, orders, or commands of a government; nonviolent resistance

Clean coal although there is no standard protocol for clean coal, this phrase usually refers to technologies capturing carbon emissions from coal combustion and storing them

Climate breakdown GHG emissions leading to global boiling and all-encompassing ecological disruption and subsequent mayhem

Climate change both the global warming driven by human emissions of greenhouse gases and the resulting large-scale shifts in weather patterns

Climate denialism dismissal or unwarranted doubt that contradicts the scientific consensus on climate change, including the extent to which it is caused by humans

Climate injustice wealthy elites reap benefits from their outsized emissions, while the rest of the world pays a price for the excesses of this greedy minority

Climate justice a framework for understanding climate change as a political issue and to transform our systems and responses to promote equity

Climate migration climate-related mobility that refers to primarily voluntary movement driven by the impact of sudden or gradual climate-exacerbated disasters

Climate Necessity Defense a political-legal tool used by climate activists to justify protest actions taken in defense of the climate

Climate rage angry responses to climate injustice and policy inaction; often cited as a motivation for environmental activism

Climate refugee a person who has been forced to leave their home as a result of the effects of climate change

Climate reparation transfers to cover loss and damage caused by climate change in coordination with payments for historical wrongdoing remaining unresolved

Climate risk potential for loss and damage from global heating

Climate science (climatology) the scientific study of Earth's climate, typically defined as weather conditions averaged over a period of at least 30 years

Climate warrior a person committed to combating climate change

Cloud storage a way for businesses and consumers to save data securely online so it can be easily shared and accessed anytime from any location over a network

Cluster a group of similar things positioned or occurring closely together

Clustered electrification block conversions from fossil energy like gas to renewable sources like heat pumps

Co-location to site two or more things together; placement of various entities in a single area

Co-pollutants unregulated contaminants treated in absentia with regulated emissions

Coal combustion residuals (CCRs) the byproducts and waste created when coal is burned

Coal exposure to be invested in companies that are significantly active in the coal sector with more than 25% of revenue from thermal coal production

Collective something done by people acting as a group; a cooperative enterprise

Collective trauma a traumatic psychological effect shared by a group of people or an impacted community

Colonialism exploitative and cruel extraction and production to benefit institutions originating from outside that violently imposed themselves on the oppressed areas from which they extract wealth

Commemoration a collective process of preserving memories of people or events

Community-based participatory action research collaborative investigation involving directly impacted communities and scientists or health practitioners

Company town spatial construct exemplifying extreme control where stores and housing are owned by a company that is also the main employer

Connectivity extent or capacity of being connected or interconnected; a topological property relating to how geographical areas are attached to one another functionally, spatially, or logically

Contested space where struggle shapes perceptions and realities of a location

Convict leasing a system of forced penal labor

Corporate social responsibility (CSR) a form of business self-regulation; may be used to boost public relations without actually accomplishing what the firm claims (see wokewashing)

Corridor a belt or strip of land linking two other areas, or lining the built environment like a road or pipeline or a natural landscape like a river

Corruption dishonest or fraudulent conduct by those in power, typically involving bribery

Counter-mapping strategic use of local knowledge and data to challenge state and corporate territorial assumptions and misrepresentations

Counter-surveillance measures undertaken to prevent or deter geospatial overwatch and other forms of monitoring

Cradle-to-grave analysis of all inputs and outputs from the start to end; life cycle assessment

Cribbing measuring each unit as less than the actual amount; a form of wage theft

Criminalization (1) turning an activity into a criminal offense by making it illegal or increasing punishments; (2) turning protesters into criminals by making activities illegal and increasing punishment

Crisis capitalism exploitative investments and financial arrangements following disruption or emergency

Critical infrastructure infrastructure sectors whose assets, systems, and networks, whether physical or virtual, are considered so vital that their incapacitation or destruction would have a debilitating effect on security; energy supply and utilities are considered critical

Crowdsourcing data collected from the public at large

Cultural appropriation adoption of an element or elements of one culture or identity by members of another culture or identity, especially controversial when members of a dominant culture appropriate from a minority group

Cumulative impact assessment innovative scientific and regulatory approaches to include cumulative, and potentially synergistic, effects of environmental and social stressors on the health of communities

Cumulative impacts from multiple sources over time magnifying over time and space; documenting this synergy creates a more realistic depiction of body burden

Deathprint the number of people killed by an energy source per kilowatt-hour (kWh)

Debt trap diplomacy an international financial relationship where a creditor country or institution extends debt to a borrowing nation partially, or solely, to increase the lender's political leverage

Debt treadmill creates unfair conditions such that in spite of paying installments on money borrowed, the debtor cannot pay off outstanding loans

Deception playbook a series of well-financed industry campaigns promoting disinformation

Decolonization involves cultural, psychological, and economic freedom so all ethnic groups can practice self-determination over their land, cultures, and political and economic systems

Deep time geological or cosmic time measured in billions of years

Deference politics submitting to existing hierarchies, whether out of respect or self-interest

Degrowth systematic reduction of consumption designed to reduce inequality and improve well-being

Deregulation the reduction or elimination of government power or repeal of rules

Derivative exploitation economic offshoots of harm which are often described or even tallied economically as "progress" though they could compound problems

Deterritorialization the severance of social, political, or cultural practices from land; territorial context or grounding is altered or weakened

Digital divide the gap between those able to access modern information and communications technology, and those that have restricted or no access

Direct action the use of strikes, demonstrations, blockages or other public forms of protest to achieve protest demands

Disaster capitalism to seize on crisis to generate profit; to capitalize on misfortune

Disavowal subtle but powerful soft denialism based in unconscious psychological defenses

Discontinuity a break from the past during which pacing and tempo must readjust

Discourses of climate delay justifications people use to avoid climate action

Displacement the forced loss of one's land, resources, and place identity

Dispossession the act of depriving someone of land or property; the severance of social, political, or cultural practices from Native inhabitants

Divest-reinvest the combined act of removing funds from negative investments and directing funds toward socially and ecologically sustainable initiatives

Divestment selling shares of stocks, bonds, or investment funds

Do-it-yourself (DIY) publishing a method of circulating texts without the direct aid of professionals or experts (i.e., zines, crowdsourcing, EJ Atlas)

Double exposure synergistic disadvantages from poverty and climate disruption

Downstream the part of the energy vector that involves refining as well as selling or distributing products; downstream facilities include petrochemical plants, oil refineries, gas distribution companies, and retail outlets (i.e., gas stations) or sales as well as byproducts and outputs

Dronescape the landscape of an uncrewed aerial vehicle (drone), whose small mobile shape and long reach makes it a popular surveillance method; free from the constraints of a human body (i.e. a pilot), drone usage enlarges the possibilities for overwatch

Dust sampling fraud skirting operation requirements of an apparatus used for collecting data on dust resulting from drilling in rock

Eco-defense tactics a range of actions intended to disrupt human activities perceived to be damaging to the environment, including civil disobedience, direct action, property damage, or sabotage, is an important discussion or debate

Ecocide destruction of the natural environment by deliberate or negligent action

Ecofeminism an equity worldview opposing gender inequality and seeking to end exploitation of the Earth and abuse of power

Ecological cascade a series of secondary extinctions that are triggered by the primary extinction of a key species in an ecosystem

Ecological debt accumulated obligation for restitution from wealthier countries after exploitation of resources and degradation of the environment in poor countries

Ecologically harmful subsidies payments from the state to inventive growth such as with tax breaks that fund for projects that damage ecosystem services such as biodiversity and water filtration or pollute the atmosphere with GHGs

Ecosocialism a political ideology recognizing intersections between social and ecological wellbeing and promoting democratic ecological planning

EJScreen the EPA's standardized, national-level screening and mapping tool used to look at characteristics such as race and poverty in relation to pollution

Elastic metrics selecting parameters with bias to warp measurement strategies to target a particular outcome or agenda

Elite capture a form of corruption whereby public resources are biased for the benefit of a few individuals of superior social status in detriment to the welfare of the larger population

Elites people who hold a disproportionate amount of wealth, privilege, and political power

Embodied carbon carbon dioxide (CO_2) emissions associated with materials and construction processes throughout the supply chain

Embodied research uniting of body and mind in investigative practices with attention to embodiment, empowerment, and praxis

Embodiment awareness and visibility of corporeal expression; how people engage the world through bodies and are perceived based on characteristics such as race, gender, size, and age

Eminent domain expropriation of private property for public and private projects deemed in the public interest and public need

Enchantment a feeling of pleasure or delight

Energy the capacity to do work; found in a number of different forms (i.e., heat, kinetic, chemical, electric)

Energy burden percentage of total household income per watt of electricity purchased

Energy colonialism a spatially uneven process with benefits exported and high costs in areas of production

Energy commoning an alternative approach that recognizes replenishable energy sources and associated technologies as a shared good rather than a commodity

Energy democracy participatory representative energy governance with the objective of sustainable, equitable production and consumption by/for communities and workers

Energy governance decision-making and oversight by the formal institutions of the state as well as non-state actors, such as the private sector and civil society

Energy justice involves fair distribution of costs and benefits with meaningful participation in all stages of decision-making

Energy ladder a simplistic traditional model which assumes that as income increases, the energy types used by households would become cleaner and more efficient

Energy poverty lack of access to utilities like electricity due to insufficient income or local unavailability with broad negative ramifications for quality of life

Energy regime a combination of state planning and a carbon-centric economy that is imposed from above and authoritarian

Energy security long-term uninterrupted availability of renewable energy at an affordable price

Energy sprawl extensive land area used per kilowatt/hour produced

Energy transformation structural change across the technical, economic, social, and political architecture of energy production and consumption

Energy transition substituting renewable energy in place of fossil fuels

Energy violence damage from the energy sector to oneself, another person, or against a group or community, that either results in or has a high likelihood of resulting in injury, death, psychological harm, maldevelopment, or deprivation

Entanglements constraints due to epochs of structural violence that impede overthrow or sweeping change

Environmental illiteracy the inability to make informed ecological decisions

Environmental justice meaningful involvement in ecological laws, policies, and decisions leading to fair, equitable, and non-discriminatory treatment so that no group has to bear a disproportionate share of negative consequences resulting from policies, decisions, and actions

Environmental racism the disproportionate impact of environmental hazards on people of color created through policies, decisions, and actions

Equator principles a financial industry benchmark for determining, assessing, and managing environmental and social risk in large projects like energy infrastructure

Erasure a form of hiding or silencing; excising certain voices from the dominant historical narrative

ESG (Environmental, Social, Governance) business sustainability index

ESG patchwork an uneven incomplete mosaic of business sustainability practices claimed under the ESG framework

Ethos of responsibility the moral duty to address oppression individually and collectively so that we may heal

Exposure duration, frequency, and proximity of chemical release determine a person's exposure

Externalized costs negative social and ecological impacts remaining outside the price of a good or service; also called externalities

Extra-territoriality the state of being exempted from the jurisdiction of local law, usually as the result of diplomatic negotiations

Extreme energy high-intensity, ecologically destructive extractive operations

Extreme heat period of high heat and humidity with temperatures above 90° for at least 2—3 days

Extreme weather a time and place in which weather, climate, or environmental conditions rank above a threshold value near the upper or lower ends of the range of historical measurements

False solutions don't address the scope or timeframe of the problem, or generate a problem of equal or greater proportions

Fascism centralized autocracy and forcible suppression of opposition

Fast fashion rapid production of inexpensive, low-quality clothing that often mimics popular styles

Fast violence rapid physical violence, i.e., homicide, occupational incidents like explosions and fires

Fiduciary responsibility the obligation that one party has in relationship with another to act in the other party's best interest

Fixity refers to a stationary position; a spatial "fix"

Flare a gas combustion device

Flashpoint a place, event, or time at which an issue or topic flares up or ignites

Floating rigs FPSO/FSO systems (floating production, storage, and offloading) facilitate production, liquefaction, and storage of gas at sea

Flow usually means the movement of a liquid, solid, or gas past a point during a time period, but can also apply to energy concepts, like an electrical supply chain or circuit

Fly in, fly out workforce a method of employing people in remote areas by flying them temporarily to the work site

Fog of enactment a gap between policy goals and actual results

Follow-the-money a research strategy to track the policy repercussions from donations and lobbying

Foreclosure wherein a person implicitly understands and internalizes what is unspeakable; self-censorship, sometimes unconsciously

Forward-facing policy government agencies asking to start off on a new foot after demonstrated structural violence, without addressing injustices in the recent case or cases that drew pressure for reforming practices

Fossil banks lending institutions that disproportionately finance climate chaos through fossil fuel investments

Fossil fascism authoritarian nationalist political ideology and movement characterized by militarism and forcible suppression of opposition

Fossil fuel divestment to sell shares of stocks, bonds, or investment funds tied to fossil fuels

Fossil fuel non-proliferation treaty an international pledge to phase out fossil fuels that recognizes climate change as a global threat and supports just transition

Fossil fuels a group of energy sources (i.e., coal, oil, gas) that were formed when ancient plants and organisms were subject to intense heat and pressure over millions of years; since fossil fuels are made up of hydrocarbons they can be depleted

Fossil gas a fossil fuel derived from unconventional horizontal drilling involving hydraulic fracturing

Frack pad drill site for hydraulic fracturing requiring associated equipment like water-delivery trucks

Frackademia gas industry's strategic corporate buyoff of academia with sponsorship of programs and funding research at higher education institutions

Fracking see hydraulic fracturing

Fracking ban a block on unconventional techniques for recovering gas and oil from shale rock

Fractured responsibility splintered decision-making among various governmental jurisdictions and agencies

Fragmentation the process of fracturing, splintering, or being broken into small or separate parts; includes forest fragmentation, community fragmentation

Free, Prior and Informed Consent (FPIC) to establish bottom-up participation and consultation of an Indigenous group prior to the beginning of development on ancestral land or using resources in order to respect right to self-determination guaranteed under the UN

Frontline proximity and exposure to harm

Fuel stacking using multiple energy sources within the same household or institution

Fugitive emissions emissions discharged outside of a confined flow; these emissions do not pass through vents, stacks, or other intentional openings

Fusion centers public—private counter-terrorism hubs

Future deathprint predicted number of fatalities, such as those due to residing near coal operations or from climate breakdown

Gas hook-ups the equipment and other infrastructure to connect to distribution lines

Gaslighting industry slandering, disparaging, or challenging the integrity of scientists, activists, etc.; gaslighting causes the target to begin doubting themselves as a result of the attack

Gastivists experienced opponents to fossil gas

Gender violence physical, sexual, verbal, emotional, and psychological abuse, threats, coercion, and deprivation directed at an individual based on biological sex or gender identity

Geographic imaginaries a construct that highlights how places are not static and are constantly undergoing a process of becoming; places have unrealized potential

Geographic Information Science (GIS) involves vector and raster spatial data analysis for problem-solving

Geopolitics power contests centered around the control of resources and territory

Geospatial overwatch using satellites and georeferencing equipment to observe and track from a distance

GHG exposure occurs with investment in companies that are super emitters or have significant revenue from high-emitting sources

GHG intensity the amount of greenhouse gas (GHG) emissions as measured in carbon dioxide equivalent (CO_2e) emitted per unit of economic activity

Ghostwriting producing text behind the scene for someone else who is the named author

Global boiling current state of our planet where many terrestrial and marine locations experience brutally hot temperatures

Global brightening more irradiance at the Earth's surface creating temperature increase from lowering of cloud cover, reflective aerosols, or absorbing aerosols, such as due to pollution controls

Global dimming pollution blocking atmospheric radiation

Global warming potential (GWP) a measure of how much energy the emissions of 1 ton of a gas will absorb over a given period of time, relative to the emissions of 1 ton of carbon dioxide (CO_2)

Grandfathering to be exempt based on a provision in which an old rule continues to apply to some existing situations, while a new rule will apply to all future cases

Gray surveillance private and corporate espionage

"Greener" capitalism incrementally improved ecological impact from regular capitalism where profit motivation is usually stronger than environmental commitments

Green grabbing dispossession for a sustainability oriented project or crop

Green halo effect using claims about sustainability to generate a positive market outlook

"Green" neoliberalism a convergence of market forces and sustainable development; "green neoliberalism" treats nature as capital and uses the market to address the climate problem through carbon markets, offset mechanisms, etc.

Green surveillance targeting environmentalists with security state and private sector

Greenhouse gases gases that trap heat in the atmosphere: carbon dioxide, methane, nitrous oxide, and fluorinated gases

Greenwashing intentional deception to embellish environmental positives or obscure negatives

Grid an interconnected network for electricity delivery, historically from producers to consumers

Growth poles regional geographies focused around a concentration of linked businesses; driven by an expansionist desire

Guaranteed return-on-investment means that whatever happens to the market, policymakers promise companies will get back what they invest, often plus interest

Hazardous air pollutants (HAPs) those known to cause cancer and other serious health impacts

Hazardscape landscape of risk and harm associated with unprecedented disruptions in the Anthropocene

Health disparity preventable differences in the burden of disease, injury, violence, or opportunities to health between high- and low-wealth areas

Heat dome created when a persistent ridge of high pressure traps heat over an area

Hidden energy debt a concept that highlights how exported industrial products generate externalized and unacknowledged costs from their energy supply that burden producing areas

High consequence area (HCA) a geographic radius zone along infrastructure where failure—like an explosion—would significantly impact persons and habitat, or result in outright loss of human life and property

Holdouts people who refuse to sell their property when facing eminent domain takings

Horizontal space runs parallel to or covers the Earth's surface

Horizontalism (*horizontalidad* in Spanish) a social relationship that advocates the creation, development, and maintenance of social structures for the equitable distribution of power

Hotspot analysis spatial analysis and mapping technique to show clustering

Hotspots clusters or concentrations of spatial phenomena (i.e., pollution)

Hub a location with concentrated facilities; the center of an activity or network

Hydraulic fracturing an unconventional drilling method to extract oil or gas by injecting water, chemicals, and sand at high pressure (i.e. fracking)

Hyperobject something that is distributed across space and time to such an extent that our minds struggle to comprehend it

Impact paradox an attempt to bring a favorable output brings unintentional harm; the intent becomes the opposite of the result

Induced seismicity typically minor earthquakes and tremors that are caused by human activity

Inequality difference in size, degree, or circumstances; inequity

Infrapolitics everyday resistance that is quiet, dispersed, disguised, or otherwise seemingly invisible to elites, the state, or mainstream society

Infrastructure distribution and transmission networks; the prefix *infra-* (i.e., beneath or within) reflects a hidden characteristic

Insecurity the state of being without protection from threat

Internal colonization exploitation of minority groups due to power inequalities within a state

Internet optimism assumes positive benefits like education will automatically follow from web access

Intersectionality intensification of harm resulting from two or more forms of discrimination, whether based on class, race, nationality, religion, age, sexual orientation, or other status

Investor-owned utilities (IOU) private enterprises acting as public utilities

Invisibility a described feeling from those sited next to oil and gas infrastructure of being overlooked or ignored; systematic violence facilitates erasure of marginalized populations

Judicial dissolution a legal procedure in which a corporation is forced to dissolve or cease to exist following the revocation of a corporation's charter due to their significant harm to society

Judicious mixture hiring policies aimed at cultural, ethnic, or racial fragmentation to discourage unionization and maintain a more controllable workforce

Just transition justice-oriented decarbonization from fossil fuels to clean, renewable energy; requires diversity, equity, inclusion, and fairness

Justice the quality of being righteous, equitable, or moral

Labor hierarchy the unequal stratification of control and authority within operations with worker rank by degree of financial security and control over their use of time

Land grabbing to dispossess of territory for the development of a project

Landscape viewshed of values as evident in the built environment and in the surrounding systems of consumption and production; ideological way of "seeing"

Lateral violence harm or loss caused by one peer to another peer within one or more oppressed groups

Leakage escape of emissions or other entities from spatial boundaries

Legacy pollution historical contamination that is not traced to a particular responsible party

Legal monkeywrenching an eco-defense tactic using a barrage of lawsuits to stop or delay harmful policies or projects

Legal violence manipulation of legislative weaknesses or loopholes from intentional use of power resulting in injury, death, psychological harm, maldevelopment, or deprivation

Life cycle the planning stages through end stages of an energy source, including waste disposal, decommissioning, and follow-up

Life cycle assessment (LCA) considering all inputs and outputs during an energy source's life cycle from the earliest extraction to final disposal

Load-shedding when utilities cut off supply to sections of the grid to rotate power during shortage

Lobbying lawfully attempting to influence the actions, policies, or decisions of government officials

Lock-in when the amount of investment in existing systems is used to justify continued use; inertia toward abandoning or switching (i.e., fuel type)

Loophole an ambiguity or inadequacy in a law to circumvent its purpose

Loss-and-damage loss refers to things that are gone forever and cannot be brought back, while damages are financial and societal costs

Maladaptation when actions intended to encourage climate adaptation create more overall harm than the benefits they portend; an impact paradox

Marginality the state or condition of being isolated from the dominant society or culture, and therefore frequently disadvantaged

Market citizenship when one's perceived economic value influences their access to rights, encouraging self-interest behavior at the expense of collective wellbeing

Market environmentalism use of economic incentives and signals to determine waste and pollution management and to filter our relationship to the Earth through a pricing lens

Market-based management a drive for continuous improvement of competitive position as the core business ideology

Martial law imposition of direct military control of normal civil functions or suspension of civil law by a government, including suspension of *habeas corpus*

Methane (CH$_4$) a flammable gas that consists of one carbon atom and four hydrogen atoms

Methane loopholes using legal ambiguity or inadequacy to circumvent regulation of the potent GHG methane (CH$_4$)

Methane rift an existential rupture between humans and the biosphere from methane's (CH$_4$) temporal and spatial impacts

Methane slip unburned methane (CH$_4$) during use of equipment

Mid-stream involves transmission networks, whether pipeline, rail, barge, tanker, or truck, storage, and wholesale marketing; part of the energy vector

Militancy the use of confrontational or violent methods in support of a cause

Miseducation teachings divorced from realities of learners such that they reproduce and extend prejudice

Mission creep change of an initiative or policy from its original scope over time

Mitigation the action of reducing the severity, seriousness, or painfulness of something

Mixed methods research using a combination of investigative methodologies, often including both qualitative and quantitative approaches

Mobility the quality or state of being mobile or movable; ability or capacity to move

Mobilizing preparing, assembling, readying to build capacity and make capable of movement

Most-affected peoples and areas (MAPA) represents people in regions most affected by climate change, who at the same time are doing the least to contribute to the problem

Mutual support horizontal assistance among equals that benefits widely

Necropolitics the power to decide who lives and dies

Negligence failure to exercise the appropriate and/or ethical rules of care that are expected to be exercised amongst specified circumstances

Neoliberal multiculturalism privileged entitlement in the marketplace for ethnic groups who support state and private sector neoliberal projects and reforms; an example of racial capitalism

Neoliberalism a contemporary phase of globalization and financialization of capitalism following the tenets of privatization, deregulation, and trade liberalization

Net-zero the rare condition of negating the amount of GHG emissions produced from an activity

Network a group or system of interconnected people or things (i.e., a trade network, a pipeline network, an environmental organizing network); networks express spatial and capillary power

Non-point source pollution contamination resulting from many diffuse releases

Non-response the absence of a verbal or written reply (uninformed nonresponse occurs due to lack of information, while informed nonresponse could suggest a collective lie)

Non-view a form of blindness to a situation that silences or stimies response

Obfuscation the intentional act of making perspectives obscure, unclear, or unintelligible

Off-grid independent living arrangements without reliance on one or more public utilities

Off-shoring a spatial process of the siting pollution from production in places that are separate from areas of consumption

Offset activities intended to counterbalance, counteract, or compensates for something else

Offset leakage a form of fugitive emission or loss of benefit

Ogoni-9 Ogoni ethnic leaders assassinated by Nigeria's military dictatorship in 1995 following the creation of the Ogoni Bill of Rights

Opposition organized resistance or dissent, expressed in action or argument

Orphaned wells abandoned wells for which the owner cannot be found or for which the owner is financially unable or unwilling to carry out clean-up

Overpressurization pipeline pressure significantly above what is usual, creating risk of explosion

Overshoot an act of going past or beyond a point, target, or limit—in this case referring to surpassing planetary limits

Participatory science local collection and analysis of data, often done in collaboration with credentialed scientists; also called citizen science or street science

Peak oil when the maximum rate of extraction of petroleum is reached, after which it is expected to enter terminal decline

People's assemblies coming together of community members to share ideas and set collective goals and agendas

Permanent surveillance constant observation continuously sending data to supercomputers

Physical violence bodily harm; intentional use of force against a person, group, or community that either results in or has a high likelihood of resulting in injury, death, psychological harm, maldevelopment, or deprivation

Place a special location with subjective meaning and significance

Placemaking collaborating to achieve the potential of a place, not only as it was but also *as it can be*

Point-source pollution any single identifiable source from which pollutants are discharged, such as a pipe, ditch, or smokestack

Policy capture when public decisions over policies are consistently or repeatedly directed away from the public interest towards to a specific special interest

Positionality how social position and power shape identity and viewpoint

Post-traumatic stress disorder a health condition triggered by a terrifying event, either experiencing it or witnessing it; symptoms include flashbacks, nightmares, and severe anxiety

Power ability to do or act; capability of doing or accomplishing something

Power lens focusing on political constructs like power relations to understand an issue or topic

Power map a useful analytical method and visual tool to show influence and identify targets to promote change

Power over the ability to make someone do something they would not do otherwise; possession of control, authority, or influence over others

Power with occurs when groups of actors pursue joint or common goals

Praxis the process by which a theory, lesson, or skill is enacted, embodied, or realized; the act of engaging, applying, exercising, realizing, or practicing ideas

Price gouging a pejorative term used to describe the situation when a seller increases the prices of goods, services, or commodities to a level much higher than is considered reasonable or fair

Productive exclusion occurs when the penetration of capital reduces access to means, whether in subsistence economies or complex supply circuits

Prosumers consumers who become involved in production; energy prosumers are innovators, owners, investors, and decision-makers, not just ratepayers

Protracted lawfare delayed decision-making in the courts with unresolved claims lingering

Proximity nearness in space, time, or relation (i.e., distance to toxic exposure)

Psychological violence anguish, threat, intimidation, stress, and other forms of trauma in the energy sector that result in emotional or mental (psychological) harm

Psychoterratic emotions related to the Earth

Qualitative methods involves collecting and analyzing non-numerical data to understand concepts, opinions, or experiences

Quantitative methods based on numbers; numerical statistics and measures

Race wedge using race or racial difference as a means to divide and conquer

Race-to-the-bottom a tendency for footloose firms to drive down prices by moving to locations without adequate regulations or benefits

Racial capitalism process of deriving social and economic value from the racial identity of another person; racialized exploitation and capital accumulation are mutually reinforcing

Racism prejudice, discrimination, or antagonism directed against a person or people on the basis of skin color and hatred towards people of color

Radical transparency an approach or act that uses abundant networked information to access previously confidential data to drastically increase the openness of organizational processes

Rebound effect also called the "take back effect"; part or all of the environmental gains recovered from a policy or action are lost once more due to gaps, miscalculations, or impact paradox

Redlining when services are withheld from potential customers who reside in neighborhoods classified as a financial risk due to the race and income of residents

Regime of measurement a structured process and architecture for measuring in order to obtain a result expressed in numbers

Regulatory capture corruption of authority that occurs when a political entity, policymaker, or regulatory agency is co-opted to serve the interests of a regulated industry

Regulatory compact US legal norm allowing monopoly status to investor-owned utilities with centralized ownership, generation, and distribution of electricity resources

Regulatory lag the delay between when a problem is known and when there is the creation of rules to address the problem

Regulatory rollback weakening an established legal norm or standard (backsliding)

Regulatory tug-of-war back and forth on polemic policy, such as for and against gas hook-up bans

Regulatory violence manipulation or obstruction of rules to avoid legal responsibility

Remote sensing use of satellite images and long-distance interpretive methods to understand change over time across space

Renewable energy a group of energy sources, including sunlight, wind, rain, tides, waves, and geothermal heat, collected from resources naturally replenished on a human timescale

Reparations the making of amends for a wrong one has done, by paying money to or otherwise helping those who have been wronged; the action of repairing something (i.e., Indigenous Land Back movements); see also climate reparations

Representation the action of speaking or acting on behalf of a group, territory, or institution

Research and development (R&D) search for innovation, and improvement in products and processes

Residual risk the amount of risk or danger associated with an action or event remaining after natural or inherent risks have been reduced by risk controls

Resiliency ability to bounce back after disaster or recuperate after hardship

Resistance the refusal to accept or comply in an attempt to prevent something

Return-on-investment (ROI) a ratio between net income and investment as a performance measure over a period used to evaluate the efficiency of an investment

Revolving-door workers rotate between jobs in the public sector and the private sector, bringing internal secrets between work spaces

Rights-of-nature (RoN) a legal and jurisprudential theory that describes inherent rights as associated with ecosystems and species; see also buen vivir and sumak kawsay

Risk exposure to danger, harm or loss

Risk assessment identifying and analyzing potential events that may negatively impact individuals, assets, and/or the environment; making judgments on the "tolerability" of risks

Risk management a set of measures aimed at avoiding, minimizing, or decreasing the causes of incidents

Ritual repeated ceremonial practices that are grounding and healing

Sacrifice zone a marginalized geographic area permanently impaired by environmental damage or economic disinvestment that have been overlooked by policy makers

Scale socially defined, jurisdictional levels of governance or institutions (i.e., household, district, region, nation, international)

Scarcity mindset preoccupation with the lack of something, usually time or money

School-to-prison pipeline school policies and procedures that drive many youth to a pathway that begins in school and ends in the criminal justice system

Scrip a substitute for government-issued legal tender or currency issued by a company to pay its employees that can only be exchanged in company stores

Security the state of being free from danger or threat

Segmentation division into separate parts or sections; can happen during environmental permitting to dissect harm by carving into more palatable chunks of the greater whole

Self-dealing when companies sell downstream products to entities that are forms of the same (i.e., sub-sidiaries, partners); sometimes done by using shell companies

Self-governance or self-rule the ability of a person or group to exercise all necessary functions of regulation without intervention from an external authority

Setback the absolute minimum distance that must be maintained between any energy facility (for example, a drilling well, a pipeline, or a gas plant) and a dwelling or public facility

Settler colonialism an imposed system of oppression; an ongoing process that destroys and replaces

Shack rouster historically one who awakened coal workers in the morning; part of the surveillance and control of mine bosses through physical violence and threat of violence

Shadow stock contracts that are paid out as the company's value increased but that did not confer any actual ownership or holdings

Shifting base syndrome in the absence of any point of comparison, you assume what you see is normal, even if it isn't

Shock absorber a device designed to dampen motion or impulse, in this case public opposition

Shock doctrine a brutal tactic of using the public's disorientation following a collective shock like a recession or natural disaster to push through pro-corporate measures

Shock therapy pro-corporate measures like disaster capitalism and structural adjustment programs

Short-termism concentration on projects or objectives for immediate profit at the expense of long-term security

Sink an area that holds or stores contamination

Sinkhole a cavity or surface depression that forms when soil erodes around pipeline and other construction activities

Site suitability locations that meet specific criteria or restrictions, such as the historical landscape of the area, competing economic impacts, and legacy pollution

Siting to fix or build something in a particular place

Slow violence everyday exploitation, risk, and pollution; structural inequality resulting in damage to oneself, another person, or against a group or community

Social license to operate when an entity receives ongoing approval within a local community and among other stakeholders; social legitimacy rather than government authority

Soft denialism contradictory acts, practices, and statements from those who believe in climate change but continue as usual as if it were not real

Source an origin area for contamination

Space a social product; not simply a neutral container waiting to be filled; a dynamic, human-constructed means of control, domination, and power; common forms of space include absolute, relative, and virtual (metaphorical)

Spatial analysis techniques using topological, geometric, or geographic patterns across area/location; using location-oriented problems (i.e., the power of where), find patterns, assess trends, and make decisions

Spatial distribution the arrangement of a phenomenon across an area or space; the study of the relationship between objects in physical space

Spatial emancipation facilitated mobility or a claim of being free from prior physical constraints

Spatial fix securing specific boundaries for regulation, investment, or another objective tied to power or profit

Spatial imbalance when need is misaligned with density

Spatial planning forethought in arrangement of spaces and scales; a material expression of power across space

Spill an unpermitted release of polluting materials; assumed to be accidental

Spillover effect when movements spread from one place to another

Spinoff industry occurs when fossil fuels influence the creation of new or additional sectors, products, or services

Split estate where property rights to surface and underground resources belong to different parties

Split incentives when subsidies and other incentives go for and against a purpose like decarbonization

Stop loss point a pre-set advanced order to sell that is used to limit loss in trade or investing

Stranded assets holdings that suffer from premature write-downs, devaluations, or conversion to liabilities; infrastructure that is abandoned early before predicted retirement due to shifts

Strategic retirement accelerated departure from fossil fuels using tools of spatial planning alongside commitment to improving justice and equity

Strikebreakers persons hired by a company to weaken unions and force the end to a strike

Structural violence intentional use of power that either results in or has a high likelihood of resulting in injury, death, psychological harm, maldevelopment, or deprivation

Subsidies money granted by the government or a public body to assist an industry or business so that the price of a commodity or service may remain low or competitive

Sumak kawsay (Quechua) translates roughly as "fullness of life of the community"; "buen vivir" in Spanish

Sunk cost money that has already been spent and cannot be recovered

Super emitter disproportionately large emissions in relation to other operations

Surveillance the monitoring of behavior, activities, or information for the purpose of information gathering, influencing, managing, or directing

Survivance resistance and worldmaking that draws attention to how Indigenous Peoples survived the genocidal ambitions of settler colonialism and continue to enliven their cultures in fluid, critical, and generative ways

Symbolic protest activism that is temporary and non-disruptive

Synergistic effect interaction or cooperation giving rise to a whole that is greater than the simple sum of its parts

Tax avoidance systematic loopholes and individual exceptions that allow privileged groups or firms to avoid paying

Teachable moment the time at which learning a particular topic or idea becomes possible or easiest because a person seems open to reflection

Technological hubris assumption that a technological "fix" can make up for poor governance

Technological optimism belief that scientific innovation can drive solutions without structural change

Technologically enhanced naturally occurring radioactive materials (TENORMs) naturally occurring radioactive materials concentrated or exposed as a result of human activities

Temporal analysis techniques for analyzing data linked to time and change over time

Territorial violence dispossession of property rights or fragmentation of collective land holdings for marginalized groups or households (privatization); land grab; consolidation of ownership

Territoriality a claim to jurisdiction over space or possession

Territory organized land or controlled division of an area of land under the jurisdiction of a ruler or state, including subterranean claims

"the Great Turning" a deep, broad transformation from exploitative, self-destructive status quo to mutualistic, life-sustaining systems that are forming and being recovered to heal vast harm

Threat multiplier something that exacerbates other drivers of insecurity or makes already precarious situations worse

Time-space compression the physical ability to cover more distance in less time as a result of modern transportation and communication infrastructure

Tipping point where a changing climate could push parts of the Earth system into abrupt or irreversible change; the idea of a tipping point is increasingly used to discuss social movements

Topographical approach pays particular attention to the forms and features of land surfaces—terrain or relief involves the vertical and horizontal dimensions of land surface

Toponym a place name, especially one derived from a topographical feature; a proper name of a geographical feature

Toponymic workspace describes how names are being changed to reflect current knowledge and values as places and institutions come to terms with their past

Transition risks emerge from climate mitigation when switching from fossil fuels to renewables

Uneven development systemic process by which the power relations are translated into spatial forms, some areas prosper while others stagnate or decline due to inequity in access to power and resources across space

Union a trade union or labor union is an organization of workers intent on maintaining or improving the conditions of their employment, such as obtaining better wages or safer working conditions

Union-busting a range of activities undertaken to disrupt or prevent the formation of trade unions or their attempts to grow their membership

Unnatural disaster a catastrophe in the Anthropocene worsened by human-induced factors like GHGs

Upcycling creative reuse of discarded objects or materials to create a product of higher quality or value than the original

Upstream exploration and production activities that involve searching for and producing crude oil or gas, such as drilling operations to bring raw resources to the surface

Vapor-speak industry rhetoric that reifies false solutions to continue use of fossil energy while promulgating doubt about science

Vector a structure or system transferring energy across space and time

Vertical space zones above and below the surface of the Earth, including atmospheric and subterranean geographies

Vested interest a person or group having a personal stake or involvement in an undertaking or state of affairs

Violence intentional use of physical force or power, threatened or actual, against oneself, another person, or against a group or community, which either results in or has a high likelihood of resulting in injury, death, psychological harm, maldevelopment, or deprivation

Virtue-signaling a pejorative term for an unearned claim to a moral stance or position

Voluntary standards non-binding commitments from states, firms, and other parties

Vulnerability propensity or predisposition to be adversely affected to hazards; can be physical or social vulnerability

Waste pit impoundments, depressions, or ponds that hold wastewater and solid waste

Wastelanding ecocide alongside genocide; for example, settler colonialists' impacts on Indigenous Peoples and lands

Watchdog a watchful guardian against waste, theft, or harmful practices

Water stress when water resources in an area are critically low; supply insufficient for need

Water-stranded assets resources or investments that are no longer useable because of water scarcity or contamination

Wet-bulb temperature a measure of heat stress that takes into consideration humidity as well, indicating combinations of heat and humidity that are dangerous and potentially fatal

Whitewashing painting a narrative of racial integration while forwarding white privilege

Wokewashing painting a narrative of equity and harmony while othering and stigmatizing

Worldmaking to imagine and build places and our planet in ways that are fairer and more sustainable than what we inherited

Zine a form of DIY publishing with roots in activism; a simple booklet created with social justice purposes

Index

US Mine Safety and Health Administration
  (MSHA), 51
US Natural Gas Act [1938], 294–295
US Pipeline and Hazardous Materials Safety
  Administration (PHMSA), 189

V

Vaca Muerta gas development, 350–351
Vapor speak, 317
Vertical space, 199
  grabs, 282–283
Vested interests, 29
Violence, 30
  energy, 30, 43–48, 91–98, 121–124, 154–166,
    195–197, 235–241, 434–437
  fast, 25, 32–33, 78–79, 121–122, 156, 272, 280,
    415–416
  legal, 137–138, 195–196
  physical, 41
  regulatory, 91, 122–123
  slow, 10, 25, 32–33, 51, 78–79, 93, 100, 113,
    417
  structural, 32–33, 43, 154
Virginia Electric and Power Company (VEPCO),
  132
Virtual and personal learning spaces, 483–484
Virtue signaling, exposing, 470–471
Volatile organic compounds (VOCs), 162–163,
  309, 327–329
Vulnerabilities, 342

W

Wasteful flaring, 206–207
Wastelanding, 246
Wastewater disposal, 282–283
Watchdogs, 438–439, 480–481
Water impoundments, 283
Water molecules, 406
Water protectors, 256–257
Watershed-wide damages, 350
Water-stranded assets, 146–147
Water stress, 78–79
Water worries, 283–284
Web resources, 467
Websites, 478
West Virginia's Manufacturing Association
  (WVMA), 92
Wet-bulb temperature, 416–417
Whitewashing, 109–111
World Health Organization (WHO), 30
Worldmaking, 34–35, 361
  Indigenous, 247
  regenerative, 90

Y

Yacimientos Petrolíferos Fiscales (YPF), 349
Youth mobilization, xii–xii

Z

Zine, 468

Printed in the United States
by Baker & Taylor Publisher Services